PERFORMANCE ANALYSIS
OF COMMUNICATIONS
NETWORKS AND SYSTEMS

PIET VAN MIEGHEM
Delft University of Technology

T0172078

CAMBRIDGE
UNIVERSITY PRESS

CAMBRIDGE UNIVERSITY PRESS
Cambridge, New York, Melbourne, Madrid, Cape Town, Singapore, São Paulo, Delhi

Cambridge University Press
The Edinburgh Building, Cambridge CB2 8RU, UK

Published in the United States of America by Cambridge University Press, New York

www.cambridge.org
Information on this title: www.cambridge.org/9780521108737

First published 2006
This digitally printed version 2009

A catalogue record for this publication is available from the British Library

ISBN 978-0-521-85515-0 hardback
ISBN 978-0-521-10873-7 paperback

Additional resources for this publication at www.cambridge.org/9780521108737

Further resources for this publication, provided by the author, are available at
www.nas.its.tudelft.nl/people/Piet/bookPA.html

Waar een wil is, is een weg.
to my father

to my wife Saskia
and my sons Vincent, Nathan and Laurens

Contents

Preface   xi

1   Introduction   1

     Part I   Probability theory   7

2   Random variables   9

     2.1   Probability theory and set theory   9
     2.2   Discrete random variables   16
     2.3   Continuous random variables   20
     2.4   The conditional probability   26
     2.5   Several random variables and independence   28
     2.6   Conditional expectation   34

3   Basic distributions   37

     3.1   Discrete random variables   37
     3.2   Continuous random variables   43
     3.3   Derived distributions   47
     3.4   Functions of random variables   51
     3.5   Examples of other distributions   54
     3.6   Summary tables of probability distributions   58
     3.7   Problems   59

4   Correlation   61

     4.1   Generation of correlated Gaussian random variables   61
     4.2   Generation of correlated random variables   67
     4.3   The non-linear transformation method   68

	4.4	Examples of the non-linear transformation method	74
	4.5	Linear combination of independent auxiliary random variables	78
	4.6	Problem	82
5		**Inequalities**	83
	5.1	The minimum (maximum) and infimum (supremum)	83
	5.2	Continuous convex functions	84
	5.3	Inequalities deduced from the Mean Value Theorem	86
	5.4	The Markov and Chebyshev inequalities	87
	5.5	The Hölder, Minkowski and Young inequalities	90
	5.6	The Gauss inequality	92
	5.7	The dominant pole approximation and large deviations	94
6		**Limit laws**	97
	6.1	General theorems from analysis	97
	6.2	Law of Large Numbers	101
	6.3	Central Limit Theorem	103
	6.4	Extremal distributions	104
		Part II Stochastic processes	113
7		**The Poisson process**	115
	7.1	A stochastic process	115
	7.2	The Poisson process	120
	7.3	Properties of the Poisson process	122
	7.4	The nonhomogeneous Poisson process	129
	7.5	The failure rate function	130
	7.6	Problems	132
8		**Renewal theory**	137
	8.1	Basic notions	138
	8.2	Limit theorems	144
	8.3	The residual waiting time	149
	8.4	The renewal reward process	153
	8.5	Problems	155
9		**Discrete-time Markov chains**	157
	9.1	Definition	157

9.2   Discrete-time Markov chain   158

9.3   The steady-state of a Markov chain   168

9.4   Problems   177

10   Continuous-time Markov chains   179

10.1   Definition   179

10.2   Properties of continuous-time Markov processes   180

10.3   Steady-state   187

10.4   The embedded Markov chain   188

10.5   The transitions in a continuous-time Markov chain   193

10.6   Example: the two-state Markov chain in continuous-time   195

10.7   Time reversibility   196

10.8   Problems   199

11   Applications of Markov chains   201

11.1   Discrete Markov chains and independent random variables   201

11.2   The general random walk   202

11.3   Birth and death process   208

11.4   A random walk on a graph   218

11.5   Slotted Aloha   219

11.6   Ranking of webpages   224

11.7   Problems   228

12   Branching processes   229

12.1   The probability generating function   231

12.2   The limit W of the scaled random variables W_k 233

12.3   The Probability of Extinction of a Branching Process   237

12.4   Asymptotic behavior of W 240

12.5   A geometric branching processes   243

13   General queueing theory   247

13.1   A queueing system   247

13.2   The waiting process: Lindley's approach   252

13.3   The Beneš approach to the unfinished work   256

13.4   The counting process   263

13.5   PASTA   266

13.6   Little's Law   267

14   Queueing models   271

14.1   The M/M/1 queue   271
14.2   Variants of the M/M/1 queue   276
14.3   The M/G/1 queue   283
14.4   The GI/D/m queue   289
14.5   The M/D/1/K queue   296
14.6   The N*D/D/1 queue   300
14.7   The AMS queue   304
14.8   The cell loss ratio   309
14.9   Problems   312

Part III   Physics of networks   317

15   General characteristics of graphs   319

15.1   Introduction   319
15.2   The number of paths with j hops   321
15.3   The degree of a node in a graph   322
15.4   Connectivity and robustness   325
15.5   Graph metrics   328
15.6   Random graphs   329
15.7   The hopcount in a large, sparse graph with unit link
       weights   340
15.8   Problems   346

16   The Shortest Path Problem   347

16.1   The shortest path and the link weight structure   348
16.2   The shortest path tree in K_N with exponential link
       weights   349
16.3   The hopcount h_N in the URT   354
16.4   The weight of the shortest path   359
16.5   The flooding time T_N 361
16.6   The degree of a node in the URT   366
16.7   The minimum spanning tree   373
16.8   The proof of the degree Theorem 16.6.1 of the URT   380
16.9   Problems   385

17   The efficiency of multicast   387

17.1   General results for $g_N(m)$ 388
17.2   The random graph $G_p(N)$ 392
17.3   The k-ary tree   401

17.4 The Chuang–Sirbu law   404
17.5 Stability of a multicast shortest path tree   407
17.6 Proof of (17.16): $g_N(m)$ for random graphs   410
17.7 Proof of Theorem 17.3.1: $g_N(m)$ for k-ary trees   414
17.8 Problem   416

18   The hopcount to an anycast group   417

18.1 Introduction   417
18.2 General analysis   419
18.3 The k-ary tree   423
18.4 The uniform recursive tree (URT)   424
18.5 Approximate analysis   431
18.6 The performance measure η in exponentially growing
      trees   432

Appendix A   Stochastic matrices   435

Appendix B   Algebraic graph theory   471

Appendix C   Solutions of problems   493

Bibliography   523

Index   529

Preface

Performance analysis belongs to the domain of applied mathematics. The major domain of application in this book concerns telecommunications systems and networks. We will mainly use stochastic analysis and probability theory to address problems in the performance evaluation of telecommunications systems and networks. The first chapter will provide a motivation and a statement of several problems.

This book aims to present methods rigorously, hence mathematically, with minimal resorting to intuition. It is my belief that intuition is often gained after the result is known and rarely before the problem is solved, unless the problem is simple. Techniques and terminologies of axiomatic probability (such as definitions of probability spaces, filtration, measures, etc.) have been omitted and a more direct, less abstract approach has been adopted. In addition, most of the important formulas are interpreted in the sense of "What does this mathematical expression teach me?" This last step justifies the word "applied", since most mathematical treatises do not interpret as it contains the risk to be imprecise and incomplete.

The field of stochastic processes is much too large to be covered in a single book and only a selected number of topics has been chosen. Most of the topics are considered as classical. Perhaps the largest omission is a treatment of Brownian processes and the many related applications. A weak excuse for this omission (besides the considerable mathematical complexity) is that Brownian theory applies more to physics (analogue fields) than to system theory (discrete components). The list of omissions is rather long and only the most noteworthy are summarized: recent concepts such as martingales and the coupling theory of stochastic variables, queueing networks, scheduling rules, and the theory of long-range dependent random variables that currently governs in the Internet. The confinement to stochastic analysis also excludes the recent new framework, called *Network Calculus* by Le Boudec and Thiran (2001). Network calculus is based on min-plus algebra and has been applied to (Inter)network problems in a deterministic setting.

As prerequisites, familiarity with elementary probability and the knowledge of the theory of functions of a complex variable are assumed. Parts in the text in small font refer to more advanced topics or to computations that can be skipped at first reading. Part I (Chapters 2–6) reviews probability theory and it is included to make the remainder self-contained. The book essentially starts with Chapter 7 (Part II) on Poisson processes. The Poisson process (independent increments and discontinuous sample paths) and Brownian motion (independent increments but continuous sample paths) are considered to be the most important basic stochastic processes. We briefly touch upon renewal theory to move to Markov processes. The theory of Markov processes is regarded as a fundament for many applications in telecommunications systems, in particular queueing theory. A large part of the book is consumed by Markov processes and its applications. The last chapters of Part II dive into queueing theory. Inspired by intriguing problems in telephony at the beginning of the twentieth century, Erlang has pushed queueing theory to the scene of sciences. Since his investigations, queueing theory has grown considerably. Especially during the last decade with the advent of the Asynchronous Transfer Mode (ATM) and the worldwide Internet, many early ideas have been refined (e.g. discrete-time queueing theory, large deviation theory, scheduling control of prioritized flows of packets) and new concepts (self-similar or fractal processes) have been proposed. Part III covers current research on the physics of networks. This Part III is undoubtedly the least mature and complete. In contrast to most books, I have chosen to include the solutions to the problems in an Appendix to support self-study.

I am grateful to colleagues and students whose input has greatly improved this text. Fernando Kuipers and Stijn van Langen have corrected a large number of misprints. Together with Fernando, Milena Janic and Almerima Jamakovic have supplied me with exercises. Gerard Hooghiemstra has made valuable comments and was always available for discussions about my viewpoints. Bart Steyaert eagerly gave the finer details of the generating function approach to the GI/D/m queue. Jan Van Mieghem has given overall comments and suggestions beside his input with the computation of correlations. Finally, I thank David Hemsley for his scrupulous corrections in the original manuscript.

Although this book is intended to be of practical use, in the course of writing it, I became more and more persuaded that mathematical rigor has ample virtues of its own.

Per aspera ad astra

January 2006   PIET VAN MIEGHEM

1

Introduction

The aim of this first chapter is to motivate why stochastic processes and probability theory are useful to solve problems in the domain of telecommunications systems and networks.

In any system, or for any transmission of information, there is always a non-zero probability of failure or of error penetration. A lot of problems in quantifying the failure rate, bit error rate or the computation of redundancy to recover from hazards are successfully treated by probability theory. Often we deal in communications with a large variety of signals, calls, source-destination pairs, messages, the number of customers per region, and so on. And, most often, precise information at any time is not available or, if it is available, deterministic studies or simulations are simply not feasible due to the large number of different parameters involved. For such problems, a stochastic approach is often a powerful vehicle, as has been demonstrated in the field of physics.

Perhaps the first impressing result of a stochastic approach was Boltzmann's and Maxwell's statistical theory. They studied the behavior of particles in an ideal gas and described how macroscopic quantities as pressure and temperature can be related to the microscopic motion of the huge amount of individual particles. Boltzmann also introduced the stochastic notion of the thermodynamic concept of entropy S,

$$S = k \log W$$

where W denotes the total number of ways in which the ensembles of particles can be distributed in thermal equilibrium and where k is a proportionality factor, afterwards attributed to Boltzmann as the Boltzmann constant. The pioneering work of these early physicists such as Boltzmann, Maxwell and others was the germ of a large number of breakthroughs in science. Shortly after their introduction of stochastic theory in classical physics, the theory of quantum mechanics (see e.g. Cohen-Tannoudji *et al.*, 1977) was established. This theory proposes that the elementary building blocks of nature, the atom and electrons, can only be described in a probabilistic sense. The conceptually difficult notion of a wave function whose squared modulus expresses the probability that a set of particles is in a certain state and the Heisenberg's uncertainty relation exclude in a dramatic way our deterministic, macroscopic view on nature at the fine atomic scale.

At about the same time as the theory of quantum mechanics was being created, Erlang applied probability theory to the field of telecommunications. Erlang succeeded to determine the number of telephone input lines m of a switch in order to serve N_S customers with a certain probability p. Perhaps his most used formula is the Erlang B formula (14.17), derived in Section 14.2.2,

$$\Pr\left[N_S = m\right] = \frac{\frac{\rho^m}{m!}}{\sum_{j=0}^{m} \frac{\rho^j}{j!}}$$

where the load or traffic intensity ρ is the ratio of the arrival rate of calls to the telephone local exchange or switch over the processing rate of the switch per line. By equating the desired blocking probability $p = \Pr\left[N_S = m\right]$, say $p = 10^{-4}$, the number of input lines m can be computed for each load ρ. Due to its importance, books with tables relating p, ρ and m were published.

Another pioneer in the field of communications that deserves to be mentioned is Shannon. Shannon explored the concept of entropy S. He introduced (see e.g. Walrand, 1998) the notion of the Shannon capacity of a channel, the maximum rate at which bits can be transmitted with arbitrary small (but non zero) probability of errors, and the concept of the entropy rate of a source which is the minimum average number of bits per symbol required to encode the output of a source. Many others have extended his basic ideas and so it is fair to say that Shannon founded the field of information theory.

A recent important driver in telecommunication is the concept of quality of service (QoS). Customers can use the network to transmit different types of information such as pictures, files, voice, etc. by requiring a specific level of service depending on the type of transmitted information. For example, a telephone conversation requires that the voice packets arrive at the receiver D ms later, while a file transfer is mostly not time critical but requires an extremely low information loss probability. The value of the mouth-to-ear delay D is clearly related to the perceived quality of the voice conversation. As long as $D < 150$ ms, the voice conversation has toll quality, which is roughly speaking, the quality that we are used to in classical telephony. When D exceeds 150 ms, rapid degradation is experienced and when $D > 300$ ms, most of the test persons have great difficulty in understanding the conversation. However, perceived quality may change from person to person and is difficult to determine, even for telephony. For example, if the test person knows a priori that the conversation is transmitted over a mobile or wireless channel as in GSM, he or she is willing to tolerate a lower quality. Therefore, quality of service is both related to the nature of the information and to the individual desire and perception. In future Internetworking, it is believed that customers may request a certain QoS for each type of information. Depending on the level of stringency, the network may either allow or refuse the customer. Since customers will also pay an amount related to this QoS stringency, the network function that determines to either accept or refuse a call for service will be of crucial interest to any network operator. Let us now state the connection admission control (CAC) problem for a voice conversation to illustrate the relation to stochastic analysis: "How many customers m are allowed in order to guarantee that the ensemble of all voice packets reaches the destination within D ms with probability p?" This problem is exceptionally difficult because it depends on the voice codecs used, the specifics of the network topology, the capacity of the individual network elements, the arrival process of calls from the customers, the duration of the conversation and other details. Therefore, we will simplify the question. Let us first assume that the delay is only caused by the waiting time of a voice packet in the queue of a router (or switch). As we will see in Chapter 13, this waiting time T of voice packets in a single queueing system depends on (a) the arrival process: the way voice packets arrive, and (b) the service process: how they are processed. Let us assume that the arrival process specified by the average arrival rate λ and the service process specified by the average service rate μ are known. Clearly, the arrival rate λ is connected to the number of customers m. A simplified statement of the CAC problem is, "What is the maximum λ allowed such that $\Pr[T > D] < \epsilon$?" In essence, the CAC problem consists in computing the tail probability of a quantity that depends on parameters of interest. We have elaborated on the CAC problem because it is a basic design problem that appears under several disguises. A related dimensioning problem is the determination of the buffer size in a router in order not to lose more than a certain number of packets with probability p, given the arrival and service process. The above mentioned problem of Erlang is a third example. Another example treated in Chapter 18 is the server placement problem: "How many replicated servers m are needed to guarantee that any user can access the information within k hops with probability $\Pr[h_N(m) > k] \leq \epsilon$", where

ϵ is certain level of stringency and $h_N(m)$ is the number of hops towards the most nearby of the m servers in a network with N routers.

The popularity of the Internet results in a number of new challenges. The traditional mathematical models as the Erlang B formula assume "smooth" traffic flows (small correlation and Markovian in nature). However, TCP/IP traffic has been shown to be "bursty" (long-range dependent, self-similar and even chaotic, non-Markovian (Veres and Boda, 2000)). As a consequence, many traditional dimensioning and control problems ask for a new solution. The self-similar and long range dependent TCP/IP traffic is mainly caused by new complex interactions between protocols and technologies (e.g. TCP/IP/ATM/SDH) and by other information transported than voice. It is observed that the content size of information in the Internet varies considerably in size causing the "Noah effect": although immense floods are extremely rare, their occurrence impacts significantly Internet behavior on a global scale. Unfortunately, the mathematics to cope with the self-similar and long range dependent processes turns out to be fairly complex and beyond the scope of this book.

Finally, we mention the current interest in understanding and modeling complex networks such as the Internet, biological networks, social networks and utility infrastructures for water, gas, electricity and transport (cars, goods, trains). Since these networks consists of a huge number of nodes N and links L, classical and algebraic graph theory is often not suited to produce even approximate results. The beginning of probabilistic graph theory is commonly attributed to the appearance of papers by Erdös and Rényi in the late 1940s. They investigated a particularly simple growing model for a graph: start from N nodes and connect in each step an arbitrary random, not yet connected pair of nodes until all L links are used. After about $N/2$ steps, as shown in Section 16.7.1, they observed the birth of a giant component that, in subsequent steps, swallows the smaller ones at a high rate. This phenomenon is called a phase transition and often occurs in nature. In physics it is studied in, for example, percolation theory. To some extent, the Internet's graph bears some resemblance to the Erdös-Rényi random graph. The Internet is best regarded as a dynamic and growing network, whose graph is continuously changing. Yet, in order to deploy services over the Internet, an accurate graph model that captures the relevant structural properties is desirable. As shown in Part III, a probabilistic approach based on random graphs seems an efficient way to learn about the Internet's intriguing behavior. Although the Internet's topology is not a simple Erdös-Rényi random graph, results such as the hopcount of the shortest path and the size of a multicast tree deduced from the simple random graphs provide a first order estimate for the Internet. Moreover, analytic formulas based on other classes of graphs than the simple random graph prove difficult to obtain. This observation is similar to queueing theory, where, beside the M/G/x class of queues, hardly closed expressions exist.

We hope that this brief overview motivates sufficiently to surmount the mathematical barriers. Skill with probability theory is deemed necessary to understand complex phenomena in telecommunications. Once mastered, the power and beauty of mathematics will be appreciated.

Part I
Probability theory

2

Random variables

This chapter reviews basic concepts from probability theory. A random variable (rv) is a variable that takes certain values by chance. Throughout this book, this imprecise and intuitive definition suffices. The precise definition involves axiomatic probability theory (Billingsley, 1995).

Here, a distinction between discrete and continuous random variables is made, although a unified approach including also mixed cases via the Stieltjes integral (Hardy *et al.*, 1999, pp. 152–157), $\int g(x)df(x)$, is possible. In general, the distribution $F_X(x) = \Pr[X \leq x]$ holds in both cases, and

$$\int g(x)dF_X(x) = \sum_k g(k)\Pr[X = k] \quad \text{where } X \text{ is a discrete rv}$$

$$= \int g(x)\frac{dF_X(x)}{dx}dx \quad \text{where } X \text{ is a continuous rv}$$

In most practical situations, the Stieltjes integral reduces to the Riemann integral, else, Lesbesgue's theory of integration and measure theory (Royden, 1988) is required.

2.1 Probability theory and set theory

Pascal (1623–1662) is commonly regarded as one of the founders of probability theory. In his days, there was much interest in games of chance[1] and the likelihood of winning a game. In most of these games, there was a finite number n of possible outcomes and each of them was equally likely. The

---

[1] "La règle des partis", a chapter in Pascal's mathematical work (Pascal, 1954), consists of a series of letters to Fermat that discuss the following problem (together with a more complex question that is essentially a variant of the probability of gambler's ruin treated in Section 11.2.1): Consider the game in which 2 dice are thrown n times. How many times n do we have to throw the 2 dice to throw double six with probability $p = \frac{1}{2}$?

probability of the event A of interest was defined as

$$\Pr[A] = \frac{n_A}{n}$$

where n_A is the number of favorable outcomes (samples points of A). If the number of outcomes of an experiment is not finite, this classical definition of probability does not suffice anymore. In order to establish a coherent and precise theory, probability theory employs concepts of group or set theory.

The set of all possible outcomes of an experiment is called the sample space Ω. A possible outcome of an experiment is called a sample point ω that is an element of the sample space Ω. An event A consists of a set of sample points. An event A is thus a subset of the sample space Ω. The *complement* A^c of an event A consists of all sample points of the sample space Ω that are not in (the set) A, thus $A^c = \Omega \backslash A$. Clearly, $(A^c)^c = A$ and the complement of the sample space is the empty set, $\Omega^c = \emptyset$ or, vice a versa, $\emptyset^c = \Omega$. A family $\mathcal{F}$ of events is a set of events and thus a subset of the sample space Ω that possesses particular events as elements. More precisely, a family $\mathcal{F}$ of events satisfies the three conditions that define a σ-field[2]: (a) $\emptyset \in \mathcal{F}$, (b) if $A_1, A_2, \ldots \in \mathcal{F}$, then $\cup_{j=1}^{\infty} A_j \in \mathcal{F}$ and (c) if $A \in \mathcal{F}$, then $A^c \in \mathcal{F}$. These conditions guarantee that $\mathcal{F}$ is closed under countable unions and intersections of events.

Events and the probability of these events are connected by a probability measure $\Pr[.]$ that assigns to each event of the family $\mathcal{F}$ of events of a sample space Ω a real number in the interval $[0, 1]$. As **Axiom 1**, we require that $\Pr[\Omega] = 1$. If $\Pr[A] = 0$, the occurrence of the event A is not possible, while $\Pr[A] = 1$ means that the event A is certain to occur. If $\Pr[A] = p$ with $0 < p < 1$, the event A has probability p to occur.

If the events A and B have no sample points in common, $A \cap B = \emptyset$, the events A and B are called *mutually exclusive events*. As an example, the event and its complement are mutually exclusive because $A \cap A^c = \emptyset$. **Axiom 2** of a probability measure is that for mutually exclusive events A and B holds that $\Pr[A \cup B] = \Pr[A] + \Pr[B]$. The definition of a probability measure and the two axioms are sufficient to build a consistent framework on which probability theory is founded. Since $\Pr[\emptyset] = 0$ (which follows from

---

[2] A field $\mathcal{F}$ posseses the properties:

   (i) $\emptyset \in \mathcal{F}$;
   (ii) if $A, B \in \mathcal{F}$, then $A \cup B \in \mathcal{F}$ and $A \cap B \in \mathcal{F}$;
   (iii) if $A \in \mathcal{F}$, then $A^c \in \mathcal{F}$.

   This definition is redundant. For, we have by (ii) and (iii) that $(A \cup B)^c \in \mathcal{F}$. Further, by De Morgan's law $(A \cup B)^c = A^c \cap B^c$, which can be deduced from Figure 2.1 and again by (iii), the argument shows that the reduced statement (ii), if $A, B \in \mathcal{F}$, then $A \cup B \in \mathcal{F}$, is sufficient to also imply that $A \cap B \in \mathcal{F}$.

Axiom 2 because $A \cap \emptyset = \emptyset$ and $A = A \cup \emptyset$), for mutually exclusive events A and B holds that $\Pr[A \cap B] = 0$.

As a classical example that explains the formal definitions, let us consider the experiment of throwing a fair die. The sample space consists of all possible outcomes: $\Omega = \{1, 2, 3, 4, 5, 6\}$. A particular outcome of the experiment, say $\omega = 3$, is a sample point $\omega \in \Omega$. One may be interested in the event A where the outcome is even in which case $A = \{2, 4, 6\} \subset \Omega$ and $A^c = \{1, 3, 5\}$.

If A and B are events, the union of these events $A \cup B$ can be written using set theory as

$$A \cup B = (A \cap B) \cup (A^c \cap B) \cup (A \cap B^c)$$

because $A \cap B$, $A^c \cap B$ and $A \cap B^c$ are mutually exclusive events. The relation is immediately understood by drawing a Venn diagram as in Fig. 2.1. Taking

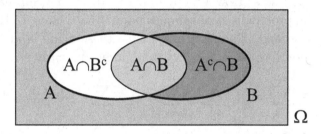

Fig. 2.1. A Venn diagram illustrating the union $A \cup B$.

the probability measure of the union yields

$$\begin{aligned}
\Pr[A \cup B] &= \Pr[(A \cap B) \cup (A^c \cap B) \cup (A \cap B^c)] \\
&= \Pr[A \cap B] + \Pr[A^c \cap B] + \Pr[A \cap B^c]
\end{aligned} \qquad (2.1)$$

where the last relation follows from Axiom 2. Figure 2.1 shows that $A = (A \cap B) \cup (A \cap B^c)$ and $B = (A \cap B) \cup (A^c \cap B)$. Since the events are mutually exclusive, Axiom 2 states that

$$\begin{aligned}
\Pr[A] &= \Pr[A \cap B] + \Pr[A \cap B^c] \\
\Pr[B] &= \Pr[A \cap B] + \Pr[A^c \cap B]
\end{aligned}$$

Substitution into (2.1) yields the important relation

$$\Pr[A \cup B] = \Pr[A] + \Pr[B] - \Pr[A \cap B] \qquad (2.2)$$

Although derived for the measure $\Pr[.]$, relation (2.2) also holds for other measures, for example, the cardinality (the number of elements) of a set.

2.1.1 The inclusion-exclusion formula

A generalization of the relation (2.2) is the *inclusion-exclusion formula*,

$$\Pr\left[\cup_{k=1}^n A_k\right] = \sum_{k_1=1}^n \Pr\left[A_{k_1}\right] - \sum_{k_1=1}^n \sum_{k_2=k_1+1}^n \Pr\left[A_{k_1} \cap A_{k_2}\right]$$

$$+ \sum_{k_1=1}^n \sum_{k_2=k_1+1}^n \sum_{k_3=k_2+1}^n \Pr\left[A_{k_1} \cap A_{k_2} \cap A_{k_3}\right]$$

$$+ \cdots + (-1)^{n-1} \sum_{k_1=1}^n \sum_{k_2=k_1+1}^n \cdots \sum_{k_n=k_{n-1}+1}^n \Pr\left[\cap_{j=1}^n A_{k_j}\right] \quad (2.3)$$

The formula shows that the probability of the union consists of the sum of probabilities of the individual events (first term). Since sample points can belong to more than one event A_k, the first term possesses double countings. The second term removes all probabilities of samples points that belong to precisely two event sets. However, by doing so (draw a Venn diagram), we also subtract the probabilities of samples points that belong to three events sets more than needed. The third term adds these again, and so on. The inclusion-exclusion formula can be written more compactly as,

$$\Pr\left[\cup_{k=1}^n A_k\right] = \sum_{j=1}^n (-1)^{j-1} \sum_{k_1=1}^n \sum_{k_2=k_1+1}^n \cdots \sum_{k_j=k_{j-1}+1}^n \Pr\left[\cap_{m=1}^j A_{k_m}\right] \quad (2.4)$$

or with

$$S_j = \sum_{1 \le k_1 < k_2 < \ldots < k_j \le n} \Pr\left[\cap_{m=1}^j A_{k_m}\right]$$

as

$$\Pr\left[\cup_{k=1}^n A_k\right] = \sum_{j=1}^n (-1)^{j-1} S_j \quad (2.5)$$

Proof of the inclusion-exclusion formula[3]: Let $A = \cup_{k=1}^{n-1} A_k$ and $B = A_n$ such that

---

[3] Another proof (Grimmett and Stirzacker, 2001, p. 56) uses the indicator function defined in Section 2.2.1. Useful indicator function relations are

$$1_{A \cap B} = 1_A 1_B$$
$$1_{A^c} = 1 - 1_A$$
$$1_{A \cup B} = 1 - 1_{(A \cup B)^c} = 1 - 1_{A^c \cap B^c} = 1 - 1_{A^c} 1_{B^c}$$
$$= 1 - (1 - 1_A)(1 - 1_B) = 1_A + 1_B + 1_A 1_B = 1_A + 1_B + 1_{A \cap B}$$

Generalizing the last relation yields

$$1_{\cup_{k=1}^n A_k} = 1 - \prod_{k=1}^n (1 - 1_{A_k})$$

Multiplying out and taking the expectations using (2.13) leads to (2.3).

$A \cup B = \cup_{k=1}^{n} A_k$ and $A \cap B = A_n \cap \left(\cup_{k=1}^{n-1} A_k\right) = \cup_{k=1}^{n-1} A_k \cap A_n$ by the distributive law in set theory, then application of (2.2) yields the recursion in n

$$\Pr\left[\cup_{k=1}^{n} A_k\right] = \Pr\left[\cup_{k=1}^{n-1} A_k\right] + \Pr\left[A_n\right] - \Pr\left[\cup_{k=1}^{n-1} A_k \cap A_n\right] \tag{2.6}$$

By direct substitution of $n \to n-1$, we have

$$\Pr\left[\cup_{k=1}^{n-1} A_k\right] = \Pr\left[\cup_{k=1}^{n-2} A_k\right] + \Pr\left[A_{n-1}\right] - \Pr\left[\cup_{k=1}^{n-2} A_k \cap A_{n-1}\right]$$

while substitution in this formula of $A_k \to A_k \cap A_n$ gives

$$\Pr\left[\cup_{k=1}^{n-1} A_k \cap A_n\right] = \Pr\left[\cup_{k=1}^{n-2} A_k \cap A_n\right] + \Pr\left[A_{n-1} \cap A_n\right] - \Pr\left[\cup_{k=1}^{n-2} A_k \cap A_n \cap A_{n-1}\right]$$

Substitution of the last two terms into (2.6) yields

$$\Pr\left[\cup_{k=1}^{n} A_k\right] = \Pr\left[A_{n-1}\right] + \Pr\left[A_n\right] - \Pr\left[A_{n-1} \cap A_n\right] + \Pr\left[\cup_{k=1}^{n-2} A_k\right]$$
$$- \Pr\left[\cup_{k=1}^{n-2} A_k \cap A_{n-1}\right] - \Pr\left[\cup_{k=1}^{n-2} A_k \cap A_n\right] + \Pr\left[\cup_{k=1}^{n-2} A_k \cap A_n \cap A_{n-1}\right] \tag{2.7}$$

Similarly, in a next iteration we use (2.6) after suitable modification in the right-hand side of (2.7) to lower the upper index in the union,

$$\Pr\left[\cup_{k=1}^{n-2} A_k\right] = \Pr\left[\cup_{k=1}^{n-3} A_k\right] + \Pr\left[A_{n-2}\right] - \Pr\left[\cup_{k=1}^{n-3} A_k \cap A_{n-2}\right]$$
$$\Pr\left[\cup_{k=1}^{n-2} A_k \cap A_{n-1}\right] = \Pr\left[\cup_{k=1}^{n-3} A_k \cap A_{n-1}\right] + \Pr\left[A_{n-2} \cap A_{n-1}\right]$$
$$- \Pr\left[\cup_{k=1}^{n-3} A_k \cap A_{n-1} \cap A_{n-2}\right]$$
$$\Pr\left[\cup_{k=1}^{n-2} A_k \cap A_n\right] = \Pr\left[\cup_{k=1}^{n-3} A_k \cap A_n\right] + \Pr\left[A_{n-2} \cap A_n\right] - \Pr\left[\cup_{k=1}^{n-3} A_k \cap A_n \cap A_{n-2}\right]$$
$$\Pr\left[\cup_{k=1}^{n-2} A_k \cap A_n \cap A_{n-1}\right] = \Pr\left[\cup_{k=1}^{n-3} A_k \cap A_n \cap A_{n-1}\right] + \Pr\left[A_{n-2} \cap A_n \cap A_{n-1}\right]$$
$$- \Pr\left[\cup_{k=1}^{n-3} A_k \cap A_n \cap A_{n-1} \cap A_{n-2}\right]$$

The result is

$$\Pr\left[\cup_{k=1}^{n} A_k\right] = \Pr\left[A_{n-2}\right] + \Pr\left[A_{n-1}\right] + \Pr\left[A_n\right] + - \Pr\left[A_{n-2} \cap A_{n-1}\right] - \Pr\left[A_{n-2} \cap A_n\right]$$
$$- \Pr\left[A_{n-1} \cap A_n\right] + \Pr\left[A_{n-2} \cap A_{n-1} \cap A_n\right] + \Pr\left[\cup_{k=1}^{n-3} A_k\right]$$
$$- \Pr\left[\cup_{k=1}^{n-3} A_k \cap A_{n-2}\right] - \Pr\left[\cup_{k=1}^{n-3} A_k \cap A_{n-1}\right] - \Pr\left[\cup_{k=1}^{n-3} A_k \cap A_n\right]$$
$$+ \Pr\left[\cup_{k=1}^{n-3} A_k \cap A_{n-1} \cap A_{n-2}\right] + \Pr\left[\cup_{k=1}^{n-3} A_k \cap A_n \cap A_{n-2}\right]$$
$$+ \Pr\left[\cup_{k=1}^{n-3} A_k \cap A_n \cap A_{n-1}\right] - \Pr\left[\cup_{k=1}^{n-3} A_k \cap A_n \cap A_{n-1} \cap A_{n-2}\right]$$

which starts revealing the structure of (2.3). Rather than continuing the iterations, we prove the validity of the inclusion-exclusion formula (2.3) via induction. In case $n = 2$, the basic expression (2.2) is found. Assume that (2.3) holds for n, then the case for $n+1$ must obey (2.6) where $n \to n+1$,

$$\Pr\left[\cup_{k=1}^{n+1} A_k\right] = \Pr\left[\cup_{k=1}^{n} A_k\right] + \Pr\left[A_{n+1}\right] - \Pr\left[\cup_{k=1}^{n} A_k \cap A_{n+1}\right]$$

Substitution of (2.3) into the above expression yields, after suitable grouping of the terms,

$$\Pr\left[\cup_{k=1}^{n+1}A_k\right]=\Pr[A_{n+1}]+\sum_{k_1=1}^{n}\Pr[A_{k_1}]-\sum_{k_1=1}^{n}\sum_{k_2=k_1+1}^{n}\Pr[A_{k_1}\cap A_{k_2}]-\sum_{k_1=1}^{n}\Pr[A_{k_1}\cap A_{n+1}]$$

$$+\sum_{k_1=1}^{n}\sum_{k_2=k_1+1}^{n}\sum_{k_3=k_2+1}^{n}\Pr[A_{k_1}\cap A_{k_2}\cap A_{k_3}]+\sum_{k_1=1}^{n}\sum_{k_2=k_1+1}^{n}\Pr[A_{k_1}\cap A_{k_2}\cap A_{n+1}]$$

$$+\cdots+(-1)^{n-1}\sum_{k_1=1}^{n}\sum_{k_2=k_1+1}^{n}\cdots\sum_{k_n=k_{n-1}+1}^{n}\Pr\left[\cap_{j=1}^{n}A_{k_j}\right]-$$

$$+\cdots+(-1)^{n}\sum_{k_1=1}^{n}\sum_{k_2=k_1+1}^{n}\cdots\sum_{k_n=k_{n-1}+1}^{n}\Pr\left[\cap_{j=1}^{n}A_{k_j}\cap A_{n+1}\right]$$

$$=\sum_{k_1=1}^{n+1}\Pr[A_k]-\sum_{k_1=1}^{n+1}\sum_{k_2=k_1+1}^{n+1}\Pr\left[A_{k_1}\cap A_{k_2}\right]$$

$$+\sum_{k_1=1}^{n+1}\sum_{k_2=k_1+1}^{n+1}\sum_{k_3=k_2+1}^{n+1}\Pr\left[A_{k_1}\cap A_{k_2}\cap A_{k_3}\right]$$

$$+\cdots+(-1)^{n}\sum_{k_1=1}^{n+1}\sum_{k_2=k_1+1}^{n+1}\cdots\sum_{k_{n+1}=k_n+1}^{n+1}\Pr\left[\cap_{j=1}^{n}A_{k_j}\cap A_{n+1}\right]$$

which proves (2.3). □

Although impressive, the inclusion-exclusion formula is useful when dealing with dependent random variables because of its general nature. In particular, if $\Pr\left[\cap_{m=1}^{j}A_{k_m}\right]=a_j$ and not a function of the specific indices k_m, the inclusion-exclusion formula (2.4) becomes more attractive,

$$\Pr\left[\cup_{k=1}^{n}A_k\right]=\sum_{j=1}^{n}(-1)^{j-1}a_j\sum_{1\le k_1<k_2<...<k_j\le n}1$$

$$=\sum_{j=1}^{n}(-1)^{j-1}\binom{n}{j}a_j$$

An application of the latter formula to multicast can be found in Chapter 17 and many others are in Feller (1970, Chapter IV). Sometimes it is useful to reason with the complement of the union $(\cup_{k=1}^{n}A_k)^c=\Omega\backslash\cup_{k=1}^{n}A_k=\cap_{k=1}^{n}A_k^c$. Applying Axiom 2 to $(\cup_{k=1}^{n}A_k)^c\cup(\cup_{k=1}^{n}A_k)=\Omega$,

$$\Pr\left[(\cup_{k=1}^{n}A_k)^c\right]=\Pr\left[\Omega\right]-\Pr\left[\cup_{k=1}^{n}A_k\right]$$

and using Axiom 1 and the inclusion-exclusion formula (2.5), we obtain

$$\Pr\left[(\cup_{k=1}^{n}A_k)^c\right]=1-\sum_{j=1}^{n}(-1)^{j-1}S_j=\sum_{j=0}^{n}(-1)^{j}S_j \qquad (2.8)$$

with the convention that $S_0 = 1$. The Boole's inequalities

$$\Pr\left[\cup_{k=1}^n A_k\right] \leq \sum_{k=1}^n \Pr\left[A_k\right] \qquad (2.9)$$

$$\Pr\left[\cap_{k=1}^n A_k\right] \geq 1 - \sum_{k=1}^n \Pr\left[A_k^c\right]$$

are derived as consequences of the inclusion-exclusion formula (2.3). Only if all events are mutually exclusive, the equality sign in (2.9) holds whilst the inequality sign follows from the fact that possible overlaps in events are, in contrast to the inclusion-exclusion formula (2.3), not subtracted.

The inclusion-exclusion formula is of a more general nature and also applies to other measures on sets than $\Pr[.]$, for example to the cardinality as mentioned above. For the cardinality of a set A, which is usually denoted by $|A|$, the inclusion-exclusion variant of (2.8) is

$$\left|(\cup_{k=1}^n A_k)^c\right| = \sum_{j=0}^n (-1)^j |S_j| \qquad (2.10)$$

where the total number of elements in the sample space is $|S_0| = N$ and

$$|S_j| = \sum_{1 \leq k_1 < k_2 < \ldots < k_j \leq n} \left|\cap_{m=1}^j A_{k_m}\right|$$

A nice illustration of the above formula (2.10) applies to the *sieve of Eratosthenes* (Hardy and Wright, 1968, p. 4), a procedure to construct the table of prime numbers[4] up to N. Consider the increasing sequence of integers

$$\Omega = \{2, 3, 4, \ldots, N\}$$

and remove successively all multiples of 2 (even numbers starting from 4, 6, ...), all multiples of 3 (starting from 3^2 and not yet removed previously), all multiples of 5, all multiples of the next number larger than 5 and still in the list (which is the prime 7) and so on, up to all multiples of the largest possible prime divisor that is equal to or smaller than $\left[\sqrt{N}\right]$. Here $[x]$ is the largest integer smaller than or equal to x. The remaining numbers in the list are prime numbers. Let us now compute the number of primes $\pi(N)$ smaller than or equal to N by using the inclusion-exclusion formula (2.10).

---

[4] An integer number p is prime if $p > 1$ and p has no other integer divisors than 1 and itself p. The sequence of the first primes are 2, 3, 5, 7, 11, 13, etc. If a and b are divisors of n, then $n = ab$ from which it follows that a and b cannot exceed both $\sqrt{n}$. Hence, any composite number n is divisible by a prime p that does not exceed $\sqrt{n}$.

The number of primes smaller than a real number x is $\pi(x)$ and, evidently, if p_n denotes the n-th prime, then $\pi(p_n) = n$. Let A_k denote the set of the multiples of the k-th prime p_k that belong to Ω. The number of such sets A_k in the sieve of Eratosthenes is equal to the largest prime number p_n smaller than or equal to $\left[\sqrt{N}\right]$, hence, $n = \pi\left(\sqrt{N}\right)$. If $q \in (\cup_{k=1}^n A_k)^c$, this means that q is not divisible by each prime number smaller than p_n and that q is a prime number lying between $\sqrt{N} < q \le N$. The cardinality of the set $(\cup_{k=1}^n A_k)^c$, the number of primes between $\sqrt{N} < q \le N$ is

$$|(\cup_{k=1}^n A_k)^c| = \pi(N) - \pi\left(\sqrt{N}\right)$$

On the other hand, if $r \in \cap_{m=1}^j A_{k_m}$ for $1 \le k_1 < k_2 < \cdots < k_j \le n$, then r is a multiple of $p_{k_1} p_{k_2} \ldots p_{k_j}$ and the number of multiples of the integer $p_{k_1} p_{k_2} \ldots p_{k_j}$ in Ω is

$$\left[\frac{N}{p_{k_1} p_{k_2} \cdots p_{k_j}}\right] = \left|\cap_{m=1}^j A_{k_m}\right|$$

Applying the inclusion-exclusion formula (2.10) with $|\Omega| = S_0 = N - 1$ and $n = \pi\left(\sqrt{N}\right)$ gives

$$\pi(N) - \pi\left(\sqrt{N}\right) = N - 1 - \sum_{j=1}^n (-1)^j \sum_{1 \le k_1 < k_2 < \ldots < k_j \le n} \left[\frac{N}{p_{k_1} p_{k_2} \cdots p_{k_j}}\right]$$

The knowledge of the prime numbers smaller than or equal to $\left[\sqrt{N}\right]$, i.e. the first $n = \pi\left(\sqrt{N}\right)$ primes, suffices to compute the number of primes $\pi(N)$ smaller than or equal to N without explicitly knowing the primes q lying between $\sqrt{N} < q \le N$.

2.2 Discrete random variables

Discrete random variables are real functions X defined on a discrete probability space Ω as $X : \Omega \to \mathbb{R}$ with the property that the event

$$\{\omega \in \Omega : X(\omega) = x\} \in \mathcal{F}$$

for each $x \in \mathbb{R}$. The event $\{\omega \in \Omega : X(\omega) = x\}$ is further abbreviated as $\{X = x\}$. A discrete probability density function (pdf) $\Pr[X = x]$ has the following properties:

(i) $0 \le \Pr[X = x] \le 1$ for real x that are possible outcomes of an experiment. The set of values x can be finite or countably infinite and constitute the discrete probability space.

(ii) $\sum_x \Pr[X = x] = 1$.

In the classical example of throwing a die, the discrete probability space $\Omega = \{1, 2, 3, 4, 5, 6\}$ and, since each of the six edges of the (fair) die is equally possible as outcome, $\Pr[X = x] = \frac{1}{6}$ for each $x \in \Omega$.

2.2.1 The expectation

An important operator acting on a discrete random variable X is the expectation, defined as

$$E[X] = \sum_x x \Pr[X = x] \tag{2.11}$$

The expectation $E[X]$ is also called the mean or average or first moment of X. More generally, if X is a discrete random variable and g is a function, then $Y = g(X)$ is also a discrete random variable with expectation $E[Y]$ equal to

$$E[g(X)] = \sum_x g(x) \Pr[X = x] \tag{2.12}$$

A special and often used function in probability theory is the indicator function 1_y defined as 1 if the condition y is true and otherwise it is zero. For example,

$$E[1_{X>a}] = \sum_x 1_{x>a} \Pr[X = x] = \sum_{x>a} \Pr[X = x] = \Pr[X > a]$$

$$E[1_{X=a}] = \Pr[X = a] \tag{2.13}$$

The higher moments of a random variable are defined as the case where $g(x) = x^n$,

$$E[X^n] = \sum_x x^n \Pr[X = x] \tag{2.14}$$

From the definition (2.11), it follows that the expectation is a linear operator,

$$E\left[\sum_{k=1}^n a_k X_k\right] = \sum_{k=1}^n a_k E[X_k]$$

The variance of X is defined as

$$\mathrm{Var}[X] = E\left[(X - E[X])^2\right] \tag{2.15}$$

The variance is always non-negative. Using the linearity of the expectation operator and $\mu = E[X]$, we rewrite (2.15) as

$$\text{Var}[X] = E\left[X^2\right] - \mu^2 \tag{2.16}$$

Since $\text{Var}[X] \geq 0$, relation (2.16) indicates that $E\left[X^2\right] \geq (E[X])^2$. Often the standard deviation, defined as $\sigma = \sqrt{\text{Var}[X]}$, is used. An interesting variational principle of the variance follows, for the variable u, from

$$E\left[(X-u)^2\right] = E\left[(X-\mu)^2\right] + (u-\mu)^2$$

which is minimized at $u = \mu = E[X]$ with value $\text{Var}[X]$. Hence, the best least square approximation of the random variable X is the number $E[X]$.

2.2.2 The probability generating function

The probability generating function (pgf) of a discrete random variable X is defined, for complex z, as

$$\varphi_X(z) = E\left[z^X\right] = \sum_x z^x \Pr[X = x] \tag{2.17}$$

where the last equality follows from (2.12). If X is integer-valued and non-negative, then the pgf is the Taylor expansion of the complex function $\varphi_X(z)$. Commonly the latter restriction applies, otherwise the substitution $z = e^{it}$ is used such that (2.17) expresses the Fourier series of $\varphi_X\left(e^{it}\right)$. The importance of the pgf mainly lies in the fact that the theory of functions can be applied. Numerous examples of the power of analysis will be illustrated. Concentrating on non-negative integer random variables X,

$$\varphi_X(z) = \sum_{k=0}^{\infty} \Pr[X = k]\, z^k \tag{2.18}$$

and the Taylor coefficients obey

$$\Pr[X = k] = \frac{1}{k!} \left.\frac{d^k \varphi_X(z)}{dz^k}\right|_{z=0} \tag{2.19}$$

$$= \frac{1}{2\pi i} \int_{C(0)} \frac{\varphi_X(z)}{z^{k+1}} dz \tag{2.20}$$

where $C(0)$ denotes a contour around $z = 0$. Both are inversion formulae[5]. Since the general form $E[g(X)]$ is completely defined when $\Pr[X = x]$ is

---

[5] A similar inversion formula for Fourier series exist (see e.g. Titchmarsh (1948)).

known, the knowledge of the pgf results in a complete alternative description,

$$E\left[g(X)\right] = \sum_{k=0}^{\infty} \frac{g(k)}{k!} \frac{d^k \varphi_X(z)}{dz^k}\Bigg|_{z=0} \tag{2.21}$$

Sometimes it is more convenient to compute values of interest directly from (2.17) rather than from (2.21). For example, n-fold differentiation of $\varphi_X(z) = E\left[z^X\right]$ yields

$$\frac{d^n \varphi_X(z)}{dz^n} = E\left[X(X-1)\cdots(X-n+1)z^{X-n}\right] = \frac{1}{n!} E\left[\binom{X}{n} z^{X-n}\right]$$

such that

$$E\left[\binom{X}{n}\right] = \frac{1}{n!} \frac{d^n \varphi_X(z)}{dz^n}\Bigg|_{z=1} \tag{2.22}$$

Similarly, let $z = e^t$, then

$$\frac{d^n \varphi_X(e^t)}{dt^n} = E\left[X^n e^{tX}\right]$$

from which the moments follow as

$$E\left[X^n\right] = \frac{d^n \varphi_X(e^t)}{dt^n}\Bigg|_{t=0} \tag{2.23}$$

and, more generally,

$$E\left[(X-a)^n\right] = \frac{d^n\left(e^{-ta}\varphi_X(e^t)\right)}{dt^n}\Bigg|_{t=0} \tag{2.24}$$

2.2.3 The logarithm of the probability generating function

The logarithm of the probability generating function is defined as

$$L_X(z) = \log\left(\varphi_X(z)\right) = \log\left(E\left[z^X\right]\right) \tag{2.25}$$

from which $L_X(1) = 0$ because $\varphi_X(1) = 1$. The derivative $L_X'(z) = \frac{\varphi_X'(z)}{\varphi_X(z)}$ shows that $L_X'(1) = \varphi_X'(1)$, while from $L_X''(z) = \frac{\varphi_X''(z)}{\varphi_X(z)} - \left(\frac{\varphi_X'(z)}{\varphi_X(z)}\right)^2$, it follows that $L_X''(1) = \varphi_X''(1) - (\varphi_X'(1))^2$. These first few derivatives are interesting because they are related directly to probabilistic quantities. Indeed, from (2.23), we observe that

$$E[X] = \varphi_X'(1) = L_X'(1) \tag{2.26}$$

and from $E[X^2] = \varphi''_X(1) + \varphi'_X(1)$

$$\text{Var}[X] = \varphi''_X(1) + \varphi'_X(1) - \left(\varphi'_X(1)\right)^2$$
$$= L''_X(1) + L'_X(1) \tag{2.27}$$

2.3 Continuous random variables

Although most of the concepts defined above for discrete random variables are readily transferred to continuous random variables, the calculus is in general more difficult. Indeed, instead of reasoning on the pdf, it is more convenient to work with the probability distribution function defined for both discrete and continuous random variables as

$$F_X(x) = \Pr[X \leq x] \tag{2.28}$$

Clearly, we have $\lim_{x \to -\infty} F_X(x) = 0$, while $\lim_{x \to +\infty} F_X(x) = 1$. Further, $F_X(x)$ is non-decreasing in x and

$$\Pr[a < X \leq b] = F_X(b) - F_X(a) \tag{2.29}$$

This relation follows from the observations $\{X \leq a\} \cup \{a < X \leq b\} = \{X \leq b\}$ and $\{X \leq a\} \cap \{a < X \leq b\} = \emptyset$. For mutually exclusive events $A \cap B = \emptyset$, Axiom 2 in Section 2.1 states that $\Pr[A \cup B] = \Pr[A] + \Pr[B]$ which proves (2.29). As a corollary of (2.29), $F_X(x)$ is continuous at the right which follows from (2.29) by denoting $a = b - \epsilon$ for any $\epsilon > 0$. Less precise, it follows from the equality sign at the right, $X \leq b$, and inequality at the left, $a < X$. Hence, $F_X(x)$ is not necessarily continuous at the left which implies that $F_X(x)$ is not necessarily continuous and that $F_X(x)$ may possess jumps. But even if $F_X(x)$ is continuous, the pdf is not necessary continuous[6].

The pdf of a continuous random variable X is defined as

$$f_X(x) = \frac{dF_X(x)}{dx} \tag{2.30}$$

---

[6] Weierstrass was the first to present a continuous non-differentiable function,

$$f(x) = \sum_{n=0}^{\infty} b^n \cos(a^n \pi x)$$

where $0 < b < 1$ and a is an odd positive integer. Since the series is uniformly convergent for any x, $f(x)$ is continuous everywhere. Titchmarsh (1964, Chapter IX) demonstrates for $ab > 1 + \frac{3\pi}{2}$ that $\frac{f(x+h)-f(x)}{h}$ takes arbitrarily large values such that $f'(x)$ does not exist. Another class of continuous non-differentiable functions are the sample paths of a Brownian motion. The Cantor function which is discussed in (Berger, 1993, p. 21) and (Billingsley, 1995, p. 407) is an other classical, noteworthy function with peculiar properties.

Assuming that $F_X(x)$ is differentiable at x, from (2.29), we have for small, positive Δx

$$\Pr\left[x < X \le x + \Delta x\right] = F_X(x + \Delta x) - F_X(x)$$
$$= \frac{dF_X(x)}{dx}\Delta x + O\left((\Delta x)^2\right)$$

Using the definition (2.30) indicates that, if $F_X(x)$ is differentiable at x,

$$f_X(x) = \lim_{\Delta x \to 0} \frac{\Pr\left[x < X \le x + \Delta x\right]}{\Delta x} \tag{2.31}$$

If $f_X(x)$ is finite, then $\lim_{\Delta x \to 0} \Pr\left[x < X \le x + \Delta x\right] = \Pr\left[X = x\right] = 0$, which means that for well-behaved (i.e. $F_X(x)$ is differentiable for most x) continuous random variables X, the event that X precisely equals x is zero[7]. Hence, for well-behaved continuous random variables where $\Pr\left[X = x\right] = 0$ for all x, the inequality signs in the general formula (2.29) can be relaxed,

$$\Pr\left[a < X \le b\right] = \Pr\left[a \le X \le b\right] = \Pr\left[a \le X < b\right] = \Pr\left[a < X < b\right]$$

If $f_X(x)$ is not finite, then $F_X(x)$ is not differentiable at x such that

$$\lim_{\Delta x \to 0} F_X(x + \Delta x) - F_X(x) = \Delta F_X(x) \ne 0$$

This means that $F_X(x)$ jumps upwards at x over $\Delta F_X(x)$. In that case, there is a probability mass with magnitude $\Delta F_X(x)$ at the point x. Although the second definition (2.31) is strictly speaking not valid in that case, one sometimes denotes the pdf at $y = x$ by $f_X(y) = \Delta F_X(x)\delta(y - x)$ where $\delta(x)$ is the Dirac impulse or delta function with basic property that $\int_{-\infty}^{+\infty} \delta(y - x)dx = 1$. Even apart from the above-mentioned difficulties for certain classes of non-differentiable, but continuous functions, the fact that probabilities are always confined to the region $[0,1]$ may suggest that $0 \le f_X(x) \le 1$. However, the second definition (2.31) shows that $f_X(x)$ can be much larger than 1. For example, if X is a Gaussian random variable with mean μ and variance σ^2 (see Section 3.2.3) then $f_X(\mu) = \frac{1}{\sqrt{2\pi}\sigma}$ can be made arbitrarily large. In fact,

$$\lim_{\sigma \to 0} \frac{\exp\left(-\frac{(x-\mu)^2}{2\sigma^2}\right)}{\sqrt{2\pi}\sigma} = \delta(x - \mu)$$

---

[7] In Lesbesgue measure theory (Titchmarsh, 1964; Billingsley, 1995), it is said that a countable, finite or enumerable (i.e. function evaluations at individual points) set is measurable, but its measure is zero.

2.3.1 Transformation of random variables

It frequently appears useful to know how to compute $F_Y(x)$ for $Y = g(X)$. Only if the inverse function g^{-1} exists, the event $\{g(X) \le x\}$ is equivalent to $\{X \le g^{-1}(x)\}$ if $\frac{dg}{dx} > 0$ and to $\{X > g^{-1}(x)\}$ if $\frac{dg}{dx} < 0$. Hence,

$$F_Y(x) = \Pr\left[g(X) \le x\right] = \begin{cases} F_X\left(g^{-1}(x)\right), & \frac{dg}{dx} > 0 \\ 1 - F_X\left(g^{-1}(x)\right), & \frac{dg}{dx} < 0 \end{cases} \quad (2.32)$$

For well-behaved continuous random variables, we may rewrite (2.31) in terms of differentials,

$$f_X(x)\, dx = \Pr\left[x \le X \le x + dx\right]$$

and, similarly for $f_Y(y)$,

$$f_Y(y)\, dy = \Pr\left[y \le Y = g(X) \le y + dy\right]$$

If g is increasing, then the event $\{y \le g(X) \le y + dy\}$ is equivalent to $\{g^{-1}(y) \le X \le g^{-1}(y + dy)\} = \{x \le X \le x + dx\}$ such that

$$f_Y(y)\, dy = f_X(x)\, dx$$

If g is deceasing, we find that $f_Y(y)\, dy = -f_X(x)\, dx$. Thus, if g^{-1} and g' exists, then the relation between the pdf of a well-behaved continuous random variable X and that of the transformed random variable $Y = g(X)$ is

$$f_Y(y) = f_X(x)\left|\frac{dx}{dy}\right| = \frac{f_X(x)}{|g'(x)|}$$

This expression also follows by straightforward differentiation of (2.32). The chi-square distribution introduced in Section 3.3.3 is a nice example of the transformation of random variables.

2.3.2 The expectation

Analogously to the discrete case, we define the expectation of a continuous random variable as

$$E[X] = \int_{-\infty}^{\infty} x f_X(x)\, dx \quad (2.33)$$

In addition for the expectation to exist[8], we require $\int_{-\infty}^{\infty} |x|\, f_X(x)\, dx < \infty$. If X is a continuous random variable and g is a continuous function, then

---

[8] This requirement is borrowed from measure theory and Lebesgue integration (Titchmarsh, 1964, Chapter X)(Royden, 1988, Chapter 4), where a measurable function is said to be integrable (in the Lebesgue sense) over A if $f^+ = \max(f(x), 0)$ and $f^- = \max(-f(x), 0)$ are both integrable over A. Although this restriction seems only of theoretical interest, in some applications (see the

$Y = g(X)$ is also a continuous random variable with expectation $E[Y]$ equal to

$$E[g(X)] = \int_{-\infty}^{\infty} g(x) f_X(x) dx \qquad (2.34)$$

It is often useful to express the expectation $E[X]$ of a non-negative random variable X in tail probabilities. Upon integration by parts,

$$E[X] = \int_0^{\infty} x f_X(x) dx = -x \int_x^{\infty} f_X(u) du \Big|_0^{\infty} + \int_0^{\infty} dx \int_x^{\infty} f_X(u) du$$

$$= \int_0^{\infty} (1 - F_X(x)) \, dx \qquad (2.35)$$

The case for a non-positive random variable X is derived analogously,

$$E[X] = \int_{-\infty}^0 x f_X(x) dx = x \int_{-\infty}^x f_X(u) du \Big|_{-\infty}^0 - \int_{-\infty}^0 dx \int_{-\infty}^x f_X(u) du$$

$$= -\int_{-\infty}^0 F_X(x) dx$$

The general case follows by addition:

$$E[X] = \int_0^{\infty} (1 - F_X(x)) \, dx - \int_{-\infty}^0 F_X(x) dx$$

A similar expression exists for discrete random variables. In general for any discrete random variable X, we can write

$$E[X] = \sum_{k=-\infty}^{\infty} k \Pr[X = k] = \sum_{k=-\infty}^{-1} k \Pr[X = k] + \sum_{k=0}^{\infty} k \Pr[X = k]$$

$$= \sum_{k=-\infty}^{-1} k \left(\Pr[X \le k] - \Pr[X \le k-1] \right) + \sum_{k=0}^{\infty} k \left(\Pr[X \ge k] - \Pr[X \ge k+1] \right)$$

$$= \sum_{k=-\infty}^{-1} k \Pr[X \le k] - \sum_{k=-\infty}^{-2} (k+1) \Pr[X \le k] + \sum_{k=1}^{\infty} k \Pr[X \ge k] - \sum_{k=1}^{\infty} (k-1) \Pr[X \ge k]$$

$$= -\Pr[X \le -1] - \sum_{k=-\infty}^{-2} \Pr[X \le k] + \sum_{k=1}^{\infty} \Pr[X \ge k]$$

Cauchy distribution defined in (3.38)) the Riemann integral may exists where the Lesbesgue does not. For example, $\int_0^{\infty} \frac{\sin x}{x} dx$ equals, in the Riemann sense, $\frac{\pi}{2}$ (which is a standard excercise in contour integration), but this integral does not exists in the Lesbesgue sense. Only for improper integrals (integration interval is infinite), Riemann integration may exist where Lesbesgue does not. However, in most other cases (integration over a finite interval), Lesbesgue integration is more general. For instance, if $f(x) = 1_{\{x \text{ is rational}\}}$, then $\int_0^1 f(u) du$ does not exist in the Riemann sense (since upper and lower sums do not converge to each other). However, $\int_0^1 f(u) du = 0$ in the Lesbesgue sense (since there is only a set of measure zero different from 0, namely all rational numbers in $[0, 1]$). In probability theory and measure theory, Lesbesgue integration is assumed.

or the mean of a discrete random variable X expressed in tail probabilities is[9]

$$E[X] = \sum_{k=1}^{\infty} \Pr[X \geq k] - \sum_{k=-\infty}^{-1} \Pr[X \leq k] \qquad (2.36)$$

2.3.3 The probability generating function

The probability generating function (pgf) of a continuous random variable X is defined, for complex z, as the Laplace transform

$$\varphi_X(z) = E\left[e^{-zX}\right] = \int_{-\infty}^{\infty} e^{-zt} f_X(t) dt \qquad (2.37)$$

Again, in some cases, it may be more convenient to use $z = iu$ in which case the double sided Laplace transform reduces to a Fourier transform. The strength of these transforms is based on the numerous properties, especially the inverse transform,

$$f_X(t) = \frac{1}{2\pi i} \int_{c-i\infty}^{c+i\infty} \varphi_X(z) e^{zt} dz \qquad (2.38)$$

where c is the smallest real variable $\text{Re}(z)$ for which the integral in (2.37) converges. Similarly as for discrete random variables, we have $e^{za}\varphi_X(z) = E\left[e^{-z(X-a)}\right]$

$$E[(X-a)^n] = (-1)^n \left.\frac{d^n\left(e^{za}\varphi_X(z)\right)}{dz^n}\right|_{z=0} \qquad (2.39)$$

The main difference with the discrete case lies in the definition $E\left[e^{-zX}\right]$ (continuous) versus $E\left[z^X\right]$ (discrete). Since the exponential is an entire

---

[9] We remark that

$$E[X] = \sum_{k=-\infty}^{\infty} k\Pr[X = k] = \sum_{k=-\infty}^{\infty} k\left(\Pr[X \geq k] - \Pr[X \geq k+1]\right)$$

$$\neq \sum_{k=-\infty}^{\infty} k\Pr[X \geq k] - \sum_{k=-\infty}^{\infty} k\Pr[X \geq k+1] = \sum_{k=-\infty}^{\infty} \Pr[X \geq k]$$

because the series in the second line are diverging. In fact, there exists a finite integer k_ϵ such that, for any real arbitrarily small $\epsilon > 0$ holds that $\Pr[X \geq k_\epsilon] = 1 - \epsilon$ and $\Pr[X \geq k_\epsilon] \leq \Pr[X \geq k]$ for all $k < k_\epsilon$. Hence,

$$E[X] = \sum_{k=-\infty}^{k_\epsilon} \Pr[X \geq k] + \sum_{k=k_\epsilon}^{\infty} \Pr[X \geq k] \geq (1 - \epsilon) \sum_{k=-\infty}^{k_\epsilon} 1 + c \to \infty$$

where $\sum_{k=k_\epsilon}^{\infty} \Pr[X \geq k] = c$ is finite. Also, even for negative X, $\sum_{k=-\infty}^{\infty} \Pr[X \geq k]$ is always positive.

function[10] with power series around $z = 0$, $e^{-zX} = \sum_{k=0}^{\infty} \frac{(-1)^k X^k}{k!} z^k$, the expectation and summation can be reversed leading to

$$E\left[e^{-zX}\right] = \sum_{k=0}^{\infty} \frac{(-1)^k E\left[X^k\right]}{k!} z^k \tag{2.40}$$

provided[11] $E\left[X^k\right] = O\left(k!\right)$ which is a necessary condition for the summation to converge for $z \neq 0$. Assuming convergence[12], the Taylor series of $E\left[e^{-zX}\right]$ around $z = 0$ is expressed as function of the moments of X, whereas in the discrete case, the Taylor series of $E\left[z^X\right]$ around $z = 0$ given by (2.18) is expressed in terms of probabilities of X. This observation has led to call $E\left[e^{-zX}\right]$ sometimes the moment generating function, while $E\left[z^X\right]$ is the probability generating function of the random variable X. On the other hand, series expansion of $E\left[z^X\right]$ around $z = 1$,

$$\varphi_X(z) = \sum_{k=0}^{\infty} \Pr\left[X = k\right] (z + 1 - 1)^k = \sum_{k=0}^{\infty} \Pr\left[X = k\right] \sum_{j=0}^{k} \binom{k}{j} (z - 1)^j$$

$$= \sum_{j=0}^{\infty} \left[\sum_{k=j}^{\infty} \binom{k}{j} \Pr\left[X = k\right]\right] (z - 1)^j$$

shows with (2.22) that

$$E\left[\binom{X}{j}\right] = \sum_{k=j}^{\infty} \binom{k}{j} \Pr\left[X = k\right]$$

If moments are desired, the substitution $z \to e^{-u}$ in $E\left[z^X\right]$ is appropriate.

2.3.4 The logarithm of the probability generating function

The logarithm of the probability generating function is defined as

$$L_X(z) = \log\left(\varphi_X(z)\right) = \log\left(E\left[e^{-zX}\right]\right) \tag{2.41}$$

---

[10] An entire (or integral) function is a complex function without singularities in the finite complex plane. Hence, a power series around any finite point has infinite radius of convergence. In other words, it exists for all finite complex values.

[11] The Landau big O-notation specifies the "order of a function" when the argument tends to some limit. Most often the limit is to infinity, but the O-notation can also be used to characterize the behavior of a function around some finite point. Formally, $f(x) = O\left(g(x)\right)$ for $x \to \infty$ means that there exist positive numbers c and x_0 for which $|f(x)| \leq c|g(x)|$ for $x > x_0$.

[12] The lognormal distribution defined by (3.43) is an example where the summation (2.40) diverges for any $z \neq 0$.

from which $L_X(0) = 0$ because $\varphi_X(0) = 1$. Further, analogous to the discrete case, we see that $L'_X(0) = \varphi'_X(0)$, $L''_X(0) = \varphi''_X(0) - (\varphi'_X(0))^2$ and

$$E[X] = -\varphi'_X(0) = -L'_X(0)$$

However, the difference with the discrete case lies in the higher moments,

$$E[X^n] = (-1)^n \left.\frac{d^n \varphi_X(z)}{dz^n}\right|_{z=0} \tag{2.42}$$

because with $E[X^2] = \varphi''_X(0)$,

$$\begin{aligned}
\mathrm{Var}[X] &= \varphi''_X(0) - (\varphi'_X(0))^2 \\
&= L''_X(0)
\end{aligned} \tag{2.43}$$

The latter expression makes $L_X(z)$ for a continuous random variable particularly useful. Since the variance is always positive, it demonstrates that $L_X(z)$ is convex (see Section 5.5) around $z = 0$. Finally, we mention that

$$E\left[(X - E[X])^3\right] = -L'''_X(0)$$

2.4 The conditional probability

The conditional probability of the event A given the event B (or on the hypothesis B) is defined as

$$\Pr[A|B] = \frac{\Pr[A \cap B]}{\Pr[B]} \tag{2.44}$$

The definition implicitly assumes that the event B has positive probability, otherwise the conditional probability remains undefined. We quote Feller (1970, p. 116):

Taking conditional probabilities of various events with respect to a particular hypothesis B amounts to choosing B as a new sample space with probabilities proportional to the original ones; the proportionality factor $\Pr[B]$ is necessary in order to reduce the total probability of the new sample space to unity. This formulation shows that all general theorems on probabilities are valid for conditional probabilities with respect to any particular hypothesis. For example, the law $\Pr[A \cup B] = \Pr[A] + \Pr[B] - \Pr[A \cap B]$ takes the form

$$\Pr[A \cup B|C] = \Pr[A|C] + \Pr[B|C] - \Pr[A \cap B|C]$$

The formula (2.44) is often rewritten in the form

$$\Pr[A \cap B] = \Pr[A|B]\Pr[B] \tag{2.45}$$

which easily generalizes to more events. For example, denote $A = A_1$ and $B = A_2 \cap A_3$, then

$$\Pr[A_1 \cap A_2 \cap A_3] = \Pr[A_1|A_2 \cap A_3]\Pr[A_2 \cap A_3]$$
$$= \Pr[A_1|A_2 \cap A_3]\Pr[A_2|A_3]\Pr[A_3]$$

Another application of the conditional probability occurs when a partitioning of the sample space Ω is known: $\Omega = \cup_k B_k$ and all B_k are mutually exclusive, which means that $B_k \cap B_j = \emptyset$ for any k and $j \neq k$. Then, with (2.45),

$$\sum_k \Pr[A \cap B_k] = \sum_k \Pr[A|B_k]\Pr[B_k]$$

The event $A_k = \{A \cap B_k\}$ is a decomposition (or projection) of the event A in the basis event B_k, analogous to the decomposition of a vector in terms of a set of orthogonal basis vectors that span the total state space. Indeed, using the associative property $A \cap \{B \cap C\} = A \cap B \cap C$ and $A \cap A = A$, the intersection $A_k \cap A_j = \{A \cap B_k\} \cap \{A \cap B_j\} = A \cap \{B_k \cap B_j\} = \emptyset$, which implies mutual exclusivity (or orthogonality). Using the distributive property $A \cap \{B_k \cup B_j\} = \{A \cap B_k\} \cup \{A \cap B_j\}$, we observe that

$$A = A \cap \Omega$$
$$= A \cap \{\cup_k B_k\} = \cup_k \{A \cap B_k\} = \cup_k A_k$$

Finally, since all events A_k are mutually exclusive, $\Pr[A] = \sum_k \Pr[A_k] = \sum_k \Pr[A \cap B_k]$. Thus, if $\Omega = \cup_k B_k$ and in addition, for any pair j, k holds that $B_k \cap B_j = \emptyset$, we have proved the *law of total probability* or decomposability,

$$\Pr[A] = \sum_k \Pr[A|B_k]\Pr[B_k] \tag{2.46}$$

Conditioning on events is a powerful tool that will be used frequently. If the conditional probability $\Pr[A|B_k]$ is known as a function $g(B_k)$, the law of total probability can also be written in terms of the expectation operator defined in (2.12) as

$$\Pr[A] = E[g(B_k)] \tag{2.47}$$

Also the important memoryless property of the exponential distribution (see Section 3.2.2) is an example of the application of the conditional probability. Another classical example is Bayes' rule. Consider again the events B_k defined above. Using the definition (2.44) followed by (2.45),

$$\Pr[B_k|A] = \frac{\Pr[B_k \cap A]}{\Pr[A]} = \frac{\Pr[A \cap B_k]}{\Pr[A]} = \frac{\Pr[A|B_k]\Pr[B_k]}{\Pr[A]} \tag{2.48}$$

Using (2.46), we arrive at Bayes' rule

$$\Pr\left[B_k | A\right] = \frac{\Pr\left[A | B_k\right] \Pr\left[B_k\right]}{\sum_j \Pr\left[A | B_j\right] \Pr\left[B_j\right]} \tag{2.49}$$

where $\Pr\left[B_k\right]$ are called the *a-priori* probabilities, while $\Pr\left[B_k | A\right]$ are the *a-posteriori* probabilities.

The conditional distribution function of the random variable Y given X is defined by

$$F_{Y|X}\left(y | x\right) = \Pr\left[Y \leq y | X = x\right] \tag{2.50}$$

for any x provided $\Pr\left[X = x\right] > 0$. This condition follows from the definition (2.44) of the conditional probability. The conditional probability density function of Y given X is defined by

$$f_{Y|X}\left(y | x\right) = \Pr\left[Y = y | X = x\right] = \frac{\Pr[X = x, Y = y]}{\Pr\left[X = x\right]}$$

$$= \frac{f_{XY}(x, y)}{f_X(x)} \tag{2.51}$$

for any x such that $\Pr\left[X = x\right] > 0$ (and similarly for continuous random variables $f_X(x) > 0$) and where $f_{XY}(x, y)$ is the joint probability density function defined below in (2.59).

2.5 Several random variables and independence

2.5.1 Discrete random variables

Two events A and B are independent if

$$\Pr\left[A \cap B\right] = \Pr\left[A\right] \Pr\left[B\right] \tag{2.52}$$

Similarly, we define two discrete random variables to be independent if

$$\Pr\left[X = x, Y = y\right] = \Pr\left[X = x\right] \Pr\left[Y = y\right] \tag{2.53}$$

If $Z = f(X, Y)$, then Z is a discrete random variable with

$$\Pr\left[Z = z\right] = \sum_{f(x,y)=z} \Pr\left[X = x, Y = y\right]$$

Applying the expectation operator (2.11) to both sides yields

$$E\left[f(X, Y)\right] = \sum_{x,y} f(x, y) \Pr\left[X = x, Y = y\right] \tag{2.54}$$

If X and Y are independent and f is separable, $f(x,y) = f_1(x)f_2(y)$, then the expectation (2.54) reduces to

$$E\left[f(X,Y)\right] = \sum_x f_1(x) \Pr\left[X = x\right] \sum_y f_2(y) \Pr\left[Y = y\right] = E\left[f_1(X)\right] E\left[f_2(Y)\right]$$

$$(2.55)$$

The simplest example of the general function is $Z = X + Y$. In that case, the sum is over all x and y that satisfy $x + y = z$. Thus,

$$\Pr\left[X + Y = z\right] = \sum_x \Pr\left[X = x, Y = z - x\right] = \sum_y \Pr\left[X = z - y, Y = y\right]$$

If X and Y are independent, we obtain the convolution,

$$\Pr\left[X + Y = z\right] = \sum_x \Pr\left[X = x\right] \Pr\left[Y = z - x\right]$$

$$= \sum_y \Pr\left[X = z - y\right] \Pr\left[Y = y\right]$$

2.5.2 The covariance

The covariance of X and Y is defined as

$$\mathrm{Cov}\left[X, Y\right] = E\left[(X - \mu_X)(Y - \mu_Y)\right] = E\left[XY\right] - \mu_X \mu_Y \qquad (2.56)$$

If $\mathrm{Cov}[X, Y] = 0$, then the variables X and Y are *uncorrelated*. If X and Y are independent, then $\mathrm{Cov}[X, Y] = 0$. Hence, independence implies uncorrelation, but the converse is not necessarily true. The classical example[13] is $Y = X^2$ where X has a normal distribution $N(0,1)$ (Section 3.2.3) because $\mu_X = 0$ and $E\left[XY^2\right] = E\left[X^3\right] = 0$ as follows from (3.23). Although X and Y are perfect dependent, they are uncorrelated. Thus, independence is a stronger property than uncorrelation. The covariance $\mathrm{Cov}[X, Y]$ measures the degree of dependence between two (or generally more) random variables. If X and Y are positively (negatively) correlated, the large values of X tend to be associated with large (small) values of Y.

As an application of the covariance, consider the problem of computing the variance of a sum S_n of random variables $X_1, X_2, \ldots, X_n$. Let $\mu_k = E\left[X_k\right]$,

---

[13] Another example: let U be uniform on $[0,1]$ and $X = \cos(2\pi U)$ and $Y = \sin(2\pi U)$. Using (2.34),

$$E\left[XY\right] = \int_0^1 \cos(2\pi u) \sin(2\pi u)\, du = 0$$

as well as $E[X] = E[Y] = 0$. Thus, $\mathrm{Cov}[X, Y] = 0$, but X and Y are perfectly dependent because $X = \cos(\arcsin Y) = \pm\sqrt{1 - Y^2}$.

then $E[S_n] = \sum_{k=1}^{n} \mu_k$ and

$$\text{Var}[S_n] = E\left[(S_n - E[S_n])^2\right] = E\left[\left(\sum_{k=1}^{n}(X_k - \mu_k)\right)^2\right]$$

$$= E\left[\sum_{k=1}^{n}\sum_{j=1}^{n}(X_k - \mu_k)(X_j - \mu_j)\right]$$

$$= E\left[\sum_{k=1}^{n}(X_k - \mu_k)^2 + 2\sum_{k=1}^{n}\sum_{j=k+1}^{n}(X_k - \mu_k)(X_j - \mu_j)\right]$$

Using the linearity of the expectation operator and the definition of the covariance (2.56) yields

$$\text{Var}[S_n] = \sum_{k=1}^{n}\text{Var}[X_k] + 2\sum_{k=1}^{n}\sum_{j=k+1}^{n}\text{Cov}[X_k, X_j] \qquad (2.57)$$

Observe that for a set of independent random variables $\{X_k\}$ the double sum with covariances vanishes.

The Cauchy-Schwarz inequality (5.17) derived in Chapter 5 indicates that

$$\left(E\left[(X - \mu_X)(Y - \mu_Y)\right]\right)^2 \leq E\left[(X - \mu_X)^2\right]E\left[(X - \mu_X)^2\right]$$

such that the covariance is always bounded by

$$|\text{Cov}[X, Y]| \leq \sigma_X \sigma_Y$$

2.5.3 The linear correlation coefficient

Since the covariance is not dimensionless, the *linear* correlation coefficient defined as

$$\rho(X, Y) = \frac{\text{Cov}[X, Y]}{\sigma_X \sigma_Y} \qquad (2.58)$$

is often convenient to relate two (or more) different physical quantities expressed in different units. The linear correlation coefficient remains invariant (possibly apart from the sign) under a linear transformation because

$$\rho(aX + b, cY + d) = \text{sign}(ac)\rho(X, Y)$$

This transform shows that the linear correlation coefficient $\rho(X, Y)$ is independent of the value of the mean μ_X and the variance σ_X^2 provided $\sigma_X^2 > 0$. Therefore, many computations simplify if we normalize the random variable properly. Let us introduce the concept of a *normalized random* variable

$X^* = \frac{X - \mu_X}{\sigma_X}$. The normalized random variable has a zero mean and a variance equal to one. By the invariance under a linear transform, the correlation coefficient $\rho(X, Y) = \rho(X^*, Y^*)$ and also $\rho(X, Y) = \text{Cov}[X^*, Y^*]$. The variance of $X^* \pm Y^*$ follows from (2.57) as

$$\text{Var}[X^* \pm Y^*] = \text{Var}[X^*] + \text{Var}[Y^*] \pm 2\,\text{Cov}[X^*, Y^*]$$
$$= 2(1 \pm \rho(X, Y))$$

Since the variance is always positive, it follows that $-1 \leq \rho(X, Y) \leq 1$. The extremes $\rho(X, Y) = \pm 1$ imply a *linear* relation between X and Y. Indeed, $\rho(X, Y) = 1$ implies that $\text{Var}[X^* - Y^*] = 0$, which is only possible if $X^* = Y^* + c$, where c is a constant. Hence, $X = \frac{\sigma_X}{\sigma_Y} Y + c'$. A similar argument applies for the case $\rho(X, Y) = -1$. For example, in curve fitting, the goodness of the fit is often expressed in terms of the correlation coefficient. A perfect fit has correlation coefficient equal to 1. In particular, in linear regression where $Y = aX + b$, the regression coefficients a_R and b_R are the minimizers of the square distance $E\left[(Y - (aX + b))^2\right]$ and given by

$$a_R = \frac{\text{Cov}[X, Y]}{\sigma_X^2}$$
$$b_R = E[Y] - a_R E[X]$$

Since a correlation coefficient $\rho(X, Y) = 1$ implies $\text{Cov}[X, Y] = \sigma_X \sigma_Y$, we see that $a_R = \frac{\sigma_X}{\sigma_Y}$ as derived above with normalized random variables.

Although the linear correlation coefficient is a natural measure of the dependence between random variables, it has some disadvantages. First, the variances of X and Y must exist, which may cause problems with heavy-tailed distributions. Second, as illustrated above, dependence can lead to uncorrelation, which is awkward. Third, linear correlation is not invariant under non-linear strictly increasing transformations T such that $\rho(T(X), T(Y)) \neq \rho(X, Y)$. Common intuition expects that dependence measures should be invariant under these transforms T. This leads to the definition of *rank* correlation which satisfies that invariance property. Here, we merely mention Sperman's rank correlation coefficient, which is defined as

$$\rho_S(X, Y) = \rho(F_X(X), F_Y(Y))$$

where ρ is the linear correlation coefficient and where the non-linear strict increasing transform is the probability distribution. More details are found in Embrechts *et al.* (2001b) and in Chapter 4.

2.5.4 *Continuous random variables*

We define the joint distribution function by $F_{XY}(x,y) = \Pr[X \leq x, Y \leq y]$ and the joint probability density function by

$$f_{XY}(x,y) = \frac{\partial^2 F_{XY}(x,y)}{\partial x \partial y} \tag{2.59}$$

Hence,

$$F_{XY}(x,y) = \Pr[X \leq x, Y \leq y] = \int_{-\infty}^{x} \int_{-\infty}^{y} f_{XY}(u,v)dudv \tag{2.60}$$

The analogon of (2.54) is

$$E[g(X,Y)] = \int_{-\infty}^{\infty} \int_{-\infty}^{\infty} g(x,y)f_{XY}(x,y)dxdy \tag{2.61}$$

Most of the difficulties occur in the evaluation of the multiple integrals. The change of variables in multiple dimensions involves the Jacobian. Consider the transformed random variables $U = g_1(X,Y)$ and $V = g_2(X,Y)$ and denote the inverse transform by $x = h_1(u,v)$ and $y = h_2(u,v)$, then

$$f_{UV}(u,v) = f_{XY}(h_1(u,v), h_1(u,v)) J(u,v)$$

where the Jacobian $J(u,v)$ is

$$J(u,v) = \det \begin{bmatrix} \frac{\partial x}{\partial u} & \frac{\partial x}{\partial v} \\ \frac{\partial y}{\partial u} & \frac{\partial y}{\partial v} \end{bmatrix}$$

If X and Y are independent and $Z = X + Y$, we obtain the convolution,

$$f_Z(z) = \int_{-\infty}^{\infty} f_X(x)f_Y(z-x)dx = \int_{-\infty}^{\infty} f_X(z-y)f_Y(y)dy \tag{2.62}$$

which is often denoted by $f_Z(z) = (f_X * f_Y)(z)$. If both $f_X(x) = 0$ and $f_Y(x) = 0$ for $x < 0$, then the definition (2.62) of the convolution reduces to

$$(f_X * f_Y)(z) = \int_0^z f_X(x)f_Y(z-x)dx$$

2.5.5 *The sum of independent random variables*

Let $S_N = \sum_{k=1}^{N} X_k$, where the random variables X_k are all independent. We first concentrate on the case where $N = n$ is a (fixed) integer. Since $S_N = S_{N-1} + X_N$, direct application of (2.62) yields the recursion

$$f_{S_N}(z) = \int_{-\infty}^{\infty} f_{S_{N-1}}(z-y)f_{X_N}(y)dy \tag{2.63}$$

which, when written out explicitly, leads to the N-fold integral

$$f_{S_N}(z) = \int_{-\infty}^{\infty} f_{X_N}(y_N)dy_N \cdots \int_{-\infty}^{\infty} f_{X_1}(y_1)f_{X_0}(z - y_N - \cdots - y_1)dy_1 \quad (2.64)$$

In many cases, convolutions are more efficiently computed via generating functions. The generating function of S_n equals

$$\varphi_{S_n}(z) = E\left[z^{S_n}\right] = E\left[z^{\sum_{k=1}^{n} X_k}\right] = E\left[\prod_{k=1}^{n} z^{X_k}\right]$$

Since all X_k are independent, (2.55) can be applied,

$$\varphi_{S_n}(z) = \prod_{k=1}^{n} E\left[z^{X_k}\right]$$

or, in terms of generating functions,

$$\varphi_{S_n}(z) = \prod_{k=1}^{n} \varphi_{X_k}(z) \quad (2.65)$$

Hence, we arrive at the important result that the generating function of a sum of independent random variables equals the product of the generating functions of the individual random variables. We also note that the condition of independence is crucial in that it allows the product and expectation operator to be reversed, leading to the useful result (2.65). Often, the random variables X_k all possess the same distribution. In this case of independent identically distributed (i.i.d.) random variables with generating function $\varphi_X(z)$, the relation (2.65) further simplifies to

$$\varphi_{S_n}(z) = (\varphi_X(z))^n \quad (2.66)$$

In the case where the number of terms N in the sum S_N is a random variable with generating function $\varphi_N(z)$, independent of the X_k, we use the general definition of expectation (2.54) for two random variables,

$$\varphi_{S_N}(z) = E\left[z^{S_N}\right] = \sum_{k=0}^{\infty}\sum_{x} z^x \Pr\left[S_N = x, N = k\right]$$

$$= \sum_{k=0}^{\infty}\sum_{x} z^x \Pr\left[S_N = x | N = k\right]\Pr\left[N = k\right]$$

where the conditional probability (2.45) is used. Since the value of S_N

depends on the number of terms N in the sum, we have $\Pr[S_N = x | N = k] = \Pr[S_k = x]$. Further, with

$$\sum_x z^x \Pr[S_N = x | N = k] = \varphi_{S_k}(z)$$

we have

$$\varphi_{S_N}(z) = \sum_{k=0}^{\infty} \varphi_{S_k}(z) \Pr[N = k] \tag{2.67}$$

The average $E[S_N]$ follows from (2.26) as

$$E[S_N] = \sum_{k=0}^{\infty} \varphi'_{S_k}(1) \Pr[N = k] = \sum_{k=0}^{\infty} E[S_k] \Pr[N = k] \tag{2.68}$$

Since $E[S_k] = E\left[\sum_{j=1}^{k} X_j\right] = \sum_{j=1}^{k} E[X_j]$ and assuming that all random variables X_j have equal mean $E[X_j] = E[X]$, we have

$$E[S_N] = \sum_{k=0}^{\infty} k E[X] \Pr[N = k]$$

or

$$E[S_N] = E[X] E[N] \tag{2.69}$$

This relation (2.69) is commonly called *Wald's identity*. Wald's identity holds for any random sum of (possibly dependent) random variables X_j provided the number N of those random variables is independent of the X_j.

In the case of i.i.d. random variables, we apply (2.66) in (2.67) so that

$$\varphi_{S_N}(z) = \sum_{k=0}^{\infty} (\varphi_X(z))^k \Pr[N = k] = \varphi_N(\varphi_X(z)) \tag{2.70}$$

This expression is a generalization of (2.66).

2.6 Conditional expectation

The generating function (2.67) of a random sum of independent random variables can be derived using the conditional expectation $E[Y | X = x]$ of two random variables X and Y. We will first define the conditional expectation and derive an interesting property.

Suppose that we know that $X = x$, the conditional density function

$f_{Y|X}(y|x)$ defined by (2.51) of the random variable $Y_c = Y|X$ can be regarded as only function of y. Using the definition of the expectation (2.33) for continuous random variables (the discrete case is analogous), we have

$$E[Y|X=x] = \int_{-\infty}^{\infty} y f_{Y|X}(y|x) \, dy \qquad (2.71)$$

Since this expression holds for any value of x that the random variable X can take, we see that $E[Y|X=x] = g(x)$ is a function of x and, in addition since $X = x$, $E[Y|X=x] = g(X)$ can be regarded as a random variable that is a function of the random variable X. Having identified the conditional expectation $E[Y|X=x]$ as a random variable, let us compute its expectation or the expectation of the slightly more general random variable $h(X)g(X)$ with $g(X) = E[Y|X=x]$. From the general definition (2.34) of the expectation, it follows that

$$E[h(X)g(X)] = \int_{-\infty}^{\infty} h(x)g(x)f_X(x)\,dx = \int_{-\infty}^{\infty} h(x)E[Y|X=x]f_X(x)\,dx$$

Substituting (2.71) yields

$$E[h(X)g(X)] = \int_{-\infty}^{\infty}\int_{-\infty}^{\infty} h(x)yf_{Y|X}(y|x)f_X(x)\,dy dx$$
$$= \int_{-\infty}^{\infty}\int_{-\infty}^{\infty} h(x)yf_{XY}(x,y)\,dy dx = E[h(X)Y]$$

where we have used (2.51) and (2.61). Thus, we find the interesting relation

$$E[h(X)E[Y|X=x]] = E[h(X)Y] \qquad (2.72)$$

As a special case where $h(x) = 1$, the expectation of the conditional expectation follows as

$$E[Y] = E_X[E_Y[Y|X=x]]$$

where the index in E_Z clarifies that the expectation is over the random variable Z. Applying this relation to $Y = z^{S_N}$ where $S_N = \sum_{k=1}^{N} X_k$ and all X_k are independent yields

$$\varphi_{S_N}(z) = E\left[z^{S_N}\right] = E_N\left[E_S\left[z^{S_N}|N=n\right]\right]$$

Since $E_S\left[z^{S_N}|N=n\right] = \varphi_{S_N}(z)$ and specified in (2.65), we end up with

$$\varphi_{S_N}(z) = E_N[\varphi_{S_N}(z)] = \sum_{k=0}^{\infty} \varphi_{S_k}(z)\Pr[N=k]$$

which is (2.67).

3

Basic distributions

This chapter concentrates on the most basic probability distributions and their properties. From these basic distributions, other useful distributions are derived.

3.1 Discrete random variables

3.1.1 The Bernoulli distribution

A Bernoulli random variable X can only take two values: either 1 with probability p or 0 with probability $q = 1 - p$. The standard example of a Bernoulli random variable is the outcome of tossing a biased coin, and, more generally, the outcome of a trial with only two possibilities, either success or failure. The sample space is $\Omega = \{0, 1\}$ and $\Pr[X = 1] = p$, while $\Pr[X = 0] = q$. From this definition, the pgf follows from (2.17) as

$$\varphi_X(z) = E\left[z^X\right] = z^0 \Pr[X = 0] + z^1 \Pr[X = 1]$$

or

$$\varphi_X(z) = q + pz \tag{3.1}$$

From (2.23) or (2.14), the n-th moment is

$$E[X^n] = p$$

which shows that $\mu = E[X] = p$. From (2.24), we find $E[(X - a)^n] = p(1 - a)^n + q(-a)^n$ such that the moments centered around the mean μ are

$$E[(X - \mu)^n] = pq^n + (-1)^n qp^n = pq\left(q^{n-1} + (-1)^n p^{n-1}\right)$$

Explicitly, with $p + q = 1$, $\mathrm{Var}[X] = pq$ and $E\left[(X - \mu)^3\right] = pq(q - p)$.

37

3.1.2 The binomial distribution

A binomial random variable X is the sum of n independent Bernoulli random variables. The sample space is $\Omega = \{0, 1, \cdots, n\}$. For example, X may represent the number of successes in n independent Bernoulli trials such as the number of heads after n-times tossing a (biased) coin. Application of (2.66) with (3.1) gives

$$\varphi_X(z) = (q + pz)^n \tag{3.2}$$

Expanding the binomial pgf in powers of z, which justifies the name "binomial",

$$\varphi_X(z) = \sum_{k=0}^{n} \binom{n}{k} p^k q^{n-k} z^k$$

and comparing to (2.18) yields

$$\Pr[X = k] = \binom{n}{k} p^k q^{n-k} \tag{3.3}$$

The alternative, probabilistic approach starts with (3.3). Indeed, the probability that X has k successes out of n trials consists of precisely k successes (an event with probability p^k) and $n - k$ failures (with probability equal to q^{n-k}). The total number of ways in which k successes out of n trials can be obtained is precisely $\binom{n}{k}$.

The mean follows from (2.23) or from the definition $X = \sum_{j=1}^{n} X_{\text{Bernoulli}}$ and the linearity of the expectation as $E[X] = np$. Higher order moments around the mean can be derived from (2.24) as

$$E[(X - \mu)^m] = \left. \frac{d^m \left(e^{-tnp} \left(q + pe^t \right)^n \right)}{dt^m} \right|_{t=0} = \left. \frac{d^m}{dt^m} \sum_{k=0}^{n} \binom{n}{k} q^k p^{n-k} e^{t(qn-k)} \right|_{t=0}$$

$$= \sum_{k=0}^{n} \binom{n}{k} q^k p^{n-k} (qn - k)^m$$

In general, this form seems difficult to express more elegantly. It illustrates that, even for simple random variables, computations may rapidly become unattractive. For $m = 2$, the above differentiation leads to $\text{Var}[X] = npq$. But, this result is more economically obtained from (2.27), since $L_X(z) = n \log(q + pz)$, $L'_X(z) = \frac{np}{q+pz}$ and $L''_X(z) = -\frac{np^2}{(q+pz)^2}$. Thus,

$$\text{Var}[X] = -np^2 + np = npq \tag{3.4}$$

3.1.3 The geometric distribution

The geometric random variable X returns the number of independent Bernoulli trials needed to achieve the first success. Here the sample space Ω is the infinite set of integers. The probability density function is

$$\Pr[X = k] = pq^{k-1} \tag{3.5}$$

because a first success (with probability p) obtained in the k-th trial is proceeded by $k - 1$ failures (each having probability $q = 1 - p$). Clearly, $\Pr[X = 0] = 0$. The series expansion of the probability generating function,

$$\varphi_X(z) = pz \sum_{k=0}^{\infty} q^k z^k = \frac{pz}{1 - qz} \tag{3.6}$$

justifies the name "geometric".

The mean $E[X] = \varphi_X'(1)$ equals $E[X] = \frac{1}{p}$. The higher-order moments can be deduced from (2.24) as

$$E[(X - \mu)^n] = p \frac{d^n}{dt^n} \left(\frac{e^{-tq/p}}{1 - qe^t} \right) \Bigg|_{t=0} = p \sum_{k=0}^{\infty} q^k (k - q/p)^n$$

Similarly as for the binomial random variable, the variance most easily follows from (2.27) with $L_X(z) = \log p + \log(z) - \log(1 - qz)$, $L_X'(z) = \frac{1}{z} + \frac{q}{1-qz}$, $L_X''(z) = -\frac{1}{z^2} + \frac{q^2}{(1-qz)^2}$. Thus,

$$\mathrm{Var}[X] = \frac{q^2}{p^2} + \frac{q}{p} = \frac{q}{p^2} \tag{3.7}$$

The distribution function $F_X(k) = \Pr[X \le k] = \sum_{j=1}^{k} \Pr[X = j]$ is obtained as

$$\Pr[X \le k] = p \sum_{j=0}^{k-1} q^j = p \frac{1 - q^k}{1 - q} = 1 - q^k$$

The tail probability is

$$\Pr[X > k] = q^k \tag{3.8}$$

Hence, the probability that the number of trials until the first success is larger than k decreases geometrically in k with rate q. Let us now consider an important application of the conditional probability. The probability that, given the success is not found in the first k trials, success does not occur within the next m trials, is with (2.44)

$$\Pr[X > k + m | X > k] = \frac{\Pr[\{X > k + m\} \cap \{X > k\}]}{\Pr[X > k]} = \frac{\Pr[X > k + m]}{\Pr[X > k]}$$

and with (3.8)

$$\Pr[X > k + m | X > k] = q^m \equiv \Pr[X > m]$$

This conditional probability turns out to be independent of the hypothesis, the event $\{X > k\}$, and reflects the famous *memoryless property*. Only because $\Pr[X > k]$ obeys the functional equation $f(x + y) = f(x)f(y)$, the hypothesis or initial knowledge does not matter. It is precisely as if past failures have never occurred or are forgotten and as if, after a failure, the number of trials is reset to 0. Furthermore, the only solution to the functional equation is an exponential function. Thus, the geometric distribution is the only discrete distribution that possesses the memoryless property.

3.1.4 The Poisson distribution

Often we are interested to count the number of occurrences of an event in a certain time interval, such as, for example, the number of IP packets during a time slot or the number of telephony calls that arrive at a telephone exchange per unit time. The Poisson random variable X with probability density function

$$\Pr[X = k] = \frac{\lambda^k e^{-\lambda}}{k!} \tag{3.9}$$

turns out to model many of these counting phenomena well as shown in Chapter 7. The corresponding generating function is

$$\varphi_X(z) = e^{-\lambda} \sum_{k=0}^{\infty} \frac{\lambda^k}{k!} z^k = e^{\lambda(z-1)} \tag{3.10}$$

and the average number of occurrences in that time interval is

$$E[X] = \lambda \tag{3.11}$$

This average determines the complete distribution. In applications it is convenient to replace the unit interval by an interval of arbitrary length t such that

$$\Pr[X = k] = \frac{(\lambda t)^k e^{-\lambda t}}{k!}$$

equals the probability that precisely k events occur in the interval with duration t. The probability that no events occur during t time units is $\Pr[X = 0] = e^{-\lambda t}$ and the probability that at least one event (i.e. one or more) occurs is $\Pr[X > 0] = 1 - e^{-\lambda t}$. The latter is equal to the exponential distribution. We will also see later in Theorem 7.3.2 that the Poisson counting process and the exponential distribution are intimately connected. The sum of n independent Poisson random variables each with mean λ_k is again a Poisson random variable with mean $\sum_{k=1}^{n} \lambda_k$ as follows from (2.65) and (3.10).

The higher-order moments can be deduced from (2.24) as

$$E\left[(X - \lambda)^n\right] = e^{-\lambda} \left.\frac{d^n \left(e^{-\lambda(t - e^t)}\right)}{dt^n}\right|_{t=0}$$

from which

$$E[X] = \mathrm{Var}[X] = E\left[(X - \lambda)^3\right] = \lambda$$

The Poisson tail distribution equals

$$\Pr\left[X > m\right] = 1 - \sum_{k=0}^{m} \frac{\lambda^k e^{-\lambda}}{k!}$$

which precisely equals the sum of m exponentially distributed variables as demonstrated below in Section 3.3.1.

The Poisson density approximates the binomial density (3.3) if $n \to \infty$ but the mean $np = \lambda$. This phenomenon is often referred to as the *law of rare events*: in an arbitrarily large number n of independent trials each with arbitrarily small success $p = \frac{\lambda}{n}$, the total number of successes will approximately be Poisson distributed.

The classical argument is to consider the binomial density (3.3) with $p = \frac{\lambda}{n}$

$$\Pr[X = k] = \frac{n!}{k!(n-k)!} \frac{\lambda^k}{n^k} \left(1 - \frac{\lambda}{n}\right)^{n-k}$$

$$= \frac{\lambda^k}{k!} \left(1 - \frac{\lambda}{n}\right)^{-k} \prod_{j=1}^{k-1}\left(1 - \frac{j}{n}\right)\left(1 - \frac{\lambda}{n}\right)^n$$

or

$$\log\left(\Pr[X = k]\right) = \log\left(\frac{\lambda^k}{k!}\right) - k\log\left(1 - \frac{\lambda}{n}\right) + \sum_{j=1}^{k-1}\log\left(1 - \frac{j}{n}\right) + n\log\left(1 - \frac{\lambda}{n}\right)$$

For large n, we use the Taylor expansion $\log\left(1 - \frac{x}{n}\right) = -\frac{x}{n} - \frac{x^2}{2n^2} + O\left(n^{-3}\right)$ to obtain up to order $O\left(n^{-2}\right)$

$$\log\left(\Pr[X = k]\right) = \log\left(\frac{\lambda^k}{k!}\right) + k\frac{\lambda}{n} + O\left(n^{-2}\right) - \frac{k(k-1)}{2n} + O\left(n^{-2}\right) - \lambda - \frac{\lambda^2}{2n} + O\left(n^{-2}\right)$$

$$= \log\left(\frac{\lambda^k}{k!}\right) - \lambda - \frac{1}{2n}\left((k - \lambda)^2 - k\right) + O\left(n^{-2}\right)$$

With $e^x = 1 + x + O(x^2)$, we finally obtain the approximation for large n,

$$\Pr[X = k] = \frac{\lambda^k e^{-\lambda}}{k!}\left(1 - \frac{1}{2n}\left[(k - \lambda)^2 - k\right] + O\left(n^{-2}\right)\right)$$

The coefficient of $\frac{1}{n}$ is negative if $k \in \left[\lambda + \frac{1}{2} - \sqrt{\lambda + \frac{1}{4}}, \lambda + \frac{1}{2} + \sqrt{\lambda + \frac{1}{4}}\right]$. In that k-interval, the Poisson density is a lower bound for the binomial density for large n and $np = \lambda$. The reverse holds for values of k outside that interval. Since for the Poisson density $\frac{\Pr[X=k]}{\Pr[X=k-1]} = \frac{\lambda}{k}$, we see that $\Pr[X = k]$ increases as $\lambda > k$ and decreases as $\lambda < k$. Thus, the maximum of the Poisson density lies around $k = \lambda = E[X]$. In conclusion, we can say that the Poisson density approximates the binomial density for large n and $np = \lambda$ from below in the region of about the standard deviation $\sqrt{\lambda}$ around the mean $E[X] = \lambda$ and from above outside this region (in the tails of the distribution).

A much shorter derivation anticipates results of Chapter 6 and starts from the probability generating function (3.2) of the binomial distribution after substitution of $p = \frac{\lambda}{n}$,

$$\lim_{n \to \infty} \varphi_X(z) = \lim_{n \to \infty} \left(1 + \frac{\lambda(z-1)}{n}\right)^n = e^{\lambda(z-1)}$$

Invoking the Continuity Theorem 6.1.3, comparison with (3.10) shows that the limit probability generating function corresponds to a Poisson distribution. The Stein–Chen (1975) Theorem[1] generalizes the law of rare events: this law even holds when the Bernoulli trials are weakly dependent.

As a final remark, let S_n be the sum of i.i.d. Bernoulli trials each with mean p, then S_n is binomially distributed as shown in Section 3.1.2. If p is a constant and independent of the number of trials n, the Central Limit Theorem 6.3.1 states that $\frac{S_n - np}{\sqrt{np(1-p)}}$ tends to a Gaussian distribution. In summary, the limit distribution of a sum S_n of Bernoulli trials depends on how the mean p varies with the number of trials n when $n \to \infty$:

$$\text{if } p = \tfrac{\lambda}{n}, \text{ then} \qquad S_n \xrightarrow{d} \frac{\lambda^k e^{-\lambda}}{k!}$$

$$\text{if } p \text{ is constant, then} \qquad \frac{S_n - np}{\sqrt{np(1-p)}} \xrightarrow{d} \frac{e^{-\frac{x^2}{2}}}{\sqrt{2\pi}}$$

---

[1] The proof (see e.g. Grimmett and Stirzacker (2001, pp. 130–132)) involves coupling theory of stochastic random variables. The degree of dependence is expressed in terms of the total variation distance. The total variation distance between two discrete random variables X and Y is defined as

$$d_{TV}(X, Y) = \sum_k |\Pr[X = k] - \Pr[Y = k]|$$

and satisfies

$$d_{TV}(X, Y) = 2 \sup_{A \subset \mathbb{Z}} |\Pr[X \in A] - \Pr[Y \in A]|$$

3.2 Continuous random variables

3.2.1 The uniform distribution

A uniform random variable X has equal probability to attain any value in the interval $[a, b]$ such that the probability density function is a constant. Since $\Pr[a \leq X \leq b] = \int_a^b f_X(x)dx = 1$, the constant value equals

$$f_X(x) = \frac{1}{b-a} 1_{x \in [a,b]} \tag{3.12}$$

where 1_y is the indicator function defined in Section 2.2.1. The distribution function then follows as

$$\Pr[a \leq X \leq x] = \frac{x-a}{b-a} 1_{x \in [a,b]} + 1_{x > b}$$

The Laplace transform (2.37) is[2]

$$\varphi_X(z) = \int_{-\infty}^{\infty} e^{-zt} f_X(t) dt = \frac{e^{-za} - e^{-zb}}{z(b-a)} \tag{3.13}$$

while the mean $\mu = E[X]$ most easily follows from

$$E[X] = \int_{-\infty}^{\infty} \frac{x\,dx}{b-a} 1_{x \in [a,b]} = \frac{a+b}{2}$$

The centered moments are obtained from (2.39) as

$$E[(X-\mu)^n] = \frac{(-1)^n}{b-a} \frac{d^n}{dz^n} \left. \frac{\left(e^{\frac{z}{2}(b-a)} - e^{-\frac{z}{2}(b-a)}\right)}{z} \right|_{z=0}$$

$$= \frac{2(-1)^n}{b-a} \frac{d^n}{dz^n} \left. \frac{\sinh(\frac{b-a}{2})z}{z} \right|_{z=0}$$

Using the power series

$$\frac{\sinh(\frac{b-a}{2})z}{z} = \sum_{k=0}^{\infty} \frac{(\frac{b-a}{2})^{2k+1}}{(2k+1)!} z^{2k}$$

leads to

$$E[(X-\mu)^{2n}] = \frac{(b-a)^{2n}}{(2n+1)2^{2n}} \tag{3.14}$$

$$E[(X-\mu)^{2n+1}] = 0$$

---

[2] Notice that $\frac{z\varphi_X(z)}{ab}$ equals the convolution $f * g$ of two exponential densities f and g with rates a and b, respectively.

Let us define U as the uniform random variable in the interval $[0, 1]$. If $W = 1 - U$ is a uniform random variable on $[0, 1]$, then W and U have the same distribution denoted as $W \overset{d}{=} U$ because $\Pr[W \le x] = \Pr[1 - U \le x] = \Pr[U \ge 1 - x] = 1 - (1 - x) = x = \Pr[U \le x]$.

The probability distribution function $F_X(x) = \Pr[X \le x] = g(x)$ whose inverse exists can be written as a function of $F_U(x) = x1_{x \in [0,1]}$. Let $X = g^{-1}(U)$. Since the distribution function is non-decreasing, this also holds for the inverse $g^{-1}(.)$. Applying (2.32) yields with $X = g^{-1}(U)$

$$F_X(x) = \Pr\left[g^{-1}(U) \le x\right] = \Pr\left[U \le g(x)\right] = F_U\left(g(x)\right) = g(x)$$

For instance, $g^{-1}(U) = -\frac{\ln(1-U)}{\alpha} \overset{d}{=} -\frac{\ln U}{\alpha}$ are exponentially random variables (3.17) with parameter α; $g^{-1}(U) = U^{1/\alpha}$ are polynomially distributed random variables with distribution $\Pr[X \le x] = x^\alpha$; $g^{-1}(U) = \cot(\pi U)$ is a Cauchy random variable defined in (3.38) below. In addition, we observe that $U = g(X) = F_X(X)$, which means that any random variable X is transformed into a uniform random variable U on $[0, 1]$ by its own distribution function.

The numbers a_k that satisfy congruent recursions of the form $a_{k+1} = (\alpha a_k + \beta) \bmod M$, where M is a large prime number (e.g. $M = 2^{31} - 1$), α and β are integers (e.g. $\alpha = 397\,204\,094$ and $\beta = 0$) are to a good approximation uniformly distributed. The scaled numbers $y_k = \frac{a_k}{M-1}$ are nearly uniformly distributed on $[0, 1]$. Since these recursions with initial value or seed $a_0 \in [0, M - 1]$ are easy to generate with computers (Press *et al.*, 1992), the above property is very useful to generate arbitrary random variables $X = g^{-1}(U)$ from the uniform random variable U.

3.2.2 The exponential distribution

An exponential random variable X satisfies the probability density function

$$f_X(x) = \alpha e^{-\alpha x} \qquad\qquad \alpha, x \ge 0 \qquad\qquad (3.15)$$

where α is the rate at which events occur. The corresponding Laplace transform is

$$\varphi_X(z) = \alpha \int_0^\infty e^{-\alpha t} e^{-zt} dt = \frac{\alpha}{z + \alpha} \qquad\qquad (3.16)$$

and the probability distribution is, for $x \ge 0$,

$$F_X(x) = 1 - e^{-\alpha x} \qquad\qquad (3.17)$$

The mean or average follows from (2.33) or from $E[X] = -\varphi'_X(0)$ as $\mu = E[X] = \frac{1}{\alpha}$. The centered moments are obtained from (2.39) as

$$E\left[\left(X - \frac{1}{\alpha}\right)^n\right] = (-1)^n \left.\frac{d^n\left(\frac{\alpha e^{z/\alpha}}{z+\alpha}\right)}{dz^n}\right|_{z=0}$$

Since the Taylor expansion of $\frac{\alpha e^{z/\alpha}}{z+\alpha}$ around $z = 0$ is

$$\frac{\alpha e^{z/\alpha}}{z+\alpha} = \sum_{k=0}^{\infty} \frac{1}{k!}\left(\frac{z}{\alpha}\right)^k \sum_{k=0}^{\infty}(-1)^k\left(\frac{z}{\alpha}\right)^k = \sum_{n=0}^{\infty}\left(\left(\frac{-1}{\alpha}\right)^n \sum_{k=0}^{n}\frac{(-1)^k}{k!}\right)z^n$$

we find that

$$E\left[\left(X - \frac{1}{\alpha}\right)^n\right] = \frac{n!}{\alpha^n}\sum_{k=0}^{n}\frac{(-1)^k}{k!} \tag{3.18}$$

For large n, the centered moments are well approximated by

$$E\left[\left(X - \frac{1}{\alpha}\right)^n\right] \simeq \frac{n!}{e\alpha^n}$$

The exponential random variable possesses, just as its discrete counterpart, the geometric random variable, the *memoryless property*. Indeed, analogous to Section 3.1.3, consider

$$\Pr[X \geq t+T | X > t] = \frac{\Pr\left[\{X \geq t+T\} \cap \{X > t\}\right]}{\Pr[X > t]} = \frac{\Pr[X \geq t+T]}{\Pr[X > t]}$$

and since $\Pr[X > t] = e^{-\alpha t}$, the memoryless property

$$\Pr[X \geq t+T | X > t] = \Pr[X > T]$$

is established. Since the only non-zero solution (proved in Feller (1970, p. 459)) to the functional equation $f(x + y) = f(x)f(y)$, which implies the memoryless property, is of the form c^x, it shows that the exponential distribution is the only continuous distribution that has the memoryless property. As we will see later, this memoryless property is a fundamental property in Markov processes.

It is instructive to show the close relation between the geometric and exponential random variable (see Feller (1971, p. 1)). Consider the waiting time T (measured in integer units of Δt) for the first success in a sequence of Bernoulli trials where only one trial occurs in a timeslot Δt. Hence, $X = \frac{T}{\Delta t}$ is a (dimensionless) geometric random variable. From (3.8), $\Pr[T > k\Delta t] = (1 - p)^k$ and the average waiting time is $E[T] = \Delta t E[X] = \frac{\Delta t}{p}$. The transition from the discrete to continuous space involves the limit process $\Delta t \to 0$ subject to a fixed average waiting time $E[T]$. Let $t = k\Delta t$, then

$$\lim_{\Delta t \to 0} \Pr[T > t] = \lim_{\Delta t \to 0} \left(1 - \frac{\Delta t}{E[T]}\right)^{t/\Delta t} = e^{-t/E[T]}$$

For arbitrary small time units, the waiting time for the first success and with average $E[T]$ turns out to be an exponential random variable.

3.2.3 The Gaussian or normal distribution

The Gaussian random variable X is defined for all x by the probability density function

$$f_X(x) = \frac{1}{\sigma\sqrt{2\pi}} \exp\left[-\frac{(x-\mu)^2}{2\sigma^2}\right] \qquad (3.19)$$

which explicitly shows its dependence on the average μ and variance σ^2. The importance of the Gaussian random variables stems from the Central Limit Theorem 6.3.1. Often a Gaussian – also called normal – random variable with average μ and variance σ^2 is denoted by $N(\mu, \sigma^2)$. The distribution function is

$$F_X(x) = \frac{1}{\sigma\sqrt{2\pi}} \int_{-\infty}^{x} \exp\left[-\frac{(t-\mu)^2}{2\sigma^2}\right] dt \equiv \Phi\left(\frac{x-\mu}{\sigma}\right) \qquad (3.20)$$

where[3] $\Phi(x) = \frac{1}{\sqrt{2\pi}} \int_{-\infty}^{x} e^{-\frac{t^2}{2}} dt$ is the normalized Gaussian distribution corresponding to $\mu = 0$ and $\sigma = 1$. The double-sided Laplace transform is

$$\varphi_X(z) = \frac{1}{\sigma\sqrt{2\pi}} \int_{-\infty}^{\infty} e^{-zt} \exp\left[-\frac{(t-\mu)^2}{2\sigma^2}\right] dt = e^{\frac{\sigma^2 z^2}{2} - \mu z} \qquad (3.22)$$

---

[3] Abramowitz and Stegun (1968, Section 7.1.1) define the error function as

$$\text{erf}(z) = \frac{2}{\sqrt{\pi}} \int_{0}^{z} e^{-t^2} dt \qquad (3.21)$$

such that (Abramowitz and Stegun, 1968, Section 7.1.22)

$$\frac{1}{\sigma\sqrt{2\pi}} \int_{-\infty}^{x} \exp\left[-\frac{(t-\mu)^2}{2\sigma^2}\right] dt = \frac{1}{2}\left(1 + \text{erf}\left(\frac{x-\mu}{\sqrt{2}\sigma}\right)\right)$$

and the centered moments (2.39) are

$$E\left[(X-\mu)^{2n}\right] = \left.\frac{d^{2n}\left(e^{\frac{\sigma^2 z^2}{2}}\right)}{dz^{2n}}\right|_{z=0} = \frac{(2n)!}{n!}\left(\frac{\sigma^2}{2}\right)^n$$

$$E\left[(X-\mu)^{2n+1}\right] = 0 \tag{3.23}$$

We note from (2.65) that a sum of independent Gaussian random variables $N(\mu_k, \sigma_k^2)$ is again a Gaussian random variable $N\left(\sum_{k=1}^n \mu_k, \sum_{k=1}^n \sigma_k^2\right)$. If $X = N(\mu, \sigma^2)$, then the scaled random variable $Y = aX$ is a $N(a\mu, (a\sigma)^2)$ random variable that is verified by computing $\Pr\left[Y \le y\right] = \Pr\left[X \le \frac{y}{a}\right]$. Similarly for translation, $Y = X + b$, then $Y = N(\mu + b, \sigma^2)$. Hence, a linear combination of Gaussian random variables is again a Gaussian random variable,

$$\sum_{k=1}^n a_k N(\mu_k, \sigma_k^2) + b = N\left(\sum_{k=1}^n a_k \mu_k + b, \sum_{k=1}^n a_k^2 \sigma_k^2\right)$$

3.3 Derived distributions

From the basic distributions, a large number of other distributions can be derived as illustrated here.

3.3.1 The sum of independent exponential random variables

By applying (2.65) and (2.38) a substantial amount of practical problems can be solved. For example, the sum of n independent exponential random variables, each with different rate $\alpha_k > 0$, has the generating function

$$\varphi_{S_n}(z) = \prod_{k=1}^n \frac{\alpha_k}{z + \alpha_k}$$

and probability density function

$$f_{S_n}(t) = \frac{\prod_{k=1}^n \alpha_k}{2\pi i} \int_{c-i\infty}^{c+i\infty} \frac{e^{zt}}{\prod_{k=1}^n (z + \alpha_k)} dz$$

The contour can be closed over the negative half plane for $t > 0$, where the integral has simple poles at $z = -\alpha_k$. From the Cauchy integral theorem, we obtain

$$f_{S_n}(t) = \left(\prod_{k=1}^n \alpha_k\right) \sum_{j=1}^n \frac{e^{-\alpha_j t}}{\prod_{k=1;k\neq j}^n (\alpha_k - \alpha_j)}$$

If all rates are equals $\alpha_k = \alpha$, the case reduces to $\varphi_{S_n}(z) = \left(\frac{\alpha}{z+\alpha}\right)^n$ with $E[S_n] = \frac{n}{\alpha}$ and with probability density

$$f_{S_n}(t) = \frac{\alpha^n}{2\pi i} \int_{c-i\infty}^{c+i\infty} \frac{e^{zt}}{(z+\alpha)^n} dz$$

Again, the contour can be closed over the negative half plane and the n-th order poles are deduced from Cauchy's relation for the n-th derivative of a complex function

$$\frac{1}{k!} \frac{d^k f(z)}{dz^k}\bigg|_{z=z_0} = \frac{1}{2\pi i} \int_{C(z_0)} \frac{f(\omega)\, d\omega}{(\omega - z_0)^{k+1}}$$

as

$$f_{S_n}(t) = \frac{\alpha^n}{(n-1)!} \frac{d^{n-1} e^{zt}}{dz^{n-1}}\bigg|_{z=-\alpha} = \frac{\alpha(\alpha t)^{n-1}}{(n-1)!} e^{-\alpha t} \qquad (3.24)$$

For integer n, this density corresponds to the $n-$Erlang random variable. When extended to real values of $n = \beta$,

$$f_X(t; \alpha, \beta) = \frac{\alpha(\alpha t)^{\beta-1}}{\Gamma(\beta)} e^{-\alpha t} \qquad (3.25)$$

it is called the Gamma probability density function, with corresponding pgf

$$\varphi_X(z; \alpha, \beta) = \left(\frac{\alpha}{z+\alpha}\right)^\beta = \left(1 + \frac{z}{\alpha}\right)^{-\beta} \qquad (3.26)$$

and distribution

$$F_X(x; \alpha, \beta) = \frac{\alpha^\beta}{\Gamma(\beta)} \int_0^x t^{\beta-1} e^{-\alpha t} dt \qquad (3.27)$$

This integral, the incomplete Gamma-function, can only be expressed in closed analytic form if β is an integer. Hence, for the n-Erlang random variable X, the distribution follows after repeated partial integration as

$$F_X(x; \alpha, n) = \frac{\alpha^n}{(n-1)!} \int_0^x t^{n-1} e^{-\alpha t} dt = 1 - \sum_{k=0}^{n-1} \frac{(\alpha x)^k}{k!} e^{-\alpha x} \qquad (3.28)$$

We observe that $\Pr[X > x] = \sum_{k=0}^{n-1} \frac{(\alpha x)^k}{k!} e^{-\alpha x}$, which equals $\Pr[Y \le n-1]$ where Y is a Poisson random variable with mean $\lambda = \alpha x$. Further, $\Pr[X > x] = \Pr[\frac{X^*}{\alpha} > x]$, where $E[X^*] = n$, or the distribution of the sum of i.i.d. exponential random variables each with rate α follows by scaling $x \to \alpha x$ from the distribution of the sum of i.i.d. exponential random variables each with unit rate (or mean 1). Moreover, (2.65) and (3.26) show that a sum of n independent Gamma random variables specified by β_k (but with same α) is again a Gamma random variable with $\beta = \sum_{k=1}^{n} \beta_k$.

At last all centered moments follow from (2.39) by series expansion around $z = 0$ as

$$E\left[(X-a)^n\right] = (-1)^n \left.\frac{d^n\left(e^{za}\left(1+\frac{z}{\alpha}\right)^{-\beta}\right)}{dz^n}\right|_{z=0}$$

$$= (-1)^n n! a^n \sum_{m=0}^{n} \binom{-\beta}{m} \frac{(\alpha a)^{-m}}{(n-m)!}$$

In particular, since $E\left[X\right] = \mu = \frac{\beta}{\alpha}$, we find with $\binom{-z}{m} = (-1)^m \frac{\Gamma(z+m)}{m!\Gamma(z)}$

$$E\left[(X-\mu)^n\right] = (-1)^n n! \frac{\beta^n}{\alpha^n} \sum_{m=0}^{n} \binom{-\beta}{m} \frac{\beta^{-m}}{(n-m)!}$$

$$= (-1)^n \frac{\beta^n}{\alpha^n} \sum_{m=0}^{n} \binom{n}{m}(-1)^m \frac{\Gamma(\beta+m)}{\Gamma(\beta)\beta^m}$$

$$= (-1)^n \frac{\beta^n}{\alpha^n} (-\beta)^\beta U\left(\beta, \beta+1+n, -\beta\right)$$

where $U(a, b, z)$ is the confluent hypergeometric function (Abramowitz and Stegun, 1968, Chapter 13). For example, if $n = 2$, the variance equals $\sigma^2 = \frac{\beta}{\alpha^2}$ and further, $E\left[(X-\mu)^3\right] = \frac{2\beta}{\alpha^3}$, $E\left[(X-\mu)^4\right] = \frac{3\beta(\beta+2)}{\alpha^4}$ and $E\left[(X-\mu)^5\right] = \frac{4\beta(5\beta+6)}{\alpha^5}$.

3.3.2 The sum of independent uniform random variables

The sum $S_k = \sum_{j=1}^{k} U_j$ of k i.i.d. uniform random variables U_j has as distribution function the k-fold convolution of the uniform density function $f_U(x) = 1_{0 \le x \le 1}$ on $[0, 1]$ denoted by $f_U^{(k*)}(x)$. The distribution function equals

$$\Pr\left[S_k \le x\right] = \sum_{j=0}^{[x]} \frac{(-1)^j}{j!(k-j)!}(x-j)^k \tag{3.29}$$

Indeed, from (2.66) and (3.13) the Laplace transform of S_k is

$$\varphi_{S_k}(z) = \left(\frac{1-e^{-z}}{z}\right)^k$$

The inverse Laplace transform determines, for $c > 0$,

$$f_U^{(k*)}(x) \triangleq \frac{d}{dx} \Pr\left[S_k \leq x\right] = \frac{1}{2\pi i} \int_{c-i\infty}^{c+i\infty} \left(\frac{1-e^{-z}}{z}\right)^k e^{zx} dz$$

Using $(1-e^{-z})^k = \sum_{j=0}^{k} \binom{k}{j}(-1)^j e^{-jz}$ and the integral $\frac{1}{2\pi i}\int_{c-i\infty}^{c+i\infty} \frac{e^{sa}}{s^{n+1}} ds = \frac{a^n}{n!} 1_{\mathrm{Re}(a)>0}$, yields

$$f_U^{(k*)}(x) = \sum_{j=0}^{k} \binom{k}{j}(-1)^j \frac{(x-j)^{k-1}}{(k-1)!} 1_{(x-j)\geq 0} \qquad (3.30)$$

from which (3.29) follows by integration.

3.3.3 The chi-square distribution

Suppose that the total error of n independent measurements X_k, each perturbed by Gaussian noise, has to be determined. In order to prevent that errors may cancel out, the sum of the squared errors $S = \sum_{k=0}^{n} e_k^2$ is preferred rather than $\sum_{k=0}^{n} |e_k|$. For simplicity, we assume that all errors $e_k = X_k - x_k$, where x_k is the exact value of quantity k, have zero mean and unit variance. The corresponding distribution of S is known as the chi-square distribution. From the χ^2-distribution, the χ^2-test in statistics is deduced which determines the goodness of a model of a distribution to a set of measurements. We refer for a discussion of the χ^2-test to Leon-Garcia (1994, Section 3.8) or Allen (1978, Section 8.4).

We first deduce the distribution of the square $Y = X^2$ of a random variable X and note that if U and V are independent so are the random variables $g(U)$ and $h(V)$. The event $\{Y \leq y\}$ or $\{X^2 \leq y\}$ is equivalent to $\{-\sqrt{y} \leq X \leq \sqrt{y}\}$ and non-existent if $y < 0$. With (2.29) and $y \geq 0$,

$$\Pr\left[Y \leq y\right] = \Pr\left[-\sqrt{y} \leq X \leq \sqrt{y}\right] = F_X(\sqrt{y}) - F_X(-\sqrt{y})$$

and, after differentiation,

$$f_{X^2}(x) = \frac{f_X(\sqrt{x}) + f_X(-\sqrt{x})}{2\sqrt{x}}$$

If X is a Gaussian random variable $N(\mu, \sigma^2)$, then is, for $x \geq 0$,

$$f_{X^2}(x) = \frac{\exp\left[-\frac{(x+\mu^2)}{2\sigma^2}\right]}{\sigma\sqrt{2\pi x}} \cosh\left(\frac{\mu\sqrt{x}}{\sigma^2}\right)$$

In particular, for $N(0,1)$ random variables where $\mu = 0$ and $\sigma = 1$, $f_{X^2}(x) =$

$\frac{e^{-\frac{x}{2}}}{\sqrt{2\pi x}}$ reduces to a Gamma distribution (3.25) with $\beta = \frac{1}{2}$ and $\alpha = \frac{1}{2}$. Since the sum of n independent Gamma random variables with (α, β) is again a Gamma random variable $(\alpha, n\beta)$, we arrive at the chi-square χ^2 probability density function,

$$f_{\chi^2}(x) = \frac{x^{\frac{n}{2}-1}}{2^{\frac{n}{2}}\Gamma\left(\frac{n}{2}\right)}e^{-\frac{x}{2}} \tag{3.31}$$

3.4 Functions of random variables

3.4.1 The maximum and minimum of a set of independent random variables

The minimum of m i.i.d. random variables $\{X_k\}_{1\leq k\leq m}$ possesses the distribution[4]

$$\Pr\left[\min_{1\leq k\leq m} X_k \leq x\right] = \Pr\left[\text{at least one } X_k \leq x\right] = \Pr\left[\text{not all } X_k > x\right]$$

or

$$\Pr\left[\min_{1\leq k\leq m} X_k \leq x\right] = 1 - \prod_{k=1}^{m} \Pr[X_k > x] \tag{3.32}$$

whereas for the maximum,

$$\Pr\left[\max_{1\leq k\leq m} X_k > x\right] = \Pr\left[\text{not all } X_k \leq x\right] = 1 - \prod_{k=1}^{m} \Pr[X_k \leq x]$$

or

$$\Pr\left[\max_{1\leq k\leq m} X_k \leq x\right] = \prod_{k=1}^{m} \Pr[X_k \leq x] \tag{3.33}$$

For example, the distribution function for the minimum of m independent exponential random variables follows from (3.17) as

$$\Pr\left[\min_{1\leq k\leq m} X_k \leq x\right] = 1 - \prod_{k=1}^{m} e^{-\alpha_k x} = 1 - \exp\left(-x\sum_{k=1}^{m}\alpha_k\right)$$

or, the minimum of m independent exponential random variables each with rate α_k is again an exponential random variable with rate $\sum_{k=0}^{m}\alpha_k$. In addition to the memoryless property, this property of the exponential distribution will determine the fundamentals of Markov chains.

---

[4] An alternative argument for independent random variables is that the event $\{\min_{1\leq k\leq m} X_k > x\}$ is only possible if and only if $\{X_k > x\}$ for each $1 \leq k \leq m$. Similarly, the event $\{\max_{1\leq k\leq m} X_k \leq x\}$ is only possible if and only if all $\{X_k \leq x\}$ for each $1 \leq k \leq m$.

3.4.2 Order statistics

The set $X_{(1)}, X_{(2)}, \ldots, X_{(m)}$ are called the order statistics of the set of random variables $\{X_k\}_{1 \le k \le m}$ if $X_{(k)}$ is the k-th smallest value of the set $\{X_k\}_{1 \le k \le m}$. Clearly, $X_{(1)} = \min_{1 \le k \le m} X_k$ while $X_{(m)} = \max_{1 \le k \le m} X_k$. If the set $\{X_k\}_{1 \le k \le m}$ consists of i.i.d. random variables with pdf f_X, the joint density function of the order statistics is, for only $x_1 < x_2 < \cdots < x_m$,

$$f_{\{X_{(j)}\}}(x_1, x_2, \ldots, x_m) = \frac{\partial^m}{\partial x_1 \ldots \partial x_m} \Pr\left[X_{(1)} \le x_1, \ldots, X_{(m)} \le x_m\right]$$

$$= m! \prod_{j=1}^{m} f_X(x_j) \tag{3.34}$$

Indeed, confining to discrete random variables for simplicity, if $x_1 < x_2 < \cdots < x_m$, then

$$\Pr\left[X_{(1)} = x_1, \ldots, X_{(m)} = x_m\right] = m! \Pr\left[X_1 = x_1, \ldots, X_m = x_m\right]$$

else

$$\Pr\left[X_{(1)} = x_1, X_{(2)} = x_2, \ldots, X_{(m)} = x_m\right] = 0$$

because there are precisely $m!$ permutations of the set $\{X_k\}_{1 \le k \le m}$ onto the given ordered sequence $\{x_1, x_2, \ldots, x_m\}$. If the sequence is not ordered such that $x_k > x_l$ for at least one couple of indices $k < l$, then the probability is zero because the event $\{X_{(k)} > X_{(l)}\}$ is, by definition, impossible. Finally, the product in (3.34) follows by independence.

If the set $\{X_k\}_{1 \le k \le m}$ is uniformly distributed over $[0, t]$, then

$$f_{\{X_{(j)}\}}(x_1, x_2, \ldots, x_m) = \frac{m!}{t^m} \qquad 0 \le x_1 < x_2 < \cdots < x_m \le t$$

$$= 0 \qquad \text{elsewhere}$$

while for exponential random variables with $f_X(x) = \alpha e^{-\alpha x}$

$$f_{\{X_{(j)}\}}(x_1, x_2, \ldots, x_m) = m! \alpha^m e^{-\alpha \sum_{j=1}^{m} x_j} \qquad 0 \le x_1 < x_2 < \cdots < x_m$$

$$= 0 \qquad \text{elsewhere}$$

The order relation between the set $X_{(1)} \le X_{(2)} \le \cdots \le X_{(m)}$ is preserved after a continuous, non-decreasing transform g, i.e. $g\left(X_{(1)}\right) \le g\left(X_{(2)}\right) \le \cdots \le g\left(X_{(m)}\right)$. If the distribution function F_X is continuous (it is always non-decreasing), the argument shows that the order statistics of a general set of i.i.d. random variable $\{X_k\}_{1 \le k \le m}$ can be reduced to a study of the order statistics of the set of i.i.d. uniform random variables $\{U_k\}_{1 \le k \le m}$ on $[0,1]$ because $U = F_X(X)$.

The event $\{X_{(k)} \leq x\}$ means that at least k among the m random variables $\{X_j\}_{1 \leq j \leq m}$ are smaller than x. Since each of the m random variables is chosen independently from a same distribution F_X, the probability that precisely n of the m random variables is smaller than x is binomially distributed with parameter $p = \Pr[X \leq x]$. Hence,

$$\Pr\left[X_{(k)} \leq x\right] = \sum_{n=k}^{m} \binom{m}{n} \left(\Pr[X \leq x]\right)^n \left(1 - \Pr[X \leq x]\right)^{m-n} \qquad (3.35)$$

The probability density function can be obtained in the usual, though cumbersome, way by differentiation,

$$f_{X_{(k)}}(x) = \frac{d\Pr\left[X_{(k)} \leq x\right]}{dx}$$

$$= \sum_{n=k}^{m} \binom{m}{n} \frac{d}{dx} \left(\Pr[X \leq x]\right)^n \left(1 - \Pr[X \leq x]\right)^{m-n}$$

$$= f_X(x) \sum_{n=k}^{m} n \binom{m}{n} \left(\Pr[X \leq x]\right)^{n-1} \left(1 - \Pr[X \leq x]\right)^{m-n}$$

$$- f_X(x) \sum_{n=k}^{m} (m-n) \binom{m}{n} \left(\Pr[X \leq x]\right)^n \left(1 - \Pr[X \leq x]\right)^{m-n-1}$$

Using $n\binom{m}{n} = m\binom{m-1}{n-1}$, $(m-n)\binom{m}{n} = m\binom{m-1}{n}$ and lowering the upper index in the last summation, we have

$$f_{X_{(k)}}(x) = m f_X(x) \sum_{n=k}^{m} \binom{m-1}{n-1} \left(\Pr[X \leq x]\right)^{n-1} \left(1 - \Pr[X \leq x]\right)^{m-n}$$

$$- m f_X(x) \sum_{n=k}^{m-1} \binom{m-1}{n} \left(\Pr[X \leq x]\right)^n \left(1 - \Pr[X \leq x]\right)^{m-n-1}$$

$$= m f_X(x) \sum_{n=k}^{m} \binom{m-1}{n-1} \left(\Pr[X \leq x]\right)^{n-1} \left(1 - \Pr[X \leq x]\right)^{m-n}$$

$$- m f_X(x) \sum_{n=k+1}^{m} \binom{m-1}{n-1} \left(\Pr[X \leq x]\right)^{n-1} \left(1 - \Pr[X \leq x]\right)^{m-n}$$

or, with $F_X(x) = \Pr[X \leq x]$,

$$f_{X_{(k)}}(x) = m f_X(x) \binom{m-1}{k-1} \left(F_X(x)\right)^{k-1} \left(1 - F_X(x)\right)^{m-k} \qquad (3.36)$$

The more elegant and faster argument is as follows: in order for $X_{(k)}$ to be equal to x, exactly $k-1$ of the m random variables $\{X_j\}_{1 \leq j \leq m}$ must be less than x, one equal to x and the other $m - k$ must all be greater than x. Abusing the notation $f_X(x) = \Pr\left[X_{(k)} = x\right]$ and observing that $m\binom{m-1}{k-1} = \frac{m!}{1!(k-1)!(m-k)!}$ is an instance of the multinomial coefficient $\frac{m!}{n_1! n_2! \cdots n_k!}$ which gives the number of ways of putting $m = n_1 + n_2 + \cdots + n_k$ different objects into k different boxes with n_j in the j-th box, leads alternatively to (3.36).

3.5 Examples of other distributions

1. The **Gumbel distribution** appears in the theory of extremes (see Section 6.4) and is defined by the distribution function

$$F_{\text{Gumbel}}(x) = e^{-e^{-a(x-b)}} \tag{3.37}$$

The corresponding Laplace transform is

$$\varphi_{\text{Gumbel}}(z) = \int_{-\infty}^{\infty} e^{-zt} e^{-e^{-a(t-b)}} a e^{-a(t-b)} dt = e^{-bz} \Gamma\left(1 + \frac{z}{a}\right)$$

from which the mean follows as $E[X] = -\frac{d}{dz} e^{-bz} \Gamma\left(1 + \frac{z}{a}\right)\big|_{z=0} = b + \frac{\gamma}{a}$, where $\gamma = 0.57721...$ is the Euler constant. The variance is best computed with (2.43) resulting in $\text{Var}[X] = \frac{\pi^2}{6a^2}$.

2. The **Cauchy distribution** has the probability density function

$$f_{\text{Cauchy}}(x) = \frac{1}{\pi(1 + x^2)}$$

and corresponding distribution,

$$F_{\text{Cauchy}}(x) = \frac{1}{\pi}\left(\frac{\pi}{2} + \arctan x\right) \tag{3.38}$$

The Laplace transform

$$\varphi_{\text{Cauchy}}(z) = \frac{1}{\pi} \int_{-\infty}^{\infty} \frac{e^{-zx} dx}{1 + x^2}$$

only converges for purely imaginary $z = i\omega$, in which case it reduces to a Fourier transform,

$$\varphi_{\text{Cauchy}}(i\omega) = \frac{1}{\pi} \int_{-\infty}^{\infty} \frac{e^{-i\omega x} dx}{1 + x^2}$$

This integral is best evaluated by contour integration. If $\omega \leq 0$, we consider a contour C consisting of the real axis and the semi-circle that encloses the negative $\text{Im}(x)$-plane,

$$\int_C \frac{e^{-i\omega x} dx}{1 + x^2} = \int_{-\infty}^{\infty} \frac{e^{-i\omega x} dx}{1 + x^2} + \lim_{r \to \infty} \int_0^\pi \frac{e^{-i\omega r e^{-i\theta}} d\left(r e^{-i\theta}\right)}{1 + r^2 e^{-2i\theta}}$$

Since $\left| e^{-i\omega r e^{-i\theta}} \right| = e^{\omega r \sin\theta} = e^{-|\omega| r \sin\theta}$ and $\sin\theta \geq 0$ for $0 \leq \theta \leq \pi$, the limit of the last integral vanishes. The contour encloses the simple pole (zero of $x^2 + 1 = (x-i)(x+i)$) at $x = -i$. Applying Cauchy's residue theorem, we obtain

$$\int_{-\infty}^{\infty} \frac{e^{-i\omega x} dx}{1 + x^2} = -2\pi i \lim_{x \to -i} \frac{e^{-i\omega x}(x+i)}{1 + x^2} = \pi e^{-\omega}$$

If $\omega \geq 0$, we close the contour over the positive $\mathrm{Im}(x)$-plane such that the contribution of the semi-circle to the contour C again vanishes. The resulting contour then encloses the simple pole at $x = i$ and

$$\int_{-\infty}^{\infty} \frac{e^{-i\omega x} dx}{1 + x^2} = 2\pi i \lim_{x \to i} \frac{e^{-i\omega x}(x-i)}{1 + x^2} = \pi e^{-\omega}$$

Combining both expressions results in

$$\varphi_{\text{Cauchy}}(i\omega) = E\left[e^{-i\omega X}\right] = e^{-|\omega|}$$

Since $|\omega|$ is not analytic around $\omega = 0$, none of the moments of the Cauchy distribution exists! Hence, the Cauchy distribution is an example of a distribution without mean (see the requirement for the existence of the expectation in Section 2.3.2), although the improper integral $\int_{-\infty}^{\infty} \frac{x dx}{1+x^2} = 0$ due to symmetry (in the Riemann sense), but both $\int_{-\infty}^{0} \frac{x dx}{1+x^2}$ and $\int_{0}^{\infty} \frac{x dx}{1+x^2}$ diverge. In addition, if $S_n = \sum_{k=1}^{n} X_k$ is the sum of i.i.d. Cauchy random variables X_k, the sample mean $\frac{S_n}{n}$ has the Fourier transform,

$$E\left[e^{-i\omega \frac{S_n}{n}}\right] = E\left[e^{-i\frac{\omega}{n} \sum_{k=1}^{n} X_k}\right] = \prod_{k=1}^{n} E\left[e^{-i\frac{\omega}{n} X_k}\right] = \left(E\left[e^{-i\frac{\omega}{n} X}\right]\right)^n = e^{-|\omega|}$$

Hence, the sample mean $\frac{S_n}{n}$ of i.i.d. Cauchy random variables is again a Cauchy random variable independent of n. This means that the law of large numbers (see Section 6.2) does not hold for the Cauchy random variable, as a consequence of the non-existence of the mean. Also, the sum S_n has Fourier transform $e^{-|n\omega|}$ and the pdf equals $f_{S_n}(x) = \frac{1}{n\pi(1+(x/n)^2)}$.

3. The **Weibull distribution** with pdf defined for $x \geq 0$ and $a, b > 0$

$$f_{\text{Weibull}}(x) = \frac{\exp\left(-\left(\frac{x}{a}\right)^b\right)}{a\Gamma\left(1 + \frac{1}{b}\right)} \tag{3.39}$$

generalizes the exponential distribution (3.17) corresponding to $b = 1$ and $a = \frac{1}{\lambda}$. It is related to the Gaussian distribution if $b = 2$. Let X be a Weibull random variable. All higher moments can be computed from (2.34) as

$$E\left[X^k\right] = \frac{1}{a\Gamma\left(1+\frac{1}{b}\right)} \int_0^\infty x^k \exp\left(-\left(\frac{x}{a}\right)^b\right) dx = \frac{a^k\Gamma\left(\frac{k+1}{b}\right)}{\Gamma\left(\frac{1}{b}\right)}$$

The generating function possesses the expansion

$$\varphi_X(z) = E\left[e^{-zX}\right] = \sum_{k=0}^\infty \frac{(-z)^k}{k!} E\left[X^k\right] = \frac{1}{\Gamma\left(\frac{1}{b}\right)} \sum_{k=0}^\infty \Gamma\left(\frac{k+1}{b}\right) \frac{(-za)^k}{k!}$$

which cannot be summed in explicit form for general b.

Sometimes an alternative definition of the Weibull distribution appears

$$f_{\text{Weibull}}(x) = abx^{b-1}e^{-ax^b} \tag{3.40}$$

$$F_{\text{Weibull}}(x) = 1 - e^{-ax^b}$$

with the advantage of a simpler expression for the distribution function $F_{\text{Weibull}}(x)$. If X possesses this probability density (3.40), the moments and variance are

$$E\left[X^k\right] = \int_0^\infty x^k f_{\text{Weibull}}(x)dx = \frac{\Gamma\left(k+\frac{1}{b}\right)}{a^{k/b}}$$

$$\text{Var}[X] = \int_0^\infty (x - E[X])^2 f_{\text{Weibull}}(x)dx = \frac{\Gamma\left(1+\frac{2}{b}\right) - \Gamma^2\left(1+\frac{1}{b}\right)}{a^{2/b}}$$

The interest of the Weibull distribution in the Internet stems from the self-similar and long-range dependence of observables (i.e. quantities that can be measured such as the delay, the interarrival times of packets, etc.). Especially if the shape factor $b \in (0,1)$, the Weibull has a sub-exponential tail that decays more slowly than an exponential, but still faster than any power law.

4. Power law behavior is often described via the **Pareto distribution** with pdf for $x \geq 0$ and $\tau > 0$,

$$f_{\text{Pareto}}(x) = \frac{\alpha}{\tau}\left(1 + \frac{x}{\tau}\right)^{-\alpha-1} \tag{3.41}$$

and with distribution function

$$F_{\text{Pareto}}(x) = \frac{\alpha}{\tau} \int_0^x \left(1 + \frac{x}{\tau}\right)^{-\alpha-1} dt = 1 - \left(1 + \frac{x}{\tau}\right)^{-\alpha} \tag{3.42}$$

Since $\lim_{x\to\infty} F(x) = 1$, the power α must exceed 0. The higher moments are Beta-functions (Abramowitz and Stegun, 1968, Section 6.2.1)

$$E\left[X^k\right] = \frac{\alpha}{\tau} \int_0^\infty \frac{x^k dx}{(1+\frac{x}{\tau})^{\alpha+1}} = k!\tau^k \frac{\Gamma(\alpha-k)}{\Gamma(\alpha)}$$

and show that $E\left[X^k\right]$ only exists if $\alpha > k$. Hence, the mean $E\left[X\right]$ only exists if $\alpha > 1$. The deep tail asymptotic for large x is $f_{\text{Pareto}}(x) = O\left(x^{-\alpha-1}\right)$ and $\Pr\left[X > x\right] = O\left(x^{-\alpha}\right)$. For example, the distribution of the nodal degree in the Internet has an exponent around $\alpha = 2.4$ (see Section 15.3).

5. Another distribution with heavy tails is the **lognormal distribution** defined as the random variable $X = e^Y$ where $Y = N\left(\mu, \sigma^2\right)$ is a Gaussian or normal random variable. From (2.32), it follows that $\Pr\left[e^Y \leq x\right] = \Pr\left[Y \leq \log x\right]$ for $x \geq 0$, and with (3.20)

$$F_{\text{lognormal}}(x) = \frac{1}{\sigma\sqrt{2\pi}} \int_{-\infty}^{\log x} \exp\left[-\frac{(t-\mu)^2}{2\sigma^2}\right] dt \qquad (3.43)$$

and, for $x > 0$,

$$f_{\text{lognormal}}(x) = \frac{\exp\left[-\frac{(\log x - \mu)^2}{2\sigma^2}\right]}{\sigma x \sqrt{2\pi}} \qquad (3.44)$$

The moments are

$$E\left[X^k\right] = \frac{1}{\sigma\sqrt{2\pi}} \int_0^\infty x^{k-1} \exp\left[-\frac{(\log x - \mu)^2}{2\sigma^2}\right] dx$$

$$= \frac{1}{\sigma\sqrt{2\pi}} \int_{-\infty}^\infty e^{ku} \exp\left[-\frac{(u-\mu)^2}{2\sigma^2}\right] du$$

or, explicitly,

$$E\left[X^k\right] = \exp\left(\mu k\right) \exp\left(\frac{k^2\sigma^2}{2}\right) \qquad (3.45)$$

and

$$\text{Var}[X] = e^{2\mu}\left(e^{2\sigma^2} - e^{\sigma^2}\right) \qquad (3.46)$$

The probability generating function is by definition (2.37)

$$\varphi_X(z; \mu, \sigma^2) = \frac{1}{\sigma\sqrt{2\pi}} \int_0^\infty e^{-zt} \frac{e^{-\frac{(\log t - \mu)^2}{2\sigma^2}}}{t} dt = \frac{1}{\sigma\sqrt{2\pi}} \int_{-\infty}^\infty e^{-ze^x} e^{-\frac{(x-\mu)^2}{2\sigma^2}} dx$$

$$(3.47)$$

The integral (3.47) indicates that $\varphi_X(z; \mu, \sigma^2)$ only exists for $\text{Re}(z) \geq 0$. This means that $\varphi_X(z; \mu, \sigma^2)$ is not analytic at any point $z = it$ on the imaginary axis because the circle with arbitrary small but non-zero radius around $z = it$ necessarily encircles points with $\text{Re}(z) < 0$ where $\varphi_X(z; \mu, \sigma^2)$ does not exist. Hence, the Taylor expansion (2.40) of the generating function around $z = 0$ does not exist, although all moments or derivatives at $z = 0$

exist. Indeed, the series

$$\sum_{k=0}^{\infty} \frac{(-1)^k E\left[X^k\right]}{k!} z^k = \sum_{k=0}^{\infty} \frac{(-ze^\mu)^k}{k!} \exp\left(\frac{k^2\sigma^2}{2}\right)$$

is a divergent series (except for $\sigma = 0$ or $z = 0$). The fact that the pgf (3.44) is not available in closed form complicates the computation of the sum of i.i.d. lognormal random variables via (2.66). This sum appears in radio communications with several transmitters and receivers.

In radio communications, the received signal levels decrease with the distance between the transmitter and the receiver. This phenomenon is called *pathloss*. Attenuation of radio signals due to pathloss has been modeled by averaging the measured signal powers over long times and over various locations with the same distances to the transmitter. The mean value of the signal power found in this way is referred to as the *area mean power* $\mathbf{P}_a$ (in Watts) and is well-modeled as $\mathbf{P}_a(r) = c \cdot r^{-\eta}$ where c is a constant and η is the pathloss exponent[5]. In reality the received power levels may vary significantly around the mean power $\mathbf{P}_a(r)$ due to irregularities in the surroundings of the receiving and transmitting antennas. Measurements have revealed that the logarithm of the mean power $\mathbf{P}(r)$ at different locations on a circle with radius r around the transmitter is *approximately* normally distributed with mean equal to the logarithm of the area mean power $\mathbf{P}_a(r)$. The lognormal shadowing model assumes that the logarithm of $\mathbf{P}(r)$ is *precisely* normally distributed around the logarithmic value of the area mean power: $\log_{10}\left(\mathbf{P}(r)\right) = \log_{10}\left(\mathbf{P}_a(r)\right) + X$, where $X = N(0, \sigma)$ is a zero-mean normal distributed random variable (in dB) with standard deviation σ (also in dB and for severe fluctuations up to 12 dB). Hence, the random variable $\mathbf{P}(r) = \mathbf{P}_a(r)10^X$ has a lognormal distribution (3.43) equal to

$$\Pr\left[\mathbf{P}(r) \le x\right] = \Pr\left[X \le \log_{10}\frac{x}{\mathbf{P}_a(r)}\right] = \frac{1}{\sigma\sqrt{2\pi}\log 10} \int_0^x \exp\left[-\frac{\left(\log_{10} u - \log_{10}\left(\mathbf{P}_a(r)\right)\right)^2}{2\sigma^2}\right] \frac{du}{u}$$

3.6 Summary tables of probability distributions

3.6.1 Discrete random variables

Name	$\Pr[X = k]$	$E[X]$	$\mathrm{Var}[X]$	$\varphi_X(z) = E\left[z^X\right]$
Bernoulli	$\Pr[X = 1] = p$	p	$p(1-p)$	$1 - p + pz$
Binomial	$\binom{n}{k}p^k(1-p)^{n-k}$	np	$np(1-p)$	$((1-p) + pz)^n$
Geometric	$p(1-p)^{k-1}$	$\frac{1}{p}$	$\frac{1-p}{p^2}$	$\frac{pz}{1-(1-p)z}$
Poisson	$\frac{\lambda^k}{k!}e^{-\lambda}$	λ	λ	$e^{\lambda(z-1)}$

---

[5] The constant c depends on the transmitted power, the receiver and the transmitter antenna gains and the wavelength. The pathloss exponent η depends on the environment and terrain structure and can vary between 2 in free space to 6 in urban areas.

3.6.2 Continuous random variables

Name	$f_X(x)$	$E[X]$	$\text{Var}[X]$	$\varphi_X(z) = E\left[e^{-zX}\right]$
Uniform	$\frac{1_{a \le x \le b}}{b-a}$	$\frac{a+b}{2}$	$\frac{(b-a)^2}{12}$	$\frac{e^{-za}-e^{-zb}}{z(b-a)}$
Exponential	$\alpha e^{-\alpha x}$	$\frac{1}{\alpha}$	$\frac{1}{\alpha^2}$	$\frac{\alpha}{z+\alpha}$
Gaussian	$\dfrac{\exp\left[-\frac{(x-\mu)^2}{2\sigma^2}\right]}{\sigma\sqrt{2\pi}}$	μ	σ^2	$\exp\left[\frac{\sigma^2 z^2}{2} - \mu z\right]$
Gamma	$\frac{\alpha(\alpha x)^{\beta-1}}{\Gamma(\beta)}e^{-\alpha x}$	$\frac{\beta}{\alpha}$	$\frac{\beta}{\alpha^2}$	$\left(\frac{\alpha}{z+\alpha}\right)^{\beta}$
Gumbel	$e^{-x}e^{e^{-x}}$	$\gamma = 0.5772...$	$\frac{\pi^2}{6}$	$\Gamma(z+1)$
Cauchy	$\frac{1}{\pi(1+x^2)}$	does not exist	does not exist	$e^{-|\operatorname{Im}(z)|}$ $(\operatorname{Re}(z)=0)$
Weibull	$abx^{b-1}e^{-ax^b}$	$\frac{\Gamma(1+\frac{1}{b})}{a^{1/b}}$	$\frac{\Gamma(1+\frac{2}{b})-\Gamma^2(1+\frac{1}{b})}{a^{2/b}}$	
Pareto	$\frac{\alpha}{\tau}\left(1+\frac{x}{\tau}\right)^{-\alpha-1}$	$\frac{\tau 1_{\{\alpha>1\}}}{\alpha-1}$	$\frac{\tau^2\alpha 1_{\{\alpha>2\}}}{(\alpha-1)^2(\alpha-2)}$	
Lognormal	$\dfrac{\exp\left[-\frac{(\log x-\mu)^2}{2\sigma^2}\right]}{\sigma x\sqrt{2\pi}}$	$\exp(\mu)\exp\left(\frac{\sigma^2}{2}\right)$	$e^{2\mu}\left(e^{2\sigma^2}-e^{\sigma^2}\right)$	

3.7 Problems

(i) If $\varphi_X(z)$ is the probability generating function of a non-zero discrete random variable X, find an expression of $E[\log X]$ in terms of $\varphi_X(z)$.

(ii) Compute the mean value of the k-th order statistic in an ensemble of (a) m i.i.d. exponentially distributed random variables with mean $\frac{1}{\alpha}$ and (b) m i.i.d. polynomially distributed random variables on $[0,1]$.

(iii) Discuss how a probability density function of a continuous random variable X can be approximated from a set $\{x_1, x_2, \ldots, x_n\}$ of n measurements or simulations.

(iv) In a circle with radius r around a sending mobile node, there are $N-1$ other mobile nodes uniformly distributed over that circle. The possible interference caused by these other mobile nodes depends on their distance to the sending node at the center. Derive for large N but constant density ν of mobile nodes the pdf of the distance of the m-th nearest node to the center.

(v) Let U and V be two independent random variables. What is the probability that the one is larger than the other?

4
Correlation

In this chapter methods to compute bi-variate correlated random variables are discussed. As a measure for the correlation, the linear correlation coefficient defined in (2.58) is used. First, the generation of n correlated Gaussian random variables is explained. The sequel is devoted to the construction of two correlated random variables with arbitrary distribution.

4.1 Generation of correlated Gaussian random variables

Due to the importance of Gaussian correlated random variables as an underlying system for generating arbitrary correlated random variables, as will be demonstrated in Section 4.3, we discuss how they can be generated in multiple dimensions. With the notation of Section 3.2.3, a Gaussian (normal) random variable with average μ and variance σ^2 is denoted by $N(\mu, \sigma^2)$. By linearly combining Gaussian random variables, we can create a new Gaussian random variable with a desired mean μ and variance σ^2.

4.1.1 Generation of two independent Gaussian random variables

The fact that a linear combination of Gaussian random variables is again a Gaussian random variable allows us to concentrate on normalized Gaussian random variables $N(0, 1)$. Let X_1 and X_2 be two independent normalized Gaussian random variables. Independent random variables are not correlated and the linear correlation coefficient $\rho = 0$. The resulting joint probability distribution is $f_{X_1 X_2}(x, y; \rho) = f_{X_1}(x) f_{X_2}(y)$ and with (3.19),

$$f_{X_1 X_2}(x, y; 0) = \frac{e^{-\frac{x^2 + y^2}{2}}}{2\pi}$$

61

It is natural to consider a polar transformation and the transformed random variables $R = X_1^2 + X_2^2$ and $\Theta = \arctan\left(\frac{X_2}{X_1}\right)$. The inverse transform is $x = \sqrt{r}\cos\theta$ and $y = \sqrt{r}\sin\theta$, which differs slightly from the usual polar transformation in that we now define $r = x^2 + y^2$ instead of $r^2 = x^2 + y^2$. The reason is that the Jacobian is simpler for our purposes,

$$J(r,\theta) = \det\begin{bmatrix} \frac{\partial x}{\partial r} & \frac{\partial x}{\partial \theta} \\ \frac{\partial y}{\partial r} & \frac{\partial y}{\partial \theta} \end{bmatrix} = \det\begin{bmatrix} \frac{\cos\theta}{2\sqrt{r}} & -\sqrt{r}\sin\theta \\ \frac{\sin\theta}{2\sqrt{r}} & \sqrt{r}\cos\theta \end{bmatrix} = \frac{1}{2}$$

whereas the usual polar transformation has the Jacobian equal to the variable r. Using the transformation rules in Section 2.5.4,

$$f_{R\Theta}(r,\theta) = \frac{e^{-\frac{r}{2}}}{4\pi}$$

which shows that $f_{R\Theta}(r,\theta)$ does not depend on θ. Hence, we can write $f_{R\Theta}(r,\theta) = f_R(r)f_\Theta(\theta)$ with $f_\Theta(\theta) = c$, where c is a constant and $f_R(r) = \frac{e^{-\frac{r}{2}}}{4\pi c}$. This implies that Θ is a uniform random variable over an interval $1/c$. We also recognize from (3.15) that $f_R(r)$ is close to an exponential random variable with rate $\alpha = \frac{1}{2}$. Therefore, it is instructive to choose the constant c such that R is precisely an exponential random variable with rate $\alpha = \frac{1}{2}$. Thus, choosing $c = \frac{1}{2\pi}$, we end up with $f_R(r) = \frac{e^{-\frac{r}{2}}}{2}$ and $f_\Theta(\theta) = \frac{1}{2\pi}$. These two independent random variables R and Θ can each be generated separately from a uniform random variable U on $[0,1]$, as discussed in Section 3.2.1, leading to

$$R = -2\ln(U_1)$$
$$\Theta = 2\pi U_2$$

and, finally, to the independent Gaussian random variables

$$X_1 = \sqrt{-2\ln(U_1)}\cos 2\pi U_2 \qquad X_2 = \sqrt{-2\ln(U_1)}\sin 2\pi U_2$$

The procedure can be used to generate a single Gaussian random variable, but also more independent Gaussians by repeating the generation procedure.

4.1.2 *The n-joint Gaussian probability distribution function*

A collection of n random variables X_i is called a *random vector* $X = (X_1, X_2, \ldots, X_n)^T$, a matrix with dimension $n \times 1$. The average of a random vector is a vector with components $E[X_i]$ for $1 \leq i \leq n$. The variance of a random vector

$$\mathrm{Var}[X] = E\left[(X - E[X])(X - E[X])^T\right] = E[XX^T] - E[X](E[X])^T$$

is a matrix Σ_X with elements $(\Sigma_X)_{i,j} = \text{Cov}[X_i, X_j]$. Since $\text{Cov}[X_i, X_j] = \text{Cov}[X_j, X_i]$, the covariance matrix Σ_X is real and symmetric, $\Sigma_X = \Sigma_X^T$. The importance of real, symmetric matrices is that they have real eigenvectors (see Appendix A.2). Moreover, Σ_X is non-negative definite because, using vector norms defined in Section A.3,

$$
\begin{aligned}
x^T \Sigma_X x &= E\left[x^T (X - E[X])(X - E[X])^T x\right] \\
&= E\left[\left((X - E[X])^T x\right)^T (X - E[X])^T x\right] \\
&= E\left[\left\|(X - E[X])^T x\right\|_2^2\right] \geq 0
\end{aligned}
$$

which implies that all real eigenvalues λ_i are non-negative. Hence, there exists an orthogonal matrix U such that

$$
\Sigma_X = U \text{diag}(\lambda_i) U^T \tag{4.1}
$$

If all random variables X_i are independent, $\text{Cov}[X_i, X_j] = 0$ for $i \neq j$ and $\text{Cov}[X_i, X_i] = \text{Var}[X_i] \geq 0$ then $\Sigma_X = \text{diag}(\text{Var}[X_i])$.

Gaussian random variables are completely determined by the mean and the variance, i.e. by the first two moments. We will now show that the existence of an orthogonal transformation for any probability distribution such that $U^T \Sigma_X U = \text{diag}(\lambda_i)$ implies that a vector of joint Gaussian random variables can be transformed into a vector of independent Gaussian random variables. Also the reverse holds, which will be used below to generate n joint correlated Gaussian random variables. The multi-dimensional generating function of a n-joint Gaussian or n-joint normal random vector X is defined for the vector $z = (z_1, z_2, \ldots, z_n)^T$ as

$$
\varphi_X(z) = E\left[e^{-zX}\right] = \exp\left(\frac{1}{2} z^T \Sigma_X z - E[X]^T z\right) \tag{4.2}
$$

Using (4.1), and the fact that U is an orthogonal matrix such that $U^{-1} = U^T$ and $UU^T = I$,

$$
\varphi_X(z) = \exp\left(\frac{1}{2}\left(U^T z\right)^T \text{diag}(\lambda_i) U^T z - \left(U^T E[X]\right)^T U^T z\right)
$$

Denote the vectors $w = U^T z$ and $m = U^T E[X]$. Then we have

$$
\varphi_X(z) = \exp\left(\frac{1}{2} w^T \text{diag}(\lambda_i) w - m^T w\right)
$$

$$
= \exp\left(\sum_{j=1}^n \frac{\lambda_j w_j^2}{2} - m_j w_j\right) = \prod_{j=1}^n e^{\frac{\lambda_j w_j^2}{2} - m_j w_j}
$$

and $e^{\frac{\lambda_j w_j^2}{2} - m_j w_j} = \varphi_{X_j}(w_j)$ is the Laplace transform (3.22) of a Gaussian random variable X_j because all λ_j are real and non-negative. With (2.65), this shows that a vector of joint Gaussian random variables can be transformed into a vector of independent Gaussian random variables. Reversing the order of the manipulations also justifies that (4.2) indeed defines a general n-joint Gaussian probability generating function. If $X_1, X_2, \ldots, X_n$ are joint normal and not correlated, then Σ_X is a diagonal matrix, which implies that $X_1, X_2, \ldots, X_n$ are independent. As discussed in Section 2.5.2, independence implies non-correlation, but the converse is generally not true. These properties make Gaussian random variables particularly suited to deal with correlations.

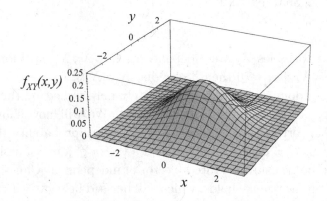

Fig. 4.1. The joint probability density function (4.4) with $\mu_X = \mu_Y = 0$ and $\sigma_X = \sigma_Y = 1$ and $\rho = 0$.

The corresponding n-joint Gaussian probability density function of the vector X can be derived after inverse Laplace transform for the vector $x = (x_1, x_2, \ldots, x_n)^T$ as

$$f_X(x) = \frac{1}{\left(\sqrt{2\pi}\right)^n \sqrt{\det \Sigma_X}} \exp\left(-\frac{1}{2}(x - E[x])^T \Sigma_X^{-1}(x - E[x])\right) \quad (4.3)$$

The inverse Laplace transform for $n = 2$ is computed in Section C.2.

After computing the inverse matrix and the determinant in (4.3) explicitly, the two-dimensional ($n = 2$) or bi-variate Gaussian probability density function is

$$f_{XY}(x, y; \rho) = \frac{\exp\left[-\dfrac{\frac{(x-\mu_X)^2}{\sigma_X^2} - \frac{2\rho}{\sigma_X \sigma_Y}(x-\mu_X)(y-\mu_Y) + \frac{(y-\mu_Y)^2}{\sigma_Y^2}}{2(1-\rho^2)}\right]}{2\pi\sigma_X\sigma_Y\sqrt{1-\rho^2}} \quad (4.4)$$

Figures 4.1–4.3 plot $f_{XY}(x, y; \rho)$ for various correlation coefficients ρ. If $\rho = 0$, we observe that $f_{XY}(x, y; 0) = f_X(x) f_Y(y)$, which indicates that uncorrelated Gaussian random variables are also independent.

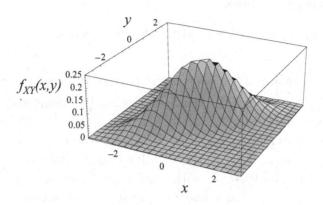

Fig. 4.2. The joint probability density function (4.4) with $\mu_X = \mu_Y = 0$ and $\sigma_X = \sigma_Y = 1$ and $\rho = 0.8$.

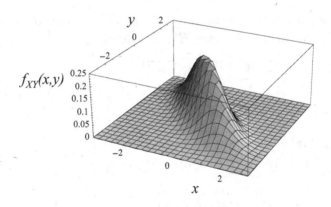

Fig. 4.3. The joint probability density function (4.4) with $\mu_X = \mu_Y = 0$ and $\sigma_X = \sigma_Y = 1$ and $\rho = -0.8$.

If we denote $x^* = \frac{(x - \mu_X)}{\sigma_X}$ and $y^* = \frac{(y - \mu_Y)}{\sigma_Y}$, the bi-variate normal density (4.4) reduces to

$$f_{XY}(x^*, y^*; \rho) = \frac{\exp\left[-\frac{(x^*)^2 - 2\rho x^* y^* + (y^*)^2}{2(1 - \rho^2)}\right]}{2\pi \sigma_X \sigma_Y \sqrt{1 - \rho^2}}$$

from which we can verify the partial differential equation

$$\frac{\partial f_{XY}(x^*, y^*; \rho)}{\partial \rho} = \frac{\partial f_{XY}(x^*, y^*; \rho)}{\partial x^* \partial y^*} \tag{4.5}$$

and the symmetry relations

$$f_{XY}(x^*, y^*; -\rho) = f_{XY}(-x^*, y^*; \rho) = f_{XY}(x^*, -y^*; \rho) \tag{4.6}$$

4.1.3 Generation of n correlated Gaussian random variables

Let $\{X_i\}_{1 \leq i \leq n}$ be a set of n independent normal random variables, where each X_i is distributed as $N(0, 1)$. The vector X is rather easily simulated. The analysis above shows that $E[X] = 0$ (the null-vector) and $\Sigma_X = \text{diag}(\text{Var}[X_i]) = \text{diag}(1) = I$, the identity matrix. We want to generate the correlated normal vector Y with a given mean vector $E[Y]$ and a given covariance matrix Σ_Y. Since linear combinations of normal random variables are normal random variables, we consider the linear transformation $Y = AX + B$ where A and B are constant matrices. We will now determine A and B. First,

$$E[Y] = E[AX] + E[B] = AE[X] + E[B] = E[B]$$

Hence, the matrix B is a vector with components equal to the given components $E[Y_i]$ of the mean vector $E[Y]$. Second,

$$\begin{aligned} \Sigma_Y &= E\left[(Y - E[Y])(Y - E[Y])^T\right] \\ &= E\left[AX(AX)^T\right] = E\left[AXX^T A^T\right] = AE\left[XX^T\right]A^T \\ &= A\Sigma_X A^T = AA^T \end{aligned}$$

From the eigenvalue decomposition of $\Sigma_Y = U\text{diag}(\lambda_i)U^T$ with real eigenvalues $\lambda_i \geq 0$ and the fact that $\text{diag}(\lambda_i) = \text{diag}(\sqrt{\lambda_i})\left(\text{diag}(\sqrt{\lambda_i})\right)^T$ such that

$$AA^T = U\text{diag}(\sqrt{\lambda_i})\left(\text{diag}(\sqrt{\lambda_i})\right)^T U^T$$

we obtain

$$A = U\text{diag}(\sqrt{\lambda_i})$$

The matrix A is also called the *square root* matrix of Σ_Y and can be found from the singular value decomposition of Σ_Y or from Cholesky factorization (Press *et al.*, 1992).

Example Generate a normal vector Y with $E[Y] = (300, 300)^T$, with standard deviations $\sigma_1 = 106.066$, $\sigma_2 = 35.355$ and correlation $\rho_Y = 0.8$.

Solution: The covariance matrix of Y is obtained using the definition of the linear correlation coefficient (2.58),

$$\Sigma_Y = \begin{pmatrix} \sigma_1^2 & \rho_Y \sigma_1 \sigma_2 \\ \rho_Y \sigma_1 \sigma_2 & \sigma_2^2 \end{pmatrix} = \begin{pmatrix} 11250 & 3000 \\ 3000 & 1250 \end{pmatrix}$$

The square root matrix A of Σ_Y is

$$A = \begin{pmatrix} 63.640 & 84.853 \\ 0 & 35.355 \end{pmatrix}$$

which is readily checked by computing $AA^T = \Sigma_Y$. It remains to generate m independent draws for X_1 and X_2 from a normal distribution with zero mean and unit variance as explained in Section 4.1.1. Each pair (X_1, X_2) out of the m pairs is transformed as $Y = AX + E[Y]$. The result, component Y_2 versus Y_1, is shown in Fig. 4.4.

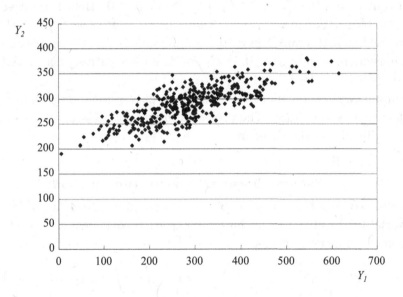

Fig. 4.4. The scatter diagram of the simulated vector Y.

4.2 Generation of correlated random variables

Let us consider the problem of generating two correlated random variables X and Y with given distribution functions F_X and F_Y. The correlation is expressed in terms of the *linear* correlation coefficient $\rho(X, Y) = \rho$ defined in (2.58). The need to generate correlated random variables often occurs in simulations. For example, as shown in Kuipers and Van Mieghem

(2003), correlations in the link weight structure may significantly increase the computational complexity of multi-constrained routing, called in brief QoS routing. The importance of measures of dependence between quantities in risk and finance is discussed in Embrechts *et al.* (2001b).

In general, given the distribution functions F_X and F_Y, not all linear correlations from $-1 \leq \rho \leq 1$ are possible. Indeed, let X and Y be positive real random variables with infinite range which means that $F_X(x) = 1$ if $x \to \infty$ and that $F_X(x) = F_Y(x) = 0$ for $x < 0$. Consider $Y = aX + b$ with $a < 0$ and $b \geq 0$. For all finite $y < 0$,

$$F_Y(y) = \Pr[Y \leq y] = \Pr[aX + b \leq y] = \Pr\left[X \geq \frac{y-b}{a}\right] \geq \Pr\left[X > \frac{y-b}{a}\right]$$

$$= 1 - F_X\left(\frac{y-b}{a}\right) > 0$$

which contradicts the fact that $F_Y(y) = 0$ for $y < 0$. Hence, positive random variables with infinite range cannot be correlated with $\rho = -1$. The requirement that the range needs to be unbounded is necessary because two uniform random variables on $[0, 1]$, U_1 and U_2, are negatively correlated with $\rho = -1$ if $U_1 = 1 - U_2$.

In summary, the set of all possible correlations is a closed interval $[\rho_{\min}, \rho_{\max}]$ for which $\rho_{\min} < 0 < \rho_{\max}$. The precise computation of $\rho_{\min}$ and $\rho_{\max}$ is, in general, difficult, as shown below.

4.3 The non-linear transformation method

The non-linear transformation approach starts from a given set of two random variables X_1 and X_2 that have a correlation coefficient $\rho_X \in [-1, 1]$. If the joint distribution function

$$F_{X_1 X_2}(x_1, x_2; \rho_X) = \Pr[X_1 \leq x_1, X_2 \leq x_2] = \int_{-\infty}^{x_1} \int_{-\infty}^{x_2} f_{X_1 X_2}(u, v; \rho_X) du\, dv$$

is known, the marginal distribution follows from (2.60) as

$$\Pr[X_1 \leq x_1] = \int_{-\infty}^{x_1} \int_{-\infty}^{\infty} f_{X_1 X_2}(u, v; \rho_X) du\, dv$$

Since for any random variable X holds that $F_X(X) = U$ where U is a uniform random variable on $[0, 1]$, it follows that $U_1 = F_{X_1}(X_1)$ and $U_2 = F_{X_2}(X_2)$ are uniformly correlated random variables with correlation coefficient ρ_U. As shown in Section 3.2.1, if U is a uniform random variable on $[0,1]$, any other random variable Y with distribution function $g(x)$ can be constructed as $g^{-1}(U)$. By combining the two transforms, we can generate

$Y_1 = g_1^{-1}(F_{X_1}(X_1))$ and $Y_2 = g_2^{-1}(F_{X_2}(X_2))$ that are correlated because X_1 and X_2 are correlated. It may be possible to construct directly the correlated random variables $Y_1 = T_1(X_1)$ and $Y_2 = T_2(X_2)$ if the transforms T_1 and T_2 are known.

The goal is to determine the linear correlation coefficient ρ_Y defined in (2.58),

$$\rho_Y = \frac{E[Y_1 Y_2] - E[Y_1] E[Y_2]}{\sqrt{\text{Var}[Y_1]}\sqrt{\text{Var}[Y_2]}}$$

as a function of ρ_X. Using (2.61),

$$E[Y_1 Y_2] = E\left[g_1^{-1}(F_{X_1}(X_1)) g_2^{-1}(F_{X_2}(X_2))\right]$$

$$= \int_{-\infty}^{\infty} \int_{-\infty}^{\infty} g_1^{-1}(F_{X_1}(u)) g_2^{-1}(F_{X_2}(v)) f_{X_1 X_2}(u, v; \rho_X) du dv \quad (4.7)$$

This relation shows that ρ_Y is a continuous function in ρ_X and that the joint distribution function of X_1 and X_2 is needed. The main difficulty lies now in the computation of the integral appearing in $E[Y_1 Y_2]$. For X_1 and X_2, Gaussian correlated random variables are most often chosen because an exact analytic expression (4.4) exists for the joint distribution function $F_{X_2 X_2}(x_1, x_2; \rho_X)$.

4.3.1 Properties of ρ_Y as a function of ρ_X

From now on, we choose Gaussian correlated random variables for X_1 and X_2.

Theorem 4.3.1 *The correlation coefficient ρ_Y is a differentiable and increasing function of ρ_X.*

Proof: From the partial differential equation (4.5) of $f_{X_1 X_2}(u, v; \rho_X)$, it follows that

$$\frac{\partial E[Y_1 Y_2]}{\partial \rho_X} = \int_{-\infty}^{\infty} \int_{-\infty}^{\infty} g_1^{-1}(F_{X_1}(u)) g_2^{-1}(F_{X_2}(v)) \frac{\partial^2 f_{X_1 X_2}(u, v; \rho_X)}{\partial u \partial v} du dv$$

Partial integration with respect to u and v yields

$$\frac{\partial E[Y_1 Y_2]}{\partial \rho_X} = \int_{-\infty}^{\infty} \int_{-\infty}^{\infty} \frac{dg_1^{-1}(F_{X_1}(u))}{du} \frac{dg_2^{-1}(F_{X_2}(v))}{du} f_{X_1 X_2}(u, v; \rho_X) du dv$$

Applying the chain rule for differentiation and $\frac{dg^{-1}(x)}{dx} = \frac{1}{g'(g^{-1}(x))}$ gives

$$\frac{dg^{-1}(F_X(u))}{du} = \frac{dg^{-1}(x)}{dx}\bigg|_{x=F_X(u)} \frac{dx}{du} = \frac{f_X(u)}{g'(g^{-1}(F_X(u)))}$$

Since $g'(x)$ and $f_X(u)$ are probability density functions and positive, $\frac{\partial E[Y_1 Y_2]}{\partial \rho_X} = \frac{\partial \rho_Y}{\partial \rho_X} > 0$. Hence, we have shown that ρ_Y is a differentiable, increasing function of ρ_X. $\square$

Since $\rho_X \in [-1, 1]$, ρ_Y increases from $\rho_{Y\,\min}$ at $\rho_X = -1$ to $\rho_{Y\,\max}$ corresponding to $\rho_X = 1$. In the sequel, we will derive expressions to compute the boundary cases $\rho_X = -1$ and $\rho_X = 1$.

Theorem 4.3.2 (of Lancaster) *For any two strictly increasing real functions T_1 and T_2 that transform the correlated Gaussian random variables X_1 and X_2 to the correlated random variables $Y_1 = T_1(X_1)$ and $Y_2 = T_2(X_2)$, it holds that*

$$|\rho_Y| \le |\rho_X|$$

If two correlated random variables Y_1 and Y_2 can be obtained by separate transformations from a bi-variate normal distribution with correlation coefficient ρ_X, the correlation coefficient ρ_Y of the transformed random variables cannot in absolute value exceed ρ_X. The interest of the proof is that it uses powerful properties of orthogonal polynomials and that ρ_Y is expanded in a power series in ρ_X in (4.12).

Proof: The proof is based on the orthogonal Hermite polynomials $H_n(x)$ (see e.g. Rainville (1960) and Abramowitz and Stegun (1968, Chapter 22)) defined by the generating function

$$\exp\left(2xt - t^2\right) = \sum_{n=0}^{\infty} \frac{H_n(x)\, t^n}{n!} \tag{4.8}$$

After expanding $\exp\left(2xt - t^2\right)$ in a Taylor series and equating corresponding powers in t, we find that

$$H_n(x) = n! \sum_{k=0}^{\left[\frac{n}{2}\right]} \frac{(-1)^k (2x)^{n-2k}}{k!\,(n-2k)!} \tag{4.9}$$

with $H_0(x) = 1$. The Hermite polynomials satisfy the orthogonality relations

$$\int_{-\infty}^{\infty} e^{-x^2} H_n(x) H_m(x)\, dx = 0 \qquad m \ne n$$

$$\int_{-\infty}^{\infty} e^{-x^2} H_n^2(x)\, dx = 2^n n!\sqrt{\pi}$$

These orthogonality relations enable us to expand functions in terms of Hermite polynomials (similar to Fourier analysis). If the expansion of a function $f(x)$,

$$f(x) = \sum_{k=0}^{\infty} a_k H_k(x)$$

converges for all x, then it follows from the orthogonality relations that

$$a_k = \frac{1}{2^k k!\sqrt{\pi}} \int_{-\infty}^{\infty} e^{-x^2} f(x) H_k(x)\, dx$$

The joint normalized Gaussian density function can be expanded (Rainville, 1960, pp. 197–198) in terms of Hermite polynomials

$$\frac{\exp\left[-\frac{x^2 - 2\rho xy + y^2}{(1-\rho^2)}\right]}{\sqrt{1-\rho^2}} = e^{-x^2 - y^2} \sum_{n=0}^{\infty} H_n(x) H_n(y) \frac{\rho^n}{2^n n!} \tag{4.10}$$

In order for the covariance $\text{Cov}[Y_1 Y_2]$ to exist, both $E\left[Y_1^2\right]$ and $E\left[Y_2^2\right]$ must be finite. Since $Y_j = T_j\left(X_j\right)$ for $j = 1, 2$, the mean is

$$E\left[Y_j\right] = \int_{-\infty}^{\infty} T_j(x) f_{X_j}(x)\, dx = \frac{1}{\sqrt{2\pi}\sigma_X} \int_{-\infty}^{\infty} T_j(x)\exp\left[-\frac{(x-\mu_X)^2}{2\sigma_X^2}\right] dx$$

$$= \frac{1}{\sqrt{\pi}} \int_{-\infty}^{\infty} T_j(\mu_X + \sqrt{2}\sigma_X u) e^{-u^2}\, du$$

Let

$$T_j\left(\mu_X + \sqrt{2}\sigma_X u\right) = \sum_{k=0}^{\infty} a_{k;j} H_k(u) \tag{4.11}$$

with

$$a_{k;j} = \frac{1}{2^k k! \sqrt{\pi}} \int_{-\infty}^{\infty} e^{-x^2} T_j\left(\mu_X + \sqrt{2}\sigma_X x\right) H_k(x)\, dx$$

then, since $H_0(x) = 1$,

$$E\left[Y_j\right] = a_{0;j}$$

The second moment $E\left[Y_j^2\right]$ follows from (2.34) as

$$E\left[Y_j^2\right] = \int_{-\infty}^{\infty} (T_j(x))^2 f_{X_j}(x)\, dx = \frac{1}{\sqrt{2\pi}\sigma_X} \int_{-\infty}^{\infty} T_j^2(x)\exp\left[-\frac{(x-\mu_X)^2}{2\sigma_X^2}\right] dx$$

$$= \frac{1}{\sqrt{\pi}} \int_{-\infty}^{\infty} T_j^2(\mu_X + \sqrt{2}\sigma_X u) e^{-u^2}\, du$$

Substituting (4.11) gives

$$E\left[Y_j^2\right] = \sum_{k=0}^{\infty}\sum_{m=0}^{\infty} a_{m;j} a_{k;j} \frac{1}{\sqrt{\pi}} \int_{-\infty}^{\infty} H_k(u) H_m(u) e^{-u^2}\, du = \sum_{k=0}^{\infty} a_{k;j}^2 2^k k!$$

which is convergent. Similarly, using (4.4),

$$E\left[Y_1 Y_2\right] = \int_{-\infty}^{\infty}\int_{-\infty}^{\infty} T_1(u) T_2(v) f_{X_1 X_2}(u, v; \rho_X)\, du\, dv$$

$$= \int_{-\infty}^{\infty}\int_{-\infty}^{\infty} T_1(u) T_2(v) \frac{\exp\left[-\dfrac{\dfrac{(u-\mu_X)^2}{\sigma_X^2} - \dfrac{2\rho_X}{\sigma_X \sigma_Y}(u-\mu_X)(v-\mu_Y) + \dfrac{(v-\mu_Y)^2}{\sigma_Y^2}}{2(1-\rho_X^2)}\right]}{2\pi\sigma_X \sigma_Y \sqrt{1-\rho^2}}\, du\, dv$$

$$= \int_{-\infty}^{\infty}\int_{-\infty}^{\infty} T_1\left(\mu_X + \sqrt{2}\sigma_X x\right) T_2\left(\mu_Y + \sqrt{2}\sigma_Y y\right) \frac{\exp\left[-\dfrac{x^2 - 2\rho_X xy + y^2}{(1-\rho_X^2)}\right]}{\pi\sqrt{1-\rho_X^2}}\, dx\, dy$$

$$= \frac{1}{\pi\sqrt{1-\rho_X^2}} \sum_{k=0}^{\infty} a_{k;1} \sum_{m=0}^{\infty} a_{m;2} \int_{-\infty}^{\infty}\int_{-\infty}^{\infty} H_k(x) H_m(y)\exp\left[-\frac{x^2 - 2\rho_X xy + y^2}{(1-\rho_X^2)}\right] dx\, dy$$

Using (4.10),

$$E\left[Y_1 Y_2\right] = \frac{1}{\pi} \sum_{k=0}^{\infty} a_{k;1} \sum_{m=0}^{\infty} a_{m;2} \sum_{n=0}^{\infty} \frac{\rho_X^n}{2^n n!} \int_{-\infty}^{\infty} e^{-x^2} H_k(x) H_n(x)\, dx \int_{-\infty}^{\infty} e^{-y^2} H_m(y) H_n(y)\, dy$$

Introducing the orthogonality relations for Hermite polynomials leads to

$$E\left[Y_1 Y_2\right] = \sum_{n=0}^{\infty} a_{n;1} a_{n;2} 2^n n! \rho_X^n$$

The correlation coefficient becomes

$$\rho_Y = \frac{\sum_{n=0}^{\infty} a_{n;1} a_{n;2} 2^n n! \rho_X^n - a_{0;1} a_{0;2}}{\sqrt{\sum_{k=0}^{\infty} a_{k;1}^2 2^k k! - a_{0;1}^2} \sqrt{\sum_{k=1}^{\infty} a_{k;2}^2 2^k k! - a_{0;2}^2}} = \frac{\sum_{n=1}^{\infty} a_{n;1} a_{n;2} 2^n n! \rho_X^n}{\sqrt{\sum_{k=1}^{\infty} a_{k;1}^2 2^k k! \sum_{k=1}^{\infty} a_{k;2}^2 2^k k!}}$$

Denote $\alpha_n = a_{n;1}\sqrt{2^n n!}$ and $\beta_n = a_{n;2}\sqrt{2^n n!}$, then $\mathrm{Var}[Y_1] = \sum_{k=1}^{\infty} \alpha_k^2$ and $\mathrm{Var}[Y_2] = \sum_{k=1}^{\infty} \beta_k^2$. Since the linear correlation coefficient $\rho(X, Y)$ equals the correlation coefficient of the corresponding normalized random variable with mean zero and variance 1, as shown in Section 2.5.3, we may choose $\sum_{k=1}^{\infty} \alpha_k^2 = \sum_{k=1}^{\infty} \beta_k^2 = 1$ such that

$$\rho_Y = \sum_{n=1}^{\infty} \alpha_n \beta_n \rho_X^n \qquad (4.12)$$

If $\alpha_1^2 = 1$ and $\beta_1^2 = 1$, then $|\rho_Y| = |\rho_X|$ because all other α_k and β_k must then vanish. In all other cases, either $\alpha_1^2 < 1$ or $\beta_1^2 < 1$ or both, such that

$$\rho_Y = \alpha_1 \beta_1 \rho_X + \sum_{n=2}^{\infty} \alpha_n \beta_n \rho_X^n$$

and

$$\left| \sum_{n=2}^{\infty} \alpha_n \beta_n \rho_X^n \right| \leq \sum_{n=2}^{\infty} |\alpha_n \beta_n| \, |\rho_X|^n \leq \sqrt{\sum_{n=2}^{\infty} \alpha_n^2 \, |\rho_X|^n} \sqrt{\sum_{n=2}^{\infty} \beta_n^2 \, |\rho_X|^n}$$

where we have used the Cauchy-Schwarz inequality $\sum ab \leq \sqrt{\sum a^2} \sqrt{\sum b^2}$ (see Section 5.5). By partial summation,

$$\sum_{n=2}^{\infty} \alpha_n^2 \, |\rho_X|^n = (1 - |\rho_X|) \sum_{n=2}^{\infty} \sum_{k=2}^{n} \alpha_k^2 \, |\rho_X|^n \leq (1 - |\rho_X|) \left(1 - \alpha_1^2\right) \frac{|\rho_X|^2}{1 - |\rho_X|} = \left(1 - \alpha_1^2\right) |\rho_X|^2$$

because $\sum_{k=2}^{n} \alpha_k^2 < \sum_{k=2}^{\infty} \alpha_k^2 = 1 - \alpha_1^2$. Thus

$$|\rho_Y| \leq |\rho_X| \left(|\alpha_1 \beta_1| + \sqrt{1 - \alpha_1^2} \sqrt{1 - \beta_1^2} \, |\rho_X| \right)$$

Finally, for $\alpha_1^2 \leq 1$ and $\beta_1^2 \leq 1$, the inequality $|\alpha_1 \beta_1| + \sqrt{1 - \alpha_1^2} \sqrt{1 - \beta_1^2} \leq 1$ holds. This proves Lancaster's theorem because $|\rho_X| \leq 1$. $\square$

4.3.2 Boundary cases

Let us investigate some cases for special values of ρ_X.

1. $\rho_X = 0$. Since uncorrelated Gaussian random variables ($\rho_X = 0$) are independent, also $Y_1 = g_1^{-1}\left(F_{X_1}(X_1)\right)$ and $Y_2 = g_2^{-1}\left(F_{X_2}(X_2)\right)$ are independent such that $\rho_Y = 0$. Hence, uncorrelated Gaussian random variables with $\rho_X = 0$ lead to uncorrelated random variables Y_1 and Y_2 with $\rho_Y = 0$.

2. $\rho_X = 1$. Perfect positively correlated Gaussian random variables $X_1 = X_2 = X$ have joint distribution

$$f_{X_1 X_2}(u, v; 1) = \frac{1}{\sqrt{2\pi}\sigma_X} \exp\left[-\frac{(u - \mu_X)^2}{2\sigma_X^2}\right] \delta(u - v)$$

which follows from $\Pr[X_1 \leq x_1, X_2 \leq x_2] = \Pr[X \leq x_1, X \leq x_2] = \Pr[X \leq x]$ with $x = \max(x_1, x_2)$. In that case,

$$E[Y_1 Y_2] = \int_{-\infty}^{\infty} g_1^{-1}(F_X(u)) g_2^{-1}(F_X(u)) dF_X(u)$$

$$= \int_0^1 g_1^{-1}(x) g_2^{-1}(x) dx \tag{4.13}$$

which may lead to $\rho_{Y\,\max} < 1$ depending on the specifics of g_1 and g_2.

By transforming $x = g_1(u)$, we obtain

$$E[Y_1 Y_2] = \int_{g_1^{-1}(0)}^{g_1^{-1}(1)} u g_2^{-1}(g_1(u)) g_1'(u) du$$

which shows that, if $g_1 = g_2 = g$,

$$E[Y_1 Y_2] = \int_{g^{-1}(0)}^{g^{-1}(1)} u^2 g'(u) du = E[Y^2]$$

Hence, if Y_1 and Y_2 have the same distribution function g as Y, the case $\rho_X = 1$ leads to

$$\rho_Y = \frac{E[Y^2] - (E[Y])^2}{\text{Var}[Y]} = 1$$

3. $\rho_X = -1$. Perfect negatively correlated Gaussian random variables $X_1 = -X_2 = X$ have joint distribution

$$f_{X_1 X_2}(u, v; -1) = \frac{1}{\sqrt{2\pi}\sigma_X} \exp\left[-\frac{(u - \mu_X)^2}{2\sigma_X^2}\right] \delta(u + v)$$

which follows from the symmetry relations (4.6). In that case,

$$E[Y_1 Y_2] = \int_{-\infty}^{\infty} g_1^{-1}(F_X(u)) g_2^{-1}(F_X(-u)) dF_X(u)$$

$$= \int_{-\infty}^{\infty} g_1^{-1}(F_X(u)) g_2^{-1}(1 - F_X(u)) dF_X(u)$$

$$= \int_0^1 g_1^{-1}(x) g_2^{-1}(1 - x) dx \tag{4.14}$$

which may lead to $\rho_{Y\,\min} > -1$, depending on the specifics of g_1 and g_2.

4.4 Examples of the non-linear transformation method

4.4.1 *Correlated uniform random variables*

Let us first focus on the relation between ρ_X and ρ_U. Since $E[U] = \frac{1}{2}$ and $\sigma_U^2 = \frac{1}{12}$, the definition of the linear correlation coefficient (2.58) gives

$$\rho_U = \frac{E[U_1 U_2] - \frac{1}{4}}{\frac{1}{12}}$$

where, using (2.61),

$$E[U_1 U_2] = E[F_{X_1}(X_1) F_{X_2}(X_2)]$$
$$= \int_{-\infty}^{\infty} \int_{-\infty}^{\infty} F_{X_1}(u) F_{X_2}(v) f_{X_1 X_2}(u, v; \rho_X) du dv$$

In the case of Gaussian correlated random variables specified by (3.20) and (4.4), we must evaluate the integral

$$E[U_1 U_2] = \int_{-\infty}^{\infty} du \int_{-\infty}^{\infty} dv \int_{-\infty}^{u} dt \, e^{-\frac{(t-\mu_{X_1})^2}{2\sigma_{X_1}^2}} \int_{-\infty}^{v} d\tau \, e^{-\frac{(\tau-\mu_{X_2})^2}{2\sigma_{X_2}^2}}$$

$$\times \frac{\exp\left[-\dfrac{\dfrac{(u-\mu_{X_1})^2}{\sigma_{X_1}^2} - \dfrac{2\rho_X}{\sigma_{X_1}\sigma_{X_2}}(u-\mu_{X_1})(v-\mu_{X_2}) + \dfrac{(v-\mu_{X_2})^2}{\sigma_{X_2}^2}}{2(1-\rho_X^2)}\right]}{(2\pi)^2 \sigma_{X_1}^2 \sigma_{X_2}^2 \sqrt{1-\rho_X^2}}$$

Substituting successively $u' = \frac{u-\mu_{X_1}}{\sigma_{X_1}}$, $t' = \frac{t-\mu_{X_1}}{\sigma_{X_1}}$, $v' = \frac{v-\mu_{X_2}}{\sigma_{X_2}}$, $\tau' = \frac{\tau-\mu_{X_2}}{\sigma_{X_2}}$, we obtain

$$E[U_1 U_2] = \frac{1}{(2\pi)^2} \int_{-\infty}^{\infty} du' \int_{-\infty}^{\infty} dv' \int_{-\infty}^{u'} e^{-\frac{t'^2}{2}} dt' \int_{-\infty}^{v'} e^{-\frac{\tau'^2}{2}} d\tau' \frac{\exp\left[-\frac{u'^2 - 2\rho_X u' v' + v'^2}{2(1-\rho_X^2)}\right]}{\sqrt{1-\rho_X^2}}$$

We now use the partial differential equation (4.5),

$$\frac{\partial}{\partial \rho} \left(\frac{\exp\left[-\frac{u'^2 - 2\rho_X u' v' + v'^2}{2(1-\rho_X^2)}\right]}{\sqrt{1-\rho_X^2}} \right) = \frac{\frac{\partial^2}{\partial u' \partial v'} \left(\exp\left[-\frac{u'^2 - 2\rho_X u' v' + v'^2}{2(1-\rho_X^2)}\right] \right)}{\sqrt{1-\rho_X^2}}$$

such that

$$\frac{\partial E[U_1 U_2]}{\partial \rho} = \int_{-\infty}^{\infty} du' \int_{-\infty}^{\infty} dv' \int_{-\infty}^{u'} e^{-\frac{t'^2}{2}} dt' \int_{-\infty}^{v'} e^{-\frac{\tau'^2}{2}} d\tau' \frac{\frac{\partial^2}{\partial u' \partial v'} \left(\exp\left[-\frac{u'^2 - 2\rho_X u' v' + v'^2}{2(1-\rho_X^2)}\right] \right)}{(2\pi)^2 \sqrt{1-\rho_X^2}}$$

Partial integration in the last integral v'

$$I_2 = \int_{-\infty}^{\infty} dv' \int_{-\infty}^{v'} e^{-\frac{\tau'^2}{2}} d\tau' \frac{\partial^2}{\partial u' \partial v'} \left(\exp\left[-\frac{u'^2 - 2\rho_X u' v' + v'^2}{2(1-\rho_X^2)}\right] \right)$$

$$= \int_{-\infty}^{\infty} dv e^{-\frac{v^2}{2}} \frac{\partial}{\partial u'} \left(\exp\left[-\frac{u'^2 - 2\rho_X u' v + v^2}{2(1-\rho_X^2)}\right] \right)$$

yields

$$\frac{\partial E\left[U_1 U_2\right]}{\partial \rho} = \int_{-\infty}^{\infty} dv e^{-\frac{v^2}{2}} \int_{-\infty}^{\infty} du' \int_{-\infty}^{u'} e^{-\frac{t'^2}{2}} dt' \frac{\frac{\partial}{\partial u'}\left(\exp\left[-\frac{u'^2 - 2\rho_X u' v' + v'^2}{2\left(1 - \rho_X^2\right)}\right]\right)}{(2\pi)^2 \sqrt{1 - \rho_X^2}}$$

and similarly in the u' integral,

$$\frac{\partial E\left[U_1 U_2\right]}{\partial \rho} = \frac{1}{(2\pi)^2 \sqrt{1 - \rho_X^2}} \int_{-\infty}^{\infty} dv \int_{-\infty}^{\infty} du \exp\left[-\frac{u^2\left(2 - \rho_X^2\right) - 2\rho_X uv + v^2\left(2 - \rho_X^2\right)}{2\left(1 - \rho_X^2\right)}\right]$$

$$= \int_{-\infty}^{\infty} dv \frac{\exp\left[-\left(\frac{\left(4 - \rho_X^2\right)}{2\left(2 - \rho_X^2\right)}\right) v^2\right]}{(2\pi)^2 \sqrt{1 - \rho_X^2}} \int_{-\infty}^{\infty} du \exp\left[-\frac{\left(2 - \rho_X^2\right)}{2\left(1 - \rho_X^2\right)}\left(u - \frac{\rho_X v}{\left(2 - \rho_X^2\right)}\right)^2\right]$$

$$= \frac{1}{(2\pi)^2 \sqrt{1 - \rho_X^2}} \sqrt{\frac{2\pi\left(2 - \rho_X^2\right)}{\left(4 - \rho_X^2\right)}} \sqrt{\frac{2\pi\left(1 - \rho_X^2\right)}{\left(2 - \rho_X^2\right)}} = \frac{1}{2\pi} \frac{1}{\sqrt{4 - \rho_X^2}}$$

Thus, we find that

$$\frac{\partial \rho_U}{\partial \rho_X} = \frac{6}{\pi} \frac{1}{\sqrt{4 - \rho_X^2}}$$

or that

$$\rho_U = \frac{6}{\pi} \int \frac{1}{\sqrt{4 - \rho_X^2}} + c = \frac{6}{\pi} \arcsin\left(\frac{\rho_X}{2}\right) + c$$

It remains to determine the constant c. We have shown in Section 4.3.2 that random variables generated from uncorrelated Gaussian random variables are also uncorrelated implying that $\rho_U = 0$ if $\rho_X = 0$ and, hence, that the constant $c = 0$. This finally results in

$$\rho_U = \frac{6}{\pi} \arcsin\left(\frac{\rho_X}{2}\right) \tag{4.15}$$

In summary, two uniform correlated random variables U_1 and U_2 with correlation coefficient ρ_U are found by transforming two Gaussian correlated random variables X_1 and X_2 with correlation coefficient $\rho_X = 2\sin\left(\frac{\pi\rho_U}{6}\right)$. Equation (4.15) further shows that $\rho_U = \pm 1$ if $\rho_X = \pm 1$, which indicates that the whole range of the correlation coefficient ρ_U is possible.

4.4.2 Correlated exponential random variables

In Section 3.2.1, we have seen that, if U is a uniform random variable on $[0,1]$, $g^{-1}(U) = -\frac{1}{\alpha}\log U$ is an exponential random variable with mean $\frac{1}{\alpha}$. The correlation coefficient for two exponential random variables, Y_1 and Y_2, with mean $\frac{1}{\alpha_1}$ and $\frac{1}{\alpha_2}$ respectively, is

$$\rho_Y = \frac{E\left[Y_1 Y_2\right] - \frac{1}{\alpha_1 \alpha_2}}{\frac{1}{\alpha_1 \alpha_2}} = E\left[\alpha_1 \alpha_2 Y_1 Y_2\right] - 1$$

As above, we generate $Y_1 = -\frac{1}{\alpha_1}\log F_{X_1}(X_1)$ and $Y_2 = -\frac{1}{\alpha_2}\log F_{X_2}(X_2)$, where X_1 and X_2 are correlated Gaussian random variables with correlation coefficient ρ_X. Then,

$$E[\alpha_1\alpha_2 Y_1 Y_2] = \frac{\alpha_1\alpha_2}{\alpha_1\alpha_2} E[\log F_{X_1}(X_1)\log F_{X_2}(X_2)]$$

$$= \int_{-\infty}^{\infty}\int_{-\infty}^{\infty}\log F_{X_1}(u)\log F_{X_2}(v)f_{X_1 X_2}(u,v;\rho_X)dudv$$

In the general case for $\rho_X \neq 0$, the previous method can be followed, which yields after substitution towards normalized variables,

$$E[\alpha_1\alpha_2 Y_1 Y_2] = \int_{-\infty}^{\infty}du\int_{-\infty}^{\infty}dv\log\left(\int_{-\infty}^{u}e^{-\frac{t^2}{2}}dt\right)\log\left(\int_{-\infty}^{v}e^{-\frac{\tau^2}{2}}d\tau\right)\frac{\exp\left[-\frac{u^2-2\rho_X uv+v^2}{2(1-\rho_X^2)}\right]}{(2\pi)^2\sqrt{1-\rho_X^2}}.$$

Unfortunately, we cannot evaluate this integral analytically.

Let us compute the upper bound $\rho_{Y\,\max}$ from (4.13) with $g_1^{-1}(x) = -\frac{1}{\alpha_1}\log x$ and $g_2^{-1}(x) = -\frac{1}{\alpha_2}\log x$,

$$E[\alpha_1\alpha_2 Y_1 Y_2; \rho_X = 1] = \int_0^1 \log^2 x\, dx = 2$$

and thus $\rho_{Y\,\max} = 1$. The lower boundary $\rho_{Y\,\min}$ follows from (4.14) as[1],

$$E[\alpha_1\alpha_2 Y_1 Y_2; \rho_X = -1] = \int_0^1 \log x\log(1-x)dx = 2 - \frac{\pi^2}{6}$$

Here, we find $\rho_{Y\,\min} = 1 - \frac{\pi^2}{6} = -0.644\,934....$

In summary, exponential correlated random variables can be generated from Gaussian correlated random variables, but the correlation coefficient

---

[1] Substituting the Taylor expansion $\log(1-x) = -\sum_{k=1}^{\infty}\frac{x^k}{k}$ gives

$$\int_0^1 \log x\log(1-x)dx = -\sum_{k=1}^{\infty}\frac{1}{k}\int_0^1 x^k\log x\,dx$$

and

$$\int_0^1 x^k\log x\,dx = -\int_0^{\infty}e^{-(k+1)u}u\,du = -\frac{1}{(k+1)^2}$$

Thus,

$$\int_0^1 \log x\log(1-x)dx = \sum_{k=1}^{\infty}\frac{1}{k(k+1)^2}$$

Since $\frac{1}{k(k+1)^2} = \frac{1}{k} - \frac{1}{k+1} - \frac{1}{(k+1)^2}$,

$$\int_0^1 \log x\log(1-x)dx = \sum_{k=1}^{\infty}\frac{1}{k} - \sum_{k=2}^{\infty}\frac{1}{k} - \sum_{k=2}^{\infty}\frac{1}{k^2} = 1 - \sum_{k=2}^{\infty}\frac{1}{k^2} = 2 - \sum_{k=1}^{\infty}\frac{1}{k^2} = 2 - \zeta(2) = 2 - \frac{\pi^2}{6}$$

ρ_Y is limited to the interval $\left[1 - \frac{\pi^2}{6}, 1\right]$. As explained in the introduction of Section 4.2, the exponential random variables are positive with infinite range for which not all negative correlations are possible. The analysis demonstrates that it is not possible to construct two exponential random variables with correlation coefficient smaller than $\rho_{Y\,\min} = 1 - \frac{\pi^2}{6} \simeq -0.645$.

4.4.3 Correlated lognormal random variables

Two correlated lognormal random variables Y_1 and Y_2 with distribution specified in (3.43) can be constructed directly from two correlated Gaussian random variables X_1 and X_2. In particular, let $Y_1 = e^{a_1 X_1}$ and $Y_2 = e^{a_2 X_2}$. The explicit scaling parameters can be used to determine the desired mean. From (4.7),

$$E[Y_1 Y_2] = \int_{-\infty}^{\infty} \int_{-\infty}^{\infty} e^{a_1 u} e^{a_2 v} f_{X_1 X_2}(u, v; \rho_X)\, du\, dv$$

$$= \exp\left[a_1 \mu_1 + a_2 \mu_2 + \frac{\sigma_1^2}{2} a_1^2 + a_1 a_2 \rho_X \sigma_1 \sigma_2 + \frac{\sigma_2^2}{2} a_2^2\right]$$

where the Laplace transform (4.2) for $n = 2$ has been used. Invoking (3.45) and (3.46) with $\mu_j \to a_j \mu_j$ and $\sigma_j^2 \to a_j^2 \sigma_j^2$, the correlation coefficient ρ_Y is

$$\rho_Y = \frac{e^{a_1 \sigma_1 a_2 \sigma_2 \rho_X} - 1}{\sqrt{\left(e^{a_1^2 \sigma_1^2} - 1\right)\left(e^{a_2^2 \sigma_2^2} - 1\right)}} \tag{4.16}$$

If at least one (but not all) of the quantities σ_1, σ_2, a_1 or a_2 grows large, ρ_Y tends to zero irrespective of ρ_X. Thus even if X_1 and X_2 and, hence also Y_1 and Y_2, have the strongest kind of dependence possible, i.e. $\rho_X = \pm 1$, the correlation coefficient ρ_Y can be made arbitrarily small. In case $a_1 \sigma_1 = a_2 \sigma_2 = \sigma$, (4.16) reduces to

$$\rho_Y = \frac{e^{\sigma^2 \rho_X} - 1}{e^{\sigma^2} - 1}$$

We observe that $\rho_{Y\,\max} = 1$, while $\rho_{Y\,\min} = -e^{-\sigma^2} > -1$ for $\sigma > 0$; again a manifestation that for positive random variables with infinite range not all negative correlations are possible.

4.5 Linear combination of independent auxiliary random variables

In spite of the generality of the non-linear transformation method, the involved computational difficulty suggests us to investigate simpler methods of construction. It is instructive to consider two independent random variables V and W with known probability generating functions $\varphi_V(z)$ and $\varphi_W(z)$ respectively. In the discussion of the uniform random variable in Section 3.2.1, it was shown how to generate by computer an arbitrary random variable from a uniform random variable. We thus assume that V and W can be constructed. Let us now write X and Y as a linear combination of V and W,

$$X = a_{11}V + a_{12}W + b_1$$
$$Y = a_{21}V + a_{22}W + b_2$$

which is specified by the matrix

$$A = \begin{bmatrix} a_{11} & a_{12} \\ a_{21} & a_{22} \end{bmatrix}$$

and compute the covariance defined in (2.56),

$$
\begin{aligned}
\mathrm{Cov}\,[X,Y] &= E\,[XY] - \mu_X \mu_Y \\
&= a_{11}a_{21}\left[E\,[V^2] - (E\,[V])^2\right] + (a_{11}a_{22} + a_{12}a_{21})\,E\,[VW] \\
&\quad - (a_{11}a_{22} + a_{12}a_{21})\,E[V]\,E[W] + a_{12}a_{22}\left[E\,[W^2] - (E[W])^2\right]
\end{aligned}
$$

Since U and V are independent, $E\,[VW] = E\,[V]\,E\,[W]$, and with the definition of the variance (2.16) and denoting $\sigma_V^2 = \mathrm{Var}[V]$ and similarly for W, we obtain

$$\mathrm{Cov}\,[X,Y] = a_{11}a_{21}\sigma_V^2 + a_{12}a_{22}\sigma_W^2$$

In a same way, we find

$$
\begin{aligned}
\sigma_X^2 &= a_{11}^2\sigma_V^2 + a_{12}^2\sigma_W^2 \\
\sigma_Y^2 &= a_{21}^2\sigma_V^2 + a_{22}^2\sigma_W^2
\end{aligned}
\tag{4.17}
$$

such that the correlation coefficient, in general, becomes

$$\rho = \frac{a_{11}a_{21}\sigma_V^2 + a_{12}a_{22}\sigma_W^2}{\sqrt{a_{11}^2\sigma_V^2 + a_{12}^2\sigma_W^2}\sqrt{a_{21}^2\sigma_V^2 + a_{22}^2\sigma_W^2}}$$

which is independent of the constants b_1 and b_2 since for a centered moment $E\left(X - E\,[X]\right)^2 = E\left(X + b - E\,[X + b]\right)^2$.

In order to achieve our goal of constructing two correlated random variables X and Y, we can choose the coefficients of the matrix A to obtain an expression as simple as possible. If we choose $X = V$ or $a_{11} = 1$, $a_{12} = b_1 = 0$, the correlation coefficient reduces to

$$\rho = \frac{a_{21}\sigma_X^2}{\sqrt{\sigma_X^2}\sqrt{a_{21}^2\sigma_X^2 + a_{22}^2\sigma_W^2}} = \frac{1}{\sqrt{1 + \frac{a_{22}^2}{a_{21}^2}\frac{\sigma_W^2}{\sigma_X^2}}}$$

By rewriting this relation, we obtain

$$a_{21} = \pm\frac{\rho\sigma_W}{\sigma_X}\frac{a_{22}}{\sqrt{1-\rho^2}}$$

If we choose $a_{22} = \sqrt{1 - \rho^2}$, the random variables X and Y are specified as

$$X = V$$
$$Y = \pm\frac{\rho\sigma_W}{\sigma_X}V + \sqrt{1-\rho^2}W + b_2$$

and the corresponding variances (4.17) are $\sigma_X^2 = \sigma_V^2$ and $\sigma_Y^2 = \sigma_W^2$. Finally, we require that $E[W] = \mu_W = 0$, which specifies

$$b_2 = E[Y] \mp \frac{\rho\sigma_Y}{\sigma_X}E[X]$$

If W is a zero mean random variable with standard deviation $\sigma_W = \sigma_Y$, the random variables X and Y are correlated with correlation coefficient ρ

$$Y = \pm\frac{\rho\sigma_Y}{\sigma_X}X + \sqrt{1-\rho^2}W + \mu_Y \mp \frac{\rho\sigma_Y}{\sigma_X}\mu_X \qquad (4.18)$$

In the sequel, we take the positive sign for ρ.

Let us now investigate what happens with the distribution functions of X and Y. Using the pgfs for continuous random variables $\varphi_X(z) = E\left[e^{-zX}\right]$ and $\varphi_Y(z) = E\left[e^{-zY}\right]$, the last relation (4.18) becomes

$$E\left[e^{-zY}\right] = e^{-z\left(\mu_Y - \frac{\rho\sigma_Y}{\sigma_X}\mu_X\right)}E\left[e^{-z\frac{\rho\sigma_Y}{\sigma_X}X}\right]E\left[e^{-z\sqrt{1-\rho^2}W}\right]$$

because $V = X$ and W are independent, or

$$\varphi_Y(z) = e^{-z\left(\mu_Y - \frac{\rho\sigma_Y}{\sigma_X}\mu_X\right)}\varphi_X\left(\frac{\rho\sigma_Y}{\sigma_X}z\right)\varphi_W\left(\sqrt{1-\rho^2}z\right) \qquad (4.19)$$

In order to produce two random variables X and Y that are correlated with correlation coefficient ρ, the pgf of the zero mean random variable W with variance σ_Y^2 must obey,

$$\varphi_W(z) = \frac{e^{\frac{\left(\mu_Y - \frac{\rho\sigma_Y}{\sigma_X}\mu_X\right)}{\sqrt{1-\rho^2}}z} \varphi_Y\left(\frac{z}{\sqrt{1-\rho^2}}\right)}{\varphi_X\left(\frac{\rho\sigma_Y}{\sigma_X\sqrt{1-\rho^2}}z\right)} \qquad (4.20)$$

which can be written in terms of the translated random variables $Y' = Y - \mu_Y$ and $X' = X - \rho\sigma_Y\mu_X$,

$$\varphi_W(z) = \frac{\varphi_{Y'}\left(\frac{z}{\sqrt{1-\rho^2}}\right)}{\varphi_{X'}\left(\frac{\rho\sigma_Y}{\sigma_X\sqrt{1-\rho^2}}z\right)}$$

This form shows that, if X' and Y' have a same distribution, W possesses, in general, a different distribution. Only the pgf of a Gaussian (with zero mean) obeys the functional equation

$$f(z) = \frac{f\left(\frac{z}{\sqrt{1-\rho^2}}\right)}{f\left(\frac{\rho}{\sqrt{1-\rho^2}}z\right)}$$

The joint probability generating function follows from (2.61) as

$$\varphi_{XY}(z_1, z_2) = E\left[e^{-z_1 X - z_2 Y}\right] = \int_{-\infty}^{\infty}\int_{-\infty}^{\infty} e^{-z_1 x - z_2 y} f_{XY}(x, y)\,dx\,dy$$

and the inverse is

$$f_{XY}(x, y) = \frac{1}{(2\pi i)^2}\int_{c_1-i\infty}^{c_1+i\infty}\int_{c_2-i\infty}^{c_2-i\infty} e^{z_1 x + z_2 y} \varphi_{XY}(z_1, z_2)\,dz_1\,dz_2 \qquad (4.21)$$

Using (4.18), we have

$$\varphi_{XY}(z_1, z_2) = e^{-z_2\left(\mu_Y - \frac{\rho\sigma_Y}{\sigma_X}\mu_X\right)} E\left[e^{-\left(z_1 + z_2\frac{\rho\sigma_Y}{\sigma_X}\right)X}\right] E\left[e^{-z_2\sqrt{1-\rho^2}W}\right]$$

$$= e^{-z_2\left(\mu_Y - \frac{\rho\sigma_Y}{\sigma_X}\mu_X\right)} \varphi_X\left(z_1 + z_2\frac{\rho\sigma_Y}{\sigma_X}\right) \varphi_W\left(z_2\sqrt{1-\rho^2}\right) \qquad (4.22)$$

Introduced into the complex double integral (4.21), the joint probability density function of the two correlated random variables can be computed.

The main deficiency of the linear combination method is the implicit assumption that any joint distribution function $f_{XY}(x, y)$ can be constructed from two independent random variables X and W. The corresponding joint pgf (4.22) possesses a product form that cannot always be made compatible with the form of an arbitrary pgf $\varphi_{XY}(z_1, z_2)$. The examples below illustrate this deficiency.

4.5.1 Correlated Gaussian random variables

If X and Y are Gaussian random variables with Laplace transform $E\left[e^{-zY}\right]$ given in (3.22), the expression (4.20) for $\varphi_W(z)$ becomes

$$\varphi_W(z) = \exp\left[\left(\frac{\sigma_Y^2}{2}\right)z^2\right]$$

which shows that W is also a Gaussian random variable with mean $\mu_W = 0$ and standard deviation $\sigma_W = \sigma_Y$. Further, the joint pgf follows from (4.22) as

$$E\left[e^{-z_1 X - z_2 Y}\right] = \exp\left[-z_2\mu_Y - \mu_X z_1 + \frac{\sigma_X^2}{2}z_1^2 + z_1 z_2 \rho\sigma_Y\sigma_X + \frac{\sigma_Y^2}{2}z_2^2\right]$$

Since

$$\frac{\sigma_X^2}{2}z_1^2 + z_1 z_2 \rho\sigma_Y\sigma_X + \frac{\sigma_Y^2}{2}z_2^2 = \frac{1}{2}\left[\begin{array}{cc} z_1 & z_2 \end{array}\right]\left[\begin{array}{cc} \sigma_X^2 & \rho\sigma_X\sigma_Y \\ \rho\sigma_X\sigma_Y & \sigma_Y^2 \end{array}\right]\left[\begin{array}{c} z_1 \\ z_2 \end{array}\right]$$

formula (4.2) indicates that $E\left[e^{-z_1 X - z_2 Y}\right]$ is the two dimensional pgf of a joint Gaussian with pdf (4.4). The linear combination method thus provides the exact results for correlated Gaussian random variables.

4.5.2 Correlated exponential random variables

Let X and Y be two correlated, exponential random variables with rate α_x and α_y. Recall that $E[X] = \sigma_X = \frac{1}{\alpha_x}$. Using the Laplace transform (3.16) in (4.20), we obtain

$$\varphi_W(z) = e^{\frac{1-\rho}{\alpha_y\sqrt{1-\rho^2}}z}\frac{1 + \dfrac{\rho z}{\alpha_y\sqrt{1-\rho^2}}}{1 + \dfrac{z}{\alpha_y\sqrt{1-\rho^2}}}$$

The corresponding probability distribution function follows from (2.38) as

$$F_W(t) = \frac{1}{2\pi i}\int_{c-i\infty}^{c+i\infty}\frac{1 + \dfrac{\rho z}{\alpha_y\sqrt{1-\rho^2}}}{1 + \dfrac{z}{\alpha_y\sqrt{1-\rho^2}}}\frac{e^{z\left(t + \frac{1-\rho}{\alpha_y\sqrt{1-\rho^2}}\right)}}{z}\,dz \qquad c > 0$$

Define the normalized time $T = \dfrac{1}{\alpha_y \sqrt{1-\rho^2}}$, then

$$F_W(t) = \frac{1}{2\pi i} \int_{c-i\infty}^{c+i\infty} \frac{1 + T\rho z}{1 + zT} \frac{e^{z(t+(1-\rho)T)}}{z} dz$$

Since $t + (1-\rho)T > 0$, the contour can be closed over the negative $\mathrm{Re}(z)$ plane encircling the poles at $z = -\frac{1}{T}$ and $z = 0$. By Cauchy's residue theorem,

$$F_W(t) = \lim_{z \to -\frac{1}{T}} \frac{(1+T\rho z)\left(z + \frac{1}{T}\right)}{(1+zT)\,z} e^{z(t+(1-\rho)T)} + \lim_{z \to 0} \frac{(1+T\rho z)\,z}{(1+zT)\,z} e^{z(t+(1-\rho)T)}$$

$$= 1 - (1-\rho)\,e^{-(1-\rho)}e^{-\frac{t}{T}}$$

Hence, for the generation of two exponential, correlated random variables, the auxiliary random variable W has an exponential distribution with an atom of size $(1-\rho)\,e^{-(1-\rho)}$ at $t = 0$, which is fortunately easily to generate with a computer. It appears that only for $\rho \geq 0$, the linear combination method leads to correct results for exponential random variables. Moreover, the method does not give an indication of the validity in the range of ρ. We have shown above that $\rho \in \left[1 - \frac{\pi^2}{6}, 1\right]$.

While the linear combination method applied to generate two exponential random variables still correctly treats a range of ρ, the application to correlated uniform random variables leads to bizarre results and definitely shows the deficiency of the method. The difficulties already encountered in this chapter in generating $n = 2$ correlated random variables with arbitrary distribution suspects that the case for $n > 2$ must be even more intractable.

4.6 Problem

(i) Show in two dimensions that (4.3), or in explicit form (4.4), is indeed the joint pdf corresponding to (4.2).

5

Inequalities

Hardy *et al.* (1999) view the most known inequalities from various angles, provide several different proofs and relate the nature of these inequalities. For example, starting from the most basic inequality between geometric and arithmetic mean[1],

$$\min(x, y) \leq \sqrt{xy} \leq \frac{x+y}{2} \leq \max(x, y) \tag{5.1}$$

which directly follows from $\left(\sqrt{x} - \sqrt{y}\right)^2 \geq 0$, they masterly extend this relation to the theorem of the arithmetic and geometric mean in several real variables x_k,

$$\prod_{k=1}^{n} x_k^{q_k} \leq \sum_{k=1}^{n} q_k x_k \tag{5.2}$$

where $\sum_{k=1}^{n} q_k = 1$. They further move to the inequalities of Cauchy-Schwarz, of Hölder, of Minkowski and many more. Only a few inequalities are reviewed here and we recommend the classic treatise on inequalities by Hardy, Littlewood and Polya for those who search for more depth, elegance and insight.

5.1 The minimum (maximum) and infimum (supremum)

Since these concepts will be frequently used, we explain the difference by concentrating on the minimum and infimum (the maximum and the supremum follow analogously). Let Υ be a non-empty subset of $\mathbb{R}$. The subset Υ

---

[1] The arithmetic-geometric mean $M(x, y)$ is the limit for $n \to \infty$ of the recursion $x_n = \frac{1}{2}(x_{n-1} + y_{n-1})$, which is an arithmetic mean, and $y_n = \sqrt{x_{n-1} y_{n-1}}$, which is a geometric mean, with initial values $x_0 = x$ and $y_0 = y$. Gauss's famous discovery on intriguing properties of $M(x, y)$ (which lead e.g. to very fast converging series for computing π) is narrated in a paper by Almkvist and Berndt (1988).

is said to be bounded from below by M if there exists a number M such that, for all $x \in \Upsilon$ holds that $x \geq M$. The largest lower bound (largest number M) is called the infimum and is denoted by $\inf(\Upsilon)$. Further, if there exists an element $m \in \Upsilon$ such that $m \leq x$ for all $x \in \Upsilon$, then this element m is called the minimum and is denoted by $\min(\Upsilon)$. If the minimum $\min(\Upsilon)$ exists, then $\min(\Upsilon) = \inf(\Upsilon)$. However, the minimum does not always exists. The classical example is the open interval (a, b), where $\inf((a, b)) = a$, but the minimum does not exist because $a \notin (a, b)$. On the other hand, for the closed interval $[a, b]$, we have that $\inf([a, b]) = \min([a, b]) = a$. This example also illustrates that every finite non-empty subset of $\mathbb{R}$ has a minimum.

5.2 Continuous convex functions

A continuous function $f(x)$ that satisfies for u and v belonging to an interval I,

$$f\left(\frac{u+v}{2}\right) \leq \frac{f(u) + f(v)}{2}$$

is called convex in that interval I. If $-f$ is convex, f is concave. Hardy *et al.* (1999, Section 3.6) demonstrate that this condition is fundamental from which the more general condition[2]

$$f\left(\sum_{k=1}^{n} q_k x_k\right) \leq \sum_{k=1}^{n} q_k f(x_k) \tag{5.3}$$

where $\sum_{k=1}^{n} q_k = 1$, can be deduced. Moreover, they show that a convex function is either very regular or very irregular and that a convex function that is not "entirely irregular" is necessarily continuous. Current textbooks, in particular the book by Boyd and Vandenberghe (2004), usually start with the definition of convexity from (5.3) in case $n = 2$ where $q_1 = 1 - q_2 = q$ and $0 \leq q \leq 1$ as

$$f(qu + (1-q)v) \leq qf(u) + (1-q)f(v) \tag{5.4}$$

where u and v can be vectors in an m-dimensional space.Fig_convex

Geometrically with $m = 1$ as illustrated in Fig. 5.1, relation (5.4) shows that each point on the chord between $(u, f(u))$ and $(v, f(v))$ lies above the

---

[2] The convexity concept can be generalized (Hardy *et al.*, 1999, Section 98) to several variables in which case the condition (5.3) becomes

$$f\left(\sum_k q_k x_k, \sum_k q_k y_k\right) \leq \sum_k q_k f(x_k, y_k)$$

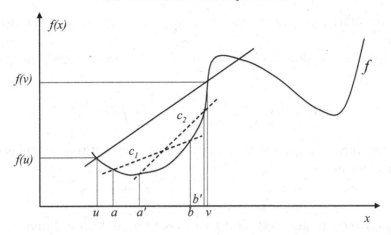

Fig. 5.1. The function f is convex between u and v.

curve f in the interval I. The more general form (5.3) asserts that the centre of gravity of any number of arbitrarily weighted points of the curve lies above or on the curve. Figure 5.1 illustrates that for any convex function f and points $a, a', b, b' \in [u, v]$ such that $a \le a' \le b'$ and $a < b \le b'$, the chord c_1 over (a, b) has a smaller slope than the chord c_2 over (a', b') or, equivalently, $\frac{f(b)-f(a)}{b-a} \le \frac{f(b')-f(a')}{b'-a'}$. Suppose that $f(x)$ is twice differentiable in the interval I, then a necessary and sufficient condition for convexity is $f''(x) \ge 0$ for each $x \in I$. This theorem is proved in Hardy *et al.* (1999, pp. 76–77). Moreover, they prove that the equality in (5.3) can only occur if $f(x)$ is linear.

Applied to probability, relation (5.3) with $q_k = \Pr[X = k]$ and $x_k = k$ is written with (2.12) as

$$f(E[X]) \le E[f(X)] \tag{5.5}$$

and is known as *Jensen's inequality*. The Jensen's inequality (5.5) also hold for continuous random variables. Indeed, if f is differentiable and convex, then $f(x) - f(y) \ge f'(y)(x - y)$. Substitute x by the random variable X and $y = E[X]$, then

$$f(X) - f(E[X]) \ge f'(E[X])(x - E[X])$$

After applying the expectation operator to both sides, we obtain (5.5). An important application of Jensen's inequality is obtained for $f(x) = e^{-zx}$ with real z as

$$e^{-zE[X]} \le E\left[e^{-zX}\right] = \varphi_X(z)$$

Any probability generating function $\varphi_X(z)$ is, for real z, bounded from below by $e^{-zE[X]}$.

A continuous analog of (5.3) with $f(x) = e^x$ (and similarly for $f(x) = -\log x$)

$$\exp\left[\frac{1}{v-u}\int_u^v f(x)dx\right] \le \frac{1}{v-u}\int_u^v e^{f(x)}dx$$

can be regarded as a generalization of the inequality between arithmetic and geometric mean.

5.3 Inequalities deduced from the Mean Value Theorem

The mean value theorem (Whittaker and Watson, 1996, p. 65) states that if $g(x)$ is continuous on $x \in [a,b]$, there exists a number $\xi \in [a,b]$ such that

$$\int_a^b g(u)du = (b-a)g(\xi)$$

or, alternatively, if $f(x)$ is differentiable on $[a,b]$, then

$$f(b) - f(a) = (b-a)f'(\xi) \tag{5.6}$$

The equivalence follows by putting $f(x) = \int_a^x g(u)du$. It is convenient to rewrite this relation for $0 \le \theta \le 1$ as

$$f(x+h) - f(x) = hf'(x+\theta h)$$

In this form, the mean value theorem is nothing else than a special case for $n = 1$ of Taylor's theorem (Whittaker and Watson, 1996, p. 96),

$$f(x+h) - f(x) = \sum_{k=1}^{n-1} \frac{f^{(k)}(x)}{k!}h^k + \frac{h^n}{n!}f^{(n)}(x+\theta h) \tag{5.7}$$

An important application of Taylor's Theorem (or of the mean value theorem) to the exponential function gives a list of inequalities. First,

$$e^x = 1 + x + \frac{x^2}{2}e^{\theta x}$$

and, since $e^{\theta x} > 0$ for any finite x, we have for any $x \ne 0$,

$$e^x > 1 + x \tag{5.8}$$

A direct generalization follows from Taylor's Theorem (5.7),

$$e^x = \sum_{k=0}^{n-1} \frac{x^k}{k!} + \frac{x^n}{n!} e^{\theta x}$$

such that, for $n = 2m$ and any x,

$$e^x > \sum_{k=0}^{2m-1} \frac{x^k}{k!}$$

and, for $n = 2m + 1$,

$$e^x > \sum_{k=0}^{2m} \frac{x^k}{k!} \qquad x > 0$$

$$e^x < \sum_{k=0}^{2m} \frac{x^k}{k!} \qquad x < 0$$

Second, estimates of the product $\prod_{k=0}^{n}(1 + a_k x)$ where $a_k x \neq 0$ are obtained from (5.8) as[3]

$$\prod_{k=0}^{n}(1 + a_k x) < \exp\left(\sum_{k=0}^{n} a_k x\right)$$

5.4 The Markov and Chebyshev inequalities

Consider first a non-negative random variable X. The expectation reads

$$E[X] = \int_0^\infty x f_X(x)dx = \int_0^a x f_X(x)dx + \int_a^\infty x f_X(x)dx$$

$$\geq \int_a^\infty x f_X(x)dx \geq a \int_a^\infty f_X(x)dx = a \Pr[X \geq a]$$

Hence, we obtain the Markov inequality

$$\Pr[X \geq a] \leq \frac{E[X]}{a} \qquad (5.9)$$

---

[3] A tighter bound is obtained if all $a_k > 0$ (e.g. a_k is a probability). The above relation indicates that $g(x) = \prod_{k=0}^{n}(1 + a_k x)$ is smaller than $f(x) = \exp\left(x \sum_{k=0}^{n} a_k\right)$ for any $x \neq 0$ and $g(0) = f(0) = 1$. Further, from $(1 + a_k x) < e^{a_k x}$ it can be verified that, for all Taylor coefficients $1 < k \leq n$ holds that $0 < g_k \leq f_k$ and $g_2 < f_2$ such that $g(x) = \sum_{k=0}^{n} g_k x^k < \sum_{k=0}^{n} f_k x^k < \sum_{k=0}^{\infty} f_k x^k$ for $x > 0$. Thus, for $x = 1$, we have $g(1) < \sum_{k=0}^{\infty} f_k$ or

$$\prod_{k=0}^{n}(1 + a_k) < \sum_{k=0}^{n} \frac{1}{k!}\left(\sum_{k=0}^{n} a_k\right)^k$$

Another proof of the Markov inequality follows after taking the expectation of the inequality $a1_{X \geq a} \leq X$ for $X \geq 0$. The restriction to non-negative random variables can be circumvented by considering the random variable $X = (Y - E[Y])^2$ and $a = t^2$ in (5.9),

$$\Pr\left[(Y - E[Y])^2 \geq t^2\right] \leq \frac{E\left[(Y - E[Y])^2\right]}{t^2} = \frac{\text{Var}[Y]}{t^2}$$

From this, the Chebyshev inequality follows as

$$\Pr\left[|X - E[X]| \geq t\right] \leq \frac{\sigma^2}{t^2} \tag{5.10}$$

The Chebyshev inequality quantifies the spread of X around the mean $E[X]$. The smaller σ, the more concentrated X is around the mean.

Further extensions of the Markov inequality use the equivalence between the events $\{X \geq a\} \Leftrightarrow \{g(X) \geq g(a)\}$ where g is a monotonously increasing function. Hence, (5.9) becomes

$$\Pr[X \geq a] \leq \frac{E[g(X)]}{g(a)}$$

For example, if $g(x) = x^k$, then $\Pr[X \geq a] \leq \frac{E[X^k]}{a^k}$. An interesting application of this idea is based on the equivalence of the events $\{X \geq E[X]+t\} \Leftrightarrow \{e^{uX} \geq e^{u(E[X]+t)}\}$ provided $u \geq 0$. For $u \geq 0$,

$$\Pr[X \geq E[X] + t] = \Pr\left[e^{uX} \geq e^{u(E[X]+t)}\right] \leq e^{-u(E[X]+t)} E\left[e^{uX}\right] \tag{5.11}$$

where in the last step Markov's inequality (5.9) has been used. If the generating function or Laplace transform $E\left[e^{uX}\right]$ is known, the sharpest bound is obtained by the minimizer u^* in u of the right-hand side because (5.11) holds for any $u > 0$. In Section 5.7, we show that this minimizer u^* obeying $\text{Re}\, u > 0$ indeed exists for probability generating functions. The resulting inequality

$$\Pr[X \geq E[X] + t] \leq e^{-u^*(E[X]+t)} E\left[e^{u^* X}\right] \tag{5.12}$$

is called the Chernoff bound.

The Chernoff bound of the binomial distribution Let X denote a binomial random variable with probability generating function given by (3.2) such that $E\left[e^{uX}\right] = E\left[(e^u)^X\right] = (q + pe^u)^n$. Then, with $E[X] = np$,

$$e^{-u(E[X]+t)} E\left[e^{uX}\right] = e^{-u(np+t)+n\log(q+pe^u)}$$

Provided $\frac{d^2}{du^2} e^{-u(E[X]+t)} E\left[e^{uX}\right]\Big|_{u=u^*} > 0$, the minimum u^* is solution of

$$\frac{d}{du} e^{-u(E[X]+t)} E\left[e^{uX}\right] = 0$$

Explicitly,

$$\frac{d}{du} e^{-u(E[X]+t)} E\left[e^{uX}\right] = e^{-u(np+t)+n\log(q+pe^u)} \left(-(np+t) + \frac{npe^u}{q+pe^u}\right)$$

from which u^* follows using $q = 1 - p$ as

$$u^* = \log\left(\frac{npq + qt}{npq - pt}\right)$$

Hence,

$$e^{-u^*(E[X]+t)} E\left[e^{u^* X}\right] = \frac{\left(1 - \frac{t}{nq}\right)^{t-nq}}{\left(1 + \frac{t}{np}\right)^{t+np}}$$

For large n, but p and t fixed, we observe[4] that $e^{-u^*(E[X]+t)} E\left[e^{u^* X}\right] = e^{-\frac{t^2}{npq}}\left(1 + O\left(\frac{1}{n}\right)\right)$. Since $\mathrm{Var}[X] = npq$ and by denoting $y^2 = \frac{t^2}{\mathrm{Var}[X]}$, we find that the asymptotic regime for large n,

$$\Pr\left[\frac{|X - E[X]|}{\sqrt{\mathrm{Var}[X]}} \geq y\right] \leq e^{-y^2} \tag{5.13}$$

is in agreement with the Central Limit Theorem 6.3.1. The corresponding Chebyshev inequality,

$$\Pr\left[\frac{|X - E[X]|}{\sqrt{\mathrm{Var}[X]}} \geq y\right] \leq \frac{1}{y^2}$$

is considerably less tight for the binomial distribution than the Chernoff bound (5.13). More advanced and sharper inequalities than that of Chebyshev are surveyed by Janson (2002).

---

[4] Write

$$e^{-u^*(E[X]+t)} E\left[e^{u^* X}\right] = \exp\left((t - nq)\log\left(1 - \frac{t}{nq}\right) - (t + np)\log\left(1 + \frac{t}{np}\right)\right)$$

and use the Taylor expansion of $\log(1 \pm x)$ around $x = 0$.

5.5 The Hölder, Minkowski and Young inequalities

The Hölder inequality is

$$\sum a^\alpha b^\beta \cdots z^\zeta \leq \left(\sum a\right)^\alpha \left(\sum b\right)^\beta \cdots \left(\sum z\right)^\zeta \qquad \alpha + \beta + \cdots \zeta = 1$$

Let $a^\alpha = X$ and $b^\beta = Y$, and further $p = \frac{1}{\alpha} > 1$ and $q = \frac{1}{\beta} > 1$ such that $\frac{1}{p} + \frac{1}{q} = 1$, then we obtain as a frequently used application,

$$E[XY] \leq (E[X^p])^{1/p} (E[Y^q])^{1/q} \tag{5.14}$$

The Hölder inequality can be deduced from the basic convexity inequality (5.4). Since $-\log x$ is a convex function for real $x > 0$, the basic convexity inequality (5.4) is with $0 \leq \theta \leq 1$,

$$\log(\theta u + (1 - \theta)v) \geq \theta \log(u) + (1 - \theta) \log(v)$$

After exponentiation, we obtain for $u, v > 0$ a more general inequality than (5.1), which corresponds to $\theta = \frac{1}{2}$,

$$u^\theta v^{1-\theta} \leq \theta u + (1 - \theta)v$$

Substitute $u = \frac{|x_j|^p}{\sum_{j=1}^n |x_j|^p}$ and $v = \frac{|y_j|^q}{\sum_{j=1}^n |y_j|^q}$, then

$$\left(\frac{|x_j|^p}{\sum_{j=1}^n |x_j|^p}\right)^\theta \left(\frac{|y_j|^q}{\sum_{j=1}^n |y_j|^q}\right)^{1-\theta} \leq \theta \frac{|x_j|^p}{\sum_{j=1}^n |x_j|^p} + (1 - \theta)\frac{|y_j|^q}{\sum_{j=1}^n |y_j|^q}$$

and summing over all j yields

$$\sum_{j=1}^n |x_j|^{p\theta} |y_j|^{q(1-\theta)} \leq \left(\sum_{j=1}^n |x_j|^p\right)^\theta \left(\sum_{j=1}^n |y_j|^q\right)^{1-\theta} \tag{5.15}$$

By choosing $p = \frac{1}{\theta}$ and $q = \frac{1}{1-\theta}$, we arrive at the Hölder inequality with $p > 1$ and $\frac{1}{p} + \frac{1}{q} = 1$,

$$\sum_{j=1}^n |x_j y_j| \leq \left(\sum_{j=1}^n |x_j|^p\right)^{\frac{1}{p}} \left(\sum_{j=1}^n |y_j|^q\right)^{\frac{1}{q}} \tag{5.16}$$

A special important case of the Hölder inequality (5.14) for $p = q = 2$ is the Cauchy–Schwarz inequality,

$$(E[XY])^2 \leq E[X^2] E[Y^2] \tag{5.17}$$

It is of interest to mention that the Hölder inequality is of a general type in the following sense (Hardy *et al.*, 1999, Theorem 101 (p. 82)). Suppose that $f(x)$ is convex (such that the inverse $g(x) = f^{-1}(x)$ is also convex) and that $f(0) = 0$. If $F(x) = \int_0^x f(u)du$ and $G(x) = \int_0^x g(u)du$, and if

$$\sum_{k=1}^n q_k a_k b_k \leq F^{-1}\left(\sum_{k=1}^n q_k F(a_k)\right) G^{-1}\left(\sum_{k=1}^n q_k G(b_k)\right)$$

with $\sum_{k=1}^n q_k = 1$ holds for all positive a_k and b_k, then $f(x) = x^r$ and the above inequality is Hölder's inequality.

The next inequalities are of a different type. For $p > 1$, the Minkowski inequality is

$$(E\left[|X + Y|^p\right])^{1/p} \leq (E\left[|X|^p\right])^{1/p} + (E\left[|Y|^p\right])^{1/p} \qquad (5.18)$$

or, written algebraically,

$$\left(\sum_{j=1}^n |x_j + y_j|^p\right)^{\frac{1}{p}} \leq \left(\sum_{j=1}^n |x_j|^p\right)^{\frac{1}{p}} + \left(\sum_{j=1}^n |y_j|^p\right)^{\frac{1}{p}}$$

Suppose that $f(x)$ is continuous and strictly increasing for $x \geq 0$ and $f(0) = 0$. Then the inverse function $g(x) = f^{-1}(x)$ satisfies the same conditions. The Young inequality states that for $a \geq 0$ and $b \geq 0$ holds that

$$ab \leq \int_0^a f(u)du + \int_0^b g(u)du \qquad (5.19)$$

with equality only if $b = f(a)$. The Young inequality follows by geometrical consideration. The first integral is the area under the curve $y = f(x)$ from $[0, a]$, while the second is the area under the curve $x = g(y) = f^{-1}(y)$ from $[0, b]$.

Applications of the Cauchy–Schwarz inequality

1. We will demonstrate that both the generating function $\varphi_X(z) = E\left[e^{-zX}\right]$ and its logarithm $L_X(z) = \log(\varphi_X(z))$ are convex functions of z. First, the second derivative is continuous and non-negative because $\varphi_X''(z) = E\left[X^2 e^{-zX}\right] \geq 0$. Further, since

$$L_X''(z) = \frac{\varphi_X(z)\varphi_X''(z) - (\varphi_X'(z))^2}{\varphi_X^2(z)}$$

it remains to show that $\varphi_X(z)\varphi_X''(z) - (\varphi_X'(z))^2 \geq 0$. From Cauchy–Schwarz inequality (5.17) applied to $\varphi_X'(z) = E\left[Xe^{-zX}\right]$ with $X \to e^{-\frac{z}{2}X}$ and $Y = Xe^{-\frac{z}{2}X}$, we obtain $(\varphi_X'(z))^2 = \left(E\left[Xe^{-zX}\right]\right)^2 \leq E\left[e^{-zX}\right]E\left[X^2 e^{-zX}\right] = \varphi_X(z)\varphi_X''(z)$. Hence, $L_X''(z) \geq 0$.

2. Let $Y = 1_{X>0}$ in (5.17) while X is a non-negative random variable, then with (2.13),

$$(E[X])^2 \leq E[X^2] E[1_{X>0}] = E[X^2] (1 - \Pr[X = 0])$$

such that an upper bound for $\Pr[X = 0]$ is obtained,

$$\Pr[X = 0] \leq 1 - \frac{(E[X])^2}{E[X^2]} \tag{5.20}$$

5.6 The Gauss inequality

In this section, we consider a continuous random variable X with even probability density function, i.e. $f_X(-x) = f_X(x)$, which is not increasing for $x > 0$. A typical example of such random variables are measurement errors due to statistical fluctuations.

In his epoch-making paper, Gauss (1821) established the method of the least squares (see e.g. Section 2.2.1 and 2.5.3). In that same paper, Gauss (1821, pp. 10-11) also stated and proved Theorem 5.6.1, which is appealing because of its generality.

We define the probability m as

$$m = \Pr[-\lambda\sigma \leq X \leq \lambda\sigma] = \int_{-\lambda\sigma}^{\lambda\sigma} f_X(u)\, du \tag{5.21}$$

where $\sigma = \sqrt{\text{Var}[X]}$ is the standard deviation.

Theorem 5.6.1 (Gauss) *If X is a continuous random variable with even probability density function, i.e. $f_X(-x) = f_X(x)$, which is not increasing for $x > 0$, then*

$$\begin{aligned}
&\text{if } m < \tfrac{2}{3} \quad \text{then } \lambda \leq m\sqrt{3} \\
&\text{if } m = \tfrac{2}{3} \quad \text{then } \lambda \leq \sqrt{\tfrac{4}{3}} \\
&\text{if } m > \tfrac{2}{3} \quad \text{then } \lambda < \tfrac{2}{3\sqrt{1-m}}
\end{aligned}$$

and, conversely,

$$\begin{aligned}
&\text{if } \lambda < \sqrt{\tfrac{4}{3}} \quad \text{then } m \geq \tfrac{\lambda}{\sqrt{3}} \\
&\text{if } \lambda > \sqrt{\tfrac{4}{3}} \quad \text{then } m \geq 1 - \tfrac{4}{\lambda^2}
\end{aligned}$$

Given a bound on the probability m, Gauss's Theorem 5.6.1 bounds the extent of the error X around its mean zero in units of the standard deviation σ or, equivalently, it provides bounds for the normalized random variable

$X^* = \frac{X-E[X]}{\sigma}$. The proof of this theorem only uses real function theory and is characteristic for the genius of Gauss.

Proof: Consider the inverse function $x = g(y)$ of the integral $y = \int_{-x}^{x} f_X(u)\, du = F_X(x) - F_X(-x)$. An interesting general property of the inverse function is

$$\int_0^1 g^2(u)\, du = \int_{-\infty}^{\infty} x^2 f_X(x)\, dx$$

which is verified by the substitution $x = g(u)$. Since $E[X] = 0$ and $\mathrm{Var}[X] = E[X^2]$, we have

$$\int_0^1 g^2(u)\, du = \sigma^2 = \mathrm{Var}[X] \qquad (5.22)$$

Beside $g(0) = 0$, the derivative

$$g'(y) = \frac{1}{F'_X(x) - F'_X(-x)} = \frac{1}{f_X(x) + f_X(-x)}$$

is increasing from $y = 0$ until $y = 1$ because $f_X(x)$ attains a maximum at $x = 0$ and is not increasing for $x > 0$. Hence, $g''(y) \geq 0$. From the differential

$$d(yg'(y)) = g'(y)\, dy + yg''(y)\, dy$$

we obtain by integration

$$yg'(y) - g(y) = \int_0^y ug''(u)\, du$$

Since $g''(y) \geq 0$, we have that $yg'(y) - g(y) \geq 0$ and since $yg'(y) > 0$ (for $y > 0$) that

$$h(y) = 1 - \frac{g(y)}{yg'(y)}$$

lies in the interval $[0, 1]$. From (5.21), it follows that $\lambda\sigma = g(m)$ and that $h(m) = 1 - \frac{\lambda\sigma}{mg'(m)}$ or

$$g'(m) = \frac{\lambda\sigma}{m(1 - h(m))}$$

With this preparation, consider now the following linear function

$$G(y) = \frac{\lambda\sigma}{m(1 - h(m))}(y - mh(m)) \qquad (5.23)$$

Clearly, we have that $G(m) = \lambda\sigma$ and that $G'(y) = \frac{\lambda\sigma}{m(1-h(m))} = g'(m)$ is independent of y. Since $g'(y)$ is non decreasing – which is the basic assumption of the theorem – the difference $g'(y) - G'(y)$ is negative if $y < m$, but positive if $y > m$. Since $g'(y) - G'(y) = \frac{d}{dy}(g(y) - G(y))$, the function $g(y) - G(y)$ is convex with minimum at $y = m$ for which $g(m) - G(m) = 0$. Hence, $g(y) - G(y) \geq 0$ for all $y \in [0, 1]$. Further, $G(y)$ is positive for $y \in (mh(m), 1]$. Especially in this interval, the inequality $g(y) \geq G(y)$ is sharp because $g(y)$ is positive in $(0, 1]$. Thus,

$$\int_{mh(m)}^1 G^2(y)\, dy \leq \int_{mh(m)}^1 g^2(y)\, dy < \int_0^1 g^2(y)\, dy$$

Using (5.22) and with (5.23), we have

$$\frac{\lambda^2\sigma^2}{m^2(1 - h(m))^2} \frac{(1 - mh(m))^3}{3} < \sigma^2$$

from which we arrive at the inequality

$$\lambda^2 < \frac{3m^2(1-z)^2}{(1-mz)^3} \tag{5.24}$$

where $z = h(m) \in [0,1]$. The derivative of the right-hand side with respect to z,

$$\frac{d}{dz}\left(\frac{3(1-z)^2}{(1-mz)^3}m^2\right) = -\frac{3m^2(1-z)}{(1-mz)^4}(2-3m+mz)$$

shows that $\frac{3m^2(1-z)^2}{(1-mz)^3}$ is monotonously decreasing for all $z \in [0,1]$ if $m < \frac{2}{3}$ with maximum at $z = 0$. Thus, if $m < \frac{2}{3}$, evaluating (5.24) at $z = 0$ yields $\lambda < \sqrt{3}m$. On the other hand, if $m > \frac{2}{3}$, then $\frac{3m^2(1-z)^2}{(1-mz)^3}$ is maximal provided $2 - 3m + mz = 0$ or for $z = 3 - \frac{2}{m}$. With that value of z, the inequality (5.24) yields $\lambda < \frac{2}{3\sqrt{1-m}}$. Both regimes $m > \frac{2}{3}$ and $m < \frac{2}{3}$ tend to a same bound $\lambda < \frac{2}{\sqrt{3}}$ if $m \to \frac{2}{3}$. The converse is similarly derived from (5.24). $\square$

If X has a symmetric uniform distribution with $f_X(x) = \frac{1}{2a}1_{x\in[-a,a]}$, then $m = \frac{\lambda\sigma}{a}$ and $\sigma = \frac{a}{\sqrt{3}}$ from which $m = \frac{\lambda}{\sqrt{3}}$. This example shows that Gauss's Theorem 5.6.1 is sharp for $m \le \frac{2}{3}$ in the sense that equality can occur in the first condition $\lambda \le \sqrt{3}m$.

5.7 The dominant pole approximation and large deviations

In this section, we relate asymptotic results of generating functions to the theory of large deviations. An asymptotic expansion in discrete-time is compared to established large deviations results.

The first approach using the generating function $\varphi_X(z)$ of the random variable X is an immediate consequence of Lemma 5.7.1.

Lemma 5.7.1 *If $\varphi_X(z)$ is meromorphic with residues r_k at the (simple) poles p_k ordered as $0 < |p_0| \le |p_1| \le |p_2| \le \cdots$ and if $\varphi_X(z) = o(z^{N+1})$ as $z \to \infty$, then holds*

$$\varphi_X(z) = \sum_{k=0}^{N}\Pr[X=k]z^k + \sum_{k=0}^{\infty}\frac{r_k}{p_k^{N+1}(z-p_k)}z^{N+1} \tag{5.25}$$

$$= \sum_{k=0}^{N}\Pr[X=k]z^k + \sum_{k=0}^{\infty}r_k\left(\frac{1}{z-p_k} + \sum_{m=0}^{N}\frac{z^m}{p_k^{m+1}}\right) \tag{5.26}$$

The normalization condition $\varphi_X(1) = 1$ implies that

$$\Pr[X>N] = 1 - \sum_{k=0}^{N}\Pr[X=k] = \sum_{k=0}^{\infty}\frac{r_k}{p_k^{N+1}(1-p_k)} \tag{5.27}$$

The Lemma follows from Titchmarsh (1964, Section 3.21). Rewriting (5.26) gives,

$$\varphi_X(z) = \sum_{k=0}^{N} \Pr\left[X = k\right] z^k - \sum_{j=N+1}^{\infty} \left(\sum_{k=0}^{\infty} \frac{r_k}{p_k^{j+1}} \right) z^j \tag{5.28}$$

and hence,

$$\Pr\left[X = j\right] = -\sum_{k=0}^{\infty} \frac{r_k}{p_k^{j+1}} \qquad (j > N) \tag{5.29}$$

The cumulative density function for $K > N$ follows from (5.29) as

$$\Pr[X > K] = \sum_{j=K+1}^{\infty} \Pr\left[X = j\right] = \sum_{k=0}^{\infty} \frac{r_k}{p_k^{K+1}(1 - p_k)} \qquad (K > N) \tag{5.30}$$

Lemma 5.7.1 means that, if the plot $\Pr\left[X = j\right]$ versus j exhibits a kink at $j = N$, then $\varphi_X(z) = O\left(z^N\right)$ as $z \to \infty$. Alternatively[5], the asymptotic regime does not start earlier than $j \geq N$. For large K, only the pole with smallest modulus, p_0, will dominate. Hence,

$$\Pr[X > K] \approx \frac{r_0}{p_0^{K+1}(1 - p_0)} \tag{5.31}$$

This approximation is called the *dominant pole* approximation with the residue at the simple pole p_0 equal to $r_0 = \lim_{z \to p_0} \varphi_X(z)(z - p_0)$.

The second approach is a large deviations approximation in discrete-time. We have

$$-\log \Pr\left[X > K\right] = -\log \sum_{j=K+1}^{\infty} \Pr\left[X = j\right]$$

$$\geq -\log \sum_{j=K+1}^{\infty} x^{j-K-1} \Pr\left[X = j\right] \qquad (x \in \mathbb{R} \text{ and } x \geq 1)$$

$$\geq -\log \left(x^{-K-1} \sum_{j=0}^{\infty} x^j \Pr\left[X = j\right] \right)$$

$$= (K + 1) \log x - \log \varphi_X(x) \tag{5.32}$$

This inequality holds for all real $x \geq 1$. To get the tightest bound, we determine the maximizer x_{max} of (5.32), thus $I(K) = \sup_{x \geq 1}[(K+1) \log x - \log \varphi_X(x)]$. There exists such a supremum on account of the convexity of

---

[5] In terms of the queue occupancy in ATM, the initial $\Pr\left[X = j\right]$-regime for $j < N$ reflects the cell scale, while the asymptotic regime $j \geq N$ refers to the burst scale.

$I(K)$ because $\varphi_X(x)$ and $\log \varphi_X(x)$ are convex for $x \geq 1$ as shown in Section 5.5. Assuming that the maximum, say $x_{\max}$ exists, then it is solution of $x_{\max} = (K+1)\frac{\varphi_X(x_{\max})}{\varphi_X'(x_{\max})}$ and the large deviations estimate becomes

$$\Pr[X > K] \leq e^{-[(K+1)\log x_{\max} - \log \varphi_X(x_{\max})]} = \varphi_X(x_{\max})\, x_{\max}^{-(K+1)} \quad (5.33)$$

Observe that (5.33) can be obtained directly from (5.11) with $K = t + E[X]$. Comparing (5.33) and (5.31) indicates, for large K, that $x_{\max} = p_0$ because

$$\lim_{K \to \infty} \frac{-\log \Pr[X > K]}{K} = \log p_0 = \log x_{\max}$$

Example A frequently appearing "dominant pole" (see, for example, the extinction probability of a Poisson branching process in Section 12.3, the M/D/1 queue in Section 14.5 and the size of the giant component in the random graph in Section 15.6.4) is the real zero ζ different from 1 of $e^{\lambda(z-1)} - z$. The trivial zero is $z = 1$. The non-trivial solution $e^{\lambda(\zeta-1)} = \zeta$ can be expressed as a Lagrange series (Markushevich, 1985, p. 94) for[6] $\lambda > 1$,

$$\zeta = e^{-\lambda} \sum_{n=0}^{\infty} \frac{(n+1)^{n-1}}{n!} \left(\lambda e^{-\lambda}\right)^n \quad (5.34)$$

An exact and fast converging expansion for ζ around $\lambda = 1$,

$$\zeta = 1 + \frac{2}{\lambda}\left[(1-\lambda) + \frac{(1-\lambda)^2}{3} + \frac{2(1-\lambda)^3}{9} + \frac{22(1-\lambda)^4}{135} + \frac{52(1-\lambda)^5}{405}\right.$$

$$+ \frac{20(1-\lambda)^6}{189} + \frac{3824(1-\lambda)^7}{42525} + \frac{1424(1-\lambda)^8}{18225} + \frac{15856(1-\lambda)^9}{229635}$$

$$\left. + \frac{11714672(1-\lambda)^{10}}{189448875} + \frac{44536288(1-\lambda)^{11}}{795685275} + o\left((1-\lambda)^{11}\right)\right] \quad (5.35)$$

is derived in Van Mieghem (1996) as the zero of $\frac{e^{\lambda(z-1)} - z}{z-1}$. The numerical data show that the approximation $\zeta \simeq \frac{1}{\lambda^2}$, which can be deduced from the series, is within 1% accurate for $0.84 < \lambda \leq 1$.

---

[6] From (14.43) we observe for $a = b = 1$ and $z = \lambda$ that in (5.34) $\zeta = 1$ for all $0 \leq \lambda < 1$.

6

Limit laws

Limit laws lie at the heart of analysis and probability theory. Solutions of problems often considerably simplify in limit cases. For example, in Section 16.5.1, the flooding time in the complete graph with N nodes and exponentially distributed link weights can be computed exactly. However, the expression is unattractive, but, fortunately, the limit result for $N \to \infty$ is appealing. Many more results and deep discussions are found in the books of Feller (1970, 1971). In this chapter, we will mainly be concerned with sums of independent random variables, $S_n = \sum_{k=1}^{n} X_k$.

6.1 General theorems from analysis

In this section, we define modes of convergence of sequences of random variables and state (without proof) some general theorems that will be used later on.

6.1.1 Summability

We will need results from the analysis on summability[1]. First the discrete case is presented and then the continuous case.

Lemma 6.1.1 *Let $\{a_n\}_{n \geq 1}$ be a sequence of numbers with $\lim_{n \to \infty} a_n = a$, then the average of the partial sums converges to a,*

$$\lim_{n \to \infty} \frac{1}{n} \sum_{m=1}^{n} a_m = a \tag{6.1}$$

---

[1] In his classical treatise on Divergent Series, Hardy (1948) discusses Césaro, Abel, Euler and Borel summability in depth.

Proof: The demonstration of (6.1) is short enough to include here. The fact that there is a limit a of the sequence $a_1, a_2, \ldots$ implies that, for an arbitrary $\varepsilon > 0$, there exist a finite number n_0 such that, for all $n > n_0$, holds that $|a_n - a| < \varepsilon$. Consider the average partial sum $s_n = \frac{1}{n} \sum_{m=1}^{n} a_m$ or, rewritten,

$$s_n - a = \frac{1}{n} \sum_{m=1}^{n_0} (a_m - a) + \frac{1}{n} \sum_{m=n_0}^{n} (a_m - a)$$

Hence,

$$|s_n - a| \leq \frac{1}{n} \sum_{m=1}^{n_0} |a_m - a| + \frac{1}{n} \sum_{m=n_0}^{n} |a_m - a| < \frac{c}{n} + \left(\frac{n - n_0}{n} \right) \varepsilon$$

$$< \frac{c}{n} + \varepsilon$$

Since c is a constant, $\frac{c}{n}$ can be made arbitrarily small for n large enough such that $|s_n - a| < \varepsilon$, which is equivalent to (6.1). $\qquad\square$

In fact, as illustrated by many examples in Hardy (1948, Chapter I and II), relation (6.1) converges in more cases than $\lim_{n \to \infty} a_n = a$ does. For example, if $a_{2n} = 1$ and $a_{2n+1} = 0$, the limit $\lim_{n \to \infty} a_n$ does not exist, but (6.1) tends to $\frac{1}{2}$. Probabilistically, the Lemma 6.1.1 is closely related to the sample mean and the Law of Large Numbers (Section 6.2).

The continuous case distinguishes between $\lim_{t \to \infty} g(t)$, which is called the pointwise limit (for sufficiently large t, all points t will be arbitrarily close to that limit) and between the limit $\lim_{t \to \infty} \frac{1}{t} \int_0^t g(u) du$, which is called the time average[2] of g.

Lemma 6.1.2 *If the pointwise limit $\lim_{t \to \infty} g(t) = g_\infty$ exists, then the time average*

$$\lim_{t \to \infty} \frac{1}{t} \int_0^t g(u) du = g_\infty.$$

Proof: The proof is analogous to that of Lemma 6.1.1 in the discrete case since $\lim_{t \to \infty} g(t) = g_\infty$ means that for an arbitrary $\varepsilon > 0$, there exist a finite number T such that, for all $t > T$ holds that $|g(t) - g_\infty| < \varepsilon$. For any $t > T$,

$$\frac{1}{t} \int_0^t g(u) du - g_\infty = \frac{1}{t} \int_0^T (g(u) - g_\infty) \, du + \frac{1}{t} \int_T^t (g(u) - g_\infty) \, du$$

---

[2] In summability theory, it is also known as the Cesaro limit of g.

and

$$\left| \frac{1}{t} \int_0^t g(u)du - g_\infty \right| \le \frac{1}{t} \left| \int_0^T (g(u) - g_\infty)\, du \right| + \frac{1}{t} \int_T^t |g(u) - g_\infty|\, du$$
$$< \frac{c}{t} + \varepsilon \frac{t - T}{t}$$

Since c is a constant, the lemma follows by letting $t \to \infty$. $\qquad \square$

Both in Markov theory (Section 9.3.2) and in Little's Law (Section 13.6) these Lemmas will be used.

6.1.2 Convergence of a sequence of random variables

A sequence $\{X_k\}_{k \ge 0}$ of random variables may converge to a random variable X in several ways. If

$$\Pr\left[\lim_{k \to \infty} |X_k - X| = 0 \right] = 1$$

then the sequence $\{X_k\}_{k \ge 0}$ converges to X with probability 1 (w.p. 1) or almost surely (a.s.). This mode of convergence is denoted by $X_k \to X$ w.p. 1 or a.s. as $k \to \infty$. If, for any $\epsilon > 0$,

$$\lim_{k \to \infty} \Pr\left[|X_k - X| > \epsilon \right] = 0$$

then it is said that the sequence $\{X_k\}_{k \ge 0}$ converges in probability or in measure to X. This mode of convergence is denoted by $X_k \overset{p}{\to} X$ as $k \to \infty$. Convergence in probability is a weaker notion of convergence than almost sure convergence. Almost sure convergence implies convergence in probability, whereas convergence in probability means that there exists a subsequence of $\{X_k\}_{k \ge 0}$ that converges almost surely. An equivalent criterion for almost surely convergence is

$$\Pr\left[|X_k - X| > \epsilon \ \text{i.o.} \right] = 0$$

where "i.o." stands for "infinitely often", thus for an infinite number of k.

If, for all x with the possible exception of a set of measure zero where $F_{X_n}(x)$ is discontinuous, the distributions

$$\lim_{k \to \infty} F_{X_n}(x) = F_X(x)$$

then the sequence $\{X_k\}_{k \ge 0}$ converges in distribution to X, denoted as $X_k \overset{d}{\to}$

X as $k \to \infty$ or, sometimes, in mixed form as $X_k \xrightarrow{d} F_X$ as $k \to \infty$. If, for $1 \le q$,

$$\lim_{k \to \infty} E\left[|X_k - X|^q\right] = 0$$

then the sequence $\{X_k\}_{k \ge 0}$ converges to X in L^q, the space of all functions f for which $\int_{-\infty}^{\infty} |f(x)|^q dx < \infty$. The most common values of q are 1, 2 and $q = \infty$. This convergence is also called convergence in norm (see Appendix A.3). The Markov inequality (5.9)

$$\Pr\left[|X_k - X| \ge \epsilon\right] \le \frac{E\left[|X_k - X|\right]}{\epsilon}$$

shows that convergence in mean ($q = 1$) implies convergence in probability.

In general, it is fair to say that the convergence of sequences belong to the most complicated topics in both analysis and probability theory. In many limit theorems, for example, the Law of Large Numbers in Section 6.2 and Little's Law in Section 13.6, the art consists in proving the theorem with the least possible number of assumptions or in its most widely applicable form.

6.1.3 List of general theorems

Theorem 6.1.3 (Continuity Theorem) *Let $\{F_n\}_{n \ge 1}$ be a sequence of distribution functions with corresponding probability generating functions $\{\varphi_n\}_{n \ge 1}$. If $\lim_{n \to \infty} \varphi_n(z) = \varphi(z)$ exists for all z, and, in addition if φ is continuous at $z = 0$, then there exists a limiting distribution function F with generating function φ for which $F_n \xrightarrow{d} F$.*

Proof: See e.g. Berger (1993, p. 51). $\square$

Theorem 6.1.4 (Dominated Convergence Theorem) *Let $\{f_n\}_{n \ge 1}$ and f be real functions and suppose that for each x*

$$\lim_{n \to \infty} f_n(x) = f(x)$$

If there exists a real function $g(x)$ such that $|f_n(x)| < g(x)$ and for which the random variable $g(X)$ has finite expectation, then

$$\lim_{n \to \infty} E\left[f_n(X)\right] = E\left[f(X)\right]$$

Proof: See e.g. Royden (1988, Chapter 4). $\square$

6.2 Law of Large Numbers

Theorem 6.2.1 (Weak Law of Large Numbers) *Let $\{X_k\}$ be a sequence of independent random variables each with distribution identical to that of the random variable X with $\mu = E[X]$. If the expectation $\mu = E[X]$ exists, then, for any $\epsilon > 0$,*

$$\lim_{n\to\infty} \Pr\left[\left|\frac{S_n}{n} - \mu\right| \geq \epsilon\right] = 0 \tag{6.2}$$

Proof[3]: Replacing X_k by $X_k - \mu$ demonstrates that, without loss of generality, we may assume that $\mu = 0$. Denote $U_n = \frac{S_n}{n}$, then $\varphi_{U_n}(z) = E\left[e^{-zU_n}\right] = E\left[e^{-zS_n/n}\right]$. Since the set $\{X_k\}$ is independent with common distribution, applying relation (2.66) yields

$$\varphi_{U_n}(z) = \left(\varphi_X\left(\frac{z}{n}\right)\right)^n$$

Since the expectation exists ($\mu = 0$), the Taylor expansion (2.40) of φ_X around $z = 0$ is $\varphi_X(z) = 1 + o(z)$ and $\varphi_{U_n}(z) = \left(1 + o\left(\frac{z}{n}\right)\right)^n$. Taking the logarithm, $\log(\varphi_{U_n}(z)) = n \log\left(1 + o\left(\frac{z}{n}\right)\right) = n.o\left(\frac{z}{n}\right) = o(z)$ for large n such that $\lim_{n\to\infty} \varphi_{U_n}(z) = 1$. By the Continuity Theorem 6.1.3, $\varphi_U(z) = E\left[e^{-zU}\right] = 1$ which implies that $U_n \xrightarrow{d} 0$. Hence, the sequence $\frac{S_n}{n}$ converges in distribution to $\mu = 0$, which is equivalent to (6.2). $\square$

The Weak Law of Large Numbers is a general result of the behavior of the sample mean $\frac{S_n}{n}$ of independent random variables with same existing expectation μ. It is weak in the sense that only convergence in probability is established. For large n, the weak law of large numbers states that the sample mean $\frac{S_n}{n}$ will be close (less than an arbitrary ϵ) to the expectation with high probability. It does not imply that $\left|\frac{S_n}{n} - \mu\right|$ remains small for all large n. In fact, large fluctuations in $\left|\frac{S_n}{n} - \mu\right|$ can happen; the Weak Law of Large Numbers only concludes that large values of $\left|\frac{S_n}{n} - \mu\right|$ occur with (very) small probability. For example, in a coin-tossing experiment with a fair coin such that $\Pr[X_k = 1] = \Pr[X_k = 0] = \mu = \frac{1}{2}$ in n-trials, the sequence of always head $\{X_k = 1\}_{1 \leq k \leq n}$ is possible with probability 2^{-n} and $\frac{S_n}{n} = 1 > \mu$. But, only for $n \to \infty$, the probability of this "always head sequence" is impossible ($\lim_{n\to\infty} 2^{-n} = 0$). For all finite n there is a non-zero probability of having a large deviation from the mean.

If we assume in addition to the existence of the expectation that also the variance $\mathrm{Var}[X]$ exists, the Weak Law follows from the Chebyshev inequality

---

[3] An alternative proof is given in Feller (1970, p. 247–248).

(5.10). This exemplifies the increasingly complexity if less restrictions in the theorems are assumed. Indeed, using (2.57) for independent random variables, $\text{Var}\left[\frac{S_n}{n}\right] = \frac{\text{Var}[X]}{n}$ and the Chebyshev inequality (5.10) gives

$$\Pr\left[\left|\frac{S_n}{n} - \mu\right| \ge \epsilon\right] \le \frac{\text{Var}[X]}{n\epsilon^2}$$

which tends to zero for any fixed ϵ and finite $\text{Var}[X]$. In fact, with the additional assumption of a finite variance $\text{Var}[X]$, a much more precise result can be proved known as the Central Limit Theorem (Section 6.3). We remark that the Weak Law of Large Numbers also holds in the case $\text{Var}[X]$ does not exist.

Theorem 6.2.2 (Strong Law of Large Numbers) *Let* $\{X_k\}$ *be a sequence of independent random variables each with distribution identical to that of the random variable* X *with* $\mu = E[X]$. *If the expectation* $\mu = E[X]$ *and variance* $\text{Var}[X]$ *exists, then,*

$$\Pr\left[\lim_{n\to\infty} \frac{S_n}{n} = \mu\right] = 1 \tag{6.3}$$

Proof: See e.g. Feller (1970, p. 259–261), Berger (1993, pp. 46–48) or Wolff (1989, pp. 40–41). Their proof is based on the Kolmogorov criterion: the convergence of $\sum_{k=1}^{\infty} \frac{\text{Var}[X_k]}{k^2}$ is a sufficient condition for the Strong Law of Large Numbers for independent random variables X_k with mean $E[X_k]$ and variance $\text{Var}[X_k]$. If the existence of $E[X^4]$ is assumed, Ross (1996, pp. 56–58) and Billingsley (1995, p. 85) provide a different proof. Wolff (1989, pp. 41–42) remarks that both the Weak and Strong Laws hold under much weaker conditions: it is only needed that the X_k are not correlated. In other words, $\frac{S_n}{n} \to \mu$ w.p. 1 implies $E[X_k] = \mu$ even if $\text{Var}[X_k] = \infty$. $\square$

The Strong Law of Large Numbers roughly states that $\left|\frac{S_n}{n} - \mu\right|$ remains small for sufficiently large n with overwhelming probability. The importance of the Law of Large Numbers is the mathematical foundation of the intuition that the sample mean is the best estimator.

Theorem 6.2.3 (Law of the Iterated Logarithm) *Let* $\{X_k\}$ *be a sequence of independent random variables each with distribution identical to that of the random variable* X *with* $\mu = E[X]$ *and, if* $\text{Var}[X]$ *exists, then,*

$$\Pr\left[\lim_{n\to\infty} \sup \frac{S_n - n\mu}{\sigma\sqrt{2n\log\log n}} = 1\right] = 1 \tag{6.4}$$

Proof: See e.g. Billingsley (1995, p. 154–156) or Feller (1970, Section VIII.5)[4].   □

In addition to the Weak and Strong Laws of Large Numbers, the Law of the Iterated Logarithm provides information about large values of $\left|\frac{S_n}{n} - \mu\right|$. Specifically, it states that the bound $\left|\frac{S_n}{n} - \mu\right| \leq \sigma\sqrt{\frac{2\log\log n}{n}}$ holds almost surely. The latter means that it is satisfied infinitely often and only for a finite number of values of n, the converse $\left|\frac{S_n}{n} - \mu\right| > \sigma\sqrt{\frac{2\log\log n}{n}}$ may occur.

6.3 Central Limit Theorem

Theorem 6.3.1 (Central Limit Theorem) *Let $\{X_k\}$ be a sequence of independent random variables each with distribution identical to that of the random variable X with finite $\mu = E[X]$ and $\sigma^2 = Var[X]$. Then $\frac{S_n - n\mu}{\sigma\sqrt{n}} \xrightarrow{d} N(0,1)$ or, explicitly,*

$$\Pr\left[\frac{S_n - n\mu}{\sigma\sqrt{n}} \leq x\right] \to \Phi(x) = \frac{1}{\sqrt{2\pi}} \int_{-\infty}^{x} e^{-\frac{t^2}{2}} dt$$

Proof: Without loss of generality, we may confine to normalized random variables – replace X_k by $\frac{X_k - \mu}{\sigma}$ – such that $\mu = 0$ and $\sigma = 1$. Consider the scaled random variable $U_n = \alpha_n S_n$, where α_n is a real number depending on n and to be determined later. Similarly as in the proof of the Weak Law of Large Numbers, we find that $\varphi_{U_n}(z) = (\varphi_X(\alpha_n z))^n$. Due to the existence of the variance, the Taylor expansion (2.40) of φ_X around $z = 0$ is known with higher precision as $\varphi_X(z) = 1 + \frac{z^2}{2} + o(z^2)$. For sufficiently small z, the logarithm

$$\log(\varphi_{U_n}(z)) = n\log\left(1 + \frac{\alpha_n^2 z^2}{2} + o(\alpha_n^2 z^2)\right) = n\frac{\alpha_n^2 z^2}{2} + o(n\alpha_n^2 z^2)$$

only converges to a finite (non-zero) number if $\alpha_n = O\left(\frac{1}{\sqrt{n}}\right)$. Choosing the simplest function that satisfies this condition, $\alpha_n = \frac{1}{\sqrt{n}}$, leads to $\lim_{n\to\infty} \log(\varphi_{U_n}(z)) = \frac{z^2}{2}$ or, since the logarithm is a continuous, increasing function, $\lim_{n\to\infty} \varphi_{U_n}(z) = \exp\left(\frac{z^2}{2}\right)$. The transform (3.22) shows that the corresponding limit random variable is a Gaussian $N(0,1)$. The theorem then follows by virtue of the Continuity Theorem 6.1.3.   □

---

[4] Feller also mentions sharper bounds.

An alternative formulation of the Central Limit Theorem is that the k-fold convolution of any probability density function converges to a Gaussian probability distribution, $f_X^{(k*)}(x) \to \frac{1}{\sigma\sqrt{2\pi}} \exp\left[-\frac{(x-\mu)^2}{2\sigma^2}\right]$ with $\mu = kE\left[X\right]$ and $\sigma^2 = k\mathrm{Var}[X]$. Both the Law of Large Numbers and the Central Limit Theorem can be shown to be valid for a surprisingly large class of sequences $\{X_k\}$ where each random variable may have a different distribution. The conditions for the extension of the Central Limit Theorem are summarized in the Lindeberg conditions (Feller, 1971, p. 263). An example where the sum of independent random variables tend to a different limit distribution than the Gaussian appears in Section 16.5.1.

If higher moments are known, the convergence to the Gaussian distribution can be bounded. Feller (1971, Chapter XVI) devotes a chapter on expansions related to the Central Limit Theorem culminating in the Berry–Esseen Theorem.

Theorem 6.3.2 (Berry–Esseen Theorem) *Let $\{X_k\}$ be a sequence of independent random variables each with distribution identical to that of the random variable X with finite $\mu = E\left[X\right]$, $\sigma^2 = \mathrm{Var}[X]$ and $\rho = E\left[\frac{|X-\mu|^3}{\sigma^3}\right]$. Then, with $C = 3$,*

$$\sup_x \left| \Pr\left[\frac{S_n - n\mu}{\sigma\sqrt{n}} \le x\right] - \Phi(x) \right| \le \frac{C\rho}{\sqrt{n}} \tag{6.5}$$

Proof: See e.g. Feller (1971, Section XVI.5). The constant C can be slightly improved to $C \le 2.05$.   □

As an example of the rate of convergence towards the Gaussian distribution, the k-fold convolutions of the uniform density given by (3.30) is plotted in Fig. 6.1 together with the Gaussian approximation (3.19).

6.4 Extremal distributions

6.4.1 Scaling laws

In this section, limit properties of the maximum and minimum of a set $\{X_k\}$ of independent random variables are discussed. For simplicity, we assume that all random variables X_k have identical distribution $F(x) = \Pr\left[X \le x\right]$

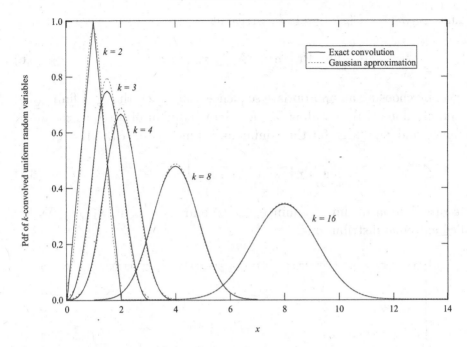

Fig. 6.1. Both the exact $f_U^{(k*)}(x)$ with $f_U(x) = 1_{0 \leq x \leq 1}$ and the Gaussian approximation for several values of k.

such that (3.33) and (3.32) simplify to

$$\Pr\left[\max_{1 \leq k \leq m} X_k \leq x\right] = F^m(x)$$

$$\Pr\left[\min_{1 \leq k \leq m} X_k > x\right] = (1 - F(x))^m$$

Consider the limit process when $m \to \infty$. Let $\{x_m\}$ be a sequence of real numbers. Then, confining to the maximum first,

$$\log\left(\Pr\left[\max_{1 \leq k \leq m} X_k \leq x_m\right]\right) = m \log F(x_m)$$

Since $0 \leq F(x_m) \leq 1$ and since the logarithm has a Taylor expansion $\log(1 - x) = -\sum_{k=1}^{\infty} \frac{x^k}{k}$ around $x = 0$ and convergent for $|x| < 1$, we rewrite the right-hand side as $\log F(x_m) = \log\left[1 - (1 - F(x_m))\right]$ and, after expansion,

$$\log\left(\Pr\left[\max_{1 \leq k \leq m} X_k \leq x_m\right]\right) = -m\left(1 - F(x_m)\right) + o\left[m\left(1 - F(x_m)\right)\right]$$

If $\lim_{m \to \infty} m\left(1 - F(x_m)\right) = \xi$, we arrive at

$$\lim_{m \to \infty} \Pr\left[\max_{1 \leq k \leq m} X_k \leq x_m\right] = e^{-\xi} \tag{6.6}$$

Hence, by choosing an appropriate sequence $\{x_m\}$ such that ξ is finite (and preferably non-zero), a scaling law for the maximum of a sequence can be obtained and, similarly, for the minimum, if $\lim_{m \to \infty} m\left(F(x_m)\right) = \zeta$,

$$\lim_{m \to \infty} \Pr\left[\min_{1 \leq k \leq m} X_k > x_m\right] = e^{-\zeta} \tag{6.7}$$

The distribution of $\lim_{m \to \infty} \min_{1 \leq k \leq m} X_k$ and $\lim_{m \to \infty} \max_{1 \leq k \leq m} X_k$ are called extremal distributions.

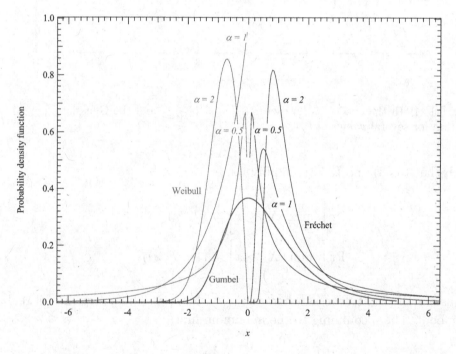

Fig. 6.2. The probability density function of the three types of extremal distributions.

6.4.2 The Law of Extremal Types

Two distribution functions F and G are said to be of the same type if there exists constants $a > 0$ and b for which $F(ax + b) = G(x)$ for all x.

Theorem 6.4.1 (Law of Extremal Types) *Any extremal distribution of a sequence of i.i.d random variables can only have one of the three types*

$$1.\ \text{Gumbel} \quad F(x) = e^{-e^{-x}}$$
$$2.\ \text{Fréchet} \quad F(x) = e^{-x^{-\alpha}} 1_{x \geq 0}$$
$$3.\ \text{Weibull} \quad F(x) = e^{-(-x)^{\alpha}} 1_{x<0} + 1_{x \geq 0}$$

where $\alpha > 0$.

Proof: See e.g. Berger (1993, pp. 65–69). $\square$

The generality of this theorem is appealing: any maximum or minimum of a set of i.i.d. random variables has (apart from the scaling constants a and b) one of the above three types. The corresponding probability density functions are plotted in Fig. 6.2.

6.4.3 Examples

1. Consider the set $\{X_k\}$ of exponentially distributed random variables with $F(x) = 1 - e^{-\alpha x}$. The condition for the maximum is $me^{-\alpha x_m} \to \xi$ or, equivalently, $x_m = \frac{1}{\alpha}(\log m - \log \xi)$ and (6.6) becomes, after putting $x = -\log \xi$,

$$\lim_{m \to \infty} \Pr \left[\max_{1 \leq k \leq m} X_k \leq \frac{1}{\alpha}(\log m + x) \right] = e^{-e^{-x}}$$

The minimum $m(1 - e^{-\alpha x_m}) \to \zeta$ is equivalent to $e^{-\alpha x_m} \sim \log m - \log \zeta$ or $x_m \sim \frac{1}{\alpha} \log(\log m - \log \zeta) \sim \frac{1}{\alpha}\left(\log \log m - \frac{\log \zeta}{\log m}\right)$. Hence, after putting $x = -\log \zeta$, the limit law for the minimum of exponential random variables is

$$\lim_{m \to \infty} \Pr \left[\min_{1 \leq k \leq m} X_k > \frac{1}{\alpha}\left(\log \log m + \frac{x}{\log m}\right) \right] = e^{-e^{-x}}$$

For both the maximum and the minimum of exponential random variables, a scaling law exists that leads to a Gumbel distribution $F_{Gumbel}(x) = e^{-e^{-x}}$. In other words, for large m, the random variables $M = \alpha \max_{1 \leq k \leq m} X_k - \log m$ and $N = \alpha \log m (\min_{1 \leq k \leq m} X_k) - \log m \log \log m$ have an identical distribution equal to the Gumbel distribution.

2. Another example is the maximum of a set of i.i.d. uniform random variables $\{U_k\}$ in $[0, 1]$ with $F(x) = x$ for $0 \leq x \leq 1$. Since $m(1 - x_m) \to \xi$ or, equivalently, $x_m \to 1 - \frac{\xi}{m}$ with $0 \leq x_m \leq 1$ we have, after putting $x = \xi$ with $x \geq 0$,

$$\lim_{m \to \infty} \Pr \left[\max_{1 \leq k \leq m} U_k \leq 1 - \frac{x}{m} \right] = e^{-x}$$

3. Consider a rectangular lattice with size z_1 and z_2 and with independent and identical, uniformly distributed link weights on $(0, 1]$ between each lattice point. The number of lattice points (nodes) equals $N = (z_1 + 1)(z_2 + 1)$ and the number of links is $L = 2z_1 z_2 + (z_1 + z_2)$. The shortest hop path between two diagonal corner points consists of $h = z_1 + z_2$ hops. The weight W_h of such a h hop path is the sum of h independent uniform random variables with distribution specified in (3.29),

$$F(x) = \Pr[W_h \le x] = \frac{1}{h!} \sum_{j=0}^{h} \binom{h}{j} (-1)^j (x-j)^h 1_{j \le x}$$

In particular, $\Pr[W_h \le h] = 1$ and for small $x < 1$ it holds that $F(x) = \frac{x^h}{h!}$. The precise computation of the minimum weight of a h hop path in a lattice is difficult due to dependence among those h hop paths and we content ourselves here with an approximate estimate. If we neglect the dependence · of the h hops paths due to possible overlap, then the minimum weight among all h hop paths can be approximated by (6.7) because the number[5] $m = \binom{z_1+z_2}{z_1} = \frac{h!}{z_1! z_2!}$ of those h hop paths is large. The limit sequence must obey $m(F(x_m)) \to \zeta$ for sufficiently large m, which implies that $F(x_m)$ must be small or, equivalently, x_m must be small. Hence, $m\frac{x_m^h}{h!} = \zeta$ or $x_m = \left(\frac{h! \zeta}{m}\right)^{\frac{1}{h}}$. The limit law (6.7) for the minimum weight $W = \min_{1 \le k \le m} W_{h,k}$ of the shortest hop path in a rectangular lattice is

$$\lim_{m \to \infty} \Pr\left[\min_{1 \le k \le m} W_{h,k} > \left(\frac{h! x}{m}\right)^{\frac{1}{h}}\right] = e^{-x}$$

In other words, the random variable $\frac{m W^h}{h!}$ tends to an exponential random variable with mean 1 for large $m = \frac{h!}{z_1! z_2!}$ or

$$\Pr[W \le y] \approx 1 - \exp\left(-m\frac{y^h}{h!}\right)$$

---

[5] Any path in a rectangular lattice can be represented by a sequence of r(ight), l(eft), u(p) and d(own), which is called an encoded path word. The encoded path word of the shortest hop path between diagonal corner points consists of z_1 r's (or l's) and z_2 d's (or u's). The total number of these paths equals $\binom{z_1+z_2}{z_1}$. Two paths coincide in a same lattice point at $g \le h$ hops from the source node if their encoded path word has the same sum of r's and d's in the first g lettres. The number of overlapping links between two paths equals the number of the same consecutive lettres (r or d) in a block after a same sum of r's and d's in the encoded path words. Checking for overlap between h hop paths requires a comparison of $\binom{z_1+z_2}{z_1}!$ permutations in the encoded path words.

From (2.35), the mean shortest weight of a h hop path equals

$$E\left[W\right] = \int_0^\infty \left(1 - F_W(x)\right) dx \approx \int_0^\infty \exp\left(-m\frac{x^h}{h!}\right) dx = \Gamma\left(1 + \frac{1}{h}\right)(z_1!z_2!)^{\frac{1}{h}}$$

For a square lattice where $z_1 = z_2 = \frac{h}{2}$, we have

$$E\left[W\right] = \Gamma\left(1 + \frac{1}{h}\right)\left(\left(\frac{h}{2}\right)!\right)^{\frac{2}{h}}$$

Using Stirling's formula (Abramowitz and Stegun, 1968, Section 6.1.38) for the factorial $h! = \sqrt{2\pi}h^{h+\frac{1}{2}}e^{-h+\frac{\theta}{12h}}$ where $0 < \theta < 1$, for large h, the mean $E\left[W\right]$ increases about linearly in the number of hops h,

$$E\left[W\right] \simeq \left(\frac{h}{2e}\right)\left(\sqrt{\pi h}e^{\frac{\theta}{12h}}\right)^{\frac{2}{h}} \approx \frac{h}{2e}$$

The average weight of a link (or 1 hop) of the shortest h hop path is roughly $\frac{1}{2e} \approx 0.184$.

In spite of the fact that path dependence (overlap) has been ignored in the computation of the minimum weight, $E\left[W\right] = O\left(h\right) = O\left(\sqrt{N}\right)$ is correct. However, the approximate analysis does *not* give the correct prefactor in $E\left[W\right]$ nor the correct limit pdf which turns out to be Gaussian. Hence, if random variables are *not* independent, Theorem 6.4.1 does not apply. Finally, a shortest h hop path is not necessarily the overall shortest path because it is possible – though with small probability – that the overall shortest path has $h + 2j$ hops with $j > 0$.

4. **The probability density function of the longest shortest path** The most commonly used process that informs each node about changes in a network topology (e.g. an autonomous domain) is called flooding: every router forwards the packet on all interfaces except for the incoming one and duplicate packets are discarded. Flooding is particularly simple and robust since it progresses, in fact, along all possible paths from the emitting node to the receiving node. Hence, a flooded packet reaches a node in the network in the shortest possible time (if overhead in routers are ignored). Therefore, the interesting problem lies in the determination of the flooding time T_N, which is the minimum time needed to inform the last node in a network with N nodes. Only after T_N, all topology databases at each router in the network are again synchronized, i.e. all routers possess the same topology information. Rather than investigating the flooding time T_N (for which we refer to Section 16.5), the largest number of traversed routers (hops) or the longest shortest path from the emitting node to the furthermost node in its shortest path tree is computed.

The number of hops, in short the hopcount H_N, along the shortest path between two arbitrary nodes in a network containing N nodes is modeled subject to the following assumptions: (a) the hopcount H_N is a Poisson random variable with mean $E\left[H_N\right] = \lambda = \alpha \log N$ with $\alpha > 0$, which is motivated in Section 16.3.1; (b) the number of nodes N is very large[6]; (c) all shortest paths from the emitting node towards any other node in the network are independent. The problem reduces to compute the pdf of the random variable $\max_{1 \le k \le N-1} H_k$. The distribution function follows from (3.9) as

$$F_{H_N}(x) = \sum_{k=0}^{x} \frac{\lambda^k e^{-\lambda}}{k!} = 1 - \sum_{k=x+1}^{\infty} \frac{\lambda^k e^{-\lambda}}{k!}$$

---

[6] The size of the Internet is currently estimated at about $N \sim 10^5$.

The condition $\lim_{m\to\infty} m\left(1 - F(x_m)\right) = \xi$ becomes

$$\lim_{N\to\infty} N^{1-\alpha} \sum_{k=x_N+1}^{\infty} \frac{\lambda^k}{k!} = \xi$$

from which we must choose the appropriate x_N as function of N. Observe that the maximum term in the series has index $k = [\lambda]$, where the latter denotes the largest integer smaller or equal to λ. For, the ratio between two consecutive (positive) terms in the k-sum equals $\frac{a_k}{a_{k-1}} = \frac{\lambda}{k}$ such that, if $\lambda > k$, then $a_k > a_{k-1}$, implying that the terms increase, while, if $\lambda < k$, the terms $a_k < a_{k-1}$ form a decreasing sequence. The series is rewritten as

$$\sum_{k=x_N+1}^{\infty} \frac{\lambda^k}{k!} = \frac{\lambda^{x_N+1}}{(x_N+1)!} \sum_{k=0}^{\infty} \frac{(x_N+1)!\lambda^k}{(x_N+1+k)!}$$

$$= \frac{\lambda^{x_N+1}}{(x_N+1)!}\left(1 + \frac{\lambda}{x_N+2} + \frac{\lambda^2}{(x_N+2)(x_N+3)} + \cdots\right)$$

We choose $x_N = [\lambda] + [\delta\lambda] \sim \lambda(1+\delta)$ for large N and, thus large λ, where δ must be related to ξ. The series then consists of decreasing terms. Moreover, for large λ,

$$\sum_{k=[\lambda]+[\delta\lambda]+1}^{\infty} \frac{\lambda^k}{k!} = \frac{\lambda^{\lambda(1+\delta)+1}}{(\lambda(1+\delta)+1)!}\left(1 + \frac{1}{(1+\delta)+2/\lambda} + \frac{1}{((1+\delta)+2/\lambda)((1+\delta)+3/\lambda)} + \cdots\right)$$

$$< \frac{\lambda^{\lambda(1+\delta)+1}}{(\lambda(1+\delta)+1)!}\left(1 + \frac{1}{(1+\delta)} + \frac{1}{(1+\delta)^2} + \cdots\right)$$

and thus,

$$\sum_{k=[\lambda]+[\delta\lambda]+1}^{\infty} \frac{\lambda^k}{k!} = \frac{\lambda^{\lambda(1+\delta)+1}}{(\lambda(1+\delta)+1)!}\frac{1+\delta}{\delta}\left(1 + O\left(\frac{1}{\lambda}\right)\right)$$

Using Stirling's formula (Abramowitz and Stegun, 1968, Section 6.1.38), $x! \sim \sqrt{2\pi}x^{x+\frac{1}{2}}e^{-x}$, for large x yields

$$\frac{\lambda^{\lambda(1+\delta)+1}}{(\lambda(1+\delta)+1)!}\frac{1+\delta}{\delta} \sim \frac{\lambda^{\lambda(1+\delta)+1}e^{\lambda(1+\delta)}}{(\lambda(1+\delta)+1)\sqrt{2\pi}\lambda^{\lambda(1+\delta)+\frac{1}{2}}(1+\delta)^{\lambda(1+\delta)+\frac{1}{2}}}\frac{1+\delta}{\delta}$$

$$\sim \frac{e^{\lambda(1+\delta)[1-\log(1+\delta)]}}{\sqrt{2\pi(1+\delta)\lambda}}\frac{1}{\delta}$$

For large N, the condition becomes

$$\xi \sim \frac{N^{\alpha(1+\delta)[1-\log(1+\delta)]+1-\alpha}}{\sqrt{2\pi(1+\delta)\alpha\log N}}\frac{1}{\delta}\left(1 + O\left(\frac{1}{\log N}\right)\right)$$

and, after taking the logarithm of both sides,

$$\log\xi \sim (\alpha(1+\delta)\left[1 - \log(1+\delta)\right] + 1 - \alpha)\log N - \frac{1}{2}\log\log N - \frac{1}{2}\log(2\pi(1+\delta)\alpha) - \log\delta + O\left(\frac{1}{\log N}\right)$$

or

$$\log\xi + (\alpha-1)\log N + \frac{1}{2}\log\log N + O\left(\frac{1}{\log N}\right) \sim (\alpha(1+\delta)\left[1 - \log(1+\delta)\right])\log N$$

$$- \frac{1}{2}\log(2\pi(1+\delta)\alpha) - \log\delta \qquad (6.8)$$

At this point, we will assume that $\delta < 1$, which justifies the expansion $\log(1+\delta) = \delta + O(\delta^2)$. This assumption will be checked later. Thus,

$$\alpha(1+\delta)\left[1 - \log(1+\delta)\right] \sim \alpha(1+\delta)\left[1 - \delta\right] \sim \alpha(1-\delta^2)$$

and δ must be solved from

$$R \sim \alpha(1 - \delta^2) \log N - \frac{\delta}{2} - \log \delta$$

with $R = \log\left(\sqrt{2\pi\alpha}\xi\right) + (\alpha - 1)\log N + \frac{1}{2}\log\log N$. The Newton–Raphson iteration can be applied with starting value δ_0 to find the solution of the equation up to the leading order in $\log N$, i.e. $R \sim \alpha(1 - \delta^2)\log N$. Hence,

$$\delta_0 = \sqrt{1 - \frac{R}{\alpha \log N}} \sim 1 - \frac{R}{2\alpha \log N} \sim 1 - \frac{\alpha - 1}{2\alpha} - \frac{\log\left(\sqrt{2\pi\alpha}\xi\right) + \frac{1}{2}\log\log N}{2\alpha \log N}$$

which demonstrates that, for $\alpha \geq 1$, the assumption $\delta < 1$ is correct for large N. The case $\alpha < 1$ requires the application of Newton–Raphson's method on (6.8), which we omit here. The second iteration in Newton–Raphson's method leads to

$$\delta_1 = \delta_0 - \frac{\frac{\delta_0}{2} + \log \delta_0}{2\alpha\delta_0 \log N + \frac{1}{2} + \frac{1}{\delta_0}}$$

and shows that the k-th iteration improves the previous with a quantity of order $O\left(\log^{-k} N\right)$. Since (6.8) is only accurate up to $O\left(\log^{-1} N\right)$, a second iteration is superfluous and we obtain the choice $x_N = \lambda(1 + \delta)$, or

$$x_N = \frac{3\alpha + 1}{2}\log N - \frac{1}{2}\log\left(\sqrt{2\pi\alpha}\xi\right) - \frac{1}{4}\log\log N$$

After substituting $x = -\log\xi$, we finally arrive for $\alpha \geq 1$ and large N at

$$\Pr\left[\max_{1 \leq k \leq N-1} H_k \leq \frac{x}{2} + \frac{3\alpha + 1}{2}\log N - \frac{1}{4}\log\log N - \frac{1}{4}\log(2\pi\alpha)\right] = e^{-e^{-x}}$$

from which the pdf of the hopcount of the longest shortest path (lsp) follows as

$$f_{lsp}(x) = 2e^{-e^{-2(x-c)}}e^{-2(x-c)} \tag{6.9}$$

with

$$c = \frac{3\alpha + 1}{2}\log N - \frac{1}{4}\log\log N - \frac{1}{4}\log(2\pi\alpha)$$

$$= \left(\frac{3}{2} + \frac{1}{2\alpha}\right)E[H_N] - \frac{1}{4}\log E[H_N] - \frac{1}{4}\log(2\pi)$$

and

$$E[lsp] = c + \frac{\gamma}{2} \approx \left(\frac{3}{2} + \frac{1}{2\alpha}\right)E[H_N] - \frac{1}{4}\log E[H_N] - 0.170$$

$$var[lsp] = \frac{\pi^2}{24} \simeq 0.4112$$

Observe that the average longest shortest path is about twice the average hopcount if $\alpha = 1$ while the variance is small, constant and independent of the scaling parameter c or λ. Figure 6.3 compares the above approximate analysis with simulations.

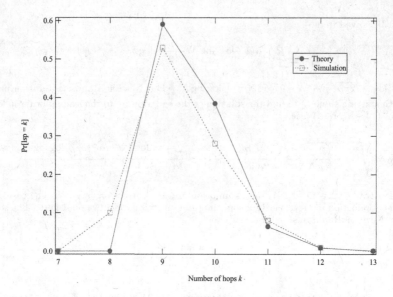

Fig. 6.3. The hopcount of the shortest path for $N = 4000$. Both simulations based on an Internet-like topology generator (with unit link weights) and theory $f_{lsp}(k)$ with $\alpha = 0.4786$ are shown.

Notes

(i) The classical theory of extremes, extremal properties of dependent sequences and extreme values in continuous-time are treated in detail in the book by Leadbetter *et al.* (1983).

(ii) A more recent book by Embrechts *et al.* (2001a) applies the theory of extremal events to problems in insurance and finance.

Part II

Stochastic processes

7

The Poisson process

The Poisson process is a prominent stochastic process, mainly because it frequently appears in a wealth of physical phenomena and because it is relatively simple to analyze. Therefore, we will first treat the Poisson process before considering the more general Markov processes.

7.1 A stochastic process

7.1.1 Introduction and definitions

A *stochastic*[1] *process*, formally denoted as $\{X(t), t \in T\}$, is a sequence of random variables $X(t)$, where the parameter t – most often the time – runs over an index set T. The *state space* of the stochastic process is the set of all possible values for the random variables $X(t)$ and each of these possible values is called the *state* of the process. If the index set T is a countable set, $X[k]$ is a discrete stochastic process. Often k is the discrete time or a time slot in computer systems. If T is a continuum, $X(t)$ is a continuous stochastic process. For example, the outcome of n tosses of a coin is a discrete stochastic process with state space {heads, tails} and the index set $T = \{0, 1, 2, \ldots, n\}$. The number of arrivals of packets in a router during a certain time interval $[a, b]$ is a continuous stochastic process because $t \in [a, b]$. Any realization of a stochastic process is called a *sample path*. For example, a sample path of the outcome of n tosses of a coin is {heads, tails, tails, ..., heads}, while a sample path of the number of arrivals in $[a, b]$ is $1_{a \leq t < a+h}, 3 \times 1_{a+h \leq t < a+4h}, 8 \times 1_{a+4h \leq t < a+5h}, \ldots, 13 \times 1_{a+(k-1)h \leq t < b}$, where $h = \frac{b-a}{k}$. Other examples are the measurement of the temperature each day, the notation of the value of a stock each minute or rolling a die and recording its value, which is illustrated in Fig. 7.1.

---

[1] The word "stochastic" is derived from $\sigma\tau o\chi\alpha\zeta\epsilon\sigma\theta\alpha i$ in Greek which means "to aim at, try to hit".

Especially in continuous stochastic processes, it is convenient to define increments as the difference $X(t) - X(u)$. The continuous time stochastic process $X(t)$ has *independent increments* if changes in the value of the process in different time intervals are independent, or, if for all $t_0 < t_1 < \cdots < t_n$, the random variables $X(t_1) - X(t_0), X(t_2) - X(t_1), \ldots, X(t_n) - X(t_{n-1})$ are independent. The continuous (time) stochastic process has *stationary increments* if $X(t+s) - X(s)$ possesses the same distribution for all s. Hence, changes in the value of the process only dependent on the distance t between process events, not on the time point s.

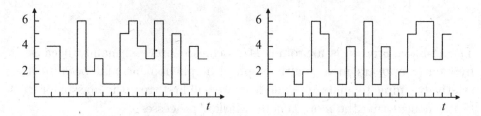

Fig. 7.1. Two different sample paths of the experiment: roll a die and record the outcome. The total number of different sample paths is 6^T where T is the number of times an outcome is recorded. The state space only contains 6 possible outcomes $\{1, 2, 3, 4, 5, 6\}$.

Stochastic processes are distinguished by (a) their state space, (b) the index set T and (c) by the dependence relations between random variables $X(t)$. For example, a standard Brownian motion (or Wiener process)[2] is defined as a stochastic process $X(t)$ having continuous sample paths, stationary independent increments and $X(t)$ has a normal distribution $N(0, t)$. A Poisson process, defined in more detail in Section 7.2, is a stochastic process $X(t)$ having discontinuous sample paths, stationary independent increments and $X(t)$ has a Poisson distribution. A generalization of the Poisson process is a counting process. A counting process is defined as a stochastic process $N(t) \geq 0$ with discontinuous sample paths, stationary independent increments, but with arbitrary distribution. A counting process $N(t)$ represents the total number of events that have occurred in a time interval $[0, t]$. Examples of a counting process are the number of telephone calls at a local exchange during an interval, the number of failures in a telecommunication network, the number of corrupted bits after transmission due to channel errors, etc.

---

[2] Harrison (1990) shows that the converse is also true: if Y is a continuous process with stationary independent increments, then Y is a Brownian motion.

7.1.2 Modeling a stochastic process from measurements

In practice, understanding observed phenomena often asks for a stochastic model that captures the main characteristics of the studied phenomena and that enables computations of diverse quantities of interest. Examples in the field of data communications networks are the determination of the arrival process at a switch or router in order to dimension the number of buffer (memory) places, the modeling of the graph of the Internet, the distribution of the duration of a telephone call or web browsing session, the number of visits to certain websites, the number of links that refer to a web page, the amount of downloaded information, the number of traversed routers by an email, etc. Accurate modeling is in general difficult and often trades off complexity against accuracy of the model.

Let us illustrate some aspects of *modeling* by considering Internet delay measurements. A motivation for obtaining an end-to-end delay model for (a part of) the Internet is the question whether massive service deployment of voice over IP (VoIP) can substitute classical telephony with a comparable quality. Specifically, classical telephony requires that the end-to-end delay of an arbitrary telephone conversation hardly exceeds 100 ms.

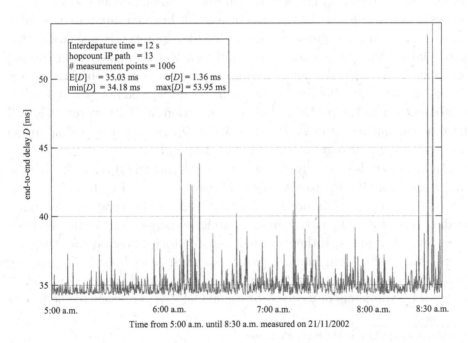

Fig. 7.2. The raw data of the end-to-end delay of IP test packets along a same path of 13 hops in the Internet measured during 3.5 hours.

The end-to-end delay along a fixed path between source and destination measured during some interval is an example of a continuous time stochastic process. We have received data of the delay measured at RIPE-NCC as illustrated in Fig. 7.2. Figure 7.2 shows a *sample path* of this continuous stochastic process. The precise details of the measurement configuration are for the present purpose not relevant. It suffices to add that Figure 7.2 shows the time difference between the departure of an IP test packet of 100 byte at the sending box and its arrival at the destination box accurate within 10 μs. The average sending rate of IP test packets is $\frac{1}{12}$ packets per second. Each IP test packet is assumed to follow the same path from sending to receiving box. The steadiness of the path is checked by trace-route measurements every 6 minutes.

Usually, in the next step, the histogram of the raw data is made. A histogram counts the number of data points that lie in an interval of ΔD ms, which is often called the bin size. Most graphical packages allow to choose the bin size. Figure 7.3 shows two different histograms with bin size $\Delta D = 0.5$ ms and $\Delta D = 0.1$ ms. In general, there is no universal rule to choose the bin size ΔD. Clearly, the bin size is bounded below by the measurement accuracy, in our case $\Delta D > 10$ μs. A finer bin size provides more detail, but the resulting histogram exhibits also more stochastic variations because there are fewer data points in a small bin and adjacent bins may possess a significantly different amount of data points. Hence, compared to one larger bin that covers a same interval, less averaging or smoothing occurs in a set of smaller bins. The normalized histogram obtained by dividing the counts per bin by the total number of data points provides a first approximation to the probability density function of D. However, it is still discrete and approximates $\Pr[k < D \leq k + \Delta D]$. A more precise description of constructing a histogram is given in Section C.1.

The histogram is generally better suited to decide whether outliners in de data points may be due to measurement errors or not. Figure 7.3 suggests to either neglect the data points with $D > 40$ ms or to measure at a higher sending rate of IP test packets in order to have more details in the intervals exceeding 38 or 40 ms. If there existed a good[3] stochastic model for the end-to-end delay along fixed Internet paths, a normal procedure[4] in engineering and physics would be to fit the histogram with that stochastic model to obtain the parameters of that stochastic model. The accuracy of the fit can

---

[3] Which is still lacking at the time of writing.

[4] Other more difficult methods in the realm of statistics must be invoked in case the measurement data are so precious and rare that any additional measurement point has a far larger cost than the cost of extensive additional computations.

be expressed in terms of the correlation coefficient ρ explained in Section 2.5.3. The closer ρ tends to 1, the better the fit, which gives confidence that the stochastic model corresponds with the real phenomenon.

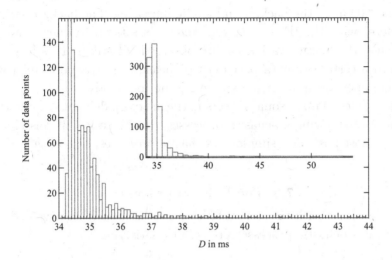

Fig. 7.3. The histogram of the end-to-end delay with a bin size of 0.1 ms (the insert has bin size of 0.5 ms).

Assuming that the presented measurement is a typical measurement along a fixed Internet path (which is true for about 80% of the investigated different paths), it demonstrates that there is a clear minimum at about 34 ms due to the propagation delay of electromagnetic waves. In addition, the end-to-end delay lies for 99% between 34 and 38 ms. However, there is insufficient data to pronounce claims in the tail behavior ($\Pr[D > x]$ for $x > 40$ ms). Just this region is of interest to compute the quality of service expressed as the probability that the end-to-end delay exceeds x ms is smaller than 10^{-a} where a specifies the stringency on the quality requirement. Toll quality in classical telephony sets x at 100 ms and a in the range of 4 to 5. The existence of a good stochastic model covering the whole possible range of the end-to-end delay D would enable us to compute tail probabilities based on the parameters that can be fitted from the measurements.

The histogram is in fact a projection of the raw measurement data onto the ordinate (end-to-end delay axis). All time information (the abscissa in Fig. 7.2) is lost. Usually, the time evolution and the dependencies or correlations over time of a stochastic phenomenon are difficult and most analyses are only tractable under certain simplifying conditions. For example, often only a steady state analysis is possible and the increments $X(t_k) - X(t_{k-1})$

of the process for all $t_0 < \cdots < t_{k-1} < t_k < \cdots < t_n$ are assumed to be independent or weakly dependent. The study of Markov processes (Chapters 9–11) basically tries to compute and analyze the process in steady state. Figure 7.2 is measured over a relatively long period of time and indicates that after 8.00 a.m. the background traffic increases. The background traffic interferes with the IP test packets and causes them to queue longer in routers such that larger variations are observed. However, it is in general difficult to ascertain that (a part of) the measurement is performed while the system operates in a certain stable regime (or steady state).

We have touched upon some aspects in the art of modeling to motivate the importance of studying stochastic processes. In the sequel of this chapter, one of the most basic and simplest stochastic processes is investigated.

7.2 The Poisson process

A Poisson process with parameter or rate $\lambda > 0$ is an integer-valued, continuous time stochastic process $\{X(t), t \geq 0\}$ satisfying

- (i) $X(0) = 0$
- (ii) for all $t_0 = 0 < t_1 < \cdots < t_n$, the increments $X(t_1) - X(t_0), X(t_2) - X(t_1), \ldots, X(t_n) - X(t_{n-1})$ are independent random variables
- (iii) for $t \geq 0$, $s > 0$ and non-negative integers k, the increments have the Poisson distribution

$$\Pr\left[X(t+s) - X(s) = k\right] = \frac{(\lambda t)^k e^{-\lambda t}}{k!} \qquad (7.1)$$

It is convenient to view the Poisson process $X(t)$ as a special counting process, where the number of events in any interval of length t is specified via condition (iii). From this definition, a number of properties can be derived:

(a) Condition (iii) implies that the increments are stationary because the right-hand side does not dependent on s. In other words, the increments only depend on the length of the interval t and not on the time s when the interval begins. Further, with (3.11), the mean $E\left[X(t+s) - X(s)\right] = \lambda t$ and because the increments are stationary, this holds for any value of s. In particular with $s = 0$ and condition 1, the expected number of events in a time interval with length t is

$$E\left[X(t)\right] = \lambda t \qquad (7.2)$$

Relation (7.2) explains why λ is called the rate of the Poisson process, namely, the derivative over time t or the number of events per time unit.

(b) The probability that exactly one event occurs in an arbitrarily small time interval of length h follows from condition (iii) as

$$\Pr\left[X(h+s) - X(s) = 1\right] = \lambda h e^{-\lambda h} = \lambda h + o(h)$$

while the probability that no event occurs in an arbitrarily small time interval of length h is

$$\Pr\left[X(h+s) - X(s) = 0\right] = e^{-\lambda h} = 1 - \lambda h + o(h)$$

Similarly, the probability that more than one event occurs in an arbitrarily small time interval of length h is

$$\Pr\left[X(h+s) - X(s) > 1\right] = o(h)$$

Example 1 A conversation in a wireless ad-hoc network is severely disturbed by interference signals according to a Poisson process of rate $\lambda = 0.1$ per minute. (a) What is the probability that no interference signals occur within the first two minutes of the conversation? (b) Given that the first two minutes are free of disturbing effects, what is the probability that in the next minute precisely 1 interfering signal disturbs the conversation?

(a) Let $X(t)$ denote the Poisson interference process, then $\Pr\left[X(2) = 0\right]$ needs to be computed. Since $X(0) = 0$ and with (7.1), we can write $\Pr\left[X(2) = 0\right] = \Pr\left[X(2) - X(0) = 0\right] = e^{-2\lambda}$, which equals $\Pr\left[X(2) = 0\right] = e^{-0.2} = 0.8187$.

(b) The events during two non-overlapping intervals of a Poisson process are independent. Thus the event $\{X(2) - X(0) = 0\}$ is independent from the event $\{X(3) - X(2) = 1\}$ which means that the asked conditional probability $\Pr\left[X(3) - X(2) = 1 | X(2) - X(0) = 0\right] = \Pr\left[X(3) - X(2) = 1\right]$. From (7.1), we obtain $\Pr\left[X(3) - X(2) = 1\right] = 0.1e^{-0.1} = 0.0905$.

Example 2 During a certain time interval $[t_1, t_1 + 10 \text{ s}]$, the number of IP packets that arrive at a router is on average $40/\text{s}$. A service provider asks us to compute the probability that there arrive 20 packets in the period $[t_1, t_1 + 1 \text{ s}]$ and 30 IP packets in $[t_1, t_1 + 3 \text{ s}]$. We may regard the arrival process as a Poisson process.

We are asked to compute $\Pr\left[X(1) = 20, X(3) = 30\right]$ knowing that $\lambda = 40$ s^{-1}. Using the independence of increments and (7.1), we rewrite

$$\begin{aligned}
\Pr\left[X(1) = 20, X(3) = 30\right] &= \Pr\left[X(1) - X(0) = 20, X(3) - X(1) = 10\right] \\
&= \Pr\left[X(1) - X(0) = 20\right]\Pr\left[X(3) - X(1) = 10\right] \\
&= \frac{(\lambda)^{20}e^{-\lambda}}{20!}\frac{(2\lambda)^{10}e^{-2\lambda}}{10!} = 10^{-26} \approx 0
\end{aligned}$$

which means that the request of the service provider does not occur in practice.

7.3 Properties of the Poisson process

The first theorem is the converse of the above property (b) that immediately followed from the definition. The Theorems presented here reveal the methodology of how stochastic processes are studied.

Theorem 7.3.1 *A counting process $N(t)$ that satisfies the conditions (i) $N(0) = 0$, (ii) the process $N(t)$ has stationary and independent increments, (iii) $\Pr[N(h) = 1] = \lambda h + o(h)$ and (iv) $\Pr[N(h) > 1] = o(h)$ is a Poisson process with rate $\lambda > 0$.*

Proof: We must show that conditions (iii) and (iv) are equivalent to condition (iii) in the definition of the Poisson process. Denote $P_n(t) = \Pr[N(t) = n]$ and consider first the case $n = 0$, then

$$P_0(t + h) = \Pr[N(t + h) = 0] = \Pr[N(t + h) - N(t) = 0, N(t) = 0]$$

Invoking independence via (ii)

$$P_0(t + h) = \Pr[N(t + h) - N(t) = 0] \Pr[N(t) = 0]$$

By definition, $P_0(t) = \Pr[N(t) = 0]$ and from (iii), (iv) and the fact that $\sum_{k=0}^{\infty} \Pr[N(h) = k] = 1$, it follows that

$$\Pr[N(h) = 0] = 1 - \lambda h + o(h) \tag{v}$$

Combining these with the stationarity in (ii), we obtain

$$P_0(t + h) = P_0(t)(1 - \lambda h + o(h))$$

or

$$\frac{P_0(t + h) - P_0(t)}{h} = -\lambda P_0(t) + \frac{o(h)}{h}$$

from which, in the limit $h \to 0$, the differential equation

$$P_0'(t) = -\lambda P_0(t)$$

is immediate. The solution is $P_0(t) = Ce^{-\lambda t}$ and the integration constant C follows from (i) and $P_0(0) = \Pr[N(0) = 0] = 1$ as $C = 1$. This establishes condition (iii) in the definition of the Poisson process for $k = 0$.

The verification for $n > 0$ is more involved. Applying the law of total probability (2.46),

$$P_n(t+h) = \Pr\left[N(t+h) = n\right]$$

$$= \sum_{j=0}^{n} \Pr\left[N(t+h) - N(t) = j | N(t) = n - j\right] \Pr\left[N(t) = n - j\right]$$

By independence (ii),

$$\Pr\left[N(t+h) - N(t) = j | N(t) = n - j\right] = \Pr\left[N(t+h) - N(t) = j\right]$$

and by definition $\Pr\left[N(t) = n - j\right] = P_{n-j}(t)$, we have

$$P_n(t+h) = \sum_{j=0}^{n} \Pr\left[N(t+h) - N(t) = j\right] P_{n-j}(t)$$

By the stationarity (ii)

$$\Pr\left[N(t+h) - N(t) = j\right] = \Pr\left[N(h) - N(0) = j\right]$$

we obtain using (i)

$$P_n(t+h) = \sum_{j=0}^{n} \Pr\left[N(h) = j\right] P_{n-j}(t)$$

while (v) and (iii) suggest to write the sum as

$$P_n(t+h) = P_n(t) \Pr\left[N(h) = 0\right] + P_{n-1}(t) \Pr\left[N(h) = 1\right]$$

$$+ \sum_{j=2}^{n} P_{n-j}(t) \Pr\left[N(h) = j\right]$$

Since $P_n(t) \leq 1$ and using (iv),

$$\sum_{j=2}^{n} P_{n-j}(t) \Pr\left[N(h) = j\right] \leq \sum_{j=2}^{n} \Pr\left[N(h) = j\right] = \Pr\left[N(h) > 1\right] = o(h)$$

we arrive with (v), (iii) at

$$P_n(t+h) = P_n(t) \left(1 - \lambda h + o(h)\right) + P_{n-1}(t) \left(\lambda h + o(h)\right) + o(h)$$

or

$$\frac{P_n(t+h) - P_n(t)}{h} = -\lambda P_n(t) + \lambda P_{n-1}(t) + \frac{o(h)}{h}$$

which leads, after taking the limit $h \to 0$, to the differential equation

$$P_n'(t) = -\lambda P_n(t) + \lambda P_{n-1}(t)$$

with initial condition $P_n(0) = \Pr[N(0) = n] = 1_{\{n=0\}}$. This differential equation is rewritten as

$$\frac{d}{dt}\left(e^{\lambda t}P_n(t)\right) = \lambda e^{\lambda t}P_{n-1}(t) \tag{7.3}$$

In case $n = 1$, the differential equation reduces with $P_0(t) = e^{-\lambda t}$ to $\frac{d}{dt}\left(e^{\lambda t}P_1(t)\right) = \lambda$. The general solution is $e^{\lambda t}P_1(t) = \lambda t + C$ and, from the initial condition $P_1(0) = 0$, we have $C = 0$ and $P_1(t) = \lambda t e^{-\lambda t}$. The general solution to (7.3) is proved by induction. Assume that $P_n(t) = \frac{(\lambda t)^n e^{-\lambda t}}{n!}$ holds for n, then the case $n+1$ follows from (7.3) as

$$\frac{d}{dt}\left(e^{\lambda t}P_{n+1}(t)\right) = \lambda\frac{(\lambda t)^n}{n!}$$

and integrating from 0 to t using $P_{n+1}(0) = 0$, yields $P_{n+1}(t) = \frac{(\lambda t)^{n+1}e^{-\lambda t}}{(n+1)!}$ which establishes the induction and finalizes the proof of the theorem. $\square$

The second theorem has very important applications since it relates the number of events in non-overlapping intervals to the interarrival time between these events.

Theorem 7.3.2 *Let $\{X(t); t \geq 0\}$ be a Poisson process with rate $\lambda > 0$ and denote by $t_0 = 0 < t_1 < t_2 < \cdots$ the successive occurrence times of events. Then the interarrival times $\tau_n = t_n - t_{n-1}$ are independent identically distributed exponential random variables with mean $\frac{1}{\lambda}$.*

Proof: For any $s \geq 0$ and any $n \geq 1$, the event $\{\tau_n > s\}$ is equivalent to the event $\{X(t_{n-1}+s) - X(t_{n-1}) = 0\}$. Indeed, the n-th interarrival time τ_n can only be longer than s time units if and only if the n-th event has not yet occurred s time units after the occurrence of the $(n-1)$-th event at t_{n-1}. Since the Poisson process has independent increments (condition (ii) in the definition of the Poisson process), changes in the value of the process in non-overlapping time intervals are independent. By the equivalence in events, this implies that the set of interarrival times τ_n are independent random variables. Further, by the stationarity of the Poisson process (deduced from condition (iii) in the definition of the Poisson process),

$$\Pr[\tau_n > s] = \Pr[X(t_{n-1}+s) - X(t_{n-1}) = 0] = e^{-\lambda s}$$

which implies that any interarrival time has an identical, exponential distribution,

$$F_{\tau_n}(x) = \Pr[\tau_n \leq x] = 1 - e^{-\lambda x}$$

This proves the theorem. □

The converse of Theorem 7.3.2 also holds: if the interarrival times $\{\tau_n\}$ of a counting process $\{N(t), t \geq 0\}$ are i.i.d. exponential random variables with mean $\frac{1}{\lambda}$, then $\{N(t), t \geq 0\}$ is a Poisson process with rate λ.

An association to the exponential distribution is the memoryless property,

$$\Pr[\tau_n > s + t | \tau_n > s] = \Pr[\tau_n > t]$$

By the equivalence of the events, for any $t, s \geq 0$,

$$\Pr[\tau_n > s + t | \tau_n > s] = \Pr[X(t_{n-1} + s + t) - X(t_{n-1}) = 0 | X(t_{n-1} + s) - X(t_{n-1}) = 0]$$
$$= \Pr[X(t_{n-1} + s + t) - X(t_{n-1} + s) = 0 | X(t_{n-1} + s) - X(t_{n-1}) = 0]$$

By the independence of increments (in non-overlapping intervals),

$$\Pr[\tau_n > s + t | \tau_n > s] = \Pr[X(t_{n-1} + s + t) - X(t_{n-1} + s) = 0]$$

and by the stationarity of the increments, the memoryless property is established,

$$\Pr[\tau_n > s + t | \tau_n > s] = \Pr[X(t_{n-1} + t) - X(t_{n-1}) = 0] = \Pr[\tau_n > t]$$

Hence, the assumption of stationary and independent increments is equivalent to asserting that, at any time s, the process probabilistically restarts again with the same distribution and is independent of occurrences in the past (before s). Thus, the process has no memory and, since the only continuous distribution that satisfies the memoryless property is the exponential distribution, exponential interarrival times τ_n are a natural consequence.

The arrival time of the n-th event or the waiting time until the n-event is $W_n = \sum_{k=1}^{n} \tau_k$. In Section 3.3.1, it is shown that the probability distribution of the sum of independent exponential random variables has a Gamma distribution or Erlang distribution (3.24). Alternatively, the equivalence of the events, $\{W_n \leq t\} \iff \{N(t) \geq n\}$, directly leads to the Erlang distribution,

$$F_{W_n}(t) = \Pr[W_n \leq t] = \Pr[N(t) \geq n] = \sum_{k=n}^{\infty} \frac{(\lambda t)^k e^{-\lambda t}}{k!}$$

The equivalence of the events, $\{W_n \leq t\} \iff \{N(t) \geq n\}$, is a general relation and a fundamental part of the theory of renewal processes, which we will study in the next Chapter 8.

Theorem 7.3.3 *Given that exactly one event of a Poisson process $\{X(t); t \geq 0\}$ has occurred during the interval $[0, t]$, the time of occurrence of this event is uniformly distributed over $[0, t]$.*

Proof: Immediate application of the conditional probability (2.44) yields for $0 \le s \le t$,

$$\Pr\left[\tau_1 \le s | X(t) = 1\right] = \frac{\Pr\left[\{\tau_1 \le s\} \cap \{X(t) = 1\}\right]}{\Pr\left[X(t) = 1\right]}$$

Using the equivalence $\{\tau_1 \le s\} \iff \{X(t_0 + s) - X(t_0) = 1\}$ and the fact that $\{X(t_0 + s) - X(t_0) = 1\} = \{X(s) = 1\}$ by the stationarity of the Poisson process gives

$$\{\tau_1 \le s\} \cap \{X(t) = 1\} = \{X(s) = 1\} \cap \{X(t) = 1\}$$
$$= \{X(s) = 1\} \cap \{X(t) - X(s) = 0\}$$

Applying the independence of increments over non-overlapping intervals and (7.1) yields

$$\Pr\left[\tau_1 \le s | X(t) = 1\right] = \frac{\Pr\left[X(s) = 1\right] \Pr\left[X(t) - X(s) = 0\right]}{\Pr\left[X(t) = 1\right]}$$

$$= \frac{(\lambda s)\, e^{-\lambda s} e^{-\lambda(t-s)}}{(\lambda t)\, e^{-\lambda t}} = \frac{s}{t}$$

which completes the proof. $\square$

Theorem 7.3.3 is immediately generalized to n events. For any set of real variables s_j satisfying $0 = s_0 < s_1 < s_2 < \cdots < s_n < t$ and given that n events of a Poisson process $\{X(t); t \ge 0\}$ have occurred during the interval $[0, t]$, the probability of the successive occurrence times $0 < t_1 < t_2 < \cdots < t_n < t$ of these n Poisson events is

$$\Pr\left[t_1 \le s_1, \ldots, t_n < s_n | X(t) = n\right] = \frac{\Pr\left[\{t_1 \le s_1, \ldots, t_n < s_n\} \cap \{X(t) = n\}\right]}{\Pr\left[X(t) = n\right]}$$

Using a similar argument as in the proof of Theorem 7.3.3,

$$p = \Pr\left[\{t_1 \le s_1, t_2 \le s_2, \ldots, t_n < s_n\} \cap \{X(t) = n\}\right]$$
$$= \Pr\left[X(s_1) - X(s_0) = 1, \ldots, X(s_n) - X(s_{n-1}) = 1, X(t) - X(s_n) = 0\right]$$

$$= \left(\prod_{j=1}^{n} \Pr\left[X(s_j) - X(s_{j-1}) = 1\right]\right) \Pr[X(t) - X(s_n) = 0]$$

$$= \left(\prod_{j=1}^{n} e^{-\lambda(s_j - s_{j-1})} \lambda\,(s_j - s_{j-1})\right) e^{-\lambda(t - s_n)}$$

$$= \lambda^n \prod_{j=1}^{n} (s_j - s_{j-1})\, e^{-\lambda \sum_{j=1}^{n}(s_j - s_{j-1}) - \lambda(t - s_n)} = \lambda^n \prod_{j=1}^{n} (s_j - s_{j-1})\, e^{-\lambda t}$$

Thus,

$$\Pr\left[t_1 \le s_1, t_2 \le s_2, \ldots, t_n < s_n | X(t) = n\right] = \frac{\lambda^n \prod_{j=1}^{n} (s_j - s_{j-1}) e^{-\lambda t}}{\frac{(\lambda t)^n e^{-\lambda t}}{n!}}$$

$$= \frac{n!}{t^n} \prod_{j=1}^{n} (s_j - s_{j-1})$$

from which the density function

$$f_{\{t_j\}}(s_1, \ldots, s_n | X(t) = n) = \frac{\partial^n}{\partial s_1 \ldots \partial s_n} \Pr\left[t_1 \le s_1, \ldots, t_n < s_n | X(t) = n\right]$$

follows as

$$f_{\{t_j\}}(s_1, s_2, \ldots, s_n | X(t) = n) = \frac{n!}{t^n}$$

which is independent of the rate λ. If $0 < t_1 < t_2 < \cdots < t_n < t$ are the successive occurrence times of n Poisson events in the interval $[0, t]$, then the random variables $t_1, t_2, \ldots, t_n$ are distributed as a set of order statistics, defined in Section 3.4.2, of n uniform random variables in $[0, t]$. In other words, if n i.i.d. uniform random variables on $[0, t]$ are assorted in increasing order, they may represent n successive occurrence times of a Poisson process. The average spacing between these n ordered i.i.d. uniform random variables is $\frac{t}{n}$ as computed in Problem (ii) of Section 3.7.

A related example is the conditional probability where $0 < s < t$ and $0 \le k \le n$,

$$\Pr\left[X(s) = k | X(t) = n\right] = \frac{\Pr\left[\{X(s) = k\} \cap \{X(t) = n\}\right]}{\Pr\left[X(t) = n\right]}$$

$$= \frac{\Pr\left[\{X(s) = k\} \cap \{X(t) - X(s) = n - k\}\right]}{\Pr\left[X(t) = n\right]}$$

$$= \frac{\Pr\left[X(s) = k\right] \Pr\left[X(t) - X(s) = n - k\right]}{\Pr\left[X(t) = n\right]}$$

$$= \frac{n!(\lambda s)^k e^{-\lambda s}}{k!(\lambda t)^n e^{-\lambda t}} \frac{(\lambda(t - s))^{n-k} e^{-\lambda(t-s)}}{(n - k)!}$$

$$= \binom{n}{k} \frac{s^k}{t^n} (t - s)^{n-k}$$

Hence, if $p = \frac{s}{t}$, the conditional probability becomes

$$\Pr\left[X(s) = k | X(t) = n\right] = \binom{n}{k} p^k (1 - p)^{n-k}$$

Given that a total number of n Poisson events have occurred in time interval $[0, t]$, the chance that precisely k events have taken place in the sub-interval $[0, s]$ is binomially distributed with parameter n and $p = \frac{s}{t}$. Observe that also this conditional probability is independent of the rate λ. In addition, since $\lim_{t \to \infty} X(t) = \infty$ such that $n \to \infty$, applying the law of rare events results in

$$\lim_{t \to \infty} \Pr[X(s) = k | X(t) = n] = \frac{s^k}{k!} e^{-s}$$

Given an everlasting Poisson process, the chance that precisely k events occur in the interval $[0, s]$ is Poisson distributed with mean equal to the length of the interval.

Application The arrival process of most real-time applications (such as telephony calls, interactive-video, ...) in a network is well approximated by a Poisson process. Suppose a measurement configuration is built to collect statistics of the arrival process of telephony calls in some region. During a period $[0, T]$, precisely 1 telephony call has been measured. What can be said of the time $x \in [0, T]$ at which the telephony call has arrived at the measurement device? Theorem 7.3.3 tells us that any time in that interval is equally probable.

Theorem 7.3.4 *If $X(t)$ and $Y(t)$ are two independent Poisson processes with rates λ_x and λ_y, then is $Z(t) = X(t) + Y(t)$ also a Poisson process with rate $\lambda_x + \lambda_y$.*

Proof: It suffices to demonstrate that the counting process $N_Z(t) = N_X(t) + N_Y(t)$ has exponentially distributed interarrival times τ_Z. Suppose that $N_Z(t_n) = n$, it remains to compute the next arrival at time $t_{n+1} = t_n + s$ for which $N_Z(t_n + s) = n + 1$. Due to the memoryless property of the Poisson process, the occurrence of an event from t_n on for each random variable X and Y is again exponentially distributed with parameter λ_x and λ_y, respectively. In other words, it is irrelevant which process X or Y has previously caused the arrival at time t_n. Further, the event that the interarrival time of the sum processes $\{\tau_Z > s\}$ is equivalent to $\{\tau_X > s\} \cap \{\tau_Y > s\}$ or

$$\Pr[\tau_Z > s] = \Pr[\tau_X > s, \tau_Y > s] = \Pr[\tau_X > s] \Pr[\tau_Y > s] = e^{-(\lambda_x + \lambda_y)s}$$

where the independence of $X(t)$ and $Y(t)$ has been used. This proves the theorem. $\square$

A direct consequence is that any sum of independent Poisson processes is also a Poisson process with aggregate rate equal to the sum of the individual rates. This theorem is in correspondence with the sum property of the Poisson distribution.

7.4 The nonhomogeneous Poisson process

As will be shown later in Section 11.3.2, the Poisson process is a special case of a birth-and-death process, which is in turn a special case of a Markov process. Hence, it seems more instructive to discuss these special processes as applications of the Markov process. Therefore, only associations to the Poisson process are treated here. In many cases, the rate is a time variant function $\lambda(t)$ and such process is termed *a nonhomogeneous or nonstationary Poisson process*. For example, the arrival rate of a large number m of individual IP-flows at a router is well approximated by a nonhomogeneous Poisson process, where the rate $\lambda(t)$ varies over the day depending on the number m and the individual rate of each flow of packets. Since the sum of independent Poisson random variables is again a Poisson random variable, we have $\lambda(t) = \sum_{j=1}^{m(t)} \lambda_j(t)$.

If $X(t)$ is a nonhomogeneous Poisson process with rate $\lambda(t)$, the increment $X(t) - X(s)$ reflects the number of events in an interval $(s, t]$ and increments of non-overlapping intervals are still independent.

Theorem 7.4.1 *If $\Lambda(t) = \int_0^t \lambda(u)du$ and $s < t$, then $X(t) - X(s)$ is Poisson distributed with mean $\Lambda(t) - \Lambda(s)$.*

The demonstration is analogous to the proof of Theorem 7.3.1.
Proof (partly): Denote by $P_n(t) = \Pr\left[N(t) - N(s) = n\right]$, then

$$P_0(t + h) = \Pr\left[N(t + h) - N(s) = 0\right]$$
$$= \Pr\left[N(t + h) - N(t) = 0, N(t) - N(s) = 0\right]$$

Invoking independence of the increments,

$$P_0(t + h) = \Pr\left[N(t + h) - N(t) = 0\right]\Pr[N(t) - N(s) = 0]$$
$$= P_0(t)(1 - \lambda(t)h + o(h))$$

or

$$\frac{P_0(t + h) - P_0(t)}{h} = -\lambda(t)P_0(t) + \frac{o(h)}{h}$$

from which, in the limit $h \to 0$, the differential equation

$$P_0'(t) = -\lambda(t)P_0(t)$$

is immediate. Rewritten as $\frac{d}{dt} \log P_0(t) = -\lambda(t)$, after integration over $(s, t]$, we find $\log P_0(t) = -(\Lambda(t) - \Lambda(s))$ since $P_0(s) = \Pr[N(s) - N(s) = 0] = 1$. Thus, for the case $n = 0$, we find $P_0(t) = \exp[-(\Lambda(t) - \Lambda(s))]$, which proves the theorem for $n = 0$.

The remainder of the proof ($n > 0$) uses the same ingredients as the proof of Theorem 7.3.1 and is omitted.   □

A nonhomogeneous Poisson process $X(t)$ with rate $\lambda(t)$ can be transformed to a homogeneous Poisson process $Y(u)$ with rate 1 by the time transform $u = \Lambda(t)$. For, $Y(u) = Y(\Lambda(t)) = X(t)$, and $Y(u + \Delta u) = Y(\Lambda(t) + \Delta\Lambda(t)) = X(t + \Delta t)$ because $\Delta\Lambda(t) = \Lambda(\Delta t)$ for small Δt such that

$$\Pr[Y(u + \Delta u) - Y(u) = 1] = \Pr[X(t + \Delta t) - X(t) = 1]$$
$$= \lambda(t)\Delta t + o(\Delta t)$$
$$= \Delta u + o(\Delta u)$$

because $\Delta u = \lambda(t)\Delta t + o(\Delta t)$. Hence, all problems concerning nonhomogeneous Poisson processes can be reduced to the homogeneous case treated above.

7.5 The failure rate function

Previous sections have shown that the Poisson process is specified by a rate function $\lambda(t)$. In this section, we consider the failure rate function of some object or system. Often it is interesting to know the probability that an object will fail in the interval $[t, t + \Delta t]$ given that the object was still functioning well up to time t. Let X denote the lifetime of an object[5], then this probability can be written with (2.44) as

$$\Pr[t \leq X \leq t + \Delta t | X > t] = \frac{\Pr[\{t \leq X \leq t + \Delta t\} \cap \{X > t\}]}{\Pr[X > t]}$$
$$= \frac{\Pr[t < X \leq t + \Delta t]}{\Pr[X > t]}$$

If $f_X(t)$ is the probability density function of X and $F_X(t) = \Pr[X \leq t]$, then for small Δt and assuming that $f_X(t)$ is well behaved[6] such that

---

[5] In medical sciences, X can represent in general the time for a certain event to occur. For example, the time it takes for an organism to die, the time to recover from illness, the time for a patient to respond to a therapy and so on.

[6] Recall the discussion in Section 2.3.

$$\Pr\left[t < X \le t + \Delta t\right] = f_X(t)\Delta t,$$

$$\Pr\left[t \le X \le t + \Delta t | X > t\right] = \frac{f_X(t)}{1 - F_X(t)}\Delta t$$

This expression shows that

$$r(t) = \frac{f_X(t)}{1 - F_X(t)} \tag{7.4}$$

can be interpreted as the intensity or rate that a t-year old object will fail. It is called *the failure rate $r(t)$* and

$$R(t) = 1 - F_X(t) = \Pr\left[X > t\right] \tag{7.5}$$

is usually termed[7] *the reliability function*. Since $r(t) = \frac{\Pr[t \le X \le t + \Delta t | X > t]}{\Delta t}$ for small Δt, the failure rate $r(t) > 0$ because $r(t) = 0$ would imply an infinite lifetime X. Using the definition (2.30) of a probability density function, we observe that

$$r(t) = -\frac{\frac{dR(t)}{dt}}{R(t)} = -\frac{d\ln R(t)}{dt} \tag{7.6}$$

Or, since $R(0) = 1$, the corresponding integrated relation is

$$R(t) = \exp\left[-\int_0^t r(u)du\right] \tag{7.7}$$

The expressions (7.6) and (7.7) are inverse relations that specify $r(t)$ as function of $R(t)$ and vice versa. The reliability function $R(t)$ is non-increasing with maximum at $t = 0$ since it is a probability distribution function. On the other hand, the failure rate $r(t)$ being a probability density function can take any positive real value. From (7.4) we obtain the density function of the lifetime X in terms of failure rate $r(t)$ as

$$f_X(t) = r(t)R(t) = r(t)\exp\left[-\int_0^t r(u)du\right]$$

with $f_X(0) = r(0)$. Using the tail relation (2.35) for the expectation of the lifetime X immediately gives the *mean time to failure*,

$$E\left[X\right] = \int_0^\infty R(t)dt \tag{7.8}$$

In case $F_X(T) = 1$ and $f_X(T) \ne 0$ for a finite time T, which is the maximum lifetime, the definition (7.4) demonstrates that $r(t)$ has a pole at

---

[7] In biology, medical sciences and physics, $R(t)$ is called the survival function and $r(t)$ is the corresponding mortality rate or hazard rate.

$t = T$. In practice, the failure rate $r(t)$ is relatively high for small t due to initial imperfections that cause a number of objects to fail early and $r(t)$ is increasing towards the maximum life time T due to aging or wear and tear. This shape of $r(t)$ as illustrated in Fig. 7.4 is called a "bath-tub" curve, which is convex.

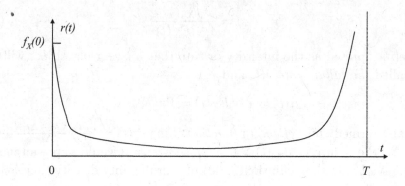

Fig. 7.4. Example of a "bath-tub" shaped failure rate function $r(t)$.

An often used model for the failure rate is $r(t) = a\lambda t^{a-1}$ with corresponding reliability function $R(t) = \exp\left[-\lambda t^a\right]$ and where the lifetime X has a Weibull distribution function $F_X(t) = 1 - R(t)$ as in (3.40). In case $a = 1$, the failure rate $r(t) = \lambda$ is constant over time, while $a > 1$ ($a < 1$) reflects an increasing (decreasing) failure rate over time. Hence, a "bath-tub" shaped (realistic) failure function as in Fig. 7.4 can be modeled by a Weibull model for $r(t)$ with $a < 1$ in the beginning, $a = 1$ in the middle and $a > 1$ at the end of the life time.

For an exponential lifetime where $f_X(t) = \lambda e^{-\lambda t}$, the failure rate (7.4) equals $r(t) = \lambda$ and is independent of time. This means that the failure rate for a t-year-old object is the same as for a new object, which is a manifestation of the memoryless property of the exponential distribution. It also explains why λ in both the exponential as Poisson process is often called a 'rate'.

7.6 Problems

(i) A series of test strings each with a variable number N of bits all equal to 1 are transmitted over a channel. Due to transmission errors, each 1-bit can be effected independently from the others and only arrives non-corrupted with probability p. The length N of the test strings (words) is a Poisson random variable with mean length λ bits. In this test, the sum Y of the bits in the arriving words is investigated to determine the channel quality via p. Compute the pdf of Y.

(ii) At a router, four QoS classes are supported and for each class packets arrive according to a Poisson process with rate λ_j for $j = 1, 2, 3, 4$. Suppose that the router had a failure at time t_1 that lasted T time units. What is the probability density function of the total number of packets of the four classes that has arrived during that period?

(iii) Let $N(t) = N_1(t) + N_2(t)$ be the sum of two independent Poisson processes with rates λ_1 and λ_2. Given that the process $N(t)$ had an arrival, what is the probability that that arrival came from the process $N_1(t)$?

(iv) Peter has been monitoring the highway for nearly his entire life and found that the cars pass his house according to a Poisson process. Moreover, he discovered that the Poisson process in one lane is independent from that in the other lanes. The rate of these independent processes differs per lane and is denoted by $\lambda_1, \lambda_2, \lambda_3$, where λ_j is expressed in the number of cars on lane j per hour.

  (a) Given that one car passed Peter, what is the probability that it passed in lane 1?
  (b) What is the probability that n cars pass Peter in 1 hour ?
  (c) What is the probability that in 1 hour n cars have passed and that they all have used lane 1?

(v) In a game, audio signals arrive in the interval $(0, T)$ according to a Poisson process with rate λ, where $T > 1/\lambda$. The player wins only if at least one audio signal arrives in that interval, and if he or she pushes a button (only one push allowed) upon the last of the signals. The player uses the following strategy: he or she pushes the button upon the arrival of the first signal (if any) after a fixed time $s \leq T$.

  (a) What is the probability that the player wins?
  (b) Which value of s maximizes the probability of winning, and what is the probability in that case?

(vi) The arrivals of voice over IP (VoIP) packets to a router is close to a Poisson process with rate $\lambda = 0.1$ packets per minute. Due to an upgrade to install weighted fair queueing as priority scheduling rule, the router is switched off for 10 minutes.

  (a) What is the probability of receiving no VoIP packets when switched off?
  (b) What is the probability that more than ten VoIP packets will arrive during this upgrade?

(c) If there was one VoIP in the meantime, what is the most probable minute of the arrival?

(vii) A link of a packet network carries on average ten packets per second. The packets arrive according to a Poisson process. A packet has a probability of 30 % to be an acknowledgment (ACK) packet independent of the others. The link is monitored during an interval of 1 second.

(a) What is the probability that at least one ACK packet has been observed?

(b) What is the expected number of all packets given that five ACK packets have been spotted on the link?

(c) Given that eight packets have been observed in total, what is the probability that two of them are ACK packets?

(viii) An ADSL helpdesk treats exclusively customer requests of one of three types: (i) login-problems, (ii) ADSL hardware and (iii) ADSL software problems. The opening hours of the helpdesk are from 8:00 until 16:00. All requests are arriving at the helpdesk according to a Poisson process with different rates: $\lambda_1 = 8$ requests with login problems/hour, $\lambda_2 = 6$ requests with hardware problems/hour, and $\lambda_3 = 6$ requests with software problems/hour. The Poisson arrival processes for different types of requests are independent.

(a) What is the expected number of requests in one day?

(b) What is the probability that in 20 minutes exactly three requests arrive, and that all of them have hardware problems?

(c) What is the probability that no requests will arrive in the last 15 minutes of the opening hours?

(d) What is the probability that one request arrives between 10:00 and 10:12 and two requests arrive between 10:06 and 10:30?

(e) If at the moment $t + s$ there are $k + m$ requests, what is the probability that there were k requests at the moment t?

(ix) Arrival of virus attacks to a PC can be modeled by a Poisson process with rate $\lambda = 6$ attacks per hour.

(a) What is the probability that exactly one attack will arrive between 1 p.m. and 2 p.m.?

(b) Suppose that at the moment the PC is turned on there were no attacks on PC, but at the shut-down time precisely 60 attacks have been observed. What is the expected amount of time that the PC has been on?

(c) Given that six attacks arrive between 1 p.m. and 2 p.m., what is the probability that the fifth attack will arrive between 1:30 p.m. and 2 p.m.?

(d) What is the expected arrival time of that fifth attack?

(x) Consider a system S consisting of n subsystems in series as shown in Fig. 7.5. The system S operates correctly only if all subsystems operate correctly. Assume that the probability that a failure in a subsystem S_i occurs is independent of that in subsystem S_j. Given the reliability functions $R_j(t)$ or each subsystem S_j, compute the reliability function $R(t)$ of the system S.

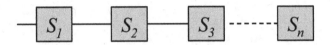

Fig. 7.5. A system consisting of n subsystems in series.

(xi) Same question as in previous exercise but applied to a system S consisting of n subsystem in parallel as shown in Fig. 7.6.

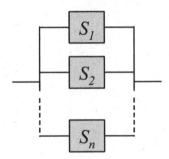

Fig. 7.6. A system consisting of n subsystems in parallel.

8

Renewal theory

A renewal process is a counting process for which the interarrival times τ_n are i.i.d. random variables with distribution $F_\tau(t)$. Hence, a renewal process generalizes the exponential interarrival times in the Poisson process (see Theorem 7.3.2) to an arbitrary distribution. Since the interarrival times are i.i.d. random variables, at each event (or renewal) the process probabilistically restarts. The classical example of a renewal process is the successive replacement of light bulbs: the first bulb is installed at time W_0, fails at time $W_1 = \tau_1$, and is immediately exchanged for a new bulb, which in turn fails at $W_2 = \tau_1 + \tau_2$, and thereafter replaced by a third bulb, and so on. How many light bulbs are replaced in a period of t time units given the life time distribution $F_\tau(t)$?

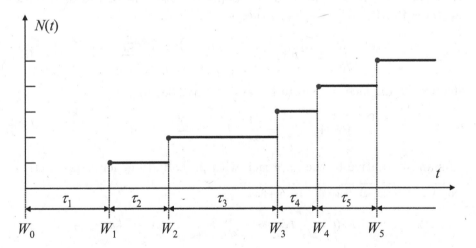

Fig. 8.1. The relation between the renewal counting process $N(t)$, the interarrival time τ_n and the waiting time W_n.

8.1 Basic notions

As illustrated in Fig. 8.1, the waiting time $W_n = \sum_{k=1}^{n} \tau_k$ (for $n \geq 1$, with $W_0 = 0$ by convention) is related to the counting process $\{N(t), t \geq 0\}$ by the equivalence $\{N(t) \geq n\} \iff \{W_n \leq t\}$: the number of events (renewals) up to time t is at least n if and only if the n-th renewal occurred on or before time t. Alternatively, the number of events by time t equals the largest value of n for which the n-th event occurs before or at time t, $N(t) = \max[n : W_n \leq t]$. The convention that $W_0 = 0$ implies that $N(0) = 0$: the counting process starts counting from zero at time 0. The main objective of renewal theory is to deduce properties of the process $\{N(t), t \geq 0\}$ as a function of the interarrival distribution $F_\tau(t) = \Pr[\tau \leq t]$.

8.1.1 The distribution of the waiting time W_n

If we assume that the interarrival times are i.i.d. having a Laplace transform

$$\varphi_\tau(z) = \int_0^\infty e^{-zt} dF_\tau(t) = \int_0^\infty e^{-zt} f_\tau(t) dt$$

the waiting time W_n is the sum of n i.i.d. random variables specified by (2.66) as

$$\varphi_{W_n}(z) = \int_0^\infty e^{-zt} f_{W_n}(t) dt = \varphi_\tau^n(z). \tag{8.1}$$

By partial integration, we find the Laplace transform of the distribution $F_{W_n}(t) = \Pr[W_n \leq t] = \int_0^t f_{W_N}(u) du$

$$\int_0^\infty e^{-zt} F_{W_n}(t) dt = \frac{\varphi_{W_n}(z)}{z} = \frac{\varphi_\tau^n(z)}{z} \tag{8.2}$$

The inverse Laplace transform follows[1] with (2.38) as

$$\Pr[W_n \leq t] = \frac{1}{2\pi i} \int_{c-i\infty}^{c+i\infty} \frac{\varphi_\tau^n(z)}{z} e^{zt} dz \tag{8.3}$$

As an alternative to the approach with probability generating functions,

---

[1] In general, by integration of (2.38), we find

$$F_X(t) = \int_0^t f_X(u) du = \frac{1}{2\pi i} \int_{c-i\infty}^{c+i\infty} \varphi_X(z) \frac{e^{zt} - 1}{z} dz$$

whose form seems different from (8.3). However, $\frac{1}{2\pi i} \int_{c-i\infty}^{c+i\infty} \frac{\varphi_X(z)}{z} dz = 0$ because the contour can be closed over the positive $\text{Re}(z) > c$ plane where $\varphi_X(z)$ is analytic and because $\lim_{R \to \infty} \varphi_X(Re^{i\theta}) = 0$ for $-\frac{\pi}{2} < \theta < \frac{\pi}{2}$, which follows from the existence of the Laplace integral $\int_0^\infty e^{-zt} f_X(t) dt$.

we can resort to the n-th convolution, which follows from (2.63) as

$$f_{W_1}(t) = f_\tau(t)$$

$$f_{W_N}(t) = \int_{-\infty}^{\infty} f_{W_{N-1}}(t-y) f_\tau(y) dy = \int_0^t f_{W_{N-1}}(t-y) f_\tau(y) dy$$

Integrated,

$$\Pr[W_n \le t] = \int_{-\infty}^t du \int_{-\infty}^{\infty} f_{W_{N-1}}(u-y) f_\tau(y) dy$$

$$= \int_{-\infty}^{\infty} \left(\int_{-\infty}^{t-y} f_{W_{N-1}}(u) du \right) f_\tau(y) dy$$

$$= \int_0^t \Pr[W_{n-1} \le t - y] f_\tau(y) dy$$

By denoting $\Pr[W_n \le t] = F_\tau^{(n*)}(t)$, we have

$$F_\tau^{(1*)}(t) = F_\tau(t)$$

$$F_\tau^{(n*)}(t) = \int_0^t F_\tau^{((n-1)*)}(t-y) f_\tau(y) dy$$

These equations also show that we can define $F_\tau^{(0*)}(t) = 1$. Let us define $U_n(t) = \sum_{k=1}^n F_\tau^{(k*)}(t)$. By summing both sides in the last equation, we obtain

$$U_n(t) = \int_0^t \sum_{k=1}^n F_\tau^{((k-1)*)}(t-y) f_\tau(y) dy = \int_0^t \sum_{k=0}^{n-1} F_\tau^{((k)*)}(t-y) f_\tau(y) dy$$

With the definition $F_\tau^{(0*)}(t) = 1$, we arrive at

$$U_n(t) = \int_0^t U_{n-1}(t-y) dF_\tau(y) + F_\tau(t) \tag{8.4}$$

or, written in terms of convolutions,

$$U_n(t) = (U_{n-1} * F_\tau)(t) + F_\tau(t)$$

Finally, we mention the interesting bound on the convolution $F_\tau^{(n*)}(t)$ for a non-negative random variable τ,

$$F_\tau^{(n*)}(t) = \int_0^t F_\tau^{((n-1)*)}(t-y) dF_\tau(y)$$

$$\le F_\tau^{((n-1)*)}(t) \int_0^t dF_\tau(y) = F_\tau^{((n-1)*)}(t) F_\tau(t)$$

which follows from the monotone increasing nature of any distribution function. By iteration on n starting from $F_\tau^{(0*)}(t) = 1$, it is immediate that

$$F_\tau^{(n*)}(t) \leq (F_\tau(t))^n \tag{8.5}$$

Since $(F_\tau(t))^n$ is the distribution of the maximum (3.33) of a set of n i.i.d. random variables $\{\tau_k\}_{1 \leq k \leq n}$, the bound (8.5) means that, for $\tau_k \geq 0$,

$$\Pr\left[\sum_{k=1}^n \tau_k \leq x\right] \leq \Pr\left[\max_{1 \leq k \leq n} \tau_k \leq x\right]$$

which is rather obvious because $\sum_{k=1}^n \tau_k \geq \max_{1 \leq k \leq n} \tau_k$. The equality sign is only possible if $n-1$ of the τ_k are zero.

8.1.2 The renewal function $m(t) = E[N(t)]$

From the equivalence $\{N(t) \geq n\} \Longleftrightarrow \{W_n \leq t\}$, we directly have

$$\Pr[N(t) \geq n] = \Pr[W_n \leq t] = F_\tau^{(n*)}(t) \tag{8.6}$$
$$\Pr[N(t) = n] = \Pr[N(t) \geq n] - \Pr[N(t) \geq n+1]$$
$$= F_\tau^{(n*)}(t) - F_\tau^{((n+1)*)}(t)$$

The expected number of events in $(0, t]$ expressed via the tail probabilities (2.36) follows with (8.6) as

$$m(t) = E[N(t)] = \sum_{k=1}^\infty F_\tau^{(k*)}(t) \tag{8.7}$$

and $m(t)$ is called *the renewal function*. According to a property of the counting process, $N(0) = 0$, the number of events in $(0, t]$ when $t \to 0$ is assumed to be zero such that $m(0) = 0$. From (8.5), it follows at each point t for which $F_\tau(t) < 1$ that

$$m(t) \leq \sum_{k=1}^\infty (F_\tau(t))^k = \frac{1}{1 - F_\tau(t)} - 1$$

Hence, for finite t where $F_\tau(t) < 1$, the renewal function $m(t)$ converges at least as fast as a geometric series and is bounded. In the limit $t \to \infty$, where $\lim_{t \to \infty} F_\tau(t) = 1$, we see that $m(t)$ is not bounded anymore. Intuitively, the number of repeated events (renewals) in an infinite time interval is clearly infinite.

The renewal function $m(t)$ completely characterizes the renewal process. Indeed, if $\varphi_m(z)$ is the Laplace transform of $m(t)$, then after taking the

Laplace transform of both sides in (8.7) and using the definition $\Pr[W_n \leq t] = F_\tau^{(n*)}(t)$ together with (8.2), we obtain

$$\varphi_m(z) = \frac{1}{z} \sum_{k=1}^{\infty} \varphi_\tau^k(z) = \frac{1}{z} \frac{\varphi_\tau(z)}{1 - \varphi_\tau(z)} \tag{8.8}$$

provided $|\varphi_\tau(z)| < 1$. From this expression, the interarrival time can be found from

$$\varphi_\tau(z) = \frac{z\varphi_m(z)}{1 + z\varphi_m(z)}$$

after inverse Laplace transform. By taking the inverse Laplace transform (2.38), $m(t)$ is written as a complex integral

$$m(t) = \frac{1}{2\pi i} \int_{c-i\infty}^{c+i\infty} \frac{\varphi_\tau(z)}{1 - \varphi_\tau(z)} \frac{e^{zt}}{z} dz$$

8.1.3 The renewal equation

After taking the inverse Laplace transform of $\varphi_m(z) = \varphi_m(z)\varphi_\tau(z) + \frac{\varphi_\tau(z)}{z}$, which is deduced from (8.8), a third relation for $m(t)$ that often occurs is

$$m(t) = \int_0^t m(t-u)dF_\tau(u) + F_\tau(t)$$

$$= \int_0^t F_\tau(t-u)dm(u) + F_\tau(t) \tag{8.9}$$

and is called *the renewal equation*. Taking the limit $n \to \infty$ in (8.4) also leads to the renewal equation. Since $m(0) = 0$, the renewal equation implies that $F_\tau(0) = \Pr[\tau \leq 0] = 0$ or that processes where a zero interarrival time is possible (e.g. in simultaneous events) are ruled out. For a Poisson process, Theorem 7.3.1 states that the occurrence of simultaneous events ($h \to 0$) is zero. The requirement $m(0) = 0$ generalizes the exclusion of simultaneous events in any renewal process.

The probabilistic argument that leads to the renewal equation is as follows. By conditioning on the first renewal for $k > 0$,

$$\Pr[N(t) = k|W_1 = s] = 0 \qquad\qquad t < s$$
$$= \Pr[N(t-s) = k-1] \qquad t \geq s$$

where in the last case for $t \geq s$ the event $\{N(t) = k\}$ is only possible if $k-1$ renewals occur in time interval $(s, t]$, which is, due to the stationarity of the renewal process, equal to $k-1$ renewals in $(0, t-s]$. By the law of total probability (2.46), we uncondition to find for $k \geq 1$,

$$\Pr[N(t) = k] = \int_0^\infty \Pr[N(t) = k|W_1 = s] \frac{d\Pr[W_1 \leq s]}{ds} ds$$

$$= \int_0^t \Pr[N(t-s) = k-1] f_\tau(s) ds \qquad (8.10)$$

Multiplying both sides by k and summing over all $k \geq 1$ gives the average at the left-hand side,

$$E[N(t)] = \sum_{k=1}^\infty k \Pr[N(t) = k]$$

The sum at the right-hand side is

$$\sum_{k=1}^\infty k \Pr[N(t-s) = k-1] = \sum_{k=0}^\infty (k+1) \Pr[N(t-s) = k]$$

$$= E[N(t-u)] + 1$$

Combining both sides yields

$$E[N(t)] = F_\tau(t) + \int_0^t E[N(t-u)] dF_\tau(s)$$

which is again the renewal equation (8.9) since $m(t) = E[N(t)]$.

8.1.4 A generalization of the renewal equation

The renewal equation (8.9) is a special case of the more general class of integral equations

$$Y(t) = h(t) + \int_0^t Y(t-u) dF(u), \qquad t \geq 0 \qquad (8.11)$$

in the unknown function $Y(t)$, where $h(t)$ is a known function and $F(t)$ is a distribution function. This equation can be written using the convolution notation as

$$Y(t) = h(t) + Y * F(t)$$

By conditioning on the first renewal as shown above, many renewal problems can be recast into the form of the general renewal equation (8.11). An example is the derivation of the residual life or waiting time given in Section 8.3. Therefore, it is convenient to present the solution to the general renewal equation (8.11).

Lemma 8.1.1 *If $h(t)$ is bounded for all t, then the unique solution of the general renewal equation (8.11) is*

$$Y(t) = h(t) + \int_0^t h(t-u)dm(u) \qquad (8.12)$$

where $m(t) = \sum_{k=1}^{\infty} F^{(k*)}(t)$ is the renewal function.

Proof: Let us first concentrate on the formal solution. In general, convolutions are best treated in the transformed domain. After taking the Laplace transform of the general renewal equation (8.11), we obtain

$$\varphi_Y(z) = \varphi_h(z) + \varphi_Y(z)\varphi_F(z)$$

such that

$$\varphi_Y(z) = \frac{\varphi_h(z)}{1 - \varphi_F(z)}$$

There always exists a region in the z-domain where $|\varphi_F(z)| < 1$ such that the geometric series applies,

$$\varphi_Y(z) = \varphi_h(z)\sum_{k=0}^{\infty}(\varphi_F(z))^k = \varphi_h(z) + \varphi_h(z)\sum_{k=1}^{\infty}(\varphi_F(z))^k$$

Back transforming and taking into account that $(\varphi_F(z))^k$ is the transform of a k-fold convolution yields

$$Y(t) = h(t) + h * \sum_{k=1}^{\infty} F^{(k*)}(t) = h(t) + h * m(t)$$

This formal manipulation demonstrates[2] that (8.12) is a solution of the general renewal equation (8.11).

Suppose now that there are two solutions $Y_1(t)$ and $Y_2(t)$. Their difference $V(t) = Y_1(t) - Y_2(t)$ obeys

$$V(t) = \int_0^t V(t-u)dF(u) = V * F(t)$$

---

[2] Alternatively, by substituting the solution into the equation, a check is

$$Y(t) = h(t) + Y * F(t) = h(t) + h * F(t) + h * m * F(t)$$

$$= h(t) + h * \left[F(t) + \sum_{k=1}^{\infty} F^{(k*)}(t) * F(t)\right]$$

$$= h(t) + h * \left[F(t) + \sum_{k=2}^{\infty} F^{(k*)}(t)\right] = h(t) + h * \left[\sum_{k=1}^{\infty} F^{(k*)}(t)\right]$$

$$= h(t) + h * m(t)$$

By convolving both sides with F and using the original equation, we deduce that $V(t) = V * F * F(t)$. Continuing this process, for each k, we have that $V(t) = V * F^{(k*)}(t)$. Since $F^{(k*)}(t) \to 0$ for all finite t and $k \to \infty$ (because $m(t)$ exists for all finite t), and if $V(t)$ is bounded, this implies that $V(t) = 0$ for all finite t. This demonstrates the uniqueness and motivates the requirement that $h(t)$ should be bounded. $\qquad \square$

8.1.5 The renewal function for a Poisson process

Before showing below that the renewal function $m(t)$ can be specified in detail as $t \to \infty$, we consider first the Poisson process where the interarrival times $\{\tau_n\}_{n \geq 1}$ are i.i.d. exponentially distributed with rate λ. Since $\varphi_\tau(z) = \frac{\lambda}{z+\lambda}$,

$$m(t) = \frac{1}{2\pi i} \int_{c-i\infty}^{c+i\infty} \frac{\lambda e^{zt}}{z^2} dz \qquad c > 0$$

The contour can be closed over the negative Re(z)-plane (because $t \geq 0$). The only singularity of the integrand is a double pole at $z = 0$ with residue $m(t) = \lambda \left. \frac{de^{zt}}{dz} \right|_{z=0} = \lambda t$. This result, of course, follows directly from the definition of the Poisson process given in (7.2). We see that the renewal function $m(t)$ for the Poisson process is linear for all t. Moreover, the Poisson process is the only continuous time renewal process with a linear renewal function $m(t)$. Indeed, if[3] $m(t) = \alpha t$, the renewal equation is

$$\alpha t = \int_0^t (\alpha(t-u))\, dF_\tau(u) + F_\tau(t) = \alpha \int_0^t F_\tau(u) du - t F_\tau(0) + F_\tau(t)$$

By differentiation with respect to t and assuming non-zero interarrival times such that $F_\tau(0) = \Pr[\tau \leq 0] = 0$, we obtain a differential equation

$$\alpha = \alpha F_\tau(t) + \frac{dF_\tau(t)}{dt}$$

whose solution is $F_\tau(t) = 1 - e^{-\alpha t}$. By Theorem 7.3.2, exponential interarrival times characterize a Poisson process with rate $\lambda = \alpha$.

8.2 Limit theorems

In the limit $t \to \infty$, the equivalence relation (8.6) indicates that, for any fixed value of n, $\Pr[N(t) \geq n] = 1$, which means that the number of events

---

[3] A linear form $m(t) = \alpha t + \beta$ with $\beta \neq 0$ is impossible because $m(0) = 0$.

$N(t) \to \infty$ as $t \to \infty$. Let us consider $\frac{W_{N(t)}}{N(t)}$, which is the sample mean of the first $N(t)$ interarrival times in the interval $(0, t]$. The Strong Law of Large Numbers (6.3) indicates that $\Pr\left[\lim_{n\to\infty} \frac{W_n}{n} = \mu\right] = 1$ and, because $N(t) \to \infty$ as $t \to \infty$, we have that $\frac{W_{N(t)}}{N(t)} \to \mu = E[\tau]$ as $t \to \infty$. Since $W_{N(t)} \le t < W_{N(t)+1}$, we obtain the inequality

$$\frac{W_{N(t)}}{N(t)} \le \frac{t}{N(t)} < \frac{W_{N(t)+1}}{N(t)}$$

Since both lower and upper bound tend to μ, we arrive at the important result that $\lim_{t\to\infty} \frac{N(t)}{t} = \frac{1}{\mu}$. The random variable counting the number of events in $(0, t]$ per interval length t, converges to the average interarrival time $\mu = E[\tau]$. Unfortunately[4], we cannot simply deduce the intuitive result that also the expectation, $E\left[\frac{N(t)}{t}\right]$ tends to $\frac{1}{\mu}$. On the other hand, the expectation of $W_{N(t)}$ is obtained from Wald's identity (2.69) as $E[W_{N(t)}] = E[N(t)]E[\tau]$. Taking the expectation in the inequality $W_{N(t)} \le t < W_{N(t)+1}$, leads to $\frac{E[N(t)]}{t} \le \frac{1}{E[\tau]} < \frac{E[N(t)]}{t} + \frac{1}{t}$ from which, after the limit $t \to \infty$, the intuitive result follows. Thus, we have proved[5] the following theorem:

Theorem 8.2.1 (Elementary Renewal Theorem) *If $\mu = E[\tau]$ is the average interarrival time of events in the renewal process, then*

$$\lim_{t\to\infty} \frac{E[N(t)]}{t} = \lim_{t\to\infty} \frac{m(t)}{t} = \frac{1}{\mu} \tag{8.13}$$

$$\lim_{t\to\infty} \frac{N(t)}{t} = \frac{1}{\mu}$$

The left-hand side in (8.13) describes the long run average number of events (renewals) per unit time. The right-hand side is the reciprocal of the average interarrival rate (or life time). For example, in the light bulb replacement process, a bulb lasts on average μ time units, then, in the long run or steady state, the light bulbs must be replaced at rate $\frac{1}{\mu}$ per time unit.

---

[4] As remarked by Ross (1996, p. 108), if U is uniformly distributed on $(0, 1)$, consider the random variables Y_n defined as $Y_n = n 1_{U \le \frac{1}{n}}$. For large n, $U > 0$ with probability 1, whence $Y_n \to 0$ if $n \to \infty$. However, $E[Y_n] = nE\left[1_{U \le \frac{1}{n}}\right] = n\frac{1}{n} = 1$, for all n. The sequence of random variables Y_n converges to 0, although the expected values of Y_n are all precisely 1.

[5] The elementary renewal theorem can be proved only by resorting to complex function theory and using Laplace–Stieltjes transforms (Cohen, 1969, p. 100). The limit argument provided by the Strong Law of Large Numbers follows then from a Tauberian theorem.

The extension[6] of the Elementary Renewal Theorem is the Key Renewal Theorem. The Key Renewal Theorem gives the limit $t \to \infty$ of the solution (8.12) of the general renewal equation (8.11).

Theorem 8.2.2 (Key Renewal Theorem) *If $g(t)$ is directly[7] Riemann integrable over $[0, \infty)$, then*

$$\lim_{t \to \infty} \int_0^t g(t - u)dm(u) = \frac{1}{\mu} \int_0^\infty g(u)du \qquad (8.14)$$

The proof[8] is more complicated, based on analysis and found in Feller (1971, Section XI.1). The essential difficulty is demonstrating that the limit at the right-hand side indeed exists. An application of the Key Renewal Theorem is presented in Section 8.3 and here we consider Blackwell's Theorem.

Blackwell's Theorem follows from the Key Renewal Theorem when choosing $h(t) = 1_{t \in [0,T)}$ in the general renewal equation (8.11). The corresponding solution (8.12) for $t > T$ is

$$Y(t) = \int_0^t 1_{t-u \in [0,T)} dm(u) = \int_{t-T}^t dm(u) = m(t) - m(t - T)$$

while the Key Renewal Theorem states that $\lim_{t \to \infty} Y(t) = \frac{1}{\mu} \int_0^\infty h(u)du = \frac{T}{\mu}$. Hence, we arrive at Blackwell's Theorem, for any fixed $T > 0$,

$$\lim_{t \to \infty} \frac{m(t) - m(t - T)}{T} = \frac{1}{\mu}$$

The interpretation of Blackwell's Theorem is that the number of expected renewals in an interval with length T sufficiently far from the origin (or in steady-state regime) is approximately equal to $\frac{T}{\mu}$. It can be shown that the reverse, i.e. the Key Renewal Theorem can be deduced from Blackwell's theorem, also holds. Hence, the Key Renewal Theorem is equivalent to Blackwell's Theorem.

Similarly to the Key Renewal Theorem the difficulty in Blackwell's Theorem is the proof that the limit exists. If the existence of the limit is proved,

---

[6] In the sequel we assume that the distribution of the interarrival times $F_\tau(t)$ is not periodic in the sense that there exists no integer d such that $\sum_{n=0}^\infty \Pr[\tau = nd] = 1$ or, the random variable τ does not only take integer units of some integer d.

[7] The concept is introduced to avoid widly oscillating functions that are still integrable over $[0, \infty)$, such as $g(t) = t1_{\{|t-n| < \frac{1}{n^2}\}}$. The precise definition is given in Feller (1971). A sufficient condition for direct Riemann integrability is (a) $g(t) \geq 0$ for all $t \geq 0$, (b) $g(t)$ is non-increasing and (c) $\int_0^\infty g(u)du < \infty$.

[8] Based on the relatively new probabilistic concept of "coupling", alternative proofs of the Key Renewal Theorem exist (see e.g. Grimmett and Stirzacker (2001, pp. 429–430)).

which means that $\lim_{t\to\infty} m(t) - m(t - T) = a(T)$ exists, the Elementary Renewal Theorem suffices to prove that the limit has value $\frac{1}{\mu}$. Following the argument of Ross (1996, p. 110), we can write, for finite x and y,

$$
\begin{aligned}
a(x + y) &= \lim_{t\to\infty} [m(t) - m(t - x - y)] \\
&= \lim_{t\to\infty} [m(t) - m(t - x)] + \lim_{t\to\infty} [m(t - x) - m(t - x - y)] \\
&= a(x) + a(y)
\end{aligned}
$$

Apart from the trivial solution $a(x) = 0$, the only other[9] solution of $a(x + y) = a(x) + a(y)$ is $a(x) = cx$, where c is a constant. Hence, given that $\lim_{t\to\infty} m(t) - m(t - T) = a(T)$ exists, this is equivalent to the fact that the sequence $\{b_n\}_{n\geq 0}$ where $b_n = \frac{m(t_n) - m(t_n - T)}{T}$ and $t_n > t_{n-1}$ converges to a constant c. The simplest sequence with this property is $\{b_n^*\}_{n\geq 0}$ where $b_n^* = m(n) - m(n - 1)$ and $T = 1$. Lemma 6.1.1 states that

$$
c = \lim_{n\to\infty} \frac{1}{n} \sum_{k=1}^{n} b_k^* = \lim_{n\to\infty} \frac{1}{n} \sum_{k=1}^{n} m(k) - m(k - 1) = \lim_{n\to\infty} \frac{m(n)}{n} = \frac{1}{\mu}
$$

where the last equality follows from the Elementary Renewal Theorem (8.13).

Theorem 8.2.3 (Asymptotic Renewal Distribution) *If the average $\mu = E[\tau]$ and variance $\sigma^2 = Var[\tau]$ of the interarrival time of the events in a renewal process exist, then*

$$
\lim_{t\to\infty} \Pr\left[\frac{N(t) - \frac{t}{\mu}}{\sigma\sqrt{\frac{t}{\mu^3}}} < x \right] = \frac{1}{\sqrt{2\pi}} \int_{-\infty}^{x} e^{-u^2/2} du \tag{8.15}
$$

Proof: The Elementary Renewal Theorem states that $N(t) \sim \frac{t}{\mu}$ for large t, which suggests to consider the random variable $U(t) = N(t) - \frac{t}{\mu}$ with $E[U(t)] \to 0$. From the equivalence $\{N(t) < n\} \iff \{W_n > t\}$, we have $\{U(t) < x_t\} \iff \{W_{x_t + \frac{t}{\mu}} > t\}$ where x_t is such that $x_t + \frac{t}{\mu}$ is a positive

---

[9] The proof is as follows: (i) if $y = 0$, we see that $a(x + 0) = a(x) + a(0)$ or $a(0) = 0$. (ii) $a(nx) = na(x)$ for integer n. (iii) Using (ii), we have that $a(nx + my) = na(x) + ma(y)$. By choosing $nx + my = 0$ it follows from (i) that $a\left(-\frac{m}{n}y\right) = -\frac{m}{n}a(y)$ such that (ii) holds for rational numbers. Thus, $a(q_1 x + q_2 y) = q_1 a(x) + q_2 a(y)$ for rational numbers q_1 and q_2. (iv) Recalling the definition $f\left(\frac{u+v}{2}\right) \leq \frac{f(u)+f(v)}{2}$ of a convex function in Section 5.2 and the fact that a function that is both concave and convex is a linear function, it follows that $a(x)$ is linear and with (i) that $a(x) = cx$.

integer. Then,

$$\Pr\left[U(t) < x_t\right] = \Pr\left[W_{x_t + \frac{t}{\mu}} > t\right]$$

$$= \Pr\left[\frac{W_{x_t + \frac{t}{\mu}} - \left(x_t + \frac{t}{\mu}\right)\mu}{\sigma\sqrt{x_t + \frac{t}{\mu}}} > \frac{t - \left(x_t + \frac{t}{\mu}\right)\mu}{\sigma\sqrt{x_t + \frac{t}{\mu}}}\right]$$

The waiting time W_n consists of a sum of i.i.d. random variables with mean μ and variance σ^2. By the Central Limit Theorem 6.3.1, there holds that

$$\lim_{n\to\infty} \Pr\left[\frac{W_n - n\mu}{\sigma\sqrt{n}} > x\right] = \frac{1}{\sqrt{2\pi}} \int_x^\infty e^{-u^2/2} du$$

which implies that

$$\lim_{t\to\infty} \Pr\left[\frac{W_{x_t + \frac{t}{\mu}} - \left(x_t + \frac{t}{\mu}\right)\mu}{\sigma\sqrt{x_t + \frac{t}{\mu}}} > \frac{t - \left(x_t + \frac{t}{\mu}\right)\mu}{\sigma\sqrt{x_t + \frac{t}{\mu}}}\right] = \frac{1}{\sqrt{2\pi}} \int_y^\infty e^{-u^2/2} du$$

provided $\lim_{t\to\infty} \frac{t - \left(x_t + \frac{t}{\mu}\right)\mu}{\sigma\sqrt{x_t + \frac{t}{\mu}}} = y$. Hence, we must determine x_t such that, for large t,

$$\frac{\mu x_t}{\sigma\sqrt{x_t + \frac{t}{\mu}}} = -y$$

which is satisfied if $x_t = \frac{y^2\sigma^2}{2\mu^2}\left(1 \pm \sqrt{1 + 4\left(\frac{\mu}{y\sigma}\right)^2 \frac{t}{\mu}}\right)$ and provided the negative sign is chosen. For large t, we see that $x_t \sim -\frac{y\sigma}{\mu}\sqrt{\frac{t}{\mu}} + O(1)$. Thus,

$$\lim_{t\to\infty} \Pr\left[U(t) < -\frac{y\sigma}{\mu}\sqrt{\frac{t}{\mu}}\right] = \frac{1}{\sqrt{2\pi}} \int_y^\infty e^{-u^2/2} du$$

which is equivalent to

$$\lim_{t\to\infty} \Pr\left[\frac{N(t) - \frac{t}{\mu}}{\sigma\sqrt{\frac{t}{\mu^3}}} < x\right] = \frac{1}{\sqrt{2\pi}} \int_{-x}^\infty e^{-u^2/2} du$$

Noting that $\int_{-x}^\infty e^{-u^2/2} du = \int_{-\infty}^x e^{-u^2/2} du$ finally proves (8.15). $\square$

Comparing Theorem 8.2.3 to the Central Limit Theorem 6.3.1 shows that the asymptotic variance of $N(t)$ behaves as

$$\lim_{t\to\infty} \frac{\text{Var}\left[N(t)\right]}{t} = \frac{\sigma^2}{\mu^3} \tag{8.16}$$

Moreover, Theorem 8.2.3 is a central limit theorem for the *dependent* random variables $N(W_n)$ where dependence is obvious from $N(W_n) = N(W_{n-1})+1$.

8.3 The residual waiting time

Suppose we inspect a renewal process at time t and ask the question "How long do we have to wait on average to see the next renewal?" This question frequently arises in renewal problems. For instance, the arrivals of taxis at a station is a renewal process and, often, we are interested to known how long we have to wait until the next taxi. Also, packets arriving at a router may find an earlier packet that is partially served. In order to compute the total time spent in the system, it is desirable to know the residual service time of that packet. In addition, this problem belongs to one of the classical examples to demonstrate how misleading intuition in probability problems can be. There are two different arguments to the question above leading to two different answers:

(i) since my inspection of the process does not alter or influence the process, the distribution of my waiting time should not depend on the time t; hence, my average waiting time equals the average interarrival time of the renewal process.

(ii) the time t of the inspection is chosen at random in (i.e. uniformly distributed over) the interval between two consecutive renewals; hence my expected waiting time should be half of the average interarrival time.

Both arguments seem reasonable although it is plain that one of them must be wrong. Let us try to sort out the correct answer to this apparent paradox, which, according to Feller (1971, pp. 12–13), has puzzled many before its solution was properly understood.

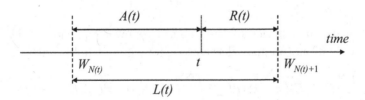

Fig. 8.2. Definition of the random variables the age $A(t)$, the lifetime $L(t)$ and the residual life (or waiting time) $R(t)$.

Figure 8.2 defines the setting of the renewal problem and the quantities of interest: $A(t)$ is the age at time t, which is the total time elapsed since the last renewal before t at time $W_{N(t)}$, the residual waiting time (or residual life or excess life) $R(t)$ is the remaining time at t until the next renewal at time $W_{N(t)+1}$ and $L(t)$ is the total waiting time (or life time). From Fig. 8.2, we verify that

$$A(t) = t - W_{N(t)}$$
$$R(t) = W_{N(t)+1} - t$$
$$L(t) = W_{N(t)+1} - W_{N(t)} = A(t) + R(t)$$

The distribution of the residual waiting time, $F_{R(t)}(x) = \Pr[R(t) \le x]$ will be derived. Similar to the probabilistic argument before, we condition on the first renewal. If $W_1 = s \le t$, then the first renewal occurs before time t and the event $\{R(t) > x | W_1 = s\}$ has the same probability as the event $\{R(t - s) > x\}$ because the renewal process restarts from scratch at time s. If $s > t$, the residual waiting time $R(t)$ lies in the first renewal interval $[0, s]$. In this case, we have either that the residual waiting time $R(t)$ is certainly shorter than x if s is contained in the interval $[t, t + x]$, else the residual waiting time $R(t)$ is surely larger than x. In summary,

$$\Pr[R(t) > x | W_1 = s] = \begin{cases} \Pr[R(t - s) > x] & \text{if } 0 \le s \le t \\ 0 & \text{if } t < s \le t + x \\ 1 & \text{if } s > x + t \end{cases}$$

Using the total law of probability (2.46),

$$\Pr[R(t) > x] = \int_0^\infty \Pr[R(t) > x | W_1 = s] \frac{d\Pr[W_1 \le s]}{ds} ds$$
$$= \int_0^t \Pr[R(t - s) > x] f_\tau(s) ds + \int_{x+t}^\infty \frac{d\Pr[\tau \le s]}{ds} ds$$
$$= \int_0^t \Pr[R(t - s) > x] dF_\tau(s) + 1 - F_\tau(x + t)$$

This relation is an instance of the general renewal equation (8.11). Since $1 - F_\tau(x + t)$ is monotonously decreasing, for all x, it holds with (2.35) that

$$\int_0^\infty (1 - F_\tau(x + t)) \, dt \le \int_0^\infty (1 - F_\tau(t)) \, dt = E[\tau] < \infty$$

which also implies that $\lim_{t \to \infty} 1 - F_\tau(x + t) = 0$. Hence, $h(t) = 1 - F_\tau(x + t)$ is bounded for all $t \ge 0$ and Lemma 8.1.1 is applicable, yielding

$$\Pr[R(t) > x] = 1 - F_\tau(x + t) + \int_0^t [1 - F_\tau(x + t - s)] \, dm(s)$$

Also, the conditions for direct Riemann integrability in the Key Renewal Theorem 8.2.2 for $g(t) = 1 - F_\tau(x + t)$ are satisfied such that

$$
\lim_{t \to \infty} \Pr[R(t) > x] = \lim_{t \to \infty} \int_0^t [1 - F_\tau(x + t - s)] \, dm(s)
$$

$$
= \frac{1}{E[\tau]} \int_0^\infty (1 - F_\tau(x + t)) \, dt \qquad \text{with (8.14)}
$$

$$
= \frac{1}{E[\tau]} \int_x^\infty (1 - F_\tau(t)) \, dt
$$

In other words, the steady-state or equilibrium distribution function for the residual waiting time equals

$$
\lim_{t \to \infty} \Pr[R(t) \le x] = \Pr[R \le x] = F_R(x) = \frac{1}{E[\tau]} \int_0^x (1 - F_\tau(t)) \, dt \quad (8.17)
$$

Similarly, for $t > y$, the event $\{A(t) > y\}$ is equivalent to the event $\{$no renewals in $[t - y, t]\}$, which is equivalent to $\{R(t - y) > y\}$. Hence,

$$
\lim_{t \to \infty} \Pr[A(t) > y] = \lim_{t \to \infty} \Pr[R(t - y) > y] = \lim_{t \to \infty} \Pr[R(t) > y]
$$

$$
= \frac{1}{E[\tau]} \int_y^\infty (1 - F_\tau(t)) \, dt
$$

or, both the residual waiting time R and the age A have the same distribution in steady state $(t \to \infty)$. Intuitively, when reversing the time axis in steady state or looking backward in time, an identically distributed renewal process is observed in which the role of the age A and the residual life R are interchanged. Thus, by a time symmetry argument, both distributions must be the same in steady state.

It is instructive to compute the average residual waiting time $E[R] = E[A]$ in steady state. Using the expression of the average in terms of tail probabilities (2.35), we have

$$
E[R] = \int_0^\infty (1 - F_R(x)) \, dx
$$

$$
= \frac{1}{E[\tau]} \int_0^\infty dx \int_x^\infty (1 - F_\tau(t)) \, dt
$$

Reversing the order of the x- and t-integration yields

$$
E[R] = \frac{1}{E[\tau]} \int_0^\infty dt \, (1 - F_\tau(t)) \int_0^t dx
$$

$$
= \frac{1}{E[\tau]} \int_0^\infty t \, (1 - F_\tau(t)) \, dt
$$

After partial integration, we end up with

$$E\left[R\right] = \frac{1}{2E\left[\tau\right]} \int_0^\infty t^2 f_\tau(t)dt = \frac{E\left[\tau^2\right]}{2E\left[\tau\right]} = \frac{\mathrm{Var}[\tau] + (E\left[\tau\right])^2}{2E\left[\tau\right]}$$

or

$$E\left[R\right] = \frac{E\left[\tau\right]}{2} + \frac{\mathrm{Var}[\tau]}{2E\left[\tau\right]} \tag{8.18}$$

This expression shows that the average remaining waiting time equals half of the average interarrival time plus the ratio of the variance over the mean of the interarrival time. The last term is always positive. Since $E\left[A\right] = E\left[R\right]$ and $E\left[L\right] = E\left[A\right] + E\left[R\right]$, we observe the curious result that

$$E\left[L\right] = E\left[\tau\right] + \frac{\mathrm{Var}[\tau]}{E\left[\tau\right]} \geq E\left[\tau\right]$$

or that the average total waiting $E\left[L\right]$ is longer than the average interarrival time $E\left[\tau\right]$, contrary to intuition. This fact is referred to as the *inspection paradox*: the steady-state interrenewal time, $L(t) = W_{N(t)+1} - W_{N(t)}$, containing the inspection point at time t, exceeds on average the generic interarrival time, say W_1. The explanation is that the inspection point at time t is uniformly chosen over the time axis and every inspection point is thus equally likely. The chance that the inspection point t lies in a renewal interval is proportional to the length of that interval. Hence, it has higher probability to fall in a long interval, which explains[10] why $E\left[L\right] \geq E\left[\tau\right]$. Only for deterministic interarrival times where $\mathrm{Var}[\tau] = 0$ holds the equality sign, $E\left[L\right] = E\left[\tau\right]$. For exponential interarrival times, application of (3.18) gives $\mathrm{Var}[\tau] = (E\left[\tau\right])^2$ and $E\left[R\right] = E\left[\tau\right]$ while $E\left[L\right] = 2E\left[\tau\right]$: the fact of being inspected at time t changes the lifetime distribution and even doubles the expected total life time for exponentially distributed failure or interoccurrence times.

Returning to the initial question, we observe that the intuitive result that my waiting time $E\left[R\right] = \frac{E[\tau]}{2}$ is only correct for deterministic processes. Thus, the variability in the interarrival process causes the paradox. We will see later, in queueing theory in Section 14.3.1, that also in queueing systems the variability in service discipline causes the average waiting time to increase. At last, Feller (1971, p. 187) remarks that an apparently unbiased inspection plan may lead to false conclusions because the actual observations are not typical of the population as a whole. When people complain that buses or trains start running irregularly, the inspection paradox shows

---

[10] A similar type of reasoning is used in the computation of the waiting of the GI/D/m queueing system in Section 14.4.2.

that above-average interarrival times are experienced more often. The inspection paradox thus implies that complaints may be erroneously based on an overestimation of the real deviations from the regular time schedule of busses or trains.

By separating each renewal interval into two non-overlapping subintervals $A(t)$ and $R(t)$, we have described an alternating renewal process. An alternating renewal process models a system that can be in on- or off-period with a repeating pattern $X_1, Y_1, X_2, Y_2, \ldots$ where each on-period X_n has a same distribution F_{on} and is followed by an off-period Y_n. Each off-period has also a same distribution F_{off}. The off-period Y_n may dependent on the on-period X_n, but the n-th renewal cycle with duration $X_n + Y_n$ is independent of any other cycle. An alternating renewal process can be used to model a data stream of packets, where the on-period reflects the time to store or process an arriving packet and the off-period a (random) delay between two packets. Another example is the modeling of the end-to-end delay from a source s to a destination d in the Internet, where the off-period describes a queueing in a router due to other interfering traffic along that path. During the on-period, a packet is not blocked by other packets. The on-period equals the propagation delay to travel from the output port of one router to the output port of the next-hop router. The end-to-end delay along a path with h hops equals the sum of h consecutive off-periods augmented by the propagation time from s to d.

8.4 The renewal reward process

The renewal reward process associates at each renewal at time W_n a certain cost or reward R_n, which may vary over time and can be negative. For example, each time a light bulb fails, it must be replaced at a certain cost (negative reward) or each customer in a restaurant pays for his meal (positive reward). The reward R_n may depend on the interarrival time τ_n or length of the n-th renewal interval, but it is independent of other renewal epochs (different from the n-th). Thus, the pairs (R_n, τ_n) are assumed to be independent and identically distributed. Most often one is interested in the total reward $R(t)$ over a period t (not to be confused with the residual life time) defined as

$$R(t) = \sum_{n=1}^{N(t)} R_n \qquad (8.19)$$

In this setting, the renewal reward process is a generalization of the counting process where $R_n = 1$.

By slightly rewriting the total reward $R(t)$ earned over an interval t as

$$\frac{R(t)}{t} = \frac{\sum_{n=1}^{N(t)} R_n}{N(t)} \frac{N(t)}{t}$$

and taking the limit $t \to \infty$, the first fraction tends with probability one to the average reward $E[R]$ per renewal period by the Strong Law of Large Numbers (Theorem 6.2.2), while the second fraction tends to $\frac{1}{\mu} = \frac{1}{E[\tau]}$ by the Elementary Renewal Theorem 8.2.1. Hence, with probability one holds that

$$\lim_{t\to\infty} \frac{R(t)}{t} = \frac{E[R]}{E[\tau]} \tag{8.20}$$

which means that the time average reward rate equals the average award per renewal period multiplied by the interarrival rate of renewals (or divided by the average length of a renewal interval).

Similarly as in the proof of the Elementary Renewal Theorem 8.2.1, the inequality for any t,

$$\sum_{n=1}^{N(t)} R_n \le R(t) \le \sum_{n=1}^{N(t)} R_n + R_{N(t)+1}$$

leads, after taking the expectations and using Wald's identity (2.69), to an inequality for the averages,

$$E[N(t)]\,E[R] \le E[R(t)] \le E[N(t)]\,E[R] + E\left[R_{N(t)+1}\right]$$

Dividing by t, the limit $t \to \infty$ becomes

$$E[R]\lim_{t\to\infty} \frac{E[N(t)]}{t} \le \lim_{t\to\infty} \frac{E[R(t)]}{t} \le E[R]\lim_{t\to\infty} \frac{E[N(t)]}{t} + \lim_{t\to\infty} \frac{E\left[R_{N(t)+1}\right]}{t}$$

Since the average reward per renewal period is finite and $E\left[R_{N(t)+1}\right] = E[R]$, we obtain by the Elementary Renewal Theorem 8.2.1 that

$$\lim_{t\to\infty} \frac{E[R(t)]}{t} = \frac{E[R]}{E[\tau]} \tag{8.21}$$

Hence, by comparing (8.21) and (8.20), the time average of the average reward rate equals the time average of the reward rate.

Example The hard disc in a network server is replaced at cost C_1 at time T. The lifetime or age of this mass storage has pdf f_A. If the hard disc fails earlier, the cost of the repair and the penalties for service disruption is C_2. What is the long run cost of the hard disc in the server per unit time?

Consider the replacement of hard discs as a renewal process with i.i.d. interarrival times τ and with distribution

$$\Pr[\tau < t] = \begin{cases} \int_0^t f_A(u)du & \text{if } t < T \\ 1 & \text{if } t \geq T \end{cases}$$

The average replacement time follows from the tail expression (2.35),

$$E[\tau] = \int_0^\infty (1 - \Pr[\tau < t])\, dt = \int_0^T (1 - \Pr[\tau < t])\, dt$$

The replacement cost $C(\tau)$ equals $C(\tau) = C_1 1_{\tau < T} + C_2 1_{\tau \geq T}$ and the average cost is with (2.13),

$$E[C] = C_1 \Pr[\tau < T] + C_2 (1 - \Pr[\tau < T])$$

The Elementary Renewal Reward Theorem (8.21) and (8.20) states that the long-run cost of replacements equals

$$\frac{E[C]}{E[\tau]} = \frac{C_1 \Pr[\tau < T] + C_2 (1 - \Pr[\tau < T])}{\int_0^T (1 - \Pr[\tau < t])\, dt}$$

Usually the replacement time T is chosen to minimize this long-run cost.

8.5 Problems

More worked examples can be found in Karlin and Taylor (1975, Chapter 5).

(i) Calculate $\Pr\left[W_{N(t)} \leq x\right]$.
(ii) Derive a recursion equation for the generating function $\varphi_{N(t)}(z) = E\left[z^{N(t)}\right]$ of the number of renewals in the interval $[0, t]$ and deduce from that equation the renewal equation (8.9) and a relation for $\text{Var}[N(t)]$.
(iii) In a TCP session from A to B, IP data packets and IP acknowledgement packets travel a distance of 2000 km over precisely the same bi-directional path. In case of congestion, the average speed is 40000 km/s and without congestion the speed is three times higher. Congestion only occurs in 20% of the travels. What is the average speed of IP packets in the TCP session?
(iv) The production of digitalized speech samples depends primarily on the codec, with an effective average rate r (bits/s). Since this rate is low compared to the ATM capacity C (bits/s), UMTS will use AAL2 mini-cells in which 1 ATM cell is occupied by N users. The financial cost of an UMTS operator increases at nc euro per unit time whenever

there are $n < N$ speech samples are waiting for transmission and an additional cost of K euro each time an ATM cell is transmitted. What is the average cost per unit time for the UMTS operator?

(v) The cost of replacing a router that has failed is A euro. However, one can decide to replace a router that has been in service for a period of time T. The advantage of this approach is that the cost of replacing a working router is only B euro, where $B < A$. The policy CHANGEROUTER consists of replacing a router either upon failure or upon reaching the age T, whichever occurs first. Replacement of the current router by a new one occurs instantaneously and at each time there can only be one router in the network. Let $\{X_j\}$ be a sequence of i.i.d. random variables, where X_j is the lifetime of a router j.

(a) Find the time average cost rate C of the policy CHANGE-ROUTER.

(b) Compute C if $T = 5$ years, the cost of replacing the failed router is $A = 10000$ euro and the cost of replacing a working router is $B = 7000$ euro. The independent random variables X_j are exponentially distributed and the average lifetime of a router is 10 years.

9

Discrete-time Markov chains

A large number of stochastic processes belong to the important class of Markov processes. The theory of Markov chains and Markov processes is well established and furnishes powerful tools to solve practical problems. This chapter will be mainly devoted to the theory of discrete-time Markov chains, while the next chapter concentrates on continuous time Markov chains. The theory of Markov processes will be applied in later chapters to compute or formulate queueing and routing problems.

9.1 Definition

A stochastic process $\{X(t), t \in T\}$ is a Markov process if the future state of the process only depends on the current state of the process and not on its past history. Formally, a stochastic process $\{X(t), t \in T\}$ is a continuous time Markov process if for all $t_0 < t_1 < \cdots < t_{n+1}$ of the index set T and for any set $\{x_0, x_1, \ldots, x_{n+1}\}$ of the state space it holds that

$$\Pr[X(t_{n+1}) = x_{n+1} | X(t_0) = x_0, \ldots, X(t_n) = x_n] = \Pr[X(t_{n+1}) = x_{n+1} | X(t_n) = x_n]$$
(9.1)

Similarly, a discrete-time Markov chain $\{X_k, k \in T\}$ is a stochastic process whose state space is a finite or countably infinite set with index set $T = \{0, 1, 2, \ldots\}$ obeying

$$\Pr[X_{k+1} = x_{k+1} | X_0 = x_0, \ldots, X_k = x_k] = \Pr[X_{k+1} = x_{k+1} | X_k = x_k]$$
(9.2)

A Markov process is called a Markov chain if its state space is discrete. The conditional probabilities $\Pr[X_{k+1} = j | X_k = i]$ are called the transition probabilities of the Markov chain. In general, these transition probabilities can depend on the (discrete) time k. A Markov chain is entirely defined by the transition probabilities (9.2) and the initial distribution of the Markov chain

$\Pr[X_0 = x_0]$. Indeed, by the definition of conditional probability (2.45), we obtain

$$\Pr[X_0 = x_0, \ldots, X_k = x_k] = \Pr[X_k = x_k | X_0 = x_0, \ldots, X_{k-1} = x_{k-1}]$$
$$\times \Pr[X_0 = x_0, \ldots, X_{k-1} = x_{k-1}]$$

and, by the definition of the Markov chain (9.2),

$$\Pr[X_0 = x_0, \ldots, X_k = x_k] = \Pr[X_k = x_k | X_{k-1} = x_{k-1}]$$
$$\times \Pr[X_0 = x_0, \ldots, X_{k-1} = x_{k-1}]$$

This recursion relation can be iterated resulting in

$$\Pr[X_0 = x_0, \ldots, X_k = x_k] = \prod_{j=1}^{k} \Pr[X_j = x_j | X_{j-1} = x_{j-1}] \Pr[X_0 = x_0]$$

$$(9.3)$$

which demonstrates that the complete information of the Markov chain is obtained if, apart from the initial distribution, all time depending transition probabilities are known.

9.2 Discrete-time Markov chain

If the transition probabilities are independent of time k,

$$P_{ij} = \Pr[X_{k+1} = j | X_k = i] \qquad (9.4)$$

the Markov chain is called stationary. In the sequel, we will confine ourselves to stationary Markov chains. Since the discrete-time Markov chain is conceptually simpler than the continuous counterpart, we start the discussion with the discrete case.

Let us consider a state space S with N states (where $N = \dim S$ can be infinite). It is convenient to introduce a vector notation[1]. Since X_k can only take N possible values, we denote the corresponding state vector at discrete-time k by $s[k] = [s_1[k] \; s_2[k] \; \cdots \; s_N[k]]$ with $s_i[k] = \Pr[X_k = i]$. Hence, $s[k]$ is a $1 \times N$ vector. Since the state X_k at discrete-time k must be in one of the N possible states, we have that $\sum_{i=1}^{N} \Pr[X_k = i] = 1$ or, in vector notation, $s[k].u = \sum_{i=1}^{N} 1.s_i[k] = 1$, where $u^T = [1 \; 1 \; \cdots \; 1]$. This fact is also written as $\|s[k]\|_1 = 1$, where $\|a\|_1$ is the $q = 1$ norm of vector a

---

[1] Unfortunately, a vector in Markov theory is represented as a single row matrix which deviates from the general theory in linear algebra, followed in Appendix A, where a vector is represented as a single column matrix. In order to be consistent with the literature on Markov processes, we have chosen to follow the notation of Markov theory here, but elsewhere we adhere to the general convention of linear algebra.

defined in the Appendix A.3. In a stationary Markov chain, the states X_{k+1} and X_k are connected via the law of total probability (2.46),

$$\Pr[X_{k+1} = j] = \sum_{i=1}^{N} \Pr[X_{k+1} = j|X_k = i]\Pr[X_k = i]$$

$$= \sum_{i=1}^{N} P_{ij}\Pr[X_k = i] \tag{9.5}$$

which holds for all j, or, in vector notation,

$$s[k+1] = s[k]P \tag{9.6}$$

where the transition probability matrix P is

$$P = \begin{bmatrix} P_{11} & P_{12} & P_{13} & \cdots & P_{1;N-1} & P_{1N} \\ P_{21} & P_{22} & P_{23} & \cdots & P_{2;N-1} & P_{2N} \\ P_{31} & P_{32} & P_{33} & \cdots & P_{3;N-1} & P_{3N} \\ \vdots & \vdots & \vdots & \cdots & \vdots & \vdots \\ P_{N-1;1} & P_{N-1;2} & P_{N-1;3} & \cdots & P_{N-1;N-1} & P_{N-1;N} \\ P_{N1} & P_{N;2} & P_{N3} & \cdots & P_{N;N-1} & P_{NN} \end{bmatrix} \tag{9.7}$$

Since (9.6) must hold for any initial state vector $s[0]$, by choosing $s[0]$ equal to a base vector $[0 \cdots 0\ 1\ 0 \cdots\ 0]$ (all columns zero except for column i) which expresses that the Markov chain starts from one of the possible states, say state i, then $s[1] = [P_{i1}\ P_{i2}\ \cdots\ P_{iN}]$. Furthermore, since $\|s[k]\|_1 = 1$ for any k, it must hold that $\sum_{j=1}^{N} P_{ij} = 1$ for any state i. The relation

$$\sum_{j=1}^{N} P_{ij} = 1 \tag{9.8}$$

means that, at discrete-time k, there certainly occurs a transition in the Markov chain, possibly to the same state as at time $k-1$. The $N \times N$ transition probability matrix P thus consists of $N^2 - N$ transition probabilities P_{ij} and at each row, one transition probability can be expressed in terms of the others, e.g. $P_{ik} = 1 - \sum_{j=1;j\neq k}^{N} P_{ij}$. A matrix with elements $0 \leq P_{ij} \leq 1$ obeying (9.8) is called a *stochastic matrix* whose properties are investigated in Appendix A. Apart from the matrix representation, Markov chains are often described by a directed graph (as illustrated in the figure below), where P_{ij} is represented by an edge from state i to j provided $P_{ij} > 0$. Especially, this feature enables to deduce structural properties of the Markov chain (such as e.g. communicating states) elegantly.

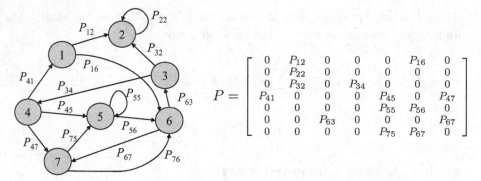

Given the initial state vector $s[0]$, the general solution of (9.6) is

$$s[k] = s[0]P^k \tag{9.9}$$

Similarly, when knowledge of the Markov chain at discrete-time k is available, we obtain from (9.6) that

$$s[k+n] = s[k]P^n$$

The elements of the matrix P^n are called the n-step transition probabilities,

$$P_{ij}^n = \Pr\left[X_{k+n} = j | X_k = i\right] \tag{9.10}$$

for $k \geq 0$ and $n \geq 0$. Since the discrete Markov chain must be surely in one of the N states n time units later given that it started at time k in state i, we obtain an extension of (9.8), for all $n \geq 1$,

$$\sum_{j=1}^{N} P_{ij}^n = 1 \tag{9.11}$$

The demonstration of (9.11) is by induction. If $n = 1$, (9.8) justifies (9.11). Assume that (9.11) holds. Any matrix element of P^{n+1} can be written as $P_{ij}^{n+1} = \sum_{k=1}^{N} P_{ik}P_{kj}^n$. Summing over all j, yields for $n \geq 0$,

$$\sum_{j=1}^{N} P_{ij}^{n+1} = \sum_{k=1}^{N} P_{ik} \sum_{j=1}^{N} P_{kj}^n$$

$$= \sum_{k=1}^{N} P_{ik} \qquad \text{(induction argument)}$$

$$= 1 \qquad \qquad (n = 1 \text{ case})$$

This proves (9.11).

9.2.1 Definitions and classification

9.2.1.1 Irreducible Markov chains

A state j in a Markov chain is said to be *reachable* from state i if it is possible to proceed from state i to state j in a finite number of transitions which is equivalent to $P_{ij}^n > 0$ for finite n. If every state is reachable from every other state, the Markov chain is said to be *irreducible*. The example of the Markov graph above is not irreducible because state 2 is absorbing. Markov theory is considerably more simplified if we know that the chain is irreducible, which justifies to investigate methods to determine irreducibility.

An equivalent requirement for the Markov chain to be irreducible is that the associated directed graph is strongly connected, i.e. if there is a path from node i to node j for any pair of distinct nodes (i, j). Let us review some basic notions from graph theory (see Appendix B.1). Denote by A the adjacency matrix of P where all non-zero elements in P are replaced by 1. A walk of length k from state i to state j is a succession of k arcs of the form $(n_0 \rightarrow n_1)(n_1 \rightarrow n_2) \cdots (n_{k-1} \rightarrow n_k)$, where $n_0 = i$ and $n_k = j$. A path is a walk in which all nodes are different, i.e. $n_l \neq n_m$ for all $0 \leq l \neq m \leq k$. Lemma B.1.1 (proved in Appendix B.1, art. **5**) states that the number of walks of length k from state i to state j is equal to the element $\left(A^k\right)_{ij}$.

A directed graph is strongly connected if and only if each non-diagonal element of the matrix $\sum_{k=1}^{N-1} P^k$ or, equivalent, of $B = \sum_{k=1}^{N-1} A^k$ is positive. Since P has N states, the longest possible path between two states consists of $N - 1$ hops. By summing over all powers of $1 \leq k \leq N - 1$, the element b_{ij} of the matrix B equals the number of all possible walks (of any possible length) between i and j. Hence, if $b_{ij} > 0$ for all $i \neq j$, there exists walks from any state i to any other state j. The converse is readily verified.

Another way to determine irreducibility follows from the definition of reducibility in the Appendix A.4. However, the methods for strongly connectivity or irreducibility are still algebraic in that they require matrix operations. A computationally more efficient method consists of applying all-pair shortest path algorithms on the Markov graph. Examples of all-pair shortest path algorithms are that of Floyd-Warshall (with computational complexity $C_{\text{Floyd-Warschall}} = O(N^3)$) or the algorithm of Johnson (complexity $C_{\text{Johnson}} = O(N^2 \log N + NL)$, where L is the total number of links in the Markov graph). These algorithms are nicely discussed in Cormen *et al.* (1991).

9.2.1.2 Communicating states

If two states i and j are reachable from one to each other, they are said to *communicate*, which is often denoted by $i \longleftrightarrow j$.

The concept of communication is an equivalence relation: (a) reflexivity: $i \longleftrightarrow i$ since $P^0 = I$ or $P_{ij} = \delta_{ij}$. (b) symmetry: $i \longleftrightarrow j$ then $j \longleftrightarrow i$ which follows from the definition of communication and (c) transitivity: if $i \longleftrightarrow j$ and $j \longleftrightarrow k$ then $i \longleftrightarrow k$. The transitivity follows from the non-negativity of P and $P^{n+m} = P^n P^m$ such that the matrix element $P_{ik}^{n+m} = \sum_{l=1}^{N} P_{il}^n P_{lk}^m \geq P_{ij}^n P_{jk}^m$. By definition of $i \longleftrightarrow j$ and $j \longleftrightarrow k$, we have $P_{ij}^n > 0$ and $P_{jk}^m > 0$ for some finite n and m. Hence, $P_{ik}^{n+m} > 0$, which implies $i \longleftrightarrow k$. As an application, the total state space can be partitioned into equivalence classes. States in one equivalence class communicate with each other. If there is a possibility to start in one class and to enter another class in which case there is no return possible to the first class (otherwise the two classes would form one class), the Markov chain is reducible. In other words, a Markov chain is irreducible if the equivalence relations result into one class.

9.2.1.3 Periodic and aperiodic Markov chains

Consider a state j in a Markov chain with $P_{jj}^n > 0$ for some $n \geq 1$. The *period* d_j is defined as the greatest common divisor of those n for which $P_{jj}^n > 0$. The figure below illustrates a Markov chain with period $d = N$.

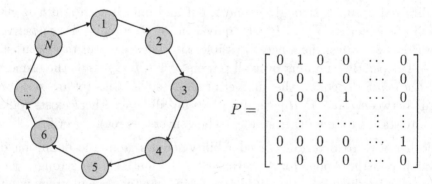

$$P = \begin{bmatrix} 0 & 1 & 0 & 0 & \cdots & 0 \\ 0 & 0 & 1 & 0 & \cdots & 0 \\ 0 & 0 & 0 & 1 & \cdots & 0 \\ \vdots & \vdots & \vdots & & \ddots & \vdots & \vdots \\ 0 & 0 & 0 & 0 & \cdots & 1 \\ 1 & 0 & 0 & 0 & \cdots & 0 \end{bmatrix}$$

Since the greatest common divisor of a set is the largest integer d that divides any integer in a set, it is smaller than the minimum element in the set. Thus,

$$1 \leq d_j \leq \min_n \{P_{jj}^n > 0\}$$

The relation $P_{jj}^{n+m} = P_{jj}^n P_{jj}^m + \sum_{l=1;l\neq j}^{N} P_{jl}^n P_{lj}^m$ deduced from matrix multiplication and the fact that all elements in P are non-negative shows that

P_{jj}^k cannot decrease with increasing $k = n + m$. Hence, if $P_{jj} > 0$, then all $P_{jj}^k > 0$ for $k > 1$ and thus $d_j = 1$.

Lemma 9.2.1 *If two states i and j communicate ($i \longleftrightarrow j$), then $d_i = d_j$.*

Proof: Let n and m be integers such that $P_{ij}^n > 0$ and $P_{ji}^m > 0$. From $P^{n+m} = P^n P^m$, the matrix element $P_{ii}^{n+m} = \sum_{k=1}^{N} P_{ik}^n P_{ki}^m \geq P_{ij}^n P_{ji}^m$. By definition of n and m, $P_{ii}^{n+m} > 0$, and, by definition of a period, $d_i | (n+m)$. Similarly, from $P^{n+l+m} = P^n P^{l+m}$, the matrix element

$$P_{ii}^{n+l+m} = \sum_{r=1}^{N} P_{ir}^n \sum_{k=1}^{N} P_{rk}^l P_{ki}^m \geq P_{ij}^n P_{jj}^l P_{ji}^m \tag{9.12}$$

Now, if $P_{jj}^l > 0$, which implies by definition that $d_j | l$, then we also have that $P_{ii}^{n+l+m} > 0$ from which $d_i | (n + m + l)$. Both conditions $d_i | (n + m)$ and $d_i | (n + m + l)$ imply that $d_i | l$. But since d_j is the largest such divisor $d_j \geq d_i$. By symmetry of the communication relation (replace $i \to j$ and $j \to i$), $d_i \geq d_j$ which proves the lemma. $\square$

The consequence of Lemma 9.2.1 is that all the states in an irreducible Markov chain have common period d. The irreducible Markov chain is periodic with period d if $d > 1$ else it is aperiodic ($d = 1$). A simple sufficient condition for an irreducible chain to be aperiodic is that $P_{ii} > 0$ for some state i. Most Markov chains of practical interest are aperiodic.

9.2.2 The hitting time

Let A be a subset of states, $A \subset S$. The hitting time T_A is the first positive time the Markov chain is in a state of the set A, thus for $k \geq 0$, $T_A = \min(k : X_k \in A)$. The hitting time[2] of a state j follows from the definition if $A = \{j\}$. For irreducible Markov chains, the hitting time T_j is finite, for any state j.

From the definition of the hitting time, the recursion

$$\Pr[T_j = m | X_0 = i] = \sum_{k \neq j} \Pr[X_1 = k | X_0 = i] \Pr[T_j = m - 1 | X_0 = k]$$

is immediate. Indeed, in order to have the transition from state i to state j at discrete-time m, it is necessary to have first a transition from state i to some other state k and to pass from that state k to state j for the first time

---

[2] The hitting time T_j is also called the first passage time into a state j.

after $m - 1$ time units. For a stationary Markov chain, we have for $m > 0$ that

$$\Pr\left[T_j = m | X_0 = i\right] = \sum_{k \neq j} P_{ik} \Pr\left[T_j = m - 1 | X_0 = k\right] \qquad (9.13)$$

and, by definition for $m \geq 0$,

$$\Pr\left[T_j = m | X_0 = j\right] = \delta_{0m} \qquad (9.14)$$

The event $\{X_n = j\}$ can be decomposed in terms of the hitting time T_j. Indeed, since the events $\{T_j = m, X_n = j\}$ are disjointed for $1 \leq m \leq n$,

$$\{X_n = j\} = \cup_{m=1}^{n}\{T_j = m, X_n = j\}$$

Applied to the n-step transition probabilities,

$$\begin{aligned}
\Pr\left[X_n = j | X_0 = i\right] &= \Pr\left[\cup_{m=1}^{n}\{T_j = m, X_n = j\} | X_0 = i\right] \\
&= \sum_{m=1}^{n} \Pr\left[T_j = m, X_n = j | X_0 = i\right] \\
&= \sum_{m=1}^{n} \Pr\left[T_j = m | X_0 = i\right] \Pr\left[X_n = j | X_0 = i, T_j = m\right]
\end{aligned}$$

By definition of the hitting time, $\{T_j = m\} = \cup_{k=1}^{m-1}\{X_k \neq j\}\{X_m = j\}$ such that

$$\begin{aligned}
\Pr\left[X_n = j | X_0 = i, T_j = m\right] &= \Pr\left[X_n = j | X_0 = i, \cup_{k=1}^{m-1}\{X_k \neq j\}\{X_m = j\}\right] \\
&= \Pr\left[X_n = j | X_m = j\right]
\end{aligned}$$

where the last step follows from the Markov property (9.2). Thus we obtain

$$\Pr\left[X_n = j | X_0 = i\right] = \sum_{m=1}^{n} \Pr\left[T_j = m | X_0 = i\right] \Pr\left[X_n = j | X_m = j\right]$$

or, written in terms of n-step transition probabilities with (9.10),

$$P_{ij}^{n} = \sum_{m=1}^{n} \Pr\left[T_j = m | X_0 = i\right] P_{jj}^{n-m} \qquad (9.15)$$

For an absorbing state j where $P_{jj} = 1$, relation (9.15) simplifies to

$$P_{ij}^{n} = \Pr\left[T_j \leq n | X_0 = i\right] \qquad (9.16)$$

9.2.3 Transient and recurrent states

The probability that a Markov chain initiated at state i will ever come into state j is denoted as

$$r_{ij} = \Pr\left[T_j < \infty | X_0 = i\right] \tag{9.17}$$

If the starting state i equals the target state j, then r_{ii} is the probability of ever returning to state i. If $r_{ii} = 1$, the state i is a recurrent state, while, if $r_{ii} < 1$, state i is a transient state. If i is a recurrent state, the Markov chain started at i will definitely (i.e. with probability 1) return to state i after some time. On the other hand, if i is a transient state, the Markov chain started at i has probability $1 - r_{ii}$ of never returning to state i. For an absorbing state i defined by $P_{ii} = 1$, we have by (9.16) that $r_{ij} = 1$, implying that an absorbing state is a recurrent state. Further, the mean return time to state j when the chain started in j is denoted by

$$m_j = E\left[T_j < \infty | X_0 = j\right] \tag{9.18}$$

Concepts of renewal theory will now be applied to a Markov process. Let $N_k(j)$ denote the number of times that the Markov chain is in state j during the time interval $[1, k]$ given the chain started in state i or $N_k(j) = \sum_{n=1}^{k} 1_{\{X_n=j|X_0=i\}}$. Using (2.13), the average number of visits to state j in the time interval $[1, k]$ is

$$E\left[N_k(j)|X_0 = i\right] = E\left[\sum_{n=1}^{k} 1_{\{X_n=j\}} | X_0 = i\right] = \sum_{n=1}^{k} E\left[1_{\{X_n=j\}} | X_0 = i\right]$$

$$= \sum_{n=1}^{k} \Pr\left[X_n = j | X_0 = i\right]$$

or, in terms of n-step transition probabilities (9.10),

$$E\left[N_k(j)|X_0 = i\right] = \sum_{n=1}^{k} P_{ij}^n \tag{9.19}$$

The average number of times that the Markov chain is ever in state j given that it started from state i, is with $N(j) = \lim_{k \to \infty} N_k(j)$,

$$E\left[N(j)|X_0 = i\right] = \sum_{n=1}^{\infty} \Pr\left[X_n = j | X_0 = i\right] = \sum_{n=1}^{\infty} P_{ij}^n$$

Hence, if state j is reachable from state i, by definition, there is some k for which $P_{ij}^k > 0$, which implies that $E\left[N(j)|X_0 = i\right] > 0$. Further, consider the probability $\Pr\left[N(j) \geq n | X_0 = i\right]$ that the number of returns to state j exceeds n, given the Markov chain started from state i. The event $\{N(j) \geq n\}$ is equivalent to the occurrence of the events $\{N(j) \geq n-1\}$ and the event that the Markov chain will return to j again given that it started from j. The probability of the latter event is precisely r_{jj}. Thus, we obtain the recursion

$$\Pr\left[N(j) \geq n | X_0 = i\right] = r_{jj} \Pr\left[N(j) \geq n-1 | X_0 = i\right]$$

with solution for $n \geq 1$,

$$\Pr\left[N(j) \geq n | X_0 = i\right] = (r_{jj})^{n-1} \Pr\left[N(j) \geq 1 | X_0 = i\right]$$

Now, $\Pr\left[N(j) \geq 1 | X_0 = i\right] = \Pr\left[T_j < \infty | X_0 = i\right] = r_{ij}$, such that

$$\Pr\left[N(j) \geq n | X_0 = i\right] = (r_{jj})^{n-1} r_{ij} \qquad (9.20)$$

The average computed with (2.36) yields,

$$E\left[N(j) | X_0 = i\right] = \frac{r_{ij}}{1 - r_{jj}} = \sum_{n=1}^{\infty} P_{ij}^n \qquad (9.21)$$

provided $r_{ij} > 0$. If $r_{ij} = 0$ then (9.20) vanishes for every n and thus $E\left[N(j) | X_0 = i\right] = 0$, which means that state j is not reachable from state i. In summary:

- For a recurrent state j for which $r_{jj} = 1$, we obtain from (9.21) that $E\left[N(j) | X_0 = i\right] \to \infty$ (if $r_{ij} \neq 0$ else $E\left[N(j) | X_0 = i\right] = 0$) and from (9.20),

$$\Pr\left[N(j) = \infty | X_0 = i\right] = \lim_{n \to \infty} \Pr\left[N(j) \geq n | X_0 = i\right] = r_{ij}$$

  A state j is recurrent if and only if $\sum_{n=1}^{\infty} P_{jj}^n \to \infty$.
- For a transient state j for which $r_{jj} < 1$, there holds that $E\left[N(j) | X_0 = i\right]$ will be finite and $\Pr\left[N(j) = \infty | X_0 = i\right] = 0$ or, equivalently,

$$\Pr\left[N(j) < \infty | X_0 = i\right] = 1$$

  A state j is transient if and only if $\sum_{n=1}^{\infty} P_{jj}^n$ is finite.

These relations explain the difference between a recurrent and a transient state. When the Markov chain starts at a recurrent state, it returns infinitely often to that state because $r_{jj} = \Pr\left[N(j) = \infty | X_0 = j\right] = 1$. If the chain starts at some other state i that is reachable from state j ($r_{ij} > 0$), then the chain will visit state j infinitely often. From this analysis, some consequences arise.

Corollary 9.2.2 *A finite-state Markov chain must have at least one recurrent state.*

Proof: Suppose the contrary that, if the state space S is finite, all states are transient states. For a transient state j it follows from (9.21) that $\sum_{k=1}^{\infty} P_{ij}^k$ is finite, which implies that $\lim_{k\to\infty} P_{ij}^k = 0$ for any other state i. If the state space is finite and all states are transient states, then

$$\sum_{j\in S} \lim_{k\to\infty} P_{ij}^k = 0$$

Since the summation has a finite number of terms, the limit and summation operator can be reversed,

$$\lim_{k\to\infty} \sum_{j\in S} P_{ij}^k = 0$$

But the total law of probability (9.11) requires that $\sum_{j\in S} P_{ij}^k = 1$ (for any time k), which leads to a contradiction. $\square$

Theorem 9.2.3 *If i is a recurrent state that leads to a state j, then the state j is also a recurrent state and $r_{ij} = r_{ji} = 1$.*

Proof: Clearly, the theorem is true if $i = j$. Suppose that, for $i \neq j$, $\Pr[T_i < \infty | X_0 = j] = r_{ji} < 1$. This implies that the Markov chain starting from state j has probability $1 - r_{ji} > 0$ of never hitting state i, which is impossible because i is a recurrent state that will be visited infinitely often. Hence $r_{ji} = 1$.

Since state $j \neq i$ is reachable from state i, by definition $r_{ij} > 0$ and there is a minimum discrete-time n such that $P_{ij}^n > 0$ and $\Pr[X_k = j | X_0 = i] = 0$ for $k < n$. Similarly, since $r_{ji} = 1$, there exists a minimum discrete-time m to have a transition from $j \to i$ given the chain started in state j, thus, $\Pr[X_m = i | X_0 = j] = P_{ji}^m > 0$. From (9.12) and the fact that $P_{ij}^n > 0$ and $P_{ji}^m > 0$ and state i is a recurrent state such that $P_{ii}^l > 0$, we have, for any $l \geq 0$,

$$P_{jj}^{n+l+m} \geq P_{ji}^m P_{ii}^l P_{ij}^n > 0$$

Summing over all l,

$$\sum_{l=1}^{\infty} P_{jj}^{n+l+m} \geq P_{ij}^n P_{ji}^m \sum_{l=1}^{\infty} P_{ii}^l$$

or

$$\sum_{l=1}^{\infty} P_{jj}^l > \sum_{l=n+m+1}^{\infty} P_{jj}^l = \sum_{l=1}^{\infty} P_{jj}^{n+l+m} \geq P_{ij}^n P_{ji}^m \sum_{l=1}^{\infty} P_{ii}^l$$

It follows from (9.21) that the right-hand side diverges. Hence, $\sum_{l=1}^{\infty} P_{jj}^l$ diverges and relation (9.21) indicates that $r_{jj} = 1$ or, that j must be a recurrent state. $\quad\square$

A non-empty set $C \subset S$ of states is said to be closed if no state $i \in C$ leads to a state $j \notin C$. Thus, $r_{ij} = 0$ for any $i \in C$ and $j \notin C$, which is equivalent to $P_{ij} = 0$. If the set C is closed, the Markov chain starting in C will remain, with probability 1, in C all the time. For example, if i is an absorbing state, $C = \{i\}$ is closed. A closed set C is irreducible if state i is reachable from state j for all $i, j \in C$. Theorem 9.2.3 together with Corollary 9.2.2 implies that, if C is a finite, irreducible closed set, all states are recurrent.

9.3 The steady-state of a Markov chain

9.3.1 The irreducible Markov chain

The steady-state vector $\pi = \lim_{k\to\infty} s[k]$ follows, after taking the limit $k \to \infty$ in (9.6), as

$$\pi = \pi.P \tag{9.22}$$

or, for each component π_j,

$$\pi_j = \sum_{k=1}^{N} P_{kj} \pi_k \tag{9.23}$$

with $\pi.u = 1$ or $\|\pi\|_1 = 1$. Equation (9.22) shows that the steady-state vector π does not depend on the initial state $s[0]$.

Alternatively, in view of (9.9), we trivially write $P^k - P^{k-1}P = 0$ or $PP^{k-1} - P^k = 0$ and if $A = \lim_{k\to\infty} P^k$ exists, then $A - AP = 0$ or $PA - A = 0$. This implies that $A(I - P) = (I - P)A = 0$ or $A = Q(1)$, where $Q(\lambda)$ is the adjoint matrix of P (see Appendix A.1, art. **7**). The non-zero columns (or rows) of the adjoint matrix $Q(\lambda)$ consist of the (unscaled) eigenvector(s) belonging to eigenvalue λ. By (9.11) the rows of P^k for any k are normalized and so must $A = Q(1)$. Since there is only one eigenvector π belonging to $\lambda = 1$ (Frobenius Theorem A.4.2), the rows of $A = \lim_{k\to\infty} P^k$ must all be the same and equal to $\pi = [a_{11} \ a_{12} \ \cdots \ a_{1N}]$. Furthermore, only if all rows of A are equal to the steady-state vector π, then the dependence on the initial state $s[0]$ vanishes since relation (9.9) becomes $\pi_j = \sum_{i=1}^{N} s_i[0] a_{ij} = a_{1j} \sum_{i=1}^{N} s_i[0] = a_{1j}$. Hence, $A = \lim_{k\to\infty} P^k = u.\pi$ or, componentwise, for all $1 \le j \le N$,

$$\lim_{k\to\infty} P_{ij}^k = \pi_j \qquad (9.24)$$

The sequence of matrices $P, P^2, P^3, \ldots, P^k$ thus converges to $A = u.\pi$ for sufficiently large k. Instead of multiplying the last matrix P^k in the sequence by P to obtain the next one P^{k+1}, with a same computational effort, the sequence $P, P^2, P^4, \ldots, P^{2^k}$, obtained by successively squaring, converges considerably faster to $A = u.\pi$ and may be useful for sparse P.

On the other hand, relation (9.22) is an eigenvalue equation with eigenvalue $\lambda = 1$ and eigenvector π. The Frobenius Theorem A.4.2 states that the transition probability matrix P has one eigenvalue $\lambda = 1$ with corresponding eigenvector π. Since in (9.22) the set $(P - I)^T \pi^T = 0$ has rank $N - 1$, the normalization condition $\|\pi\|_1 = 1$ furnishes the (last) remaining equation. Except for the trivial case where P is the identity matrix I, the solution of π is obtained from

$$
\begin{bmatrix}
P_{11}-1 & P_{21} & P_{31} & \cdots & P_{N-1;1} & P_{N1} \\
P_{12} & P_{22}-1 & P_{32} & \cdots & P_{N-1;2} & P_{N2} \\
P_{13} & P_{23} & P_{33}-1 & \cdots & P_{N-1;3} & P_{N3} \\
\vdots & \vdots & \vdots & \ddots & \vdots & \vdots \\
P_{1;N-1} & P_{2;N-1} & P_{3;N-1} & \cdots & P_{N-1;N-1}-1 & P_{N;N-1} \\
1 & 1 & 1 & \cdots & 1 & 1
\end{bmatrix}
\cdot
\begin{bmatrix}
\pi_1 \\ \pi_2 \\ \pi_3 \\ \vdots \\ \pi_{N-1} \\ \pi_N
\end{bmatrix}
=
\begin{bmatrix}
0 \\ 0 \\ 0 \\ \vdots \\ 0 \\ 1
\end{bmatrix}
$$
$$(9.25)$$

In practice, this method is used, especially if the number of states N is large and the transition probability matrix P does not exhibit a special matrix structure.

In summary, for irreducible Markov chains, there are in general two ways[3] of computing the state distribution π: via the limiting process (9.24) or via solving the set of linear equations (9.25). Recall that we have invoked the Frobenius Theorem A.4.2, which is only applicable for irreducible Markov chains. There exist cases of practical interest where (9.24) fails to hold. For example, in the two-state Markov chain studied in Section 9.3.3, there is a chain where the limit state bounces back and forth between state 0 and state 1. It is of importance to know whether the steady-state distribution π exists in the sense that $\pi_j \ne 0$ for at least one j. If $\pi_j = 0$ for all j, then there is no stationary (or equilibrium or steady-state) probability distribution.

---

[3] A third method consists of a directed graph solution of linear, algebraic equations discussed by Chen (1971, Chapter 3) and applied to the steady state equation (9.22) by Hooghiemstra and Koole (2000).

9.3.2 *The average number of visits to a recurrent state*

A direct application of Lemma 6.1.1 to the steady-state of a Markov chain is that, if (9.24) holds, then

$$\lim_{n \to \infty} \frac{1}{n} \sum_{m=1}^{n} P_{ij}^m = \pi_j$$

Invoking (9.19) where $\frac{N_n(j)}{n}$ is the fraction of the time the chain is in state j during the interval $[1, n]$, the relation is equivalent to

$$\lim_{n \to \infty} \frac{E\left[N_n(j) | X_0 = i\right]}{n} = \pi_j \tag{9.26}$$

The time average of the average number of visits to state j given the Markov chain started in state i converges to the steady-state distribution. In other words, the long run mean fraction of time that the chain spends in state j equals π_j and is independent of the initial state i. From (9.21), it immediately follows that, if j is a transient state, $\pi_j = 0$. Only recurrent states j have a non-zero probability π_j that the steady-state is in recurrent state j. Lemma 6.1.1 and its consequence (9.26) suggests to investigate $\frac{N_n(j)}{n}$ for recurrent states j.

If the Markov chain starts in a recurrent state j, we know from (9.21) that the chain returns to state j infinitely often. Let $W_k(j)$ denote the time of the k-th visit of the Markov chain to state j. Then,

$$W_k(j) = \min_{m \geq 1}(N_m(j) = k)$$

The interarrival time between the k-th and $(k-1)$-th visit is $\tau_k(j) = W_k(j) - W_{k-1}(j)$. The interarrival times $\{\tau_k(j)\}_{k \geq 1}$ are independent and identically distributed random variables as follows from the Markov property. Indeed, every time t the Markov chain returns to state j, it behaves from that time onwards as if the Markov process would have started from state j, ignoring the past before time t. Moreover, they have a common mean $E\left[\tau(j)\right] = E\left[\tau_1(j)\right]$ equal to the mean return time to j given by $E\left[T_j | X_0 = j\right] = m_j$ because the hitting time is $T_j = \tau_1(j)$. In other words, just as in renewal theory in Chapter 8, we have a counting process $\{N_m(j), m \geq 1\}$ with associated waiting times $W_k(j)$ and i.i.d. interarrival times $\tau_k(j)$, specified by the equivalence

$$\{N_m(j) < k\} \Longleftrightarrow \{W_k(j) > m\}$$

Invoking the Elementary Renewal Theorem (8.13), we obtain with $m_j =$

$E\left[T_j|X_0 = j\right]$

$$\lim_{n\to\infty} \frac{N_n(j)}{n} = \frac{1}{m_j} \tag{9.27}$$

Thus, the chain returns to state j on average every m_j time units and, hence, the fraction of time the chain is in state j is roughly $\frac{1}{m_j}$. These results are summarized as follows:

Theorem 9.3.1 (Limit Law of Markov Chains) *If j is a recurrent state and the Markov chain starts in state i, then, with probability 1,*

$$\lim_{n\to\infty} \frac{N_n(j)}{n} = \frac{1_{\{T_j<\infty\}}}{m_j} \tag{9.28}$$

and

$$\lim_{n\to\infty} \frac{E\left[N_n(j)|X_0 = i\right]}{n} = \frac{r_{ij}}{m_j} \tag{9.29}$$

Proof: Above we have proved the case (9.27) where the initial state $X_0 = j$. In that case $1_{\{T_j<\infty\}} = 1$. For an arbitrary initial distribution, it is possible that the chain will never reach the recurrent state j. In that case $1_{\{T_j<\infty\}} = 0$ given $X_0 = i$. It remains to proof (9.29). By definition, $0 \le N_n(j) \le n$ or, $0 \le \frac{N_n(j)}{n} \le 1$, which demonstrates that, for any n, $\frac{N_n(j)}{n}$ is bounded. From the Dominated Convergence Theorem 6.1.4, we have

$$\lim_{n\to\infty} E\left[\frac{N_n(j)}{n}\middle|X_0 = i\right] = E\left[\lim_{n\to\infty} \frac{N_n(j)}{n}\middle|X_0 = i\right]$$

$$= E\left[\frac{1_{\{T_j<\infty\}}}{m_j}\middle|X_0 = i\right]$$

$$= \frac{\Pr\left[T_j < \infty|X_0 = i\right]}{m_j} = \frac{r_{ij}}{m_j}$$

which completes the proof. □

Theorem 9.3.1 introduces the need for an additional definition. A recurrent state j is called *null recurrent* if $m_j = \infty$, in which case (9.29) reduces to

$$\lim_{n\to\infty} \frac{E\left[N_n(j)|X_0 = i\right]}{n} = 0 \tag{9.30}$$

By Tauberian theorems (which investigate conditions for the converse of Lemma 6.1.1 but which are far more difficult, as illustrated in the book by Hardy (1948)), it can be shown that, for null recurrent states, the stronger result $\lim_{n\to\infty} P_{ij}^n = 0$ also holds. A recurrent state j is called *positive recurrent* if $m_j < \infty$. The difference between a transient and a null recurrent state that both obey (9.30) lies in the fact that, for a transient state, the limit $\lim_{n\to\infty} E\left[N_n(j)|X_0 = j\right]$ is finite while, for a null recurrent state, $\lim_{n\to\infty} E\left[N_n(j)|X_0 = j\right] = \infty$. Relation (9.30) indicates that for a null recurrent state

$$E\left[N_n(j)|X_0 = j\right] = O\left(n^a\right)$$

where $0 < a < 1$, while for a positive recurrent state

$$E\left[N_n(j)|X_0 = j\right] = \pi_j n + o\left(n\right)$$

The strength of the increase of $E\left[N_n(j)|X_0 = j\right]$ leads to term positive recurrent states also as *strongly ergodic* states while null recurrent states are called *weakly ergodic*. Figure 9.1 sketches the classification of states in a Markov process.

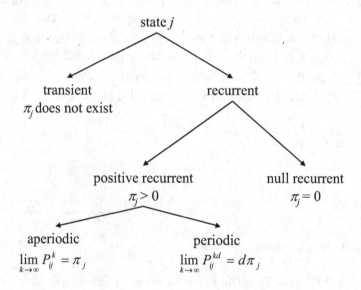

Fig. 9.1. Classification of the states in a Markov process with the corresponding steady state vector π_j.

With these additional definitions, Corollary 9.2.2 can be sharpened as follows:

Corollary 9.3.2 *A finite-state Markov chain must have at least one positive recurrent state.*

Proof: By summing (9.11) over n and dividing by n, we find

$$\frac{1}{n}\sum_{j=1}^{N}\sum_{m=1}^{n} P_{ij}^{m} = 1$$

Using (9.19) yields

$$\sum_{j=1}^{N} \frac{E\left[N_n(j)|X_0 = i\right]}{n} = 1$$

When taking the limit $n \to \infty$ of both sides, the summation and limit operator can be reversed because the summation involves a finite number of terms. Hence,

$$\sum_{j=1}^{N} \lim_{n\to\infty} \frac{E\left[N_n(j)|X_0 = i\right]}{n} = 1$$

which is only possible if at least one state j is positive recurrent because transient and null recurrent states obey (9.30).   □

Similarly, Theorem 9.2.3 and the combined consequence can be sharpened:

Theorem 9.3.3 *If i is a positive recurrent state that leads to a state j, then the state j is also a positive recurrent state.*

Theorem 9.3.4 *An irreducible Markov chain with finite-state space is positive recurrent.*

Alternatively, a Markov chain with a finite number of states has no null recurrent states. Thus, finite-state Markov chains appear to have simpler behavior than infinite-state Markov chains.

Theorem 9.3.5 *For an irreducible, positive recurrent Markov chain (even with an infinite-state space), the steady-state is unique.*

Proof: The steady-state of a positive recurrent irreducible Markov chain satisfies both (9.23) and (9.24), even for an infinite-state Markov chain. Suppose that $a \neq \pi$ is a second steady-state vector which satisfies $\|a\|_1 = 1$ and

$$a_j = \sum_{k=1}^{N} P_{kj} a_k \tag{9.31}$$

Multiplying both sides by P_{ji} and summing over all j

$$\sum_{j=1}^{N} P_{ji} a_j = \sum_{j=1}^{N} P_{ji} \sum_{k=1}^{N} P_{kj} a_k$$

$$= \sum_{k=1}^{N} a_k \sum_{j=1}^{N} P_{kj} P_{ji} = \sum_{k=1}^{N} a_k P_{ki}^2$$

The reversal in j- and k-summation is always allowed (even for $N \to \infty$) by absolute convergence. Using (9.31),

$$a_i = \sum_{k=1}^{N} a_k P_{ki}^2$$

Repeating this process leads, for any $n \geq 1$ and $i \geq 1$ to

$$a_i = \sum_{k=1}^{N} a_k P_{ki}^n$$

In the limit for $n \to \infty$, application of (9.24) yields

$$a_i = \pi_i \sum_{k=1}^{N} a_k = \pi_i$$

which demonstrates uniqueness. $\qquad\square$

Theorem 9.3.6 *For an irreducible, positive recurrent Markov chain holds*

$$\lim_{n \to \infty} \frac{E\left[N_n(j)|X_0 = i\right]}{n} = \lim_{n \to \infty} \frac{N_n(j)}{n} = \frac{1}{m_j} = \pi_j \qquad (9.32)$$

and

$$\frac{N_n(j) - n\pi_j}{\sigma_j \pi_j^{3/2} \sqrt{n}} \xrightarrow{d} N(0,1) \qquad (9.33)$$

where $\sigma_j^2 = Var[T_j|X_0 = j]$

Proof: For an irreducible, finite-state Markov chain where $r_{ij} = 1$, Theorem 9.3.1 and Theorem 9.3.4 together with (9.26) lead to the fundamental relation (9.32). Relation (9.33) is an application of the Asymptotic Renewal Distribution Theorem 8.2.3. We have shown that the interarrival times $\{\tau_k(j)\}_{k \geq 1}$ are i.i.d. with mean $E\left[\tau(j)\right] = E\left[T_j|X_0 = j\right] = m_j$ and (assumed finite) variance $Var[\tau(j)] = Var[T_j|X_0 = j] = \sigma_j^2$. $\qquad\square$

As a corollary, from (8.16), we have

$$\lim_{n\to\infty} \frac{\operatorname{Var}\left[N_n(j)|X_0 = i\right]}{n} = \sigma_j^2 \pi_j^3 \qquad (9.34)$$

Moreover, since $\|\pi\|_1 = 1$, it must hold from (9.32) that

$$\sum_{j=1}^{N} \frac{1}{m_j} = 1$$

and, from (9.22), that

$$\frac{1}{m_j} = \sum_{i=1}^{N} \frac{P_{ij}}{m_i}$$

A Markov chain that is irreducible and for which all states are positive recurrent is said to be *ergodic*. Ergodicity implies that both the steady-state distribution π and the long-run probability distribution $\lim_{k\to\infty} s[k]$ are the same. Ergodic Markov chains are basic stochastic processes in the study of queueing theory.

9.3.3 Example: the two-state Markov chain

The two-state Markov chain is defined by

$$P = \begin{bmatrix} 1-p & p \\ q & 1-q \end{bmatrix}$$

and illustrated in Fig. 9.2. A matrix computation of the two-state Markov

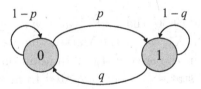

Fig. 9.2. A two-state Markov chain.

chain is presented in Appendix A.4.2. Here, we follow a probabilistic approach. Since there are only two states, at any discrete-time k, there holds that $\Pr[X_k = 0] = 1 - \Pr[X_k = 1]$. Hence, it suffices to compute $\Pr[X_k = 0]$. By the total law of probability and the Markov property (9.2), we have

$$\Pr[X_{k+1} = 0] = \Pr[X_{k+1} = 0|X_k = 1]\Pr[X_k = 1]$$
$$+ \Pr[X_{k+1} = 0|X_k = 0]\Pr[X_k = 0]$$

or, from Fig. 9.2, the Markov chain can only be in state 0 at time $k + 1$, if it is in state 0 at time k and the next event at time $k + 1$ brings it back to that same state 0, or if it is in state 1 at time k and the next event at time $k + 1$ induces a transfer to state 0. Introducing the transition probabilities,

$$\Pr[X_{k+1} = 0] = q \Pr[X_k = 1] + (1 - p) \Pr[X_k = 0]$$
$$= q(1 - \Pr[X_k = 0]) + (1 - p) \Pr[X_k = 0]$$
$$= (1 - p - q) \Pr[X_k = 0] + q$$

This recursion can be iterated back to $k = 0$,

$$\Pr[X_k = 0] = (1 - p - q)^k \Pr[X_0 = 0] + q \sum_{j=0}^{k-1} (1 - p - q)^j$$

Using the finite geometric series $\sum_{j=0}^{k-1} x^j = \frac{1-x^k}{1-x}$ for any $x \neq 1$ else $\sum_{j=0}^{k-1} x^j = k$,

$$\Pr[X_k = 0] = \frac{q}{p+q} + (1 - p - q)^k \left(\Pr[X_0 = 0] - \frac{q}{p+q} \right) \quad (9.35)$$

With $\Pr[X_k = 1] = 1 - \Pr[X_k = 0]$,

$$\Pr[X_k = 1] = \frac{p}{p+q} + (1 - p - q)^k \left(\Pr[X_0 = 1] - \frac{p}{p+q} \right) \quad (9.36)$$

If $|1 - p - q| < 1$, the state $\pi = \begin{bmatrix} \Pr[X_\infty = 0] & \Pr[X_\infty = 1] \end{bmatrix}$ directly follows as

$$\pi = \begin{bmatrix} \frac{q}{p+q} & \frac{p}{p+q} \end{bmatrix} \quad (9.37)$$

Observe from (9.35) and (9.36) that, if $\Pr[X_0 = 0] = \frac{q}{p+q} = \Pr[X_\infty = 0]$ and $\Pr[X_0 = 1] = \frac{p}{p+q} = \Pr[X_\infty = 1]$, the Markov chain starts and remains the whole time (for all k) in the steady-state. In addition, the probability of a particular sequence of states can be computed from (9.3) or directly from Fig. 9.2. For example,

$$\Pr[X_0 = 1, X_1 = 0, X_2 = 1, X_3 = 1] = qp(1 - q) \Pr[X_0 = 1]$$

We distinguish three cases:

(i) $p = q = 0$: The Markov chain consists of two separate states that do not communicate. Each state can be considered as a single state, irreducible, Markov chain. Any real number belonging to $[0, 1]$ is a steady-state solution of each separate set. Also, $P = I$ and, hence, $P^\infty = \lim_{k \to \infty} P^k = I$.

(ii) $0 < p + q < 2$: The Markov chain is aperiodic irreducible positive recurrent with steady-state π given in (9.37). This is the regular case.

(iii) $p = q = 1$: The Markov chain is periodic with period 2, but still irreducible positive recurrent with steady-state $\pi = \left[\frac{1}{2} \ \frac{1}{2}\right]$ given above. However, $P^{2n} = I$ and $P^{2n+1} = P$ such that $\lim_{k\to\infty} P^k$ does not exist, but

$$\lim_{k\to\infty} \frac{1}{k} \sum_{j=1}^{k} P^j = \begin{bmatrix} \frac{1}{2} & \frac{1}{2} \\ \frac{1}{2} & \frac{1}{2} \end{bmatrix}$$

9.4 Problems

(i) Given the transition probability matrix P,

$$P = \begin{bmatrix} 0.8 & 0.2 & 0.0 \\ 0.8 & 0.0 & 0.2 \\ 0.0 & 0.8 & 0.2 \end{bmatrix}$$

(a) draw the Markov chain, (b) compute the steady-state vector in three different ways.

(ii) Consider the discrete-time Markov chain with N states and with transition probabilities at each state j,

$$P_{j,j+1} = 1 - \frac{1}{j}$$

$$P_{j1} = \frac{1}{j}$$

(a) draw the Markov chain, (b) show that the drift is positive, but that the Markov chain is nevertheless recurrent.

(iii) Assume that trees in a forest fall into four age groups. Let $b[k]$, $y[k]$, $m[k]$ and $u[k]$ denote the number of baby trees, young trees, middle-aged trees and old trees, respectively, in the forest at a given time period k. A time period lasts 15 years. During a time period, the total number of trees remains constant, but a certain percentage of trees in each age group dies and is replaced with baby trees. All surviving trees in the baby, young and middle-aged group enter into the next age group. Surviving old trees remain old. Let $0 < p_b, p_y, p_m, p_o < 1$ denote the loss rates in each age group in percent.

(a) Make a discrete Markov chain presentation of the process of aging and replacement in the forest.

(b) The distribution of tree population amongst different age categories in time period k is represented by

$$x[k] = \begin{bmatrix} b[k] & y[k] & m[k] & u[k] \end{bmatrix}^T$$

If $x[k+1] = Px[k]$, what is the transition probability matrix
P?

(c) Let $p_b = 0.1$, $p_y = 0.2$, $p_m = 0.3$, $p_o = 0.4$ and suppose that
$x[k] = \begin{bmatrix} 5000 & 0 & 0 & 0 \end{bmatrix}^T$. What is the number of trees in
each category after 15 and after 30 years?

(d) What is the steady-state situation?

(iv) A faulty digital video conferencing system shows a clustered error
pattern. If a bit is received correctly, then the chance to receive the
next bit correctly is 0.999. If a bit is received incorrectly, then the
next bit is incorrect with probability 0.95.

(a) Model the error pattern of this system using the discrete-time
Markov chain.

(b) How many communicating classes does the Markov chain have?
Is it irreducible?

(c) In the long run, what is the fraction of correctly received bits
and the fraction of incorrectly received bits?

(d) After the system is repaired, it works properly for 99.9% of
the time. A test sequence after repair shows that, when always starting with a correctly received bit, the next 10 bits
are correctly received with probability 0.9999. What is the
probability now that a correctly (and analogously incorrectly)
received bit is followed by another correct (incorrect) bit?

10

Continuous-time Markov chains

Just as it was convenient in Chapter 2 to treat discrete and continuous random variables distinctly, the same recipe is advised for discrete-time and continuous-time Markov chains. Here also, it appears that the continuous case is more intricate than the discrete counterpart.

10.1 Definition

For the continuous-time Markov chain $\{X(t), t \geq 0\}$ with N states, the Markov property (9.1) can be written as

$$\Pr[X(t+\tau)=j|X(\tau)=i, X(u)=x(u), 0 \leq u < \tau] = \Pr[X(t+\tau)=j|X(\tau)=i]$$

and reflects the fact that the future state at time $t + \tau$ only depends on the current state at time τ. Similarly as for the discrete-time Markov chain, we assume that the transition probabilities for the continuous-time Markov chain $\{X(t), t \geq 0\}$ are stationary, i.e. independent of a point τ in time,

$$P_{ij}(t) = \Pr\left[X(t+\tau) = j|X(\tau) = i\right] = \Pr\left[X(t) = j|X(0) = i\right] \quad (10.1)$$

Analogous to (9.5) and (9.6), the state vector $s(t)$ in continuous-time with components $s_k(t) = \Pr\left[X(t) = k\right]$ obeys

$$s(t + \tau) = s(\tau)P(t) \quad (10.2)$$

Immediately, it follows from (10.2) that

$$s(t + u + \tau) = s(\tau)P(t + u)$$
$$s(t + u + \tau) = s(\tau + u)P(t) = s(\tau)P(u)P(t)$$
$$= s(\tau + t)P(u) = s(\tau)P(t)P(u)$$

such that, for all $t, u \geq 0$, the $N \times N$ transition probability matrix $P(t)$

satisfies

$$P(t + u) = P(u)P(t) = P(t)P(u) \tag{10.3}$$

This fundamental relation[1] (10.3) is called the Chapman-Kolmogorov equation. Furthermore, since the Markov chain must be at any time in one of the N states, the analogon of (9.8) is, for any state i,

$$\sum_{j=1}^{N} P_{ij}(t) = 1 \tag{10.4}$$

For continuous-time Markov chains, it is convenient to postulate the initial condition of the transition probability matrix

$$P(0) = I \tag{10.5}$$

where $P(0) = \lim_{t \downarrow 0} P(t)$. The relations (10.1), (10.3), (10.4) and (10.5) are sufficient to describe the continuous-time Markov process completely.

10.2 Properties of continuous-time Markov processes

We will now concentrate on typical properties of a continuous-time Markov process.

10.2.1 The infinitesimal generator Q

Lemma 10.2.1 *The transition probability matrix $P(t)$ is continuous for all $t \geq 0$.*

Proof: Continuity is proved if $\lim_{h \to 0} P(t+h) = \lim_{h \to 0} P(t-h) = P(t)$. From (10.3) and (10.5), we have for $h > 0$,

$$\lim_{h \to 0} P(t + h) = P(t) \lim_{h \to 0} P(h) = P(t)I = P(t)$$

Similarly, the other limit follows for $t > 0$ and $0 < h < t$ from $P(t) = P(t-h)P(h)$. $\square$

If a function is differentiable, it is continuous. However, the converse is not generally true. Therefore, we include the additional assumption that

---

[1] On a higher level of abstraction, $P(t)$ can be viewed as a linear operator acting upon the vector space defined by all possible state vectors $s(t)$. Relation $P(t + u) = P(u)P(t)$ is known as the semigroup property. The family of these commuting operators possesses an interesting algebraic structure (see e.g. Schoutens (2000)).

the matrix

$$\lim_{h\downarrow 0} \frac{P(h) - I}{h} = P'(0) = Q \tag{10.6}$$

exists. This matrix Q is called the *infinitesimal generator* of the continuous-time Markov process and it plays an important role as shown below. The infinitesimal generator Q corresponds to $P - I$ in discrete-time. From (10.4),

$$\sum_{j=1,j\neq i}^{N} P_{ij}(h) = 1 - P_{ii}(h)$$

and, dividing both sides by h and letting h approach zero, we find for each i with the definition of Q that

$$\sum_{j=1,j\neq i}^{N} q_{ij} = -q_{ii} \geq 0 \tag{10.7}$$

Hence, the sum of the rows in Q is zero, $q_{ij} = \lim_{h\downarrow 0} \frac{P_{ij}(h)}{h} \geq 0$ and $q_{ii} \leq 0$. The elements q_{ij} of Q are derivatives of probabilities and reflect a change in transition probability from state i towards state j, which suggests us to call them "rates". Usually, one defines $q_i = -q_{ii} \geq 0$. Then, $\sum_{j=1}^{N} |q_{ij}| = 2q_i$, which demonstrates that Q is bounded if and only if the rates q_i are bounded. Karlin and Taylor (1981, p. 140) show that q_{ij} is always finite. For finite-state Markov processes, q_j are finite (since q_{ij} are finite), but, in general, q_j can be infinite. If $q_j = \infty$, the state is called *instantaneous* because when the process enters this state, it immediately leaves the state. In the sequel, we confine the discussion to non-instantaneous states, thus $0 \leq q_j < \infty$. Continuous-time Markov chains with all states non-instantaneous are coined *conservative*.

Probabilistically, (10.1) indicates that, for small h,

$$\Pr\left[X(t+h) = j | X(t) = i\right] = q_{ij}h + o(h) \qquad (i \neq j)$$
$$\Pr\left[X(t+h) = i | X(t) = i\right] = 1 - q_i h + o(h) \tag{10.8}$$

which clearly generalizes the Poisson process (see Theorem 7.3.1) and motivates us to call q_i the rate corresponding to state i.

Lemma 10.2.2 *Given the infinitesimal generator Q, the transition probability matrix $P(t)$ is differentiable for all $t \geq 0$,*

$$P'(t) = P(t)Q \tag{10.9}$$
$$= QP(t) \tag{10.10}$$

These equations are called the forward (10.9) and backward (10.10) equation.

Proof: For $t = 0$, the lemma follows from the existence of $Q = P'(0)$. The derivative $P'(t)$ is defined, for $t > 0$, as

$$P'(t) = \lim_{h \to 0} \frac{P(t+h) - P(t)}{h}$$

where the derivative of the matrix has elements $P'_{ij}(t) = \frac{dP_{ij}(t)}{dt}$. Using (10.3),

$$P(t+h) - P(t) = P(t)P(h) - P(t) = P(t)\left(P(h) - I\right)$$
$$= P(h)P(t) - P(t) = \left(P(h) - I\right)P(t)$$

we obtain

$$P'(t) = P(t) \lim_{h \to 0} \frac{P(h) - I}{h} = P(t)Q$$
$$= \lim_{h \to 0} \frac{P(h) - I}{h} P(t) = QP(t)$$

which proves the lemma. $\square$

Suppose we are interested in the probabilities $s_k(t) = \Pr[X(t) = k]$ of finding the system in state k at time t. Each component of the state vector $s(t)$ is determined by (10.2) as

$$s_k(t+h) = \sum_{j=1}^{N} s_j(t) P_{jk}(h)$$

from which

$$\frac{s_k(t+h) - s_k(t)}{h} = s_k(t) \frac{P_{kk}(h) - 1}{h} + \sum_{j=1, j \neq k}^{N} s_j(t) \frac{P_{jk}(h)}{h}$$

In the limit $h \downarrow 0$, we find with $q_{jk} = \lim_{h \downarrow 0} \frac{P_{jk}(h)}{h}$ and $q_k = \lim_{h \downarrow 0} \frac{1 - P_{kk}(h)}{h}$ the differential equation for $s_k(t)$,

$$s'_k(t) = -q_k s_k(t) + \sum_{j=1, j \neq k}^{N} q_{jk} s_j(t) \tag{10.11}$$

which, together with the initial condition $s_k(0)$, completely determines the probability $s_k(t)$ that the Markov process is in state k at time t.

10.2.2 Algebraic properties of the infinitesimal generator Q

Equation (10.10) is a matrix differential equation in t that can be similarly solved as the scalar differential equation $f'(t) = qf(t)$. With the initial condition (10.5), the solution is

$$P(t) = e^{Qt} \qquad (10.12)$$

which demonstrates the importance of the infinitesimal generator Q, explicitly given by

$$
Q = \begin{bmatrix}
-q_1 & q_{12} & q_{13} & \cdots & q_{1;N-1} & q_{1N} \\
q_{21} & -q_2 & q_{23} & \cdots & q_{2;N-1} & q_{2N} \\
q_{31} & q_{32} & -q_3 & \cdots & q_{3;N-1} & q_{3N} \\
\vdots & \vdots & \vdots & \ddots & \vdots & \vdots \\
q_{N-1;1} & q_{N-1;2} & q_{N-1;3} & \cdots & -q_{N-1} & q_{N-1;N} \\
q_{N1} & q_{N;2} & q_{N3} & \cdots & q_{N;N-1} & -q_N
\end{bmatrix} \qquad (10.13)
$$

Moreover, if all eigenvalues λ_k of Q are distinct, art. **4** and art. **8** in Appendix A.1 indicate that

$$P(t) = e^{Qt} = X \operatorname{diag}(e^{\lambda_k t}) Y^T \qquad (10.14)$$

where X and Y contain as columns the right- and left-eigenvectors of Q respectively. Written explicitly in terms of the right-eigenvectors x_k and left-eigenvectors y_k (which both are an $1 \times N$ matrices or column vectors as common in vector algebra), (10.14) reads

$$P(t) = \sum_{k=1}^{N} e^{\lambda_k t} x_k y_k^T$$

where the inner or scalar vector product $y_k^T x_k = 1$ while the outer product $x_k y_k^T$ is an $N \times N$ matrix,

$$
x_k y_k^T = \begin{bmatrix}
x_{k1} y_{k1} & x_{k1} y_{k2} & x_{k1} y_{k3} & \cdots & x_{k1} y_{kN} \\
x_{k2} y_{k1} & x_{k2} y_{k2} & x_{k2} y_{k3} & \cdots & x_{k2} y_{kN} \\
x_{k3} y_{k1} & x_{k3} y_{k2} & x_{k3} y_{k3} & \cdots & x_{k3} y_{kN} \\
\vdots & \vdots & \vdots & \vdots & \vdots \\
x_{kN} y_{k1} & x_{kN} y_{k2} & x_{kN} y_{k3} & \cdots & x_{kN} y_{kN}
\end{bmatrix}
$$

If we further assume (thus omitting pathological cases) that $P(t)$ is a stochastic, irreducible matrix for any time t, Frobenius' Theorem A.4.2 indicates that all eigenvalues $\left| e^{\lambda_k t} \right| < 1$ and that only the largest one is precisely equal to 1, say $e^{\lambda_1 t} = 1$, which corresponds to the steady-state eigenvector $y_1^T = \pi$ and $x_1 = u$, where $u^T = [1 \ 1 \ \cdots \ 1]$. Frobenius' Theorem A.4.2

implies that all eigenvalues of Q have a negative real part, except for the steady-state eigenvalue $\lambda_1 = 0$. Hence, we may write

$$P(t) = u\pi + \sum_{k=2}^{N} e^{-|\operatorname{Re}\lambda_k t| + \operatorname{Im}\lambda_k t} x_k y_k^T \tag{10.15}$$

where $P_\infty = u\pi$ is the $N \times N$ matrix with each row containing the steady-state vector π. The expression (10.15) is called the spectral or eigen decomposition of the transition probability matrix $P(t)$.

Apart from the eigen decomposition method and the Taylor expansion

$$e^{Qt} = \sum_{k=0}^{\infty} \frac{(Qt)^k}{k!} \tag{10.16}$$

the matrix equivalent of $e^x = \lim_{n \to \infty} (1 + x/n)^n$ can be used,

$$P(t) = e^{Qt} = \lim_{n \to \infty} \left(I + \frac{Qt}{n}\right)^n \tag{10.17}$$

Since Q has negative diagonal elements and positive off-diagonal elements, computing the powers Q^k as required in (10.16) suffers from numerical rounding-off error propagation. Relation (10.17) circumvents this problem by choosing n sufficiently high, $\max_i q_i t < n$, such that $I + \frac{Qt}{n}$ has non-negative elements smaller than 1 everywhere. For stochastic matrices P, the sequence $P, P^2, P^4, \ldots, P^{2^k}$ rapidly converges. Yet another useful representation (10.24) of $P(t)$ is discussed in Section 10.4.1.

10.2.3 Exponential sojourn times

We end this section on properties by proving a remarkable and important characteristic of continuous-time Markov processes.

Theorem 10.2.3 *The sojourn times τ_j of a continuous-time Markov process in a state j are independent, exponential random variables with mean $\frac{1}{q_j}$.*

Proof: The independence of the sojourn times follows from the Markov property (see the renewal argument in Section 9.3.2). The exponential sojourn time is proved in two different ways.

1. The proof consists in demonstrating that the sojourn times τ_j satisfy the memoryless property. In Section 3.2.2, it has been shown that the only continuous distribution that satisfies the memoryless property is the exponential distribution.

The event $\{\tau_j \geq t + T | \tau_j > T\}$ for any $T \geq 0$ and $t \geq 0$ is equivalent to the event $\{X(t + T + u) = j | X(T + u) = j, X(u) = j\}$. According to the Markov property (9.1) and with (10.1),

$$
\begin{aligned}
\Pr\left[\tau_j \geq t + T | \tau_j > T\right] &= \Pr\left[X(t + T + u) = j | X(T + u) = j, X(u) = j\right] \\
&= \Pr\left[X(t + T + u) = j | X(T + u) = j\right] \\
&= P_{jj}(t)
\end{aligned}
$$

which is independent of T illustrating the memoryless property. Using the definition of conditional probability (2.44),

$$
\Pr\left[\tau_j \geq t + T | \tau_j > T\right] = \frac{\Pr\left[\tau_j \geq t + T\right]}{\Pr\left[\tau_j > T\right]} = P_{jj}(t)
$$

which holds for any T and thus also for $T = 0$, where $\Pr\left[\tau_j > 0\right] = 1$. The distribution of the sojourn time at state j satisfies

$$
\Pr\left[\tau_j \geq t\right] = e^{-\alpha_j t} = P_{jj}(t)
$$

After differentiation evaluated at $t = 0$, we find $\alpha_j = q_j$.

2. An alternative demonstration of the exponential sojourn times starts by considering for an initial state j, the probability H_n that the process remains in state j during an interval $[0, t]$. The idea is to first sample the continuous-time interval with step $\frac{t}{n}$ and afterwards proceed to the limit $n \to \infty$, which corresponds to a sampling with infinitesimally small step,

$$
\begin{aligned}
H_n &= \Pr\left[X(0) = j, X\left(\frac{t}{n}\right) = j, X\left(\frac{2t}{n}\right) = j, \dots, X(t) = j\right] \\
&= \prod_{m=0}^{n-1} \Pr\left[X\left(\frac{(m+1)t}{n}\right) = j \middle| X\left(\frac{mt}{n}\right) = j\right] \Pr\left[X(0) = j\right] \\
&= \left(\Pr\left[X\left(\frac{t}{n}\right) = j \middle| X(0) = j\right]\right)^n \Pr\left[X(0) = j\right] \\
&= \left[P_{jj}\left(\frac{t}{n}\right)\right]^n \Pr\left[X(0) = j\right]
\end{aligned}
$$

where (9.3) and (10.1) are used. For large n, $P_{jj}\left(\frac{t}{n}\right)$ can be expanded in a Taylor series around the origin,

$$
\begin{aligned}
P_{jj}\left(\frac{t}{n}\right) &= P_{jj}(0) + P_{jj}'(0)\frac{t}{n} + O\left(\frac{1}{n^2}\right) \\
&= 1 - q_j \frac{t}{n} + O\left(\frac{1}{n^2}\right)
\end{aligned}
$$

such that

$$\left[P_{jj}\left(\frac{t}{n}\right)\right]^n = \exp\left[n\log\left(1 - q_j\frac{t}{n} + O\left(\frac{1}{n^2}\right)\right)\right]$$

For large n the logarithm can be expanded to first order as

$$\log\left(1 - q_j\frac{t}{n} + O\left(\frac{1}{n^2}\right)\right) = -q_j\frac{t}{n} + O\left(\frac{1}{n^2}\right)$$

which shows that

$$\lim_{n\to\infty}\left[P_{jj}\left(\frac{t}{n}\right)\right]^n = e^{-q_j t}$$

On the other hand,

$$\lim_{n\to\infty} H_n = \Pr\left[X(u) = j, 0 \le u \le t\right]$$

Hence, the probability that the process remains in state j at least for a duration t equals

$$\Pr\left[X(u) = j, 0 \le u \le t\right] = e^{-q_j t}\Pr\left[X(0) = j\right]$$

Conditioned to the initial state with (2.44),

$$\Pr\left[X(u) = j, 0 \le u \le t | X(0) = j\right] = \Pr\left[\tau_j \ge t\right] = e^{-q_j t} \qquad (10.18)$$

Without resorting to the memoryless property, Theorem 10.2.3 has been proved.   □

In summary, the continuous-time Markov process $\{X(t), t \in T\}$ can be described in two equivalent ways, either by the transition probability matrix $P(t)$ or by the infinitesimal generator Q. In the first description, the process starts at time $t = t_0 = 0$ in state x_0, where it stays until a transition occurs at $t = t_1$, which makes the process jump to state x_1. In state x_1, the process stays until $t = t_2$ at which time it jumps to state x_2, and so on. The sequence of states $x_0, x_1, x_2, \ldots$ is a discrete Markov process and is called the *embedded* Markov chain. The embedded Markov chain is further discussed in Section 10.4. The infinitesimal description based on Q formulates the evolution of the process in terms of rates. The process waits in a state j until a jump or trigger occurs with rate q_j and the average waiting time in state j is $\frac{1}{q_j}$. If $q_j = 0$, the Markov process stays infinitely long in state j, implying that state j is an absorbing state.

10.3 Steady-state

Theorems 9.3.4 and 9.3.6 demonstrate that, when a finite-state Markov chain is irreducible (all states communicate and $P_{ij}(t) > 0$), the steady-state π exists. Since, by definition, the steady-state does not change over time, or $\lim_{t\to\infty} P'(t) = 0$, it follows from (10.9) and (10.10) that

$$QP_\infty = P_\infty Q = 0$$

where $\lim_{t\to\infty} P(t) = P_\infty$. This relation implies that P_∞ is the adjoint matrix of Q belonging to eigenvalue $\lambda = 0$, which plays a role analogous to $\lambda = 1$ in the discrete case. By the same arguments as in the discrete case and as shown in Section 10.2.2, all rows of P_∞ are proportional to the eigenvector of Q belonging to $\lambda = 0$. Thus, the steady-state (row) vector π is solution of

$$\pi Q = 0 \tag{10.19}$$

which means that π is orthogonal to any column vector of Q such that necessarily $\det Q = 0$ in order for a non-zero solution to exist. A single component of π in (10.19) obeys, using (10.7),

$$\pi_i q_i = \sum_{j=1, j\neq i}^{N} \pi_j q_{ji} \tag{10.20}$$

This equation has a continuity or conservation law interpretation. The left-hand side reflects the long-run rate at which the process leaves state i. The right-hand side is the sum of the long-run rates of transitions towards the state i from other states $i \neq j$ or the aggregate long-run rate towards state i. Both in- and outwards flux at any state i are in steady-state precisely in balance. Therefore relations (10.20) are called the *balance equations*. The balance equation (10.20) directly follows from the differential equation (10.11) of the state probabilities $s_k(t)$ since $\lim_{t\to\infty} s_k(t) = \pi_k$ and $\lim_{t\to\infty} s'_k(t) = 0$.

Alternatively, the steady-state vector π obeys (10.2) or

$$\pi = s(0)P_\infty = \lim_{t\to\infty} s(0)e^{Qt}$$

which, together with (10.14), implies that all eigenvalues of Q must have negative real part such that only $\lambda = 0$ determines the steady-state. This stability condition on the eigenvalues corresponds to that in a linear, time-variant system. Since all rows in P_∞ are equal (see also (10.15)), the dependence of the steady-state vector π on the initial state drops out. For, analogous to the discrete-time case and recalling the normalization $\|s(0)\|_1 = 1$, a single component becomes

$$\pi_j = \sum_{k=1}^{N} s_k(0) \, (P_\infty)_{kj} = (P_\infty)_{1j} \sum_{k=1}^{N} s_k(0) = (P_\infty)_{1j}$$

10.4 The embedded Markov chain

The main difference between discrete and continuous-time Markov chains lies, apart from the concept of time, in the determination of the number of transitions. The sojourn time in a discrete chain is deterministic and all times are equal to 1. In other words, if $F_{ij}(t)$ denotes the distribution function of the time until a transition from state i to state j occurs, then it is plain that, for a discrete-time process,

$$F_{ij}(t) = 1_{t \geq 1}$$

Even though the process remains in state j with probability P_{jj}, there has been a transition precisely after $t = 1$ units.

On the other hand, Theorem 10.2.3 demonstrates that the sojourn times in state j are exponential distributed with mean $\frac{1}{q_j}$. After, on average $\frac{1}{q_j}$ time units, a transition from state j to another state occurs. In contrast to discrete-time Markov chains, after the exponentially distributed sojourn time in state j the process makes a transition to other states $i \neq j$. Let us investigate this fact in more detail. Let us denote

$$V_{ij}(h) = \Pr\left[X(h) = j | X(h) \neq i, X(0) = i\right]$$

which describes the probability that, if a transition occurs, the process moves from state i to a different state $j \neq i$. Using the definition of conditional probability (2.44),

$$V_{ij}(h) = \frac{\Pr\left[\{X(h) = j\} \cap \{X(h) \neq i\} | X(0) = i\right]}{\Pr\left[X(h) \neq i | X(0) = i\right]} = \frac{P_{ij}(h)}{1 - P_{ii}(h)}$$

In the limit $h \downarrow 0$, we have

$$V_{ij} = \lim_{h \downarrow 0} V_{ij}(h) = \lim_{h \downarrow 0} \frac{\frac{P_{ij}(h)}{h}}{\frac{1 - P_{ii}(h)}{h}} = \frac{q_{ij}}{q_i}$$

By (10.7), we see that $\sum_{j=1, j \neq i} V_{ij} = 1$, demonstrating that, given a transition, it is a transition out of state i to another state j. The quantities V_{ij} correspond to the transition probabilities of the embedded Markov chain.

Alternatively, we can write the rate q_{ij} in terms of the transition probabilities V_{ij} of the embedded Markov chain as

$$q_{ij} = q_i V_{ij} \qquad (10.21)$$

Since q_i is the rate (i.e. the number of transitions per unit time) of the process in state i, relation (10.21) shows that the transition rate q_{ij} from state i to state j equals the rate of transitions in state i multiplied by the probability that a transition from state i to state j occurs. By definition, $V_{ii} = 0$. For, if we assume that $V_{ij} > 0$, relation (10.21) would result in $q_{ii} = V_{ii}q_i > 0$ which contradicts the definition $q_{ii} = -q_i$. Hence, in the embedded Markov chain specified by the transition probability matrix V, there are no self-transitions ($V_{ii} = 0$), which is equivalent to the fact that the sum of the eigenvalues of V is zero (A.7), since trace(V) $= 0$.

From the steady-state equation or balance equation (10.20), (10.21) and $V_{ii} = 0$, we observe that

$$\pi_i q_i = \sum_{j=1}^{N} \pi_j q_j V_{ji}$$

On the other hand, the embedded Markov chain has a steady-state vector v that obeys (9.22) or (9.23)

$$v_i = \sum_{j=1}^{N} v_j V_{ji}$$

and $\|v\|_1 = 1$. The relations between the steady-state vectors of the continuous-time Markov chain π and of its corresponding embedded discrete-time Markov chain v, are

$$v_i = \frac{\pi_i q_i}{\sum_{j=1}^{N} \pi_i q_i} \qquad (10.22)$$

$$\pi_i = \frac{v_i / q_i}{\sum_{j=1}^{N} v_j / q_j} \qquad (10.23)$$

The classification in the discrete-time case into transient and recurrent can be transferred via the embedded Markov chain to continuous Markov processes.

10.4.1 Uniformization

The restriction $V_{ii} = 0$ or $q_{ii} = 0$, which means that there are no self-transitions from a state into itself, can be removed. Indeed, we can rewrite the basic relation (10.12) between the transition probability matrix $P(t)$ and the infinitesimal generator Q for all β as

$$P(t) = \exp\left[-\beta I t + \beta t \left(I + \frac{Q}{\beta}\right)\right] = e^{-\beta t} \exp\left[\beta t \left(I + \frac{Q}{\beta}\right)\right]$$

Defining $T(\beta) = I + \frac{Q}{\beta}$ and $\beta \geq \max_i q_i$, a description, alternative to (10.15), (10.16) and (10.17), appears

$$P_{ij}(t) = e^{-\beta t} \sum_{k=0}^{\infty} \frac{(\beta t)^k}{k!} T_{ij}^k(\beta) \tag{10.24}$$

where $T(\beta)$ is a stationary transition probability matrix and, hence, a stochastic matrix.

We also observe that $\beta T(\beta) = Q + \beta I$ can be regarded as a rate matrix, with the property that, for each state i,

$$\sum_{j=1}^{N} \beta T_{ij}(\beta) = \sum_{j=1}^{N} Q_{ij} + \beta \sum_{j=1}^{N} \delta_{ij} = \beta$$

the transition rate in any state i is precisely the same, equal to β. Whereas the embedded Markov chain defined by (10.21) has no self-transitions ($V_{ii} = 0$), we see for any i and j, that $T_{ii}(\beta) = 1 - \frac{1}{\beta}\sum_{j=1;j\neq i}^{N} q_{ij} = 1 - \frac{q_i}{\beta} \geq 0$ while $T_{ij}(\beta) = \frac{q_{ij}}{\beta}$. Hence, $T(\beta)$ can be interpreted as an embedded Markov chain that allows self-transitions. In view of (10.21), the embedded structure of $T(\beta)$ is summarized as

$$q_{ij} = \beta T_{ij}(\beta) \qquad \text{for } i \neq j$$
$$q_{ii} = 1 - \beta T_{ii}(\beta)$$

where the constant rate $q_i = \beta$ for any state i is, besides self-transitions $q_{ii} \neq 0$, the characterizing property. These properties also reveal that, starting from the embedded chain V where $V_{ii} = 0$, we can add self-transitions $q_{ii} > 0$ with the effect that, on (10.7), the transition rate $q_i \to q_i + q_{ii}$. The opposite figure illustrates an embedded Markov chain with self-transitions and the corresponding transition probability matrix, where the transition rates q_i follow from $\sum_{j=1}^{6} V_{ij} = 1$ with $V_{ij} = \frac{q_{ij}}{q_i}$. This change in transition rate will change the steady-state vector since the balance equations (10.20) change. However, the Markov process $\{X(t), t \geq 0\}$ is not modified because a self-transition does not change $X(t)$ nor the distribution of the time until the next transition to a different state. But self-transitions clearly change the number of transitions during some period of time. When the transition rate q_j at each state j are the same, the embedded Markov chain $T(\beta)$ is called a *uniformized* chain.

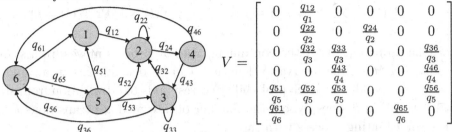

$$V = \begin{bmatrix} 0 & \frac{q_{12}}{q_1} & 0 & 0 & 0 & 0 \\ 0 & \frac{q_{22}}{q_2} & 0 & \frac{q_{24}}{q_2} & 0 & 0 \\ 0 & \frac{q_{32}}{q_3} & \frac{q_{33}}{q_3} & 0 & 0 & \frac{q_{36}}{q_3} \\ 0 & 0 & \frac{q_{43}}{q_4} & 0 & 0 & \frac{q_{46}}{q_4} \\ \frac{q_{51}}{q_5} & \frac{q_{52}}{q_5} & \frac{q_{53}}{q_5} & 0 & 0 & \frac{q_{56}}{q_5} \\ \frac{q_{61}}{q_6} & 0 & 0 & 0 & \frac{q_{65}}{q_6} & 0 \end{bmatrix}$$

In addition, in a uniformized chain, the steady-state vector $t(\beta)$ of $T(\beta)$ is the same as the steady-state vector π. Indeed, from (9.23),

$$t_j(\beta) = \sum_{k=1}^{N} T_{kj}(\beta) t_k(\beta)$$

we have, with $T(\beta) = I + \frac{Q}{\beta}$,

$$t_j(\beta) = \sum_{k=1}^{N} \left(\delta_{kj} + \frac{q_{kj}}{\beta} \right) t_k(\beta)$$

$$= t_j(\beta) + \frac{1}{\beta} \sum_{k=1}^{N} t_k(\beta) q_{kj}$$

or,

$$t_k(\beta) q_k = \sum_{k=1;k\neq j}^{N} t_k(\beta) q_{kj}$$

where $t_k(\beta) = \pi_k$ (independent of β) since it satisfies the balance equation (10.20) and Theorem 9.3.5 assures that the steady-state of a positive recurrent chain is unique.

We will now interpret (10.24) probabilistically. Let $N(t)$ denote the total number of transitions in $[0, t]$ in the uniformized (discrete) process $\{X_k(\beta)\}$. Since the transition rates $q_i = \beta$ are all the same, $N(t)$ is a Poisson process with rate β because, for any continuous-time Markov chain, the inter-transition or sojourn times are i.i.d. exponential random variables. Thus, $\Pr[N(t) = k] = e^{-\beta t} \frac{(\beta t)^k}{k!}$ is recognized as the probability that the number of transitions that occur in $[0, t]$ in the uniformized Markov chain with rate β equals k. With (9.10), $T_{ij}^k(\beta) = \Pr[X_k(\beta) = j | X_0(\beta) = i]$ is the k-step transition probability of that discrete $\{X_k(\beta)\}$ uniformized Markov process. Relation (10.24) can be interpreted as

$$P_{ij}(t) = \sum_{k=0}^{\infty} \Pr\left[X_k(\beta) = j | X_0(\beta) = i, N(t) = k\right] \Pr\left[N(t) = k\right]$$

or, the probability that the continuous Markov process moves from state i to state j in a time interval of length t, can be decomposed in an infinite sum of probabilities. Each probability corresponds to a transition from state i to state j in k-steps, where the number of intermediate transitions k is a Poisson counting process with rate β.

10.4.2 A sampled-time Markov chain

The sampled-time Markov chain approximates the continuous Markov process in that the transition probabilities $P_{ij}(t)$ are expanded to first order as in (10.8) with fixed step $h = \Delta t$. The transition probabilities of the sampled-time Markov chain are

$$P_{ij} = q_{ij}\Delta t \qquad (i \neq j)$$
$$P_{ii} = 1 - q_i\Delta t$$

Clearly, the sampled-time Markov chain also allows self-transitions, as illustrated in Fig. 10.1.

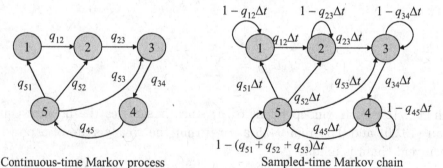

Continuous-time Markov process   Sampled-time Markov chain

Fig. 10.1. A continuous-time Markov process and its corresponding sampled-time Markov chain.

From (10.8), we observe that the approximation lies in two facts: (a) Δt is fixed such that $q_{ij}\Delta t \approx \Pr\left[X(t + \Delta t) = j | X(t) = i\right]$ is increasingly accurate as $\Delta t \to 0$ and (b) transitions occur at discrete times every Δt time units. The sampling step Δt should be chosen such that the transition probabilities obey $0 \leq P_{ij} \leq 1$, from which we find that $\Delta t \leq \frac{1}{\max_i q_i}$.

Let $\varkappa$ denote the steady-state vector of the sampled-time Markov chain with $\|\varkappa\|_1 = 1$. Being a discrete Markov chain, the steady-state vector components $\varkappa_j$ satisfy (9.23) for each component j,

$$\varkappa_j = \sum_{k=1}^{N} P_{kj}\varkappa_k = \Delta t \sum_{k=1; k\neq j}^{N} q_{kj}\varkappa_k + (1 - q_j\Delta t)\,\varkappa_j$$

or

$$q_j\varkappa_j = \sum_{k=1; k\neq j}^{N} q_{kj}\varkappa_k$$

By comparing with the balance equation (10.20) and on the uniqueness of the steady-state (Theorem 9.3.5), we observe that $\varkappa = \pi$ or, the steady-state of the sampled-time Markov chain is *exactly* (not approximately) equal to the steady-state of the continuous Markov chain for any sampling step $\Delta t \leq \frac{1}{\max_i q_i}$. Although we can possibly miss by sampling every Δt time units the smaller-scale dynamics of the continuous Markov chain, the long-run behavior or steady-state is exactly captured!

10.5 The transitions in a continuous-time Markov chain

Based on the embedded Markov chain, there exists a framework that deduces all properties of the continuous Markov chain. In particular, the exponential sojourn times of a continuous-time Markov (Theorem 10.2.3) chain is postulated as a defining characteristic.

Theorem 10.5.1 *Let V_{ij} denote the transition probabilities of the embedded Markov chain and q_{ij} the rates of the infinitesimal generator. The transition probabilities of the corresponding continuous-time Markov chain are found as*

$$P_{ij}(t) = \delta_{ij}e^{-q_i t} + q_i \sum_{k\neq i} V_{ik} \int_0^t e^{-q_i u} P_{kj}(t - u)du \qquad (10.25)$$

Proof: If i is an absorbing state ($q_i = 0$), then, by definition, $P_{ij}(t) = \delta_{ij}$ for all $t \geq 0$. For a non-absorbing state i and a process starting from state i, the event $\{\tau \leq t, X(\tau) = k\} \cap \{X(t) = j\}$ is possible if and only if the first transition from i to k occurs at some time $u \in [0, t]$ and the next transition from k to j takes place in the remaining time $t - u$. The probability density function of the sojourn time is $f_{\tau_i}(t) = \frac{d}{dt}\Pr[\tau_i \leq t] = q_i e^{-q_i t}$ for $t \geq 0$ and for infinitesimally small ϵ, we have

$$p = \Pr\left[\tau \le t, X(\tau) = k, X(t) = j | X(0) = i\right]$$

$$= \int_0^t du \, \Pr\left[\tau = u, X(u - \epsilon) = i | X(0) = i\right]$$

$$\times \Pr\left[X(u) = k | X(u - \epsilon) = i\right] \Pr\left[X(t) = j | X(u) = k\right]$$

$$= \int_0^t du \, f_{\tau_i}(u) V_{ik} P_{kj}(t - u)$$

$$= V_{ik} \int_0^t q_i e^{-q_i u} P_{kj}(t - u) du$$

Furthermore,

$$\Pr[\tau \le t \text{ and } X(t) = j | X(0) = i] = \sum_{k \ne i} \Pr[t \ge \tau, X(\tau) = k, X(t) = j | X(0) = i]$$

and

$$\Pr\left[\tau > t \text{ and } X(t) = j | X(0) = i\right] = \delta_{ij} \Pr\left[\tau_i > t\right] = \delta_{ij} e^{-q_i t}$$

Finally,

$$P_{ij}(t) = \Pr\left[X(t) = j | X(0) = i\right]$$

$$= \Pr[\tau \le t \text{ and } X(t) = j | X(0) = i] + \Pr[\tau > t \text{ and } X(t) = j | X(0) = i]$$

Combining all above relations into the last one proves the theorem. $\qquad\square$

By a change of variable $s = t - u$ in (10.25), we have

$$P_{ij}(t) = \delta_{ij} e^{-q_i t} + q_i \sum_{k \ne i} V_{ik} e^{-q_i t} \int_0^t e^{q_i s} P_{kj}(s) ds$$

and, after differentiation with respect to t, we find for $t \ge 0$,

$$P'_{ij}(t) = -q_i \delta_{ij} e^{-q_i t} - q_i \sum_{k \ne i} q_k V_{ik} e^{-q_i t} \int_0^t e^{q_i s} P_{kj}(s) ds + q_i \sum_{k \ne i} V_{ik} P_{kj}(t)$$

$$= -q_i \delta_{ij} e^{-q_i t} - q_i \left(P_{ij}(t) - \delta_{ij} e^{-q_i t}\right) + q_i \sum_{k \ne i} V_{ik} P_{kj}(t)$$

$$= -q_i P_{ij}(t) + q_i \sum_{k \ne i} V_{ik} P_{kj}(t)$$

Evaluated at $t = 0$, recalling that $P'(0) = Q$ and $P(0) = I$,

$$P'_{ij}(0) = -q_i P_{ij}(0) + q_i \sum_{k \neq i} V_{ik} P_{kj}(0)$$

$$q_{ij} = -q_i \delta_{ij} + q_i \sum_{k \neq i} V_{ik} \delta_{kj} = -q_i \delta_{ij} + q_i V_{ij}$$

which is precisely relation (10.21). With $q_i = -q_{ii}$ and (10.21), we arrive at

$$P'_{ij}(t) = \sum_{k=1}^{N} q_{ik} P_{kj}(t)$$

which is precisely the backward equation (10.10). Hence, (10.25) can be interpreted as an integrated form of the backward equation and thus of the entire continuous-time Markov process.

10.6 Example: the two-state Markov chain in continuous-time

The continuous-time two-state Markov chain is defined by the infinitesimal generator

$$Q = \begin{bmatrix} -\lambda & \lambda \\ \mu & -\mu \end{bmatrix}$$

where $\lambda, \mu \geq 0$. We will solve $P(t)$ from the forward equation (10.9),

$$\begin{bmatrix} P'_{11}(t) & P'_{12}(t) \\ P'_{21}(t) & P'_{22}(t) \end{bmatrix} = \begin{bmatrix} P_{11}(t) & P_{12}(t) \\ P_{21}(t) & P_{22}(t) \end{bmatrix} \begin{bmatrix} -\lambda & \lambda \\ \mu & -\mu \end{bmatrix}$$

which actually contains two independent transition probabilities because $P_{12}(t) = 1 - P_{11}(t)$ and $P_{21}(t) = 1 - P_{22}(t)$. The forward equation simplifies to

$$P'_{11}(t) = -(\lambda + \mu)P_{11}(t) + \mu$$
$$P'_{22}(t) = -(\lambda + \mu)P_{22}(t) + \lambda$$

Only the first equation needs to be solved since, by symmetry, the solution of $P_{11}(t)$ equals that of $P_{22}(t)$ after changing the role of $\lambda \to \mu$ and $\mu \to \lambda$. The linear, first-order, non-homogeneous differential equation consists of the solution to the corresponding homogeneous differential equation and a particular solution. The solution of the homogeneous differential equation, $P'_{11}(t) = -(\lambda + \mu)P_{11}(t)$, is $P_{11}(t) = Ce^{-(\lambda+\mu)t}$. The particular solution is generally found by variation of the constant C, which proposes $P_{11}(t) =$

$C(t)e^{-(\lambda+\mu)t}$ as general solution, where $C(t)$ needs to satisfy the original differential equation. Hence,

$$C'(t) = \mu e^{(\lambda+\mu)t}$$

or, after integration, $C(t) = \frac{\mu}{\lambda+\mu}e^{(\lambda+\mu)t} + c$. The integration constant c follows from the initial condition (10.5), $P_{11}(0) = 1$. Finally, we arrive at

$$P_{11}(t) = \frac{\mu}{\lambda+\mu} + \frac{\lambda}{\lambda+\mu}e^{-(\lambda+\mu)t}$$

$$P_{22}(t) = \frac{\lambda}{\lambda+\mu} + \frac{\mu}{\lambda+\mu}e^{-(\lambda+\mu)t}$$

from which the steady-state vector is immediate,

$$\pi = \left[\begin{array}{cc} \frac{\mu}{\lambda+\mu} & \frac{\lambda}{\lambda+\mu} \end{array} \right]$$

10.7 Time reversibility

In this section, we consider only ergodic Markov chains that have a non-zero steady-state distribution π. Suppose the Markov process operates already in the steady-state, or, in other words, the Markov process is stationary. We are interested in the time-reversed process defined by the sequence $X_n, X_{n-1}, \ldots$ We will show that this reversed time sequence again constitutes a Markov process.

Theorem 10.7.1 *The time-reversed Markov process is a Markov chain.*

Proof: It suffices to demonstrate that the time-reversed process satisfies the Markov property

$$\Pr\left[X_n = x_n | X_{n+1} = x_{n+1}, \ldots, X_{n+k} = x_{n+k}\right] = \Pr\left[X_n = x_n | X_{n+1} = x_{n+1}\right]$$

By definition of the conditional probability (2.44),

$$R = \Pr\left[X_n = x_n | X_{n+1} = x_{n+1}, X_{n+2} = x_{n+2}, \ldots, X_{n+k} = x_{n+k}\right]$$

$$= \Pr\left[X_n = x_n | \cap_{m=1}^{k}\{X_{n+m} = x_{n+m}\}\right]$$

$$= \frac{\Pr\left[\cap_{m=0}^{k}\{X_{n+m} = x_{n+m}\}\right]}{\Pr\left[\cap_{m=1}^{k}\{X_{n+m} = x_{n+m}\}\right]}$$

Since the intersection is commutative $A \cap B = B \cap A$, the indices can be reversed,

$$R = \frac{\Pr\left[\cap_{m=k}^{0}\{X_{n+m} = x_{n+m}\}\right]}{\Pr\left[\cap_{m=k}^{1}\{X_{n+m} = x_{n+m}\}\right]}$$

$$= \frac{\Pr\left[X_{n+k} = x_{n+k} | \cap_{m=k-1}^{0}\{X_{n+m} = x_{n+m}\}\right]\Pr\left[\cap_{m=k-1}^{0}\{X_{n+m} = x_{n+m}\}\right]}{\Pr\left[\cap_{m=k}^{1}\{X_{n+m} = x_{n+m}\}\right]}$$

The original stationary process is a Markov process that satisfies (9.2). Using (9.2) and (9.3) we have

$$\Pr\left[X_{n+k} = x_{n+k} \middle| \cap_{m=k-1}^{0} \{X_{n+m} = x_{n+m}\}\right] = \Pr[X_{n+k} = x_{n+k} | X_{n+k-1} = x_{n+k-1}]$$

and

$$\Pr\left[\cap_{m=k-1}^{0} \{X_{n+m} = x_{n+m}\}\right] = \Pr[X_n = x_n] \prod_{m=1}^{k-1} \Pr[X_{n+m} = x_{n+m} | X_{n+m-1} = x_{n+m-1}]$$

and, similarly,

$$\Pr\left[\cap_{m=k}^{1} \{X_{n+m} = x_{n+m}\}\right] = \Pr[X_{n+1} = x_{n+1}] \prod_{m=2}^{k} \Pr[X_{n+m} = x_{n+m} | X_{n+m-1} = x_{n+m-1}]$$

Hence,

$$\begin{aligned} R &= \frac{\Pr\left[X_{n+k} = x_{n+k} | X_{n+k-1} = x_{n+k-1}\right] \Pr\left[X_{n+1} = x_{n+1} | X_n = x_n\right] \Pr\left[X_n = x_n\right]}{\Pr\left[X_{n+k} = x_{n+k} | X_{n+k-1} = x_{n+k-1}\right] \Pr\left[X_{n+1} = x_{n+1}\right]} \\ &= \frac{\Pr\left[X_{n+1} = x_{n+1} | X_n = x_n\right] \Pr\left[X_n = x_n\right]}{\Pr\left[X_{n+1} = x_{n+1}\right]} \end{aligned}$$

Applying Bayes' rule (2.48) to the last relation finally proves the theorem. $\square$

Consider the transition probability of the time-reversed Markov process

$$R_{ij} = \Pr\left[X_n = j | X_{n+1} = i\right]$$

With Bayes' rule (2.48),

$$\Pr\left[X_n = j | X_{n+1} = i\right] = \frac{\Pr\left[X_{n+1} = i | X_n = j\right] \Pr\left[X_n = j\right]}{\Pr\left[X_{n+1} = i\right]}$$

and, since the process is stationary,

$$\Pr\left[X_n = j\right] = \pi_j, \quad \Pr\left[X_{n+1} = i\right] = \pi_i$$

the transition probability of the time-reversed process is

$$R_{ij} = \frac{\pi_j P_{ji}}{\pi_i} \tag{10.26}$$

A Markov chain is said to be *time reversible* if, for all i and j, $P_{ij} = R_{ij}$. From (10.26), the condition for time reversibility is

$$\pi_i P_{ij} = \pi_j P_{ji} \tag{10.27}$$

This condition means that, for all states i and j, the rate $\pi_i P_{ij}$ from state $i \to j$ equals the rate $\pi_j P_{ji}$ from state $j \to i$. An interesting property of time reversible Markov chains is that any vector x satisfying $\|x\|_1 = 1$ and

$x_i P_{ij} = x_j P_{ji}$ is a steady-state vector of a time-reversible Markov chain. Indeed, summing over all i,

$$\sum_i x_i P_{ij} = x_j \sum_i P_{ji} = x_j$$

Theorem 9.3.5 indicates that the steady-state is unique and, thus, $x = \pi$. As a side remark, we note that a transition matrix is only equal to its transpose $P = P^T$ if the Markov process is time reversible and doubly stochastic (i.e. $\pi_i = \frac{1}{N}$ for all i, as shown in Appendix A.5.1).

The continuous-time analogon can be immediately deduced from the discrete-time embedded Markov chain defined by the transition probabilities V_{ij}. Let U_{ij} denote the transition probabilities of the time-reversed embedded Markov chain and r_{ij} the rates of the corresponding continuous Markov chain, then by (10.21)

$$r_{ij} = r_i U_{ij} \tag{10.28}$$

We will now show that the rates r_i of the time-reversed continuous Markov process are indeed exponential random variables. Assume that the time-reversed process is in state i at time t. The probability that the process is still in state i at reversed time $t - u$ is, using Theorem 10.2.3,

$$\Pr\left[X(\tau) = i, t - u \leq \tau \leq t | X(t) = i\right] = \frac{\Pr\left[X(\tau) = i, t - u \leq \tau \leq t\right]}{\Pr\left[X(t) = i\right]}$$

$$= \frac{\Pr\left[X(t-u) = i\right] e^{-q_j t}}{\Pr\left[X(t) = i\right]} = e^{-q_j t}$$

because, in steady-state $t \to \infty$, $\Pr\left[X(t-u) = i\right] = \Pr\left[X(t) = i\right] = \pi_i$ for any finite u. Thus, the sojourn time in state i of the time-reversed process is exponentially distributed with precisely the same rate $r_i = q_i$ as the forward time process. The steady-state vector v of the embedded Markov chain can be written in terms of the steady-state vector of the continuous Markov chain via (10.22). By (10.26), we obtain

$$U_{ij} = \frac{v_j V_{ji}}{v_i} = \frac{\pi_j q_j V_{ji}}{\pi_i q_i}$$

With (10.21) and (10.28)

$$\frac{\pi_i r_{ij}}{r_i} = \frac{\pi_j q_{ji}}{q_i}$$

but, since $r_i = q_i$, we finally arrive at

$$\pi_i r_{ij} = \pi_j q_{ji} \tag{10.29}$$

Comparing (10.29) with the discrete case (10.26), we see that the transition probabilities P_{ij} and R_{ij} are changed for the rates q_{ij} and r_{ij}. We know that π_j is the portion of time the process (both forward and reversed) spend in state j and that q_{ij} is the rate at which the process makes transitions from state i to state j. Equation (10.29) has again a balance interpretation: $\pi_j q_{ji}$ is the rate at which the forward process moves from state j to i, while $\pi_i r_{ij}$ is the rate of the time-reversed process from state i to j and both rates are equal. Intuitively, when a process jumps from state $i \to j$ in forward time, it is plain that the process makes, in reversed time, just the opposite transition from $j \to i$. Similarly as above, a continuous-time Markov chain is time reversible if, for all i and j, it holds that $r_{ij} = q_{ij}$. For these processes (which occur often in practice, as demonstrated in the chapters on queueing), the rate from $i \to j$ is equal to the rate from $j \to i$ since $\pi_i q_{ij} = \pi_j q_{ji}$.

10.8 Problems

(i) Consider a computer that has two identical and independent processors. The time between failures has an exponential distribution. The mean value of this distribution is 1000 hours. The repair time for a damaged processor is exponentially distributed as well, with a mean value of 100 hours. We assume damaged processors can be repaired in parallel. There are clearly three states for this computer: (1) both processors work, (2) one processor is damaged and (3) both processors are damaged.

  (a) Make a continuous Markov chain presentation of these states.
  (b) What is the infinitesimal generator matrix Q for this Markov chain? Give the relation between the state probability at time t and its derivative.
  (c) Calculate the steady-state of this process.
  (d) What is the availability of the computer if (i) both processors are required to work, or (ii) at least one processor should work.

(ii) Consider two identical servers that are working in parallel. When one server fails, the other has to do the whole job alone under a higher load. The failure times of servers are exponentially distributed: $\mu_H = 3 \times 10^{-4}\ h^{-1}$, when the servers are equally loaded and $\mu_F = 7 \times 10^{-4}\ h^{-1}$, when one of the servers works under the "full load". In addition, both servers may fail at the same time with a failure rate of $\mu_E = 6 \times 10^{-5}\ h^{-1}$.

  As soon as one of the servers fails, the repair is initiated. The average downtime of a server is $\lambda^{-1} = 10$ hours. However, if both servers are damaged, the whole system must be shut down. The average time needed to repair both damaged servers is $\lambda_B^{-1} = 20$ hours.

(a) Draw the Markov chain for this system.
(b) Determine the infinitesimal generator matrix Q.
(c) Determine the steady-state probabilities.
(d) Determine the average lifetime of different states.
(e) What is the average number of server repairs needed during a period of one year?

11

Applications of Markov chains

This chapter illustrates the theory of Markov chains with several examples. Examples of queueing problems are deferred to later chapters. Generally, Markov processes can be solved explicitly provided the transition probability matrix P or the infinitesimal generator Q has a special structure. Only in a very small number of problems is the entire time dependence of the process available in analytic form.

11.1 Discrete Markov chains and independent random variables

This section illustrates examples of some simple Markov chains. Consider a set $\{Y_n\}_{n\geq 1}$ of positive integer, independent random variables that are identically distributed with $\Pr[Y = k] = a_k$.

The discrete-time Markov process, defined by $X_n = Y_n$ for $n \geq 1$, possesses the (infinite) transition probability matrix,

$$P = \begin{bmatrix} a_1 & a_2 & a_3 & a_4 & \cdots \\ a_1 & a_2 & a_3 & a_4 & \cdots \\ a_1 & a_2 & a_3 & a_4 & \cdots \\ a_1 & a_2 & a_3 & a_4 & \cdots \\ \cdots & \cdots & \cdots & \cdots & \cdots \end{bmatrix}$$

All rows are identical and $\Pr[X_{n+1} = j | X_n = i] = a_j$ shows that the states X_{n+1} and X_n are independent from each other.

Another, more interesting, discrete-time Markov process is defined by

$$X_n = \max[Y_1, Y_2, Y_3, \ldots, Y_n].$$

Hence, the process X_n mirrors the maxima of the first n random variables. Clearly, $X_{n+1} = \max[X_n, Y_{n+1}]$ reflects the Markov property: the next state is only dependent on the previous state of the process. From $X_{n+1} =$

$\max[X_n, Y_{n+1}]$, we observe that $\Pr[X_{n+1} = j | X_n = i] = 0$ if $j < i$ because the maximum does not decrease by adding a new random variable in the list. If $j > i$, the state j is determined by $Y_{n+1} = j$ with probability a_j. If $j = i$, then $Y_{n+1} \leq j$, which has probability

$$\Pr[Y_{n+1} \leq j] = \sum_{k=1}^{j} \Pr[Y_{n+1} = k] = \sum_{k=1}^{j} a_k = A_j$$

The corresponding probability matrix is

$$P = \begin{bmatrix} A_1 & a_2 & a_3 & a_4 & \cdots \\ 0 & A_2 & a_3 & a_4 & \cdots \\ 0 & 0 & A_3 & a_4 & \cdots \\ 0 & 0 & 0 & A_4 & \cdots \\ \cdots & \cdots & \cdots & \cdots & \cdots \end{bmatrix}$$

A related discrete-time Markov chain is $X_n = \sum_{k=1}^{n} Y_k$ which obeys $X_{n+1} = X_n + Y_{n+1}$. Furthermore, if $j \leq i$, $\Pr[X_{n+1} = j | X_n = i] = 0$ because the random variables Y_n are non-negative such that the sum cannot decrease by adding a new member. If $j > i$, then

$$\Pr[X_{n+1} = j | X_n = i] = \Pr[X_n + Y_{n+1} = j | X_n = i]$$
$$= \Pr[Y_{n+1} = j - i] = a_{j-i}$$

The corresponding probability matrix has a Toeplitz structure possessing the same elements on diagonal lines

$$P = \begin{bmatrix} 0 & a_1 & a_2 & a_3 & \cdots \\ 0 & 0 & a_1 & a_2 & \cdots \\ 0 & 0 & 0 & a_1 & \cdots \\ 0 & 0 & 0 & 0 & \cdots \\ \cdots & \cdots & \cdots & \cdots & \cdots \end{bmatrix}$$

This list can be extended by considering other integer functions of the set $\{Y_n\}_{n \geq 1}$ (such as $X_{n+1} = \min[X_n, Y_{n+1}]$ or $X_n = \prod_{k=1}^{n} Y_k$ etc.).

11.2 The general random walk

The general random walk is an important model that describes the motion of an item that is constrained to moving either one step forwards, stay at the position where it currently is or move one step backwards. In general, this "three-possibility motion" has transition probabilities that depend on the position j as depicted in Fig. 11.1. Figure 11.1 illustrates that, if the process is in state j, it has three possible choices: remain in state j with probability

$r_j = \Pr[X_{k+1} = j | X_k = j]$, move to the next state $j + 1$ with probability $p_j = \Pr[X_{k+1} = j + 1 | X_k = j]$ or jump back to state $j - 1$ with probability $q_j = \Pr[X_{k+1} = j - 1 | X_k = j]$. A general random walk is defined by the $(N + 1) \times (N + 1)$ band matrix

$$P = \begin{bmatrix} r_0 & p_0 & 0 & 0 & \cdots & 0 & 0 & 0 \\ q_1 & r_1 & p_1 & 0 & \cdots & 0 & 0 & 0 \\ 0 & q_2 & r_2 & p_2 & \cdots & 0 & 0 & 0 \\ \vdots & \vdots & \vdots & \vdots & \vdots & \vdots & \vdots & \vdots \\ 0 & 0 & 0 & 0 & \cdots & q_{N-1} & r_{N-1} & p_{N-1} \\ 0 & 0 & 0 & 0 & \cdots & 0 & q_N & r_N \end{bmatrix} \tag{11.1}$$

where $p_j > 0$, $q_j > 0$, $r_j \geq 0$ and $q_j + r_j + p_j = 1$ for all $0 \leq j \leq N$. The bordering states zero and N are special: $p_0 \geq 0$, $r_0 \geq 0$ and $p_0 + r_0 = 1$ and $q_N \geq 0$, $r_N \geq 0$ and $q_N + r_N = 1$.

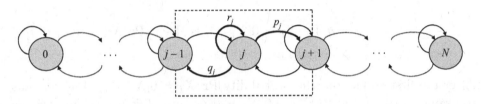

Fig. 11.1. A transition graph of the general random walk.

The general random walk serves as model for a number of practical phenomena:

- The one-dimensional motion of physical particles, electrons that hop from one atom to another. In this case, the number of states N can be very large.
- The gambler's ruin problem: a state j reflects the capital of a gambler whereas the p_j is the chance that the gambler wins while q_j is the probability that he looses. The gambler achieves his target when he reaches state N, but he is ruined at state 0. In that case are both states absorbing states with $r_0 = r_N = 1$. In games, most often the probabilities are independent of the state and simplify to $p_j = p$, $q_j = q$ and $r_j = 1 - p - q$.
- The continuous-time counterpart, the birth and death process (Section 11.3), has applications to queueing processes. For a wealth of examples and applications of the random walk, we refer to the classical treatise of Feller (1970, Chapter III, XIV).

11.2.1 The probability of gambler's ruin

The probability of gambler's ruin is defined as $u_j = \Pr[X_T = 0|X_0 = j]$ where $T = \min_k\{X_k = 0\}$ is the hitting time to state 0, which is equivalent to $u_j = \Pr[T_0 < \infty|X_0 = j]$. By definition, $u_0 = 1$ and since the gambler achieves his goal at state N, he stops and never gets ruined, $u_N = 0$. The law of total probability (2.46) gives the situation after the first transition,

$$\Pr[X_T = 0|X_0 = j] = \sum_{k=0}^{N} \Pr[X_T = 0|X_0 = j, X_1 = k]\Pr[X_1 = k|X_0 = j]$$

$$= \sum_{k=0}^{N} \Pr[X_T = 0|X_1 = k]\Pr[X_1 = k|X_0 = j]$$

(Markov property)

$$= \sum_{k=0}^{N} P_{jk}\Pr[X_T = 0|X_1 = k]$$

$$= q_j\Pr[X_T = 0|X_1 = j-1] + r_j\Pr[X_T = 0|X_1 = j]$$
$$+ p_j\Pr[X_T = 0|X_1 = j+1]$$

After the first transition, the probability $\Pr[X_T = 0|X_1 = j] = u_j$ remains the same as the initial $\Pr[X_T = 0|X_0 = j]$ because T is a random variable depending on the state and not on the discrete-time. Hence, we obtain the equations

$$u_0 = 1$$
$$u_j = q_j u_{j-1} + r_j u_j + p_j u_{j+1} \qquad (1 \le j < N)$$

which are different from the corresponding steady-state equations (11.5). The difference lies in left-multiplication of P, yielding $u = Pu$ instead of right-multiplication in $\pi = \pi P$. Substituting $r_j = 1 - p_j - q_j$ gives after some modification

$$u_{j+1} = -\frac{q_j}{p_j}u_{j-1} + \left(1 + \frac{q_j}{p_j}\right)u_j \qquad (11.2)$$

Iteration on j for the first few values using $u_0 = 1$ yields

$$u_2 = -\frac{q_1}{p_1} + \left(1 + \frac{q_1}{p_1}\right)u_1$$

$$u_3 = \frac{q_2}{p_2}u_1 + \left(1 - \frac{q_2}{p_2}\right)u_2 = -\left(\frac{q_1}{p_1} + \frac{q_1 q_2}{p_1 p_2}\right) + \left(1 + \frac{q_1}{p_1} + \frac{q_1 q_2}{p_1 p_2}\right)u_1$$

which suggests

$$u_j = -\sum_{k=1}^{j-1} \prod_{m=1}^{k} \frac{q_m}{p_m} + \left(1 + \sum_{k=1}^{j-1} \prod_{m=1}^{k} \frac{q_m}{p_m}\right) u_1$$

as readily verified by substitution in (11.2). The unknown u_1 is determined by the last relation, $u_N = 0$. Finally, the probability of gambler's ruin is

$$\Pr\left[X_T = 0 | X_0 = j\right] = \frac{\sum_{k=j}^{N-1} \prod_{m=1}^{k} \frac{q_m}{p_m}}{1 + \sum_{k=1}^{N-1} \prod_{m=1}^{k} \frac{q_m}{p_m}} \tag{11.3}$$

Similarly, the mean hitting time $\eta_j = E\left[T | X_0 = j\right]$ follows by reasoning on the possible transitions. From state j, there is a transition to state $j-1$ with probability q_j. In case of a transition from state j to state $j-1$, the hitting time T consists of the first transition plus the remaining time from state $j-1$ on which is η_{j-1}. Using the law of total probability or reading all possible transitions from the transition graph (see Fig. 11.1), we find

$$\eta_j = 1 + q_j \eta_{j-1} + r_j \eta_j + p_j \eta_{j+1}$$

with the boundary equations for the state 0 and state N (which in the gambler's ruin problem are absorbing states), $\eta_0 = \eta_N = 0$. With $r_j = 1 - p_j - q_j$,

$$\eta_{j+1} = -\frac{1}{p_j} - \frac{q_j}{p_j} \eta_{j-1} + \left(1 + \frac{q_j}{p_j}\right) \eta_j$$

By iteration,

$$\eta_2 = -\frac{1}{p_1} + \left(1 + \frac{q_1}{p_1}\right) \eta_1$$

$$\eta_3 = -\frac{1}{p_2} - \frac{q_2}{p_2} \eta_1 + \left(1 + \frac{q_2}{p_2}\right) \eta_2 = -\frac{1}{p_1} - \frac{1}{p_2}\left(1 + \frac{q_2}{p_1}\right) + \left(1 + \frac{q_1}{p_1} + \frac{q_1 q_2}{p_1 p_2}\right) \eta_1$$

$$\eta_4 = -\frac{1}{p_3} - \frac{q_3}{p_3} \eta_2 + \left(1 + \frac{q_3}{p_3}\right) \eta_3$$

$$= -\frac{1}{p_1} - \frac{1}{p_2}\left(1 + \frac{q_2}{p_1}\right) - \frac{1}{p_3}\left(1 + \frac{q_3}{p_2} + \frac{q_2 q_3}{p_1 p_2}\right) + \left(1 + \frac{q_1}{p_1} + \frac{q_1 q_2}{p_1 p_2} + \frac{q_1 q_2 q_3}{p_1 p_2 p_3}\right) \eta_1$$

or

$$\eta_j = -\sum_{k=1}^{j-1} \frac{1}{p_k}\left(1 + \sum_{n=1}^{k-1} \prod_{m=1}^{n} \frac{q_{k-m+1}}{p_{k-m}}\right) + \left(1 + \sum_{k=1}^{j-1} \prod_{m=1}^{k} \frac{q_m}{p_m}\right) \eta_1$$

Eliminating η_1 from $\eta_N = 0$, finally leads to mean hitting time to ruin or the mean duration of the game

$$\eta_j = -\sum_{k=1}^{j-1} \frac{1}{p_k}\left(1 + \sum_{n=1}^{k-1}\prod_{m=1}^{n}\frac{q_{k-m+1}}{p_{k-m}}\right)$$

$$+ \left(1 + \sum_{k=1}^{j-1}\prod_{m=1}^{k}\frac{q_m}{p_m}\right)\frac{\sum_{k=1}^{N-1}\frac{1}{p_k}\left(1 + \sum_{n=1}^{k-1}\prod_{m=1}^{n}\frac{q_{k-m+1}}{p_{k-m}}\right)}{1 + \sum_{k=1}^{N-1}\prod_{m=1}^{k}\frac{q_m}{p_m}} \qquad (11.4)$$

In spite of the relatively simple difference equations, the solution rapidly grows unattractive. The particular case of the Markov chain where $q_k = q$ and $p_k = p$ simplifies considerably. The probability of gambler's ruin (11.3) becomes

$$\Pr\left[X_T = 0 | X_0 = j\right] = \frac{\sum_{k=j}^{N-1}\left(\frac{q}{p}\right)^k}{\sum_{k=0}^{N-1}\left(\frac{q}{p}\right)^k} = \frac{\left(\frac{q}{p}\right)^j - \left(\frac{q}{p}\right)^N}{1 - \left(\frac{q}{p}\right)^N}$$

and, if $p = q = \frac{1}{2}$, via de l'Hospital's rule, $\Pr\left[X_T = 0 | X_0 = j\right] = \frac{N-j}{N}$. If the target fortune N (at which the game ends) is infinitely large,

$$\Pr\left[X_T = 0 | X_0 = j\right] = 1 \qquad \text{if } q \geq p$$

$$= \left(\frac{q}{p}\right)^j \qquad \text{if } q < p$$

which demonstrates that the gambler surely will loose all his money if his chances p on winning are smaller than those q on losing. Even in a fair game where $p = q$, he will be defeated surely. In a favorable game $(p > q)$ and with start capital j, ruin is possible with probability $\left(\frac{q}{p}\right)^j$. Another interpretation is a game with two players a and b in which player a starts with capital j and has winning chance of p, while player b starts with capital $N - j$ and wins with probability $q = 1 - p$.

Similarly, the mean duration of the game (11.4) simplifies to

$$\eta_j = -\sum_{k=1}^{j-1}\left(\frac{1 - \left(\frac{q}{p}\right)^k}{p - q}\right) + \left(\frac{1 - \left(\frac{q}{p}\right)^j}{1 - \left(\frac{q}{p}\right)^N}\right)\sum_{k=1}^{N-1}\frac{1 - \left(\frac{q}{p}\right)^k}{p - q}$$

or

$$E\left[T | X_0 = j\right] = \frac{1}{p - q}\left[N\left(\frac{1 - \left(\frac{q}{p}\right)^j}{1 - \left(\frac{q}{p}\right)^N}\right) - j\right]$$

11.2.2 *The steady-state*

The steady-state equation (9.23) for the vector component π_j becomes, for $1 \leq j < N$,

$$\pi_j = p_{j-1}\pi_{j-1} + r_j\pi_j + q_{j+1}\pi_{j+1} \tag{11.5}$$

and, for $j = 0$ and $j = N$,

$$\pi_0 = r_0\pi_0 + q_1\pi_1$$
$$\pi_N = p_{N-1}\pi_{N-1} + r_N\pi_N$$

We rewrite these equations using $r_j = 1 - q_j - p_j$, $r_0 = 1 - p_0$ and $r_N = 1 - q_N$ as

$$p_0\pi_0 = q_1\pi_1 \tag{11.6}$$
$$p_j\pi_j = (p_{j-1}\pi_{j-1} - q_j\pi_j) + q_{j+1}\pi_{j+1} \tag{11.7}$$
$$p_{N-1}\pi_{N-1} = q_N\pi_N$$

Explicitly, for a few values of j, we observe that

$$p_1\pi_1 = (p_0\pi_0 - q_1\pi_1) + q_2\pi_2 = q_2\pi_2$$
$$p_2\pi_2 = (p_1\pi_1 - q_2\pi_2) + q_3\pi_3 = q_3\pi_3$$
$$\cdots$$

or, in general, for all j,

$$p_j\pi_j = q_{j+1}\pi_{j+1}$$

By iteration of $\pi_{j+1} = \frac{p_j}{q_{j+1}}\pi_j$ starting at $j = 0$, we find

$$\pi_{j+1} = \frac{p_j}{q_{j+1}} \frac{p_{j-1}}{q_j} \cdots \frac{p_0}{q_1}\pi_0 = \pi_0 \prod_{m=0}^{j} \frac{p_m}{q_{m+1}}$$

The normalization $\|\pi\|_1 = 1$ yields a condition for π_0,

$$\pi_0 = \left(1 + \sum_{j=1}^{N} \prod_{m=0}^{j-1} \frac{p_m}{q_{m+1}}\right)^{-1}$$

which determines the complete steady-state vector π for the general random walk as

$$\pi_j = \frac{\prod_{m=0}^{j-1} \frac{p_m}{q_{m+1}}}{1 + \sum_{j=1}^{N} \prod_{m=0}^{j-1} \frac{p_m}{q_{m+1}}} \tag{11.8}$$

These relations remain valid even when the number of states N tends to infinity provided the infinite sum converges.

In the simple case where $p_k = p$ and $q_k = q$, we obtain with $\rho = \frac{p}{q}$,

$$\pi_j = \frac{(1 - \rho)\,\rho^j}{1 - \rho^{N+1}} \tag{11.9}$$

11.3 Birth and death process

A birth and death process[1] is defined by the infinitesimal generator matrix

$$Q = \begin{bmatrix} -\lambda_0 & \lambda_0 & 0 & 0 & 0 & 0 & 0 & \cdots \\ \mu_1 & -(\lambda_1 + \mu_1) & \lambda_1 & 0 & 0 & 0 & 0 & \cdots \\ 0 & \mu_2 & -(\lambda_2 + \mu_2) & \lambda_2 & 0 & 0 & 0 & \cdots \\ 0 & 0 & \mu_3 & -(\lambda_3 + \mu_3) & \lambda_3 & 0 & 0 & \cdots \\ 0 & 0 & 0 & \mu_4 & -(\lambda_4 + \mu_4) & \lambda_4 & 0 & \cdots \\ \vdots & \vdots & \vdots & \vdots & \vdots & \vdots & \vdots & \vdots \end{bmatrix}$$

The transition graph is shown in Fig. 11.2. Although the theory in the previous chapter was derived for finite-state Markov chains, the birth and death process is a generalization to an infinite number of states. The general random walk (Section 11.2) forms the embedded Markov chain of the birth and death process with transition probabilities specified by (10.21) resulting in $V_{i,i-1} = \frac{\mu_i}{\lambda_i + \mu_i}$, $V_{i,i+1} = \frac{\lambda_i}{\lambda_i + \mu_i}$ and $V_{ik} = 0$ for $k \neq i - 1 \neq i + 1$. The transition probability matrix is a tri-band diagonal matrix which is irreducible if all $\lambda_j > 0$ and $\mu_j > 0$.

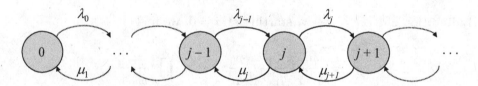

Fig. 11.2. The transition graph of a birth and death process.

The basic system of differential equations that completely describes the birth and death process follows from the general state probability equations (10.11) for $s_k(t) = \Pr[X(t) = k]$ as

$$s_0'(t) = -\lambda_0 s_0(t) + \mu_1 s_1(t) \tag{11.10}$$

$$s_k'(t) = -(\lambda_k + \mu_k)\,s_k(t) + \lambda_{k-1} s_{k-1}(t) + \mu_{k+1} s_{k+1}(t) \tag{11.11}$$

with initial condition $s_k(0) = \Pr[X(0) = k]$. Exact analytic solutions for

---

[1] Kleinrock (1975, p. ix) mentions that William Feller was the father of the birth and death process.

any λ_k and μ_k are not possible. Indeed, let us denote the Laplace transform of $s_k(t)$ by

$$S_k(z) = \int_0^\infty e^{-zt} s_k(t) dt. \tag{11.12}$$

Since $s_k(t)$ is a continuous and bounded function ($|s_k(t)| \le 1$ for all $t > 0$), the Laplace transform exists for $\operatorname{Re}(z) > 0$. The Laplace transform of (11.10) and (11.11) becomes,

$$(\lambda_0 + z) S_0(z) = s_0(0) + \mu_1 S_1(z) \tag{11.13}$$

$$(\lambda_k + \mu_k + z) S_k(z) = s_k(0) + \lambda_{k-1} S_{k-1}(z) + \mu_{k+1} S_{k+1}(z) \tag{11.14}$$

which is a set of difference equations more complex due to the initial condition $s_k(0)$ than the set (A.51) in Appendix A.5.2.3. That set (A.51) appears in the general random walk whose solution is shown to be intractable in general. This infinite set of differential equations has been thoroughly studied over years under several simplifying conditions for λ_k and μ_k, for example, $\lambda_k = \lambda$ and $\mu_k = \mu$ for all k. As shown in Chapter 13, they form the basis for the simplest set of queueing models of the family M/M/m/K.

11.3.1 The steady-state

The steady-state follows from (10.19) as solution of the set

$$-\lambda_0 \pi_0 + \mu_1 \pi_1 = 0$$

$$\lambda_{j-1} \pi_{j-1} - (\lambda_j + \mu_j) \pi_j + \mu_{j+1} \pi_{j+1} = 0$$

This set is identical to (11.6) and (11.7) provided p_j and q_j are changed for λ_j and μ_j. After this modification in (11.8), the steady-state of the birth and death process is

$$\pi_0 = \frac{1}{1 + \sum_{j=1}^\infty \prod_{m=0}^{j-1} \frac{\lambda_m}{\mu_{m+1}}} \tag{11.15}$$

$$\pi_j = \frac{\prod_{m=0}^{j-1} \frac{\lambda_m}{\mu_{m+1}}}{1 + \sum_{j=1}^\infty \prod_{m=0}^{j-1} \frac{\lambda_m}{\mu_{m+1}}} \qquad j \ge 1 \tag{11.16}$$

Theorem 9.3.4 states that an irreducible Markov chain with a finite number of states is necessarily recurrent. However, it is in general difficult to decide whether an irreducible Markov chain with an infinite number of states is recurrent or transient. In case of the birth and death process, it is possible to determine when the process is transient or recurrent. The process is transient if and only if the embedded Markov chain (determined above) is transient. Section 9.2.3 discusses that, for a recurrent chain, $r_{ij} = \Pr[T_j < \infty | X_0 = i]$ equals 1: a finite hitting time means every state j is certainly visited starting from initial state i. Applied to the embedded Markov chain, it follows from the gambler's ruin (11.3) that

$$\Pr[T_0 < \infty | X_0 = j] = 1 - \frac{\sum_{k=0}^{j-1} \prod_{m=1}^{k} \frac{q_m}{p_m}}{\sum_{k=0}^{N-1} \prod_{m=1}^{k} \frac{q_m}{p_m}}$$

Thus, for any fixed initial state j, the condition for a recurrent chain

$$\Pr[T_j < \infty | X_0 = i] = 1$$

is only possible in the limit $N \to \infty$ if $\lim_{N \to \infty} \sum_{k=0}^{N-1} \prod_{m=1}^{k} \frac{q_m}{p_m} = \infty$. Transformed to the birth and death rates, the condition for recurrence becomes $\Sigma_2 = \sum_{j=1}^{\infty} \prod_{m=0}^{j-1} \frac{\mu_m}{\lambda_m} = \infty$. Furthermore, we observe from (11.16) that the infinite series $\Sigma_1 = \sum_{j=1}^{\infty} \prod_{m=0}^{j-1} \frac{\lambda_m}{\mu_{m+1}}$ must converge to have a stationary or steady-state distribution.

In summary, if $\Sigma_1 < \infty$ and $\Sigma_2 = \infty$ the birth and death process is positive recurrent. If $\Sigma_1 = \infty$ and $\Sigma_2 = \infty$, it is null recurrent. If $\Sigma_2 < \infty$, the birth and death process is transient.

11.3.2 A pure birth process

A pure birth process is defined as a process $\{X(t), t \geq 0\}$ for which in any state i it holds that $\mu_i = 0$. It follows from Fig. 11.2 that a birth process can only jump to higher states such that $P_{ij}(t) = 0$ for $j < i$. Similarly, in a pure death process $\{X(t), t \geq 0\}$ all birth rates $\lambda_i = 0$.

11.3.2.1 The Poisson process

Let us first consider the simplest case where all birth rates are equal $\lambda_i = \lambda$ and where $P_{jj}(t) = \Pr[\tau_j > t | X(0) = j] = e^{-\lambda t}$. Using either the back or forward equation or (10.25) with $V_{i,i+1} = \delta_{i,i+1}$, yields

$$P_{ij}(t) = \delta_{ij} e^{-\lambda t} + \lambda \int_0^t e^{-\lambda u} P_{i+1,j}(t - u) du$$

or, for $j = i + k$ with $k > 0$

$$P_{i,i+k}(t) = \lambda e^{-\lambda t} \int_0^t e^{\lambda u} P_{i+1,i+k}(u) du$$

Explicitly, for $k = 1$,

$$P_{i,i+1}(t) = \lambda e^{-\lambda t} \int_0^t e^{\lambda u} e^{-\lambda u} du = \lambda t e^{-\lambda t}$$

which is independent of i and, thus, it holds for any $i \geq 1$. For $k = 2$,

$$P_{i,i+2}(t) = \lambda e^{-\lambda t} \int_0^t e^{\lambda u} P_{i+1,i+2}(u) du = \lambda e^{-\lambda t} \int_0^t e^{\lambda u} \lambda u e^{-\lambda u} du = \frac{(\lambda t)^2}{2} e^{-\lambda t}$$

This suggests us to propose for any $i \geq 0$,

$$P_{i,i+k}(t) = \frac{(\lambda t)^k}{k!} e^{-\lambda t} \tag{11.17}$$

which is verified inductively as

$$P_{i,i+k}(t) = \lambda e^{-\lambda t} \int_0^t e^{\lambda u} P_{i+1,i+k}(u) du = \lambda e^{-\lambda t} \int_0^t e^{\lambda u} P_{i,i+k-1}(u) du$$

$$= \lambda e^{-\lambda t} \int_0^t e^{\lambda u} \frac{(\lambda u)^{k-1}}{(k-1)!} e^{-\lambda u} du = \frac{(\lambda t)^k}{k!} e^{-\lambda t}$$

Hence, the transition probabilities of a pure birth process have a Poisson distribution (11.17) and are only function of the difference in states $k = j - i \geq 0$ for any $t \geq 0$. Moreover, for $0 \leq u \leq t$, consider the increment $X(t) - X(u)$,

$$\Pr[X(t) - X(u) = k] = \sum_{i \geq 0} \Pr[X(u) = i, X(t) = i + k]$$

$$= \sum_{i \geq 0} \Pr[X(u) = i] \Pr[X(t) = i + k | X(u) = i]$$

$$= \sum_{i \geq 0} \Pr[X(u) = i] P_{i,i+k}(t - u)$$

$$= \frac{(\lambda(t-u))^k}{k!} e^{-\lambda(t-u)} \sum_{i \geq 0} \Pr[X(u) = i]$$

Thus, the increment $X(t) - X(u)$ has a Poisson distribution,

$$\Pr[X(t) - X(u) = k] = \frac{(\lambda(t-u))^k}{k!} e^{-\lambda(t-u)} \tag{11.18}$$

and, since $X(0) = 0$ and the increments are independent (Markov property), we conclude that the pure birth process is a Poisson process (Section 7.2).

11.3.2.2 The general birth process

In case the birth rates λ_k depend on the actual state k, the pure birth process can be regarded as the simplest generalization of the Poisson. The Laplace transform difference equations (11.13) and (11.14) reduce to the set

$$S_0(z) = \frac{s_0(0)}{\lambda_0 + z}$$

$$S_k(z) = \frac{s_k(0)}{\lambda_k + z} + \frac{\lambda_{k-1}}{\lambda_k + z} S_{k-1}(z)$$

which, by the usual iteration, has the solution,

$$S_k(z) = \sum_{j=0}^{k} \frac{s_j(0) \prod_{m=j}^{k-1} \lambda_m}{\prod_{m=j}^{k} (\lambda_m + z)} \tag{11.19}$$

with the convention that $\prod_{m=a}^{b} f(m) = 1$ if $a > b$. The validity of this general solution is verified by substitution into the difference equation for $S_k(z)$. The form of $S_k(z)$ is a ratio that can always be transformed back to the time-domain provided that λ_k is known. If all $\lambda_k > 0$ are distinct, using (2.38) with $c > 0$, we find

$$s_k(t) = \sum_{j=0}^{k} s_j(0) \prod_{m=j}^{k-1} \lambda_m \frac{1}{2\pi i} \int_{c-i\infty}^{c+i\infty} \frac{e^{zt}}{\prod_{m=j}^{k} (\lambda_m + z)} dz$$

By closing the contour over the negative real plane ($\mathrm{Re}(z) < 0$), only simple poles at $z = -\lambda_n$ are encountered,

$$\frac{1}{2\pi i} \int_{c-i\infty}^{c+i\infty} \frac{e^{zt}}{\prod_{m=j}^{k} (\lambda_m + z)} dz = -\sum_{n=j}^{k} \frac{e^{-\lambda_n t}}{\prod_{m=j; m \neq n}^{k} (\lambda_m - \lambda_n)}$$

resulting in

$$s_k(t) = -\sum_{j=0}^{k} s_j(0) \sum_{n=j}^{k} \frac{e^{-\lambda_n t} \prod_{m=j}^{k-1} \lambda_m}{\prod_{m=j; m \neq n}^{k} (\lambda_m - \lambda_n)} \tag{11.20}$$

If some $\lambda_k = \lambda_j$, multiple poles occur and a slightly more complex result appears that still can be computed in exact analytic form.

11.3.2.3 The Yule process

A classical example of a process with distinct birth rates is the Yule process, where $\lambda_k = k\lambda$. In that case, (11.20) can be simplified. With

$$\frac{\prod_{m=j}^{k-1} \lambda_m}{\prod_{m=j; m \neq n}^{k} (\lambda_m - \lambda_n)} = \frac{(k-1)!}{(j-1)! \prod_{m=j}^{n-1} (m-n) \prod_{m=n+1}^{k} (m-n)}$$

and with $\prod_{m=j}^{n-1} (m-n) = (-1)^{n-1-j} \prod_{l=1}^{n-j} l = (-1)^{n-1-j}(n-j)!$ and $\prod_{m=n+1}^{k} (m-n) = \prod_{l=1}^{k-n} l = (k-n)!$ we find

$$\frac{\prod_{m=j}^{k-1} \lambda_m}{\prod_{m=j;m\neq n}^{k} (\lambda_m - \lambda_n)} = \frac{(-1)^{n-1-j}(k-1)!}{(j-1)!(n-j)!(k-n)!}$$

such that

$$\sum_{n=j}^{k} \frac{e^{-\lambda_n t} \prod_{m=j}^{k-1} \lambda_m}{\prod_{m=j;m\neq n}^{k} (\lambda_m - \lambda_n)} = \frac{(k-1)!}{(j-1)!} \sum_{n=0}^{k-j} \frac{e^{-(n+j)\lambda t}(-1)^{n-1}}{n!(k-j-n)!}$$

$$= -\frac{(k-1)!e^{-j\lambda t}}{(k-j)!(j-1)!} \sum_{n=0}^{k-j} \binom{k-j}{n} \left(-e^{-\lambda t}\right)^n$$

$$= -\binom{k-1}{j-1} e^{-j\lambda t} \left(1 - e^{-\lambda t}\right)^{k-j}$$

Finally, for the Yule process, we obtain from (11.20) the evolution of the state probabilities over time

$$s_k(t) = \sum_{j=0}^{k} s_j(0) \binom{k-1}{j-1} e^{-j\lambda t} \left(1 - e^{-\lambda t}\right)^{k-j} \tag{11.21}$$

In practice, $s_j(0) = \delta_{jn}$ if the process starts from state n (implying $s_k(t) = 0$ for $k < n$ because the process moves to the right for $t \geq 0$) and the general form simplifies to

$$s_k(t) = \binom{k-1}{n-1} e^{-n\lambda t} \left(1 - e^{-\lambda t}\right)^{k-n} \tag{11.22}$$

The Yule process has been used as a simple model for the evolution of a population in which each individual gives birth at exponential rate λ and $X(t)$ denotes the number of individuals in the population (that never decreases as there are no deaths) as a function of time t. At each state k the population has precisely k individuals that each generate births such that $\lambda_k = k\lambda$, the birth rate of the population. If the population starts at $t = 0$ with one individual $n = 1$, the evolution over time has the distribution $s_k(t) = e^{-\lambda t} \left(1 - e^{-\lambda t}\right)^{k-1}$, which is recognized from (3.5) as a geometric distribution with mean $e^{\lambda t}$. Since the sojourn times of a Markov process are i.i.d. exponential random variables, the average time T_k to reach k individuals from one ancestor equals $E[T_k] = \sum_{j=1}^{k} \frac{1}{\lambda_j} = \frac{1}{\lambda} \sum_{j=1}^{k} \frac{1}{j}$ which is well approximated (Abramowitz and Stegun, 1968, Section 6.3.18) as $E[T_k] \approx \frac{\log(k+1)+\gamma}{\lambda}$, where $\gamma = 0.577\,215\ldots$ is Euler's constant. If the population starts with n individuals, the distribution (11.22) at time t consists of a sum of n i.i.d. geometric random variables, which is a negative binomial distribution. The Yule process has been employed for example as a crude model to estimate the spread of a disease or epidemic and the split of molecules in new species by cosmic rays.

11.3.3 Constant rate birth and death process

In a constant rate birth and death process, both the birth rate $\lambda_k = \lambda$ and death rate $\mu_k = \mu$ are constant for any state k. From (11.16), the steady-state for all states j with $\rho = \frac{\lambda}{\mu} < 1$,

$$\pi_j = (1 - \rho)\,\rho^j \qquad\qquad j \geq 0 \qquad\qquad (11.23)$$

only depends on the ratio of birth over death rate. The time-dependent constant rate birth and death process can still be computed in analytic form. In this case, the matrix form of the infinitesimal generator Q has the tri-band Toeplitz structure, which can be diagonalized in analytic form as shown in Appendix A.5.2.1. In this section, we present an alternative approach. Instead of dealing with an infinite set of difference equations, a generating function approach seems more convenient. Let us denote the generating function of the Laplace transforms $S_k(z)$ by

$$\varphi(x, z) = \sum_{k=0}^{\infty} S_k(z)x^k \qquad\qquad (11.24)$$

Using (11.12) into (11.24) gives

$$\varphi(x, z) = \sum_{k=0}^{\infty} \int_0^{\infty} e^{-zt} s_k(t)x^k dt = \int_0^{\infty} e^{-zt} \sum_{k=0}^{\infty} s_k(t)x^k dt$$

where the reversal of summation and integration is allowed because all terms are positive. Since $0 \leq s_k(t) \leq 1$, the sum is at least convergent for $|x| < 1$, which shows that $\varphi(x, z)$ is analytic inside the unit circle $|x| < 1$ for any $\mathrm{Re}(z) > 0$.

After multiplying (11.14) by x^k and summing over all k, we obtain

$$(\lambda + z)\,S_0(z) + (\lambda + \mu + z)\sum_{k=1}^{\infty} S_k(z)x^k = \sum_{k=0}^{\infty} s_k(0)x^k + \lambda\sum_{k=1}^{\infty} S_{k-1}(z)x^k$$

$$+ \mu S_1(z) + \mu\sum_{k=1}^{\infty} S_{k+1}(z)x^k$$

and, written in terms of $\varphi(x, z)$,

$$\varphi(x, z) = \frac{-\sum_{k=0}^{\infty} s_k(0) x^{k+1} + \mu(1 - x) S_0(z)}{\lambda x^2 - x(\lambda + \mu + z) + \mu}$$

Note that in the general case λ_k and μ_k, an expression in terms of $\varphi(x, z)$ is not possible. Before continuing with the computations, we make the additional simplification that $s_k(0) = \delta_{kj}$: we assume that the constant rate birth and death process starts in state j. With this initial condition, the generating function

$$\varphi(x, z) = \frac{\mu(1 - x) S_0(z) - x^{j+1}}{\lambda x^2 - x(\lambda + \mu + z) + \mu} \tag{11.25}$$

still depends on the unknown function $S_0(z)$. The following derivation involving the theory of complex functions demonstrates a standard procedure that will also be useful in other queuing problems.

The denominator in (11.25) has two roots,

$$x_1 = \frac{\lambda + \mu + z}{2\lambda} + \frac{1}{2\lambda}\sqrt{(\lambda + \mu + z)^2 - 4\lambda\mu}$$

$$x_2 = \frac{\lambda + \mu + z}{2\lambda} - \frac{1}{2\lambda}\sqrt{(\lambda + \mu + z)^2 - 4\lambda\mu}$$

We need the powerful theorem of Rouché (Titchmarsh, 1964, p. 116) to deduce more on the location of x_1 and x_2.

Theorem 11.3.1 (Rouché) *If $f(z)$ and $g(z)$ are analytic inside and on a closed contour C, and $|g(z)| < |f(z)|$ on C, then $f(z)$ and $f(z) + g(z)$ have the same number of zeros inside C.*

Choose $f(x) = \mu - x(\lambda + \mu + z)$ and $g(x) = \lambda x^2$ such that $f(x) + g(x) = \lambda x^2 - x(\lambda + \mu + z) + \mu$, the denominator in (11.25). Since both $f(x)$ and $g(x)$ are polynomials, they are analytic everywhere in the complex x-plane. We know that $\varphi(x, z)$ is analytic inside the unit disk. If the roots x_1 or x_2 lie inside the unit disk, the numerator in (11.25) must have zeros at precisely the same place in order for $\varphi(x, z)$ to be analytic inside the unit disk. Hence, we consider as contour C in Rouché's Theorem, the unit circle $|x| = 1$. Clearly, $f(x)$ has one single zero $\frac{\mu}{\lambda + \mu + z}$ inside the unit circle (because $\lambda > 0$, $\mu > 0$ and $\mathrm{Re}(z) > 0$). Furthermore, on the unit circle $|x| = 1$,

$$|\mu - x(\lambda + \mu + z)| \geq |\mu - |x||(\lambda + \mu + z)|| = |\mu - |(\lambda + \mu + z)|| > \lambda = \lambda|x^2|$$

which shows that $|g(z)| < |f(z)|$ on the unit circle. Rouché's Theorem then tells us that $f(x) + g(x)$ has precisely one zero inside the unit circle. This implies that $|x_1| > 1$ and $|x_2| < 1$ and that the numerator in (11.25) has a zero x_2,

$$\mu (1 - x_2) S_0(z) - x_2^{j+1} = 0$$

This relation determines the unknown function $S_0(z)$ as

$$S_0(z) = \frac{x_2^{j+1}}{\mu (1 - x_2)}$$

such that (11.25) becomes

$$\varphi(x, z) = \frac{x_2^{j+1} \frac{1-x}{1-x_2} - x^{j+1}}{\lambda (x - x_1)(x - x_2)} = \frac{x_2^{j+1}(1 - x) - x^{j+1}(1 - x_2)}{\lambda (1 - x_2)(x - x_1)(x - x_2)}$$

We know that the numerator can be divided by $(x - x_2)$, or explicitly,

$$x_2^{j+1}(1 - x) - x^{j+1}(1 - x_2) = x_2^{j+1} - x^{j+1} + x_2 x(x^j - x_2^j)$$

$$= (x - x_2)\left[-\sum_{k=0}^{j} x_2^{j-k}x^k + x_2 x \sum_{k=0}^{j-1} x_2^{j-1-k}x^k \right]$$

$$= (x - x_2)\left[-x_2^j + (x_2 - 1)\sum_{k=1}^{j} x_2^{j-k}x^k \right]$$

Finally,

$$\varphi(x, z) = \frac{-x_2^j + (x_2 - 1)\sum_{k=1}^{j} x_2^{j-k}x^k}{\lambda (1 - x_2)(x - x_1)} \tag{11.26}$$

By expanding the denominator in a Taylor series around $x = 0$, and denoting

$$a_0 = -x_2^j$$
$$a_k = (x_2 - 1)x_2^{j-k}$$
$$a_k = 0 \qquad\qquad k > j$$

we have

$$\varphi(x, z) = \frac{1}{\lambda (1 - x_2)}\left(\sum_{k=0}^{\infty} a_k x^k \right)\left(-\sum_{m=0}^{\infty} x_1^{-m}x^m \right)$$

$$= \frac{-1}{\lambda (1 - x_2)} \sum_{k=0}^{\infty}\left(\sum_{m=0}^{k} \frac{a_{k-m}}{x_1^m} \right) x^k$$

Comparing with (11.24) and equating the corresponding powers in x, we find an explicit form of the Laplace transforms of the probabilities that the birth and death process is in state k,

$$S_k(z) = \frac{-1}{\lambda(1-x_2)} \sum_{m=0}^{k} \frac{a_{k-m}}{x_1^m}$$

$$= \frac{x_2^{j-k}}{\lambda} \left(\frac{1}{(1-x_2)} \frac{x_2^k}{x_1^k} + \sum_{m=0}^{k-1} \frac{x_2^m}{x_1^m} \right) 1_{k<j}$$

$$= \frac{1}{\lambda} \left(\frac{1}{(1-x_2)} \frac{x_2^j}{x_1^k} + \frac{x_2^{j-k} x_1 - x_2^j x_1^{1-k}}{x_1 - x_2} \right) 1_{k<j}$$

This expression can be put in different forms by using relations among the zeros x_1 and x_2, such as $x_1 + x_2 = \frac{\lambda+\mu+z}{\lambda}$ and $x_1 x_2 = \frac{\mu}{\lambda}$. This ingenuity is required to recognize in $S_k(z)$ a known Laplace transform. Otherwise, one has to proceed by computing the inverse Laplace transform by contour integration via (2.38). In any case, the computation needs advanced skills in complex function theory and we content ourselves here to present the result without derivation (see e.g. Cohen (1969, pp. 80–82)),

$$s_k(t) = e^{-(\lambda+\mu)t} \left[\rho^{(k-j)/2} I_{k-j}(at) + \rho^{(k-j-1)/2} I_{k+j+1}(at) \right] \tag{11.27}$$

$$+ e^{-(\lambda+\mu)t} (1-\rho)\rho^k \sum_{m=k+j+2}^{\infty} \rho^{-m/2} I_m(at)$$

where $\rho = \frac{\lambda}{\mu}$, $a = 2\mu\sqrt{\rho}$ and where $I_s(z)$ denotes the modified Bessel function (Abramowitz and Stegun, 1968, Section 9.6.1). Using the asymptotic formulas for the modified Bessel function, the behavior of $s_k(t)$ for large t can be derived (see e.g. Cohen (1969, p. 84)),

$$s_k(t) = (1-\rho)\rho^k + \frac{\rho^{(k-j)/2} e^{-(1-\sqrt{\rho})^2 \mu t}}{2\sqrt{\pi} \left(\sqrt{\rho}\mu t\right)^{3/2}} \left[\left(k - \frac{\sqrt{\rho}}{1-\sqrt{\rho}} \right) \left(j - \frac{\sqrt{\rho}}{1-\sqrt{\rho}} \right) + O(t^{-1}) \right]$$

$$= \frac{1}{\sqrt{\pi\mu t}} \left[1 + O(t^{-1}) \right] \qquad \text{only if } \rho = 1$$

This expression demonstrates that the constant rate birth death process converges to the steady-state $(1-\rho)\rho^k$ with a relaxation rate $\left(1-\sqrt{\rho}\right)^2 \mu$. Clearly, the higher ρ, the lower the relaxation rate and the slower the process tends to equilibrium as illustrated in Fig. 11.3. Intuitively, two effects play a role. Since the probability that states with large k are visited increases with increasing ρ, the built-up time for this occupation will be larger. In addition, the variability of the number of visited states (further derived for the M/M/1 queue in Section 14.1) increases with increasing ρ, which

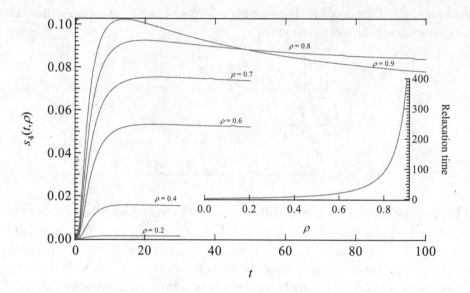

Fig. 11.3. The probability $s_4(t)$ that the process is in state 4 given that it started from state 0 as function of time (in units of average death time, $\mu = 1$) for various $\rho = \lambda$. The insert shows the relaxation time $\tau = (1 - \sqrt{\rho})^{-2}$ (in units of average death time, $\mu = 1$). The corresponding steady state probability π_4 are 0.0012, 0.015, 0.051, 0.072, 0.082, 0.065 for $\rho = 0.2, 0.4, 0.6, 0.7, 0.8, 0.9$ respectively. Observe that for $\rho = 0.9$, the plotted 100 time units are smaller than the relaxation time, which is 379 time units.

suggests that larger oscillations of the sample paths around the steady-state are likely to occur, enlarging the convergence time.

11.4 A random walk on a graph

Let $G(N, L)$ denotes a graph with N nodes and L links. Suppose that the link weight $w_{ij} = w_{ji}$ of an edge from node $i \rightarrow j$ (or vice versa) is proportional to the transition probability P_{ij} that a packet at node i decides to move to node j. Clearly, $w_{ii} = 0$. Specifically, with (9.8),

$$P_{ij} = \frac{w_{ij}}{\sum_{k=1}^{N} w_{ik}}$$

This constraint (9.8) destroys the symmetry in link weight structure ($w_{ij} = w_{ji}$) because, in general, $P_{ij} \neq P_{ji}$ since $\sum_{k=1}^{N} w_{ik} \neq \sum_{k=1}^{N} w_{jk}$. The sequence of nodes (or links) visited by that packet resembles a random walk on the graph $G(N, L)$ and constitutes a Markov chain. Moreover, the steady-state of this Markov process is readily obtained by observing that the chain is time reversible. Indeed, the condition for time reversibility (10.27) becomes

$$\frac{\pi_i w_{ij}}{\sum_{k=1}^{N} w_{ik}} = \frac{\pi_j w_{ji}}{\sum_{k=1}^{N} w_{jk}}$$

or, since $w_{ij} = w_{ji}$,

$$\frac{\pi_i}{\sum_{k=1}^{N} w_{ik}} = \frac{\pi_j}{\sum_{k=1}^{N} w_{jk}}$$

This implies that $\pi_i = \alpha \sum_{k=1}^{N} w_{ik}$ and using the normalization $\|\pi\|_1 = 1$, we obtain the steady-state probabilities for all nodes i,

$$\pi_i = \frac{\sum_{k=1}^{N} w_{ik}}{\sum_{i=1}^{N} \sum_{k=1}^{N} w_{ik}} = \frac{\sum_{k=1}^{N} w_{ik}}{2 \sum_{i=1}^{N} \sum_{k=i+1}^{N} w_{ik}}$$

This Markov process can model an active packet that monitors the network by collecting state information (number of packets, number of lost or retransmitted packets, etc.) in each router. Of course, the link weight structure w_{ij} for the active packet is decisive and requires additional information to be chosen efficiently. For example, for traffic monitoring, the distribution of the number of packets forwarded by each router must be obtained. For the collection of these data, the active packet should in steady-state visit all nodes about equally frequently or $\pi_i = \frac{1}{N}$, implying that the Markov transition matrix P must be doubly stochastic (see Appendix A.5.1).

11.5 Slotted Aloha

The Aloha protocol is a basic example of a multiple access communication scheme of which Ethernet[2] is considered as the direct descendant. Aloha – which means "hello" in the Hawaiian language – was invented by Norman Abramson at the university of Hawaii in the beginning of 1970s to provide packet-switched radio communication between a central computer and various data terminals at the campus. Slotted Aloha is a discrete-time version of the pure Aloha protocol, where all transmitted packets have equal length and where each packet requires one timeslot for transmission.

Consider a network consisting of N nodes that can communicate with each other via a shared communication channel (e.g. a radio channel) using the slotted Aloha protocol. The simplest arrival process A of packets at each node is a Poisson process. We assume that these Poisson arrivals at a

---

[2] The essential difference with the Ethernet's CSMA/CD (carrier sense multiple access with collision detection) is that Aloha does not use carrier sensing and does not stop transmitting when collisions are detected. Carrier sensing is only adequate if the nodes are near to each other (as in a local area network) such that collisions can be detected before the completion of transmission. Only then is a timely reaction possible.

node are independent from the Poisson arrivals at another node and that all Poisson arrivals at a node have the same rate $\frac{\lambda}{N}$ where λ is the overall arrival rate at the network of N nodes. The idea of the Aloha protocol is that, upon receipt of a packet, the node transmits that newly arrived packet in the next timeslot. In case two nodes happen to transmit a packet at the same timeslot, a collision occurs, which results in a retransmission of the packets. A node with a packet that must be retransmitted is said to be *backlogged*. Even if new packets arrive at a backlogged node, the retransmitted packet is the first one to be transmitted and, for simplicity (to ignore queueing of packets at a node), we assume that those new packets are discarded. If backlogged nodes retransmit the packet in the next timeslot, surely a new collision would occur. Therefore, backlogged nodes wait for some random number of timeslots before retransmitting. We assume, for simplicity, that p_r is the probability (which is the same for all backlogged nodes) that a successful transmission occurs in the next time slot. Moreover, the probability p_r of retransmission is the same for each timeslot. The number of time slots between the occurrence of a collision and a successful transmission is a geometric random variable T_r (see Section 3.1.3) with parameter p_r such that $\Pr\left[T_r = k\right] = p_r \left(1 - p_r\right)^{k-1}$.

11.5.1 The Markov chain

The slotted Aloha protocol constitutes a discrete-time Markov chain $X_k \in \{0, 1, \ldots, N\}$, where a state j counts the number of backlogged nodes out of the N nodes in total and the subscript k refers to the k-th timeslot. Each of the j backlogged nodes retransmits a packet in the next time slot with probability p_r, while each of the $N - j$ unbacklogged nodes will transmit surely a packet in the next time slot provided a packet arrives in the current timeslot. The latter event (at least one arrival A) occurs with probability $p_a = \Pr\left[A > 0\right] = 1 - \Pr\left[A = 0\right]$. If we assume that the arrival process is Poissonean, then $p_a = 1 - \exp\left(-\frac{\lambda}{N}\right)$, but the computations in this section are more generally valid.

The probability that n backlogged nodes in state j retransmit in the next time slot is binomially distributed

$$b_n(j) = \binom{j}{n} p_r^n \left(1 - p_r\right)^{j-n}$$

and, similarly, the probability that n unbacklogged nodes in state j transmit in the next time slot is

$$u_n(j) = \binom{N-j}{n} p_a^n (1-p_a)^{N-j-n}$$

A packet is transmitted successfully if and only if (a) one new arrival and no backlogged packet or (b) no new arrival and one backlogged packet is transmitted. The probability of successful transmission in state j and per time slot equals

$$p_s(j) = u_1(j)b_0(j) + u_0(j) b_1(j)$$

The transition probability $P_{j,j+m}$ equals

$$P_{j,j+m} = \begin{cases} u_m(j) & 2 \leq m \leq N-j \\ u_1(j)(1-b_0(j)) & m=1 \\ u_1(j) b_0(j) + u_0(j)(1-b_1(j)) & m=0 \\ u_0(j) b_1(j) & m=-1 \end{cases}$$

The state with j backlogged nodes jumps to the state $j-1$ with one backlogged node less if no new packets are sent and there is precisely 1 successful retransmission. The state j remains in the state j if there is 1 new arrival and there are no retransmission or if there are no new retransmissions and none or more than 1 retransmission. The state j jumps to state $j+1$ if there is 1 new arrival from a non-backlogged node and at least 1 retransmission because then there are surely collisions and the number of backlogged nodes increases by 1. Finally, the state j jumps to state $j+m$ if m new packets arrive from m different non-backlogged nodes, which always causes collisions irrespective of how many backlogged nodes also retransmit in the next time slot.

The Markov chain is illustrated in Fig. 11.4, which shows that the state can only decrease by at most 1.

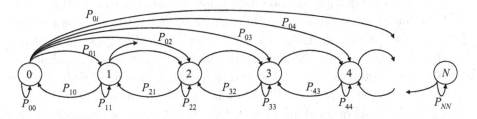

Fig. 11.4. Graph of the Markov chain for slotted Aloha. Each state j counts the number of backlogged nodes.

The transition probability matrix P has the structure

$$
P = \begin{bmatrix}
P_{00} & P_{01} & P_{02} & \cdots & & \cdots & & P_{0N} \\
P_{10} & P_{11} & P_{12} & \cdots & & \cdots & & P_{1N} \\
0 & P_{21} & P_{22} & \cdots & & \cdots & & P_{2N} \\
\vdots & \vdots & \vdots & \ddots & & & \vdots & \vdots \\
0 & 0 & \cdots & P_{N-1,N-2} & P_{N-1;N-1} & P_{N-1;N} \\
0 & 0 & \cdots & 0 & P_{N;N-1} & P_{NN}
\end{bmatrix}
$$

whose eigenstructure is computed in Appendix A.5.3.

In the asymptotic regime when $N \to \infty$, slotted Aloha has the peculiar property that the steady-state vector π does not exist. Although for a small number of nodes N, the steady-state equations can be solved, when the number N grows, Slotted Aloha turns out to be instable. It seems difficult to prove that $\lim_{N \to \infty} \pi = 0$, but there is another argument that suggests the truth of this awkward Aloha property. The expected change in backlog per time slot is equivalent to

$$
E[X_{k+1} - X_k | X_k = j] = (N - j)\, p_a - p_s(j) \tag{11.28}
$$

and equals the expected number of new arrivals minus the expected number of successful transmissions. This quantity $E[X_{k+1} - X_k | X_k = j]$ is often called the *drift*. If the drift is positive for all timeslots k, the Markov chain moves (on average) to higher states or to the right in Fig. 11.4. Since $p_s(j) \le 1$ and $p_a = 1 - \exp\left(-\frac{\lambda}{N}\right)$, it follows that

$$
\lim_{N \to \infty} E[X_{k+1} - X_k | X_k = j] = \infty
$$

Thus, the drift tends to infinity, which means that, on average, the number of backlogged nodes increases unboundedly and suggests (but does not prove, a counter example is given in problem (ii) of Section 9.4) that the Markov chain is transient for $N \to \infty$.

A more detailed discussion and engineering approaches to cure this instability are found in Bertsekas and Gallager (1992, Chapter 4). The interest of the analysis of slotted Aloha lies in the fact that other types of multiple access protocols, such as the important class of carrier sense multiple access (CSMA) protocols, can be deduced in a similar manner. Of the CSMA class with collision detection, Ethernet is by far the most important because it is the basis of local area networks. Multiple access protocols of the CSMA/CD type are discussed in our book *Data Communications Networking* (Van Mieghem, 2004a).

11.5.2 *Efficiency of slotted Aloha and the offered traffic* G

We now investigate the probability of a successful transmission in state j in more detail,

$$
\begin{aligned}
p_s\,(j) &= u_1(j)b_0\,(j) + u_0\,(j)\,b_1\,(j)\\
&= (N-j)\,p_a\,(1-p_a)^{N-j-1}\,(1-p_r)^j + jp_r\,(1-p_r)^{j-1}\,(1-p_a)^{N-j}\\
&= \left[\frac{(N-j)\,p_a}{1-p_a} + \frac{jp_r}{1-p_r}\right](1-p_a)^{N-j}\,(1-p_r)^j
\end{aligned}
$$

For small arrival probability p_a and small retransmission probability p_r, the probability of successful transmission in state j can be approximated by using the Taylor expansions of $(1-x)^\alpha = e^{\alpha\ln(1-x)} = e^{-\alpha x}\,(1+o\,(1))$ and $\frac{x}{1-x} = x + o\,(x^2)$ as

$$
\begin{aligned}
p_s\,(j) &= \left[(N-j)\,p_a + jp_r + o\,(p_a^2 + p_r^2)\right]e^{-[(N-j)p_a+jp_r]}\,(1+o\,(1))\\
&= \left[(N-j)\,p_a + jp_r\right]\exp\left(-\left[(N-j)\,p_a + jp_r\right]\right)(1+o\,(1))
\end{aligned}
$$

Similarly, the probability that no packet is transmitted in state j equals

$$
\begin{aligned}
p_{no}\,(j) &= u_0\,(j)\,b_0\,(j) = (1-p_r)^j\,(1-p_a)^{N-j}\\
&= \exp\left(-\left[(N-j)\,p_a + jp_r\right]\right)(1+o\,(1))
\end{aligned}
$$

Hence, for small p_a and small p_r, the probability of successful transmission and of no transmission in state j is well approximated by

$$
p_s\,(j) \simeq t\,(j)\,e^{-t(j)}
$$
$$
p_{no}\,(j) \simeq e^{-t(j)}
$$

Now, $t\,(j) = (N-j)\,p_a + jp_r$ is the expected number of arrivals and retransmissions in state j or, equivalently, the total rate of transmission attempts in state j. That total rate of transmissions in state j, $t(j)$, is also called the offered traffic G. The analysis shows that, for small p_a and small p_r, $p_s\,(j)$ and $p_{no}\,(j)$ are closely approximated in terms of a Poisson random variable with rate $t\,(j)$. Moreover, the probability of successful transmission $p_s\,(j)$ can be interpreted as the departure rate from state j or the throughput $S_{\mathrm{SAloha}} = Ge^{-G}$, which is maximized if $G = t\,(j) = 1$. By controlling p_r to achieve $t\,(j) = (N-j)\,p_a + jp_r = 1$, slotted Aloha performs with highest throughput. The *efficiency* η_{SAloha} of slotted Aloha with many nodes $N > 1$ is defined as the maximum fraction of time during which packets are transmitted successfully which is $\max p_s\,(j) = e^{-1}$. Hence, $\eta_{\mathrm{SAloha}} = 36\%$.

Pure Aloha (Bertsekas and Gallager, 1992), where the nodes can start transmitting at arbitrary times instead of only at the beginning of time slots, only performs half as efficiently as slotted Aloha with $\eta_{\text{PAloha}} = 18\%$. Recall that each packet is assumed to have an equal length that corresponds with the length of one timeslot. In pure Aloha, a transmitted packet at time t is successful if no other packet is sent during $(t - 1, t + 1)$. This time interval is precisely equal to two timeslots in slotted Aloha which explains why $\eta_{\text{PAloha}} = \frac{1}{2}\eta_{\text{SAloha}}$. The same observation tells us that, in pure Aloha, $p_{no}(j) \simeq e^{-2t(j)}$ because in the successful interval the expected number of arrivals and retransmissions is twice that in slotted Aloha. The throughput S roughly equals the total rate of transmission attempts G (which is the same as in slotted Aloha) multiplied by $p_{no}(j) \simeq e^{-2t(j)}$, hence, $S_{\text{PAloha}} = Ge^{-2G}$.

11.6 Ranking of webpages

To retrieve webpages related to a user's query, current websearch engines first perform a search similar to that in text processors to find all webpages containing the query terms. Due to the massive size of the World Wide Web, this first action can result in a huge number of retrieved webpages. Several thousands of webpages related to a query are not uncommon. To reduce the list of webpages, many websearch engines apply a ranking criterion to sort this list. In this section, we discuss PageRank, the hyperlink-based ranking system used by the Google search engine. PageRank elegantly exploits the power of discrete Markov theory.

11.6.1 A Markov model of the web

The hyperlink structure of the World Wide Web can be viewed as a directed graph with N nodes. Each node in the webgraph represents a certain webpage and the directed edges represent hyperlinks. Let us consider a small collection of webpages as in Fig. 11.5 to illustrate the underlying idea of PageRank, invented by Brin and Page, the founders of Google.

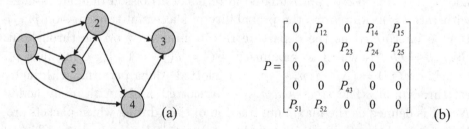

$$P = \begin{bmatrix} 0 & P_{12} & 0 & P_{14} & P_{15} \\ 0 & 0 & P_{23} & P_{24} & P_{25} \\ 0 & 0 & 0 & 0 & 0 \\ 0 & 0 & P_{43} & 0 & 0 \\ P_{51} & P_{52} & 0 & 0 & 0 \end{bmatrix}$$

(a)   (b)

Fig. 11.5. A subgraph of the World Wide Web (a) and the corresponding transition probability matrix P (b).

The topology of any graph is determined by an adjacency matrix (see Appendix B.1). A reasonable criterion to assess the importance of a webpage is the number of times that this webpage is visited. This criterion suggests us to consider a discrete Markov chain whose transition probability matrix P corresponds to adjacency matrix of the webgraph (shown in (b) in Fig. 11.5). The element P_{ij} of the Markov transition probability matrix is the probability of moving from webpage i (state i) to webpage j (state j) in one time step. The components $s_i[k]$ of the corresponding state vector $s[k]$ denotes the probability that at time k the webpage i is visited. The long run mean fraction of time that webpage i is visited equals the steady-state probability π_i of the Markov chain. This probability π_i is the ranking measure of the importance of webpage i used in Google. The basic idea is indeed simple, but we have not shown yet how to determine the elements P_{ij} nor whether the steady-state probability vector π exists. In particular, we will demonstrate that guaranteeing that the steady-state vector π exists and that π can be computed for a Markov chain containing some billion of states – the order of magnitude of the number N of webpages – requires a deeper knowledge of discrete Markov chains.

To start determining the elements P_{ij}, we assume that, given we are on webpage i, any hyperlink on that webpage has equal probability to be clicked on. This assumption implies that $P_{ij} = \frac{1}{d_i}$ where the degree d_i of a node i (see also Section 15.3) equals the number of adjacent neighbors of node i in the webgraph. This number d_i is thus equal the number of hyperlinks on webpage i. The transition probability matrix in Fig. 11.5 then becomes

$$P = \begin{bmatrix} 0 & \frac{1}{3} & 0 & \frac{1}{3} & \frac{1}{3} \\ 0 & 0 & \frac{1}{3} & \frac{1}{3} & \frac{1}{3} \\ 0 & 0 & 0 & 0 & 0 \\ 0 & 0 & 1 & 0 & 0 \\ \frac{1}{2} & \frac{1}{2} & 0 & 0 & 0 \end{bmatrix}$$

The uniformity assumption is in most cases the best we can make if no additional information is available. If, for example, web usage information is available showing that a random surfer accessing page 2 is twice as likely to jump to page 4 than to any other neighboring webpage of 2, then the second row can be replaced by $\begin{pmatrix} 0 & 0 & \frac{1}{4} & \frac{1}{2} & \frac{1}{4} \end{pmatrix}$.

When solely adopting the adjacency matrix of the webgraph as underlying structure of the Markov transition probability matrix P, we cannot assure that P is a stochastic matrix. For example, it often occurs that a node such as node 3 in our example in Fig. 11.5 does not contain outlinks. Such nodes are called *dangling nodes*. For example, many webpages may point to an important document on the web, which itself does not refer to any other webpage. The corresponding row in P possesses only zero elements, which violates the basic law (9.8) of a stochastic matrix. To rectify the deviation from a stochastic matrix, each zero row must be replaced by a particular non-zero row vector[3] v^T that obeys (9.8), i.e. $\|v\|_1 = v^T u = 1$ where $u^T = [1 \; 1 \; \cdots \; 1]$. Again, the simplest recipe is to invoke uniformity and to replace any zero row by $v^T = \frac{u^T}{N}$. In our example, we replace the third row by $\left(\frac{1}{5} \; \frac{1}{5} \; \frac{1}{5} \; \frac{1}{5} \; \frac{1}{5} \right)$ and obtain

$$\bar{P} = \begin{bmatrix} 0 & \frac{1}{3} & 0 & \frac{1}{3} & \frac{1}{3} \\ 0 & 0 & \frac{1}{3} & \frac{1}{3} & \frac{1}{3} \\ \frac{1}{5} & \frac{1}{5} & \frac{1}{5} & \frac{1}{5} & \frac{1}{5} \\ 0 & 0 & 1 & 0 & 0 \\ \frac{1}{2} & \frac{1}{2} & 0 & 0 & 0 \end{bmatrix}$$

However, this adjustment is not sufficient to insure the existence of a steady-state vector π. In Section 9.3.1, we have shown that, if the Markov chain is irreducible, the steady-state vector exists. In an irreducible Markov chain any state is reachable from any other (Section 9.2.1.1). By its very nature, the World Wide Web leads almost surely to a reducible Markov chain. In order to create an irreducible matrix, Brin and Page have considered

$$\bar{\bar{P}} = \alpha \bar{P} + (1 - \alpha) \frac{u.u^T}{N}$$

where $0 < \alpha < 1$, $u.u^T$ is a $N \times N$ matrix with each element equal to 1 and $\bar{P}$ is the previously adjusted matrix without zero-rows. The linear combination of the stochastic matrix $\bar{P}$ and a stochastic perturbation matrix $u.u^T$ ensures that $\bar{\bar{P}}$ is an irreducible stochastic matrix. Every node is now directly connected (reachable in one step) to any other (because of $u.u^T$), which makes the Markov chain irreducible with aperiodic, positive recurrent states (see Fig. 9.1). Slightly more general, we can replace the matrix $\frac{u.u^T}{N}$ by uv^T, where v is a probability vector as above but where we must additionally require that each component of v is non-zero in order to guarantee reachability. Brin and Page have called v^T the *personalization* vector which enables to deviate from non-uniformity. Hence, we arrive at the Brin and Page Markov transition probability matrix

$$\bar{\bar{P}} = \alpha \bar{P} + (1 - \alpha) uv^T \qquad (11.29)$$

---

[3] We use the normal vector algebra convention, but remark that the stochastic vectors π and $s[k]$ are *also* row vectors (without the transpose sign)!

For $v^T = \begin{bmatrix} \frac{1}{16} & \frac{4}{16} & \frac{6}{16} & \frac{4}{16} & \frac{1}{16} \end{bmatrix}$ and $\alpha = \frac{4}{5}$, the probability transition matrix in our example becomes

$$\bar{\bar{P}} = \begin{bmatrix} \frac{1}{80} & \frac{19}{60} & \frac{3}{40} & \frac{19}{60} & \frac{67}{240} \\ \frac{1}{80} & \frac{1}{20} & \frac{41}{120} & \frac{19}{60} & \frac{67}{240} \\ \frac{1}{16} & \frac{1}{4} & \frac{3}{8} & \frac{1}{4} & \frac{1}{16} \\ \frac{1}{80} & \frac{1}{20} & \frac{7}{8} & \frac{1}{20} & \frac{1}{80} \\ \frac{33}{80} & \frac{9}{20} & \frac{3}{40} & \frac{1}{20} & \frac{1}{80} \end{bmatrix}$$

If the presented method were implemented, the initially very sparse matrix P would be replaced by the dense matrix $\bar{\bar{P}}$, which for the size N of the web would increase storage dramatically. Therefore, a more effective way is to define a special vector r whose component $r_j = 1$ if row j in P is a zero-row or node j is dangling node. Then, $\bar{P} = P + rv^T$ is a rank-one update of P and so is $\bar{\bar{P}}$ because

$$\bar{\bar{P}} = \alpha \left(P + rv^T \right) + (1 - \alpha)u.v^T = \alpha P + (\alpha r + (1 - \alpha)u)\, v^T$$

11.6.2 Computation of the PageRank steady-state vector

The steady-state vector π obeys the eigenvalue equation (9.22), thus $\pi = \pi \bar{P}$. Rather than solving this equation, Brin and Page propose to compute the steady-state vector from $\pi = \lim_{k\to\infty} s[k]$. Specifically, for any starting vector $s[0]$ (usually $s[0] = \frac{u^T}{N}$), we iterate the equation (9.6) m-times and choose m sufficiently large such that $\|s[m] - \pi\| \le \epsilon$ where ϵ is a prescribed tolerance. Before turning to the convergence of the iteration process that actually computes powers of $\bar{\bar{P}}$ as observed from (9.9), we first concentrate on the basic iteration (9.6),

$$s\,[k+1] = s[k]\bar{\bar{P}} = s[k]\left(\alpha P + (\alpha r + (1 - \alpha)u)\, v^T \right)$$

Since $s[k]u = 1$, we find

$$s\,[k+1] = \alpha s[k]P + (\alpha s[k]r + (1 - \alpha))\, v^T \qquad (11.30)$$

This formula indicates that only the product of $s[k]$ with the (extremely) sparse matrix P needs to be computed and that $\bar{P}$ and $\bar{\bar{P}}$ are never formed nor stored. As shown in Appendix A.4.3, the rate of convergence of a Markov chain towards the steady-state is determined by the second largest eigenvalue. Furthermore, Lemma A.4.4 demonstrates that, for any personalization vector v^T, the second largest eigenvalue of $\bar{\bar{P}}$ is $\alpha\lambda_2$, where λ_2 is the second largest eigenvalue of $\bar{P}$. Lemma A.4.4 thus shows that by choosing

α in (11.29) appropriately, the convergence of the iteration (11.30) tends at least as α^k (since $\lambda_2 < 1$ for irreducible and $\lambda_2 = 1$ for reducible Markov chains) towards the steady-state vector π. Brin and Page report that only 50 to 100 iterations of (11.30) for $\alpha = 0.85$ are sufficient. Clearly, a fast convergence is found for small α, but then (11.29) shows that the true characteristics of the webgraph are suppressed.

This brings us to a final remark concerning the irreducibility approach. The original method of Brin and Page that resulted in (11.29) by enforcing that each node is connected to each other alters the true nature of the Webgraph even though the "connectivity strength" to create irreducibility is extremely small, $\frac{1}{N}$ in the case $v^T = \frac{u^T}{N}$. Instead of maximally connecting all nodes, an other irreducibility approach of minimally connecting nodes investigated by Langville and Meyer (2005) consists of creating one dummy node that is connected to all other nodes and to which all other nodes are connected to ensure overall reachability. Such approach changes the webgraph less. The large size N of the web introduces several challenges such as storage, stability and updating of the PageRank vector, choosing the personalization vector v and other implementational considerations for which we refer to Langville and Meyer (2005).

11.7 Problems

(i) Determine the steady-state probability distribution for the birth-death processes with following transition intensities

    (a) $\lambda_i = \lambda$ and $\mu_i = i\mu$,

    (b) $\lambda_i = \frac{\lambda}{i+1}$ and $\mu_i = \mu$

    where λ and μ are constants.

(ii) Consider a slotted ALOHA in Section 11.5. There are eight stations that compete for slots by transmitting with probability 0.12 each in one slot. Assume that the stations always have packets to transmit. Compute the average time for one station to transmit seven packets.

12

Branching processes

A branching process is an evolutionary process that starts with an initial set of items that produce several other items with a certain probability distribution. These generated items in turn again produce new items and so on. If we denote by X_k the number of items in the k-th generation and by $Y_{k,j}$ the number of items produced by the j-th item in generation k, then the basic law between the number of items in k-th and $k+1$-th generation is, for $k \geq 0$,

$$X_{k+1} = \sum_{j=1}^{X_k} Y_{k,j} \tag{12.1}$$

Figure 12.1 illustrates the basic law (12.1) of a branching process.

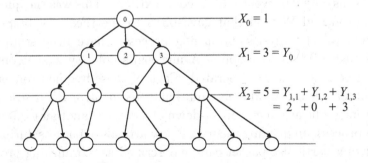

Fig. 12.1. A branching process with one root ($X_0 = 1$) drawn as a tree in which all nodes of generation k lie at a same distance k from the root (label 0).

In general, the production process in each generation k can be different, but most often and in the sequel it is assumed that all generations produce items with the same probability distribution such that all random variables $Y_{k,j}$ are independent and have the same distribution as Y. The branching

229

process is entirely defined by the basic law (12.1) and the distribution of the initial set X_0. The basic law (12.1) indicates that the number of items X_{k+1} in generation $k+1$ is only dependent on the number of items X_k in the previous generation k. The Markov property (9.2)

$$\Pr\left[X_{k+1} = x_{k+1} | X_0 = x_0, \ldots, X_k = x_k\right] = \Pr\left[X_{k+1} = x_{k+1} | X_k = x_k\right]$$

$$= \Pr\left[\sum_{j=1}^{x_k} Y = x_{k+1}\right]$$

$$= \Pr\left[x_k Y = x_{k+1}\right]$$

is obeyed, which shows that the branching process $\{X_k\}_{k\geq 0}$ is a Markov chain with transition probabilities $P_{ij} = \Pr\left[iY = j\right]$. The discrete branching process can be extended to a continuous-time branching process in which items are produced continuously in time, rather than by generations. Since continuous-time Markov processes are mathematically more difficult than their discrete counterpart, we omit the continuous-time branching processes but refer to the book of Harris (1963) and to a simple example, the Yule process, in Section 11.3.2.3.

There are many examples of branching processes and we briefly describe some of the most important. In biology, a certain species generates offsprings and the survival of that species after n generations is studied as a branching process. In the same vein, what is the probability that a family name that is inherited by sons only will eventually become extinct? This was the question posed by Galton and Watson that gave birth to the theory of branching processes in 1874. In physics, branching processes have been studied to understand nuclear chain reactions. A nucleus is split by a neutron and several new free neutrons are generated. Each of these free neutrons again may hit another nucleus producing additional free neutrons and so on. In micro-electronics, the avalanche break-down of a diode is another example of a branching process. In queuing theory, all new arrivals of packets during the service time of a particular packet can be described as a branching process. The process continues as long as the queue lasts. The number of duplicates generated by a flooding process in a communications network is a branching process: a flooded packet is sent on all interfaces of a router except for the incoming interface. The spread of computer viruses in the Internet can be modeled approximately as a branching process. The application of a branching process to compute the hopcount of the shortest path between two arbitrary nodes in a network is discussed in Section 15.7.

12.1 The probability generating function

Since $Y_{k,j}$ are independent, identically distributed random variables with same distribution as Y and independent of the random variable X_k, the probability generating function $\varphi_{X_{k+1}}(z)$ of X_{k+1} follows from (2.70) and the basic law (12.1) as

$$\varphi_{X_{k+1}}(z) = \varphi_{X_k}(\varphi_Y(z)) \tag{12.2}$$

with $\varphi_{X_0}(z) = E\left[z^{X_0}\right] = f(z)$ where $f(z)$ is a given probability generating function. Iterating the general relation (12.2) gives,

$$\varphi_{X_{k+1}}(z) = \varphi_{X_k}(\varphi_Y(z) = \varphi_{X_{k-1}}(\varphi_Y(\varphi_Y(z)))$$
$$= f(\varphi_Y(\varphi_Y(\dots(\varphi_Y(z))))) $$

where the last relation consists of k nested repeated functions $\varphi_Y(.)$, called the iterates of $\varphi_Y(.)$.

The expectation can be derived from the probability generating function by derivation and setting $z = 1$. More elegantly, by taking the expectation of the basic law (12.1) and recalling that X_k and $Y_{k,j}$ are independent, we have with $\mu = E[Y]$

$$E[X_{k+1}] = E\left[\sum_{j=1}^{X_k} Y_{k,j}\right] = E\left[\sum_{j=1}^{X_k} Y\right] = E[YX_k] = \mu E[X_k]$$

Iteration starting from a given average $E[X_0]$ of the initial population gives

$$E[X_k] = \mu^k E[X_0] \tag{12.3}$$

Using (2.27), the variance of X_{k+1} follows from (12.2) with

$$\varphi'_{X_{k+1}}(z) = \varphi'_{X_k}(\varphi_Y(z))\varphi'_Y(z)$$
$$\varphi''_{X_{k+1}}(z) = \varphi''_{X_k}(\varphi_Y(z))\left(\varphi'_Y(z)\right)^2 + \varphi'_{X_k}(\varphi_Y(z))\varphi''_Y(z)$$

as

$$\begin{aligned}
\text{Var}[X_{k+1}] &= \varphi''_{X_{k+1}}(1) + \varphi'_{X_{k+1}}(1) - \left(\varphi'_{X_{k+1}}(1)\right)^2 \\
&= \varphi''_{X_k}(1)\left(\varphi'_Y(1)\right)^2 + \varphi'_{X_k}(1)\varphi''_Y(1) + \varphi'_{X_k}(1)\varphi'_Y(1) - \left(\varphi'_{X_k}(1)\varphi'_Y(1)\right)^2 \\
&= \varphi''_{X_k}(1)\left(E[Y]\right)^2 + E[X_k]\varphi''_Y(1) + E[X_k]E[Y] - (E[X_k]E[Y])^2 \\
&= \mu^2\text{Var}[X_k] + E[X_k]\text{Var}[Y]
\end{aligned}$$

Iteration starting from a given variance $\text{Var}[X_0]$ of the initial set of items and employing the expression for the average (12.3) yields

$$\text{Var}\,[X_1] = \mu^2 \text{Var}\,[X_0] + E\,[X_0]\,\text{Var}\,[Y]$$

$$\text{Var}\,[X_2] = (E\,[Y])^2\,\text{Var}\,[X_1] + E\,[X_1]\,\text{Var}\,[Y]$$
$$= \mu^4 \text{Var}\,[X_0] + (\mu^2 + \mu)\,E\,[X_0]\,\text{Var}\,[Y]$$

$$\text{Var}\,[X_3] = (E\,[Y])^2\,\text{Var}\,[X_2] + E\,[X_2]\,\text{Var}\,[Y]$$
$$= \mu^6 \text{Var}\,[X_0] + (\mu^4 + \mu^3 + \mu^2)\,E\,[X_0]\,\text{Var}\,[Y]$$

from which we deduce

$$\text{Var}\,[X_k] = \mu^{2k}\text{Var}\,[X_0] + \sum_{j=k-1}^{2(k-1)} \mu^j E\,[X_0]\,\text{Var}\,[Y]$$

or

$$\text{Var}\,[X_k] = \mu^{2k}\text{Var}\,[X_0] + E\,[X_0]\,\text{Var}\,[Y]\,\mu^{k-1}\frac{1-\mu^k}{1-\mu} \qquad (12.4)$$

Substitution into the recursion for $\text{Var}[X_k]$ justifies the correctness of (12.4).

The relations for the expectation (12.3) and the variance (12.4) of the number of items in generation k imply that, if the average production per generation is $E\,[Y] = \mu = 1$, $E\,[X_k] = E\,[X_0]$ and that $\text{Var}[X_k] = \text{Var}[X_0] + kE\,[X_0]\text{Var}[Y]$. In the case that the average production $E\,[Y] = \mu > 1$ $(E\,[Y] < 1)$, the average population per generation grows (decreases) exponentially in k with rate $\log\mu$ and, similarly for large k, the standard deviation $\sqrt{\text{Var}\,[X_k]}$ grows (decreases) exponentially in k with the same rate $\log\mu$. Hence, the most important factor in the branching process is the average production $E\,[Y] = \mu$ per generation. The variance terms and $E\,[X_0]$ only play a role as prefactor. A branching process is called *critical* if $\mu = 1$, *subcritical* if $\mu < 1$ and *supercritical* if $\mu > 1$. In the sequel, we will only consider supercritical $(\mu > 1)$ branching processes.

Often, the initial set of items consists of only one item. In that case, $X_0 = 1$ and $\varphi_{X_0}(z) = f(z) = z$ and

$$E\,[X_k] = \mu^k$$

$$\text{Var}\,[X_k] = \text{Var}\,[Y]\,\mu^{k-1}\frac{1-\mu^k}{1-\mu}$$

while the explicit nested form of the probability generating function indicates that

$$\varphi_{X_{k+1}}(z) = \varphi_Y(\varphi_{X_k}(z)) \qquad (12.5)$$

This relation is only valid if $f(z) = z$ or, equivalently, only if $X_0 = 1$. In case $f(z) = z$, $\varphi_{X_k}(z)$ is the k-th iterate of $\varphi_Y(z)$.

Example Due to the nested structure (12.2), closed form expressions for the k-th generation probability generating function $\varphi_{X_k}(z)$ are rare. Assume that $X_0 = 1$. A simple case that allows explicit computation occurs in a deterministic production of m offsprings in each generation for which $\varphi_Y(z) = z^m$. We have from (12.5) that $\varphi_{X_1}(z) = \varphi_Y(z) = z^m$ and

$$\varphi_{X_k}(z) = z^{km}$$

This branching process evolves as an m-ary tree shown in Fig. 17.7. A second example that can be computed exactly is the geometric branching process studied in Section 12.5.

12.2 The limit W of the scaled random variables W_k

The conditional expectation, defined in Section 2.6,

$$E\left[X_{k+1}|X_k, X_{k-1}, \ldots, X_0\right] = E\left[X_{k+1}|X_k\right] \qquad \text{(Markov property)}$$

$$= E\left[\sum_{j=1}^{X_k} Y_{k,j}\,\middle|\, X_k\right] = E\left[Y X_k|X_k\right] = \mu X_k$$

is a random variable, which suggests us to consider the scaled random variable $W_k = \frac{X_k}{\mu^k}$ because

$$E\left[W_{k+1}|W_k, W_{k-1}, \ldots, W_1\right] = W_k$$

while (12.3) shows that $E\left[W_k\right] = E\left[X_0\right]$ for all k. The stochastic process $\{W_k\}_{k\geq 1}$ is a martingale process, which is a generalization of a fair game with characteristic property that at each step k in the process $E\left[W_k\right]$ is a constant (independent of k). From (12.4), the variance of the scaled random variables $W_k = \frac{X_k}{\mu^k}$ is

$$\text{Var}\left[W_k\right] = \text{Var}\left[X_0\right] + E\left[X_0\right]\frac{\text{Var}\left[Y\right]}{\mu^2 - \mu}\left(1 - \mu^{-k}\right) \qquad (12.6)$$

which geometrically tends, provided $E\left[Y\right] = \mu > 1$, to a constant independent of k. The expression for the variance (12.6) indicates that

$$\text{Var}\left[W\right] = \lim_{k\to\infty} \text{Var}\left[W_k\right] = \text{Var}\left[X_0\right] + E\left[X_0\right]\frac{\text{Var}\left[Y\right]}{\mu^2 - \mu} \qquad (12.7)$$

exists provided $E\left[Y\right] = \mu > 1$. We now show that the limit variable $W = \lim_{k\to\infty} W_k$ exists if $E\left[Y\right] > 1$.

Theorem 12.2.1 *If $E\left[Y\right] = \mu > 1$, the scaled random variables $W_n \to W$ a.s.*

Proof: Consider

$$E\left[(W_{k+n} - W_n)^2\right] = E\left[W_{k+n}^2\right] + E\left[W_n^2\right] - 2E\left[W_{k+n}W_n\right]$$

Using (2.72) with $h(x) = x$ and the Markov property,

$$E\left[W_{k+n}W_n\right] = E\left[E\left[W_{k+n}|W_n\right]W_n\right] = E\left[W_n^2\right]$$

we have, with (2.16), $E\left[W_{k+n}\right] = E\left[W_k\right] = E\left[X_0\right]$ and (12.6),

$$E\left[(W_{k+n} - W_n)^2\right] = \text{Var}\left[W_{k+n}\right] - \text{Var}\left[W_n\right] = E\left[X_0\right]\frac{\text{Var}\left[Y\right]\mu^{-n}}{\mu^2 - \mu}\left(1 - \mu^{-k}\right)$$

In the limit $k \to \infty$,

$$E\left[(W - W_n)^2\right] = O(\mu^{-n})$$

which means that the sequence $\{W_k\}_{k\geq 1}$ converges to W in L^2 or in mean square (see Section 6.1.2). Moreover,

$$\sum_{n=1}^{\infty} E\left[(W - W_n)^2\right] = E\left[\sum_{n=1}^{\infty}(W - W_n)^2\right] = O(1)$$

which means that the series has finite expectation and that $\sum_{n=1}^{\infty}(W - W_n)^2$ is finite with probability 1. The convergence of this series implies for large n that $(W - W_n)^2 \to 0$ with probability 1 or that $W_n \to W$ a.s. $\qquad\square$

Theorem 12.2.1 means that the number of items in generation k is, for large k, well approximated by $X_k \sim W\mu^k$. Hence, an asymptotic analysis of a branching process crucially relies on the properties of the limit random variable W.

The generating function $\varphi_W(z) = E\left[z^W\right]$ of this limit random variable can be deduced as the limit of the sequence of generating functions

$$\varphi_{W_k}(z) = E\left[z^{W_k}\right] = E\left[z^{\frac{X_k}{\mu^k}}\right] = \varphi_{X_k}(z^{\mu^{-k}}) \qquad (12.8)$$

Using (12.5) in case[1] $X_0 = 1$ with $z \to z^{\mu^{-k-1}}$

$$E\left[z^{\frac{X_{k+1}}{\mu^{k+1}}}\right] = \varphi_Y\left(E\left[z^{\frac{X_k}{\mu^k\mu}}\right]\right)$$

leads with $W_k = \frac{X_k}{\mu^k}$ to the recursion of the pgf of the scaled random variables W_k,

$$\varphi_{W_{k+1}}(z) = \varphi_Y\left(\varphi_{W_k}\left(z^{\frac{1}{\mu}}\right)\right)$$

---

[1] The use of the general equation (12.2) is inadequate.

In the limit $k \to \infty$ where $W_k \to W$ a.s., we can apply the Continuity Theorem 6.1.3 which results in the functional equation for the pgf of the *continuous* limit random variable W,

$$\varphi_W(z) = \varphi_Y\left(\varphi_W\left(z^{\frac{1}{\mu}}\right)\right) \tag{12.9}$$

Since W is a continuous random variable except at $W = 0$ as explained below (see (12.19)), it is more convenient to define the moment generating function $\chi_W(t) = E\left[e^{-tW}\right]$. Obviously, the relation between the two generating functions is with $z = e^{-t}$

$$\chi_W(t) = \varphi_W(e^{-t})$$

With $z = e^{-t}$ in (12.9) the functional equation of $\chi_W(t)$ is for $t \geq 0$ and $E[W] = E[X_0] = 1$,

$$\chi_W(t) = \varphi_Y\left(\chi_W\left(\frac{t}{\mu}\right)\right) \tag{12.10}$$

The functional equation (12.10) is simpler than (12.9) and $\chi_W(t)$ is convex for all t, while $\varphi_W(z)$ is not convex for all z. In particular, $\varphi_W(z) = \chi_W(-\log z)$ is not analytic at $z = 0$ and appears[2] to have a concave regime near $z \downarrow 0$ where $\chi_W'(-\log z) + \chi_W''(-\log z) < 0$.

Lemma 12.2.2 $\chi_W(t)$ *is the only probability generating function satisfying the functional equation (12.10).*

Proof: Let $\psi_{W^*}(t) = E\left[e^{-tW^*}\right]$ and $\chi_W(t) = E\left[e^{-tW}\right]$ be two probability generating functions that satisfy both (12.10). Then $\psi_{W^*}(t) - \chi_W(t)$ is continuous for $\operatorname{Re} t \geq 0$ and, since $E[W] = E[W^*] = 1$, the Taylor series (2.40) around $t = 0$ is

$$\psi_{W^*}(t) - \chi_W(t) = (-t)E[W^* - W] + E\left[\sum_{k=2}^{\infty} \frac{(-t)^k}{k!}\left((W^*)^k - W^k\right)\right]$$

$$= tE\left[\sum_{k=1}^{\infty} \frac{(-t)^k}{(k+1)!}\left(W^{k+1} - (W^*)^{k+1}\right)\right]$$

from which $\psi_{W^*}(t) - \chi_W(t) = t\,h(t)$ and $h(0) = 0$. Since $|\varphi_Y'(z)| \leq \mu$ for $|z| \leq 1$, equation (5.6) of the Mean Value Theorem implies $|\varphi_Y(a) - \varphi_Y(b)| \leq \mu|a - b|$ for any $|a|, |b| \in [0, 1]$. Since $|\chi_W(t)| \leq 1$ and $|\psi_{W^*}(t)| \leq 1$ for

---

[2] This fact is observed for both a geometric and Poisson production distribution function.

$\mathrm{Re}(t) \geq 0$, we obtain

$$
|t\,h(t)| = \left| \varphi_Y\left(\psi_{W^*}\left(\frac{t}{\mu} \right) \right) - \varphi_Y\left(\chi_W\left(\frac{t}{\mu} \right) \right) \right|
$$

$$
\leq \mu \left| \psi_{W^*}\left(\frac{t}{\mu} \right) - \chi_W\left(\frac{t}{\mu} \right) \right| = \mu \left| \frac{t}{\mu} \, h\left(\frac{t}{\mu} \right) \right|
$$

or

$$
|h(t)| \leq \left| h\left(\frac{t}{\mu} \right) \right|
$$

After K iterations, we have that $|h(t)| \leq \left| h\left(\frac{t}{\mu^K} \right) \right|$ which hold for any integer K. Hence, for any finite t and since $h(t)$ is continuous which allows that $\lim_{K \to \infty} h\left(\frac{t}{\mu^K} \right) = h\left(\lim_{K \to \infty} \frac{t}{\mu^K} \right)$,

$$
|h(t)| \leq \lim_{K \to \infty} \left| h\left(\frac{t}{\mu^K} \right) \right| = h(0) = 0
$$

which proves the Lemma.   □

Lemma 12.2.2 is important because solving the functional equation, for example by Taylor expansion, is one of the primary tools to determine $\chi_W(t)$. If $\varphi_Y(z)$ is analytic inside a circle with radius $R_Y > 0$ centered at $z = 1$, then the Taylor series around $z_0 = 1$,

$$
\varphi_Y(z) = 1 + \sum_{k=0}^{\infty} u_k (z - 1)^k
$$

converges for all $|z - 1| < R_Y$. The definition $\chi_W(t) = E\left[e^{-tW} \right]$ implies that the maximum value of $|\chi_W(t)|$ inside and on a circle with radius r around the origin is attained at $\chi_W(-r)$. The functional equation (12.10) then shows that $\chi_W(t)$ is analytic inside a circle around $t = 0$ with radius R_W for which $\chi_W\left(-\frac{R_W}{\mu} \right) < 1 + R_Y$. Since $\chi_W(0) = 1$, $\chi_W(t)$ is convex and decreasing for real t and $R_Y > 0$, there exists such a non-zero value of R_W. This implies that the Taylor series

$$
\chi_W(t) = 1 + \sum_{k=1}^{\infty} \omega_k t^k \tag{12.11}
$$

converges around $t = 0$ for $|t| < R_W$. There exists a recursion to compute ω_k for any $k \geq 1$ as shown in Van Mieghem (2005). If $\chi_W(t)$ is not known in closed form, the interest of the Taylor series (12.11) lies in the fast convergence for small values of $|t| < 1$. The recursion for the Taylor coefficients ω_k

enables the computation of $\chi_W(t)$ for $|t| < 1$ to any desired degree of accuracy. The functional equation $\chi_W(t) = \varphi_Y\left(\chi_W\left(\frac{t}{\mu}\right)\right)$ extends the t-range to the entire complex plane. For large values of t and in particular for negative real t, $\chi_W(t)$ is best computed from $\chi_W\left(\frac{t}{\mu^{[\log_\mu |t|]+1}}\right)$ after $[\log_\mu |t|] + 1$ functional iteratives of (12.10). Indeed, since $\mu > 1$ such that $\frac{t}{\mu^{[\log_\mu |t|]+1}} < 1$, the Taylor series (12.11) provides an accurate start value $\chi_W\left(\frac{t}{\mu^{[\log_\mu |t|]+1}}\right)$ for this iterative scheme.

12.3 The Probability of Extinction of a Branching Process

In many applications the probability that the process will eventually terminate and which parameters influence this extinction probability are of interest. For instance, a nuclear reaction will only lead to an explosion if critical starting conditions are obeyed. The branching process terminates if, for some generation $n > 0$, $X_n = 0$ and, of course, $X_m = 0$ for all $m > n$. Let us denote

$$q_k = \Pr[X_k = 0] = \varphi_{X_k}(0)$$

If we assume that $X_0 = 1$, the analysis simplifies because the more specific version (12.5) holds. Hence, only if the initial set consists of a single item $X_0 = 1$,

$$q_{k+1} = \varphi_{X_{k+1}}(0) = \varphi_Y(\varphi_{X_k}(0)) = \varphi_Y(q_k) \tag{12.12}$$

and with $q_0 = \varphi_{X_0}(0) = 0$, $q_1 = \varphi_Y(0) = \Pr[Y = 0] \geq 0$. Obviously, if there is no production, $\Pr[Y = 0] = 1$, or always production, $\Pr[Y = 0] = 0$, extinction never occurs. By its definition (2.18), a probability generating function of a non-negative discrete random variable is strict increasing along the positive real z-axis. When excluding the extreme cases such that $0 < \Pr[Y = 0] < 1$, by the strict increase of $\varphi_Y(x)$ for $x = \operatorname{Re} z \geq 0$, we observe that

$$0 = q_0 < q_1 = \varphi_Y(0) < q_2 = \varphi_Y(q_1) < q_3 = \varphi_Y(q_2) < \ldots$$

The series $q_0, q_1, q_2, \ldots$ is a monotone increasing sequence bounded by 1 because $\varphi_Y(1) = 1$. Hence, the probability of extinction

$$\pi_0 = \lim_{k \to \infty} \Pr[X_k = 0] = \Pr[W = 0]$$

exists and $0 < \pi_0 \leq 1$. The existence of a limiting process and the fact that the probability generating function is analytic for $|z| < 1$ and hence, continuous, which allows us to interchange $\lim_{k \to \infty} \varphi_Y(q_k) = \varphi_Y(\lim_{k \to \infty} q_k)$

yields the equation for the extinction probability π_0,

$$\pi_0 = \varphi_Y(\pi_0) \tag{12.13}$$

It demonstrates that the extinction probability π_0 is a root of $\varphi_Y(x) - x$ in the interval $x \in [0, 1]$.

Since $\varphi_W(0) = \Pr[W = 0] = \pi_0$, this equation (12.13) follows more directly from (12.9). Notice, however, that in the functional equation (12.9) the function z^μ is not analytic at $z = 0$ if $\mu \notin \mathbb{N}$ which may cause that $f_W(x)$ is possibly not continuous at $x = 0$, although the limit $\lim_{z \to 0} \varphi_W(0) = \pi_0$ exists. On the other hand, since $\varphi_W(z) = \chi_W(-\log z)$, the extinction probability is found as

$$\lim_{t \to \infty} \chi_W(t) = \pi_0$$

and the convexity of $\chi_W(t)$ implies that, for any real value of t, $\chi_W(t) \geq \pi_0$.

An alternative, more probabilistic derivation of equation (12.13) is as follows. Applying the law of total probability (2.46) to the definition of the extinction probability

$$\pi_0 = \Pr[X_n = 0 \text{ for some } n > 0]$$

$$= \sum_{j=0}^{\infty} \Pr[X_n = 0 \text{ for some } n > 0 | X_1 = j] \Pr[X_1 = j]$$

Only if $X_0 = 1$, relation (12.5) indicates that $\varphi_{X_1}(z) = \varphi_Y(z)$ which implies that $\Pr[X_1 = j] = \Pr[Y = j]$. In addition, given the first generation consists of j items, the branching process will eventually terminate if and only if each of the j sets of items generated by the first generation eventually dies out. Since each set evolves independently and since the probability that any set generated by a particular ancestor in the first generation becomes extinct is π_0, we arrive at

$$\pi_0 = \sum_{j=0}^{\infty} \pi_0^j \Pr[Y = j] = \varphi_Y(\pi_0)$$

The different viewpoints thus lead to a same result summarized by:

Theorem 12.3.1 *If $X_0 = 1$ and $0 < \Pr[Y = 0] < 1$, the extinction probability π_0 is (a) the smallest positive real root of $x = \varphi_Y(x)$ and (b) $\pi_0 = 1$ if and only if $E[Y] \leq 1$ and $\Pr[Y = 0] + \Pr[Y = 1] < 1$.*

Proof: (a) Suppose that x_o is the smallest positive real root obeying $\varphi_Y(x_o) = x_o > 0$. Then, $q_1 = \varphi_Y(0) < \varphi_Y(x_o) = x_o$. Assume (induction hypothesis) that $q_n < x_o$. The recursion (12.12) and the strict increase of $\varphi_Y(x)$ then shows that $q_{n+1} = \varphi_Y(q_n) < \varphi_Y(x_o) = x_o$. Hence, the principle of induction demonstrates that $q_n < x_o$ for all (finite) n and, hence, that $\pi_0 \leq x_o$.

(b) First, the condition $\Pr[Y = 0] + \Pr[Y = 1] < 1$ implies that $\Pr[Y > 1] > 0$ and that there exists at least one integer $j > 1$ such that $\Pr[Y = j] > 0$. In that case, for real $x > 0$ but x smaller than the radius R of convergence which is at least $R = 1$, the second derivative $\varphi_Y''(x) = \sum_{j=2}^{\infty} j(j-1) \Pr[Y = j] x^{j-2}$ is positive, which implies that $\varphi_Y(x)$ is strict convex in $(0, 1)$. Since $x = 1$ obeys $x = \varphi_Y(x)$ and $\varphi_Y(0) = \Pr[Y = 0] \in (0, 1)$, the strict convex function $y = \varphi_Y(x)$ can only intersect the line $y = x$ in some point $x \in (0, 1)$ if $\varphi_Y(x)$ is below that line near their intersection at $x = 1$ or if $\varphi_Y'(1) = E[Y] > 1$. In the other case, if $E[Y] < 1$, the only intersection is at $x = 1$. □

The two possibilities are drawn in Fig. 12.2.

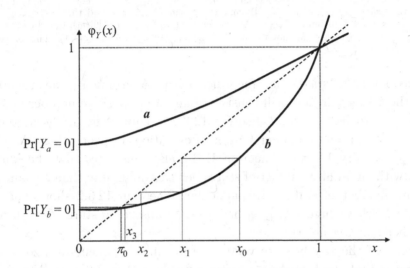

Fig. 12.2. The generating function $\varphi_Y(x)$ along the positive real axis x. The two possible cases are shown: curve a corresponds to $E[Y] < 1$ and curve b to $E[Y] > 1$. The fast convergence towards the zero π_0 is exemplified by the sequence $x_0 > x_1 = \varphi_Y(x_0) > x_2 = \varphi_Y(x_1) > x_3 = \varphi_Y(x_2)$.

A root equation such as (12.13) also appears in queuing models such as the M/G/1 (Section 14.3) and GI/D/1 (Section 14.4) and reflects the asymptotic behavior as explained in Section 5.7. The extinction probability π_0 can be expressed explicitly as a Lagrange series as demonstrated in Van Mieghem (1996).

The branching process with infinitely many generations $k \to \infty$ can be viewed as an infinite directed tree where each node has a finite degree a.s. The fact that $\pi_0 < 1$ if $E[Y] > 1$ implies that, in infinite directed trees, there exists an infinitely long path starting from the root with probability $1 - \pi_0$.

Theorem 12.3.2 *The limiting branching process with $X_0 = 1$ obeys for*

$k \to \infty$

$$\Pr[X_k = 0] \to \pi_0$$

$$\Pr[X_k = j] \to 0 \qquad \text{for any } j > 0$$

Proof: First, if $E[Y] < 1$, then Theorem 12.3.1 states that $\pi_0 = 1$. For any probability generating function $\varphi(z)$, it holds that $|\varphi(z)| \le 1$ for $|z| \le 1$. Hence, $\varphi_{X_k}(x) \le 1$ for real $x \in [0, 1]$. Moreover, $q_k = \varphi_{X_k}(0) \le \varphi_{X_k}(x)$. In the limit $k \to \infty$, $q_k \to \pi_0 = 1$, which implies that for all $x \in [0, 1]$ it holds that $\varphi_{X_k}(x) \to \pi_0 = 1$. The fact that a probability generating function, a Taylor series around $z = 0$, converges to a constant π_0 for $0 \le x \le 1$ implies that $\Pr[X_k = j] \to 0$ for any $j > 0$ and $\Pr[X_k = 0] \to \pi_0$.

The second case $E[Y] > 1$ possesses an extinction probability $\pi_0 < 1$. For $x \in (\pi_0, 1)$, Fig. 12.2 shows that $\pi_0 < \varphi_Y(x) < x < 1$. By induction using (12.5), we find that $\pi_0 < \varphi_{X_k}(x) < \varphi_{X_{k-1}}(x) < \cdots < 1$ or $\lim_{k \to \infty} \varphi_{X_k}(x) = \pi_0$ for $x \in (\pi_0, 1)$. For $x \in [0, \pi_0)$, the same argument $q_k = \varphi_{X_k}(0) \le \varphi_{X_k}(x) \le \pi_0$ shows that $\lim_{k \to \infty} \varphi_{X_k}(x) = \pi_0$ for $x \in [0, 1)$. This proves the theorem. $\square$

Theorem 12.3.2 states that, regardless of the value of $E[Y]$, the probability that the k-th generation will consists of any finite positive number of items tends to zero if $k \to \infty$. Theorem 12.3.2 is equivalent to the statement that, after an infinite number $k \to \infty$ of evolutions or generations, $X_k \to \infty$ with probability $1 - \pi_0$. Theorem 12.3.2 also illustrates that the Markov chain with an infinite number of states behaves differently than a chain with a finite number of states. In particular, Theorem 12.3.2 shows that the infinite Markov chain $\{X_k\}_{k \ge 0}$ has a single absorbing state $X_k = 0$ while all other states j are transient ($\lim_{n \to \infty} P_{ij}^n = 0$ for $1 \le i, j < \infty$). The existence of the steady-state vector π (not all components are zero) does not imply that the branching process with $X_0 = 1$ and $0 < \Pr[Y = 0] < 1$ and infinitely many states is an irreducible Markov chain.

12.4 Asymptotic behavior of W

The convexity of $\chi_W(t)$ implies that $\chi_W'(t) \le 0$ for all real t and that $\chi_W'(t)$ is decreasing in t. We know that $\chi_W'(0) = -1$. Since $\lim_{t \to \infty} \chi_W(t) = \pi_0$, it follows that $\lim_{t \to \infty} \chi_W'(t) = 0$. The following Lemma 12.4.1 is a little more precise.

Lemma 12.4.1 $\chi_W'(t) = o(t^{-1})$ *for* $t \to \infty$.

Proof: The derivative of the functional equation (12.10) is $\mu\chi_W'(\mu t) = \varphi_Y'(\chi_W(t))\chi_W'(t)$. By iteration, we have

$$\mu^K \chi_W'\left(\mu^K t\right) = \chi_W'(t) \prod_{j=0}^{K-1} \varphi_Y'\left(\chi_W(\mu^j t)\right)$$

Since $\chi_W(t) \in [\pi_0, 1]$ for real $t \ge 0$, then $\varphi_Y'(\pi_0) \le \varphi_Y'(\chi_W(\mu^j t)) \le \mu$ for any j. Theorem 12.3.1 states that if $\mu = \varphi_Y'(1) > 1$, then there are two zeros π_0 and 1 of $f(z) = \varphi_Y(z) - z$ in $z \in [0, 1]$. By Rolle's Theorem applied to the continuous function $f(z) = \varphi_Y(z) - z$, there exists an $\xi \in (\pi_0, 1)$

for which $f'(\xi) = 0$. Equivalently, $\varphi'_Y(\xi) = 1$ and $\xi > \pi_0$. Since $\varphi'_Y(z)$ is monotonously increasing in $z \in [0,1]$, we have that $\varphi'_Y(0) = \Pr[Y = 1] \leq \varphi'_Y(\pi_0) < 1$. Since $\chi_W(t)$ is continuous and monotone decreasing, there exists an integer K_0 such that $\varphi'_Y(\chi_W(\mu^j t)) < 1$ for $j > K_0$ and any $t > 0$. Hence,

$$\lim_{K \to \infty} \prod_{j=0}^{K-1} \varphi'_Y(\chi_W(\mu^j t)) = \prod_{j=0}^{K_0-1} \varphi'_Y(\chi_W(\mu^j t)) \prod_{j=K_0}^{\infty} \varphi'_Y(\chi_W(\mu^j t)) \to 0$$

and, for any finite $t > 0$, $\mu^K \chi'_W(\mu^K t) \to 0$ for $K \to \infty$ which implies the lemma. $\square$

Lemma 12.4.1 is, for large t, equivalent to $|\chi'_W(t)| \leq Ct^{-1-\beta}$ for some real $\beta > 0$ and where C is a finite positive real number. Lemma 12.4.1 thus suggests that

$$\chi'_W(t) = -g_\mu(t) t^{-\beta-1} \tag{12.14}$$

where $0 < g_\mu(t) \leq C$ on the real positive t–axis.

Lemma 12.4.2 *If $\varphi'_Y(\pi_0) > 0$ and $\mu > 1$, then*

$$F_\mu = \lim_{t \to \infty} g_\mu(t) \tag{12.15}$$

exists, is finite and strictly positive.

Proof: We first use (a) the convexity of any pgf $\chi_W(t)$ implying that $\chi''_W(t) \geq 0$ for all t and we then invoke (b) the functional equation (12.10) of $\chi_W(t)$.

(a) The function $g_\mu(t) = -\chi'_W(t) t^{\beta+1}$ is differentiable, thus continuous, and has for real $t > 0$ only one extremum at $t = \tau$ obeying

$$\tau = \frac{-\chi'_W(\tau)}{\chi''_W(\tau)} (\beta + 1) > 0$$

Since $\chi'_W(0) = -1$, implying that $g_\mu(t) = t^{\beta+1}(1 + o(t)$ as $t \downarrow 0$ or that $g_\mu(t)$ is initially monotone increasing in t, the extremum at $t = \tau$ is a maximum. The derivative of $g_\mu(t) = -\chi'_W(t) t^{\beta+1}$ is, with (12.14),

$$g'_\mu(t) = \frac{\beta+1}{t} g_\mu(t) - \chi''_W(t) t^{\beta+1}$$

such that, for τ finite, $\max g_\mu(t) = \frac{\tau^{\beta+2}}{\beta+1} \chi''_W(\tau)$. Since $\chi''_W(t) \geq 0$ for all t, we also obtain the inequality for $t \geq 0$

$$g'_\mu(t) \leq \frac{\beta+1}{t} g_\mu(t) \leq \frac{\beta+1}{t} C$$

from which $\lim_{t \to \infty} g'_\mu(t) \leq 0$. Hence, $g_\mu(t)$ is not increasing for $t \to \infty$.

(b) Substitution of (12.14) in the derivative of the functional equation (12.10) yields

$$g_\mu(t) = \varphi'_Y\left(\chi_W\left(\frac{t}{\mu}\right)\right) g_\mu\left(\frac{t}{\mu}\right) \mu^\beta \tag{12.16}$$

Since $\varphi'_Y\left(\chi_W\left(\frac{t}{\mu}\right)\right) \geq \varphi'_Y(\pi_0) > 0$ (the restriction of this Lemma), there holds with $A = \varphi'_Y(\pi_0) \mu^\beta > 0$ for all $t > 0$ that

$$g_\mu(t) \geq A g_\mu\left(\frac{t}{\mu}\right)$$

For $t < \tau$, $g_\mu(t)$ is shown in (a) to be monotone increasing, which requires that $A \geq 1$ for $\mu > 1$. But, since the inequality with $A \geq 1$ holds for all $t > 0$, we must have that $\tau \to \infty$. Hence, $g_\mu(t)$ is continuous and strict increasing for all $t \geq 0$ with a maximum at infinity, which proves the existence of a unique limit $F_\mu \leq C$.

If $F_\mu = 0$, the suggestion (12.14) is not correct implying that $\chi'_W(t)$ decreases faster than any power of t^{-1}. The proof of Lemma 12.4.1 indicates that his case can occur if $\varphi'_Y(\pi_0) = 0$. $\qquad\square$

In fact, $A = 1$. For, when passing to the limit $t \to \infty$ in (12.16) using Lemma 12.4.2, we obtain

$$\mu^{-\beta} = \varphi'_Y(\pi_0)$$

which determines the exponent $\beta \geq 1$ as

$$\beta = -\frac{\log \varphi'_Y(\pi_0)}{\log \mu} \tag{12.17}$$

After integration of (12.14), we have that

$$\chi_W(t) = \pi_0 + \int_t^\infty g_\mu(u)\, u^{-\beta-1} du \tag{12.18}$$

Approximating $g_\mu(u)$ by its limit F_μ for large t, we obtain the asymptotic form

$$\chi_W(t) \sim \pi_0 + \frac{F_\mu}{\beta} t^{-\beta}$$

Beside β and the extinction probability π_0, the parameter F_μ appears as additional characterizing quantity of a branching process. The behavior of the Laplace transform (2.37) for large t reflects the behavior of the probability density function for small x. Hence, using $\frac{1}{2\pi i} \int_{c-i\infty}^{c+i\infty} \frac{e^{xt}}{t^s} dt = \frac{x^{s-1}}{\Gamma(s)}$ for $\mathrm{Re}\, s > 0$, the probability density function is, for small x,

$$f_W(x) \approx \pi_0 \delta(x) + \frac{F_\mu}{\Gamma(\beta+1)} x^{\beta-1} \tag{12.19}$$

The probability density function $f_W(x)$ is not continuous at $x = 0$ if $\pi_0 > 0$ and reflects the two different regimes: (a) $W = 0$ implying that the branching process extincts, $X_k = 0$, from some generation k on and (b) $W > 0$ implying $X_k \sim W\mu^k$ for large k, the number of items per generation grows exponentially with prefactor W. If two sample paths of a same branching process are generated, $X_{k,1} \sim W_1\mu^k$ and $X_{k,2} \sim W_2\mu^k$ may be largely different for large k, because of the random nature of W: although the prefactor W_1 and W_2 both have the same probability density function $f_W(x)$, W_1 can differ substantially from W_2 as illustrated by the pdf $f_W(x)$ in Fig. 12.3.

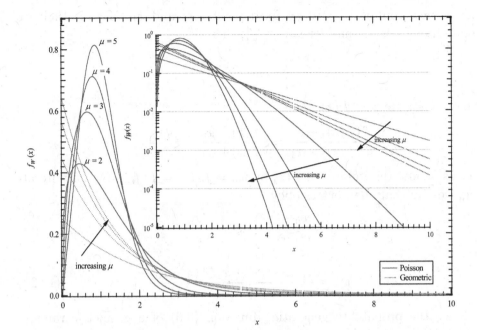

Fig. 12.3. The probability density function of the limit random variable W for both a geometric and a Poisson production process for a same set of values of the average $\mu = E[Y]$.

12.5 A geometric branching processes

Consider a production generation function $\varphi_Y(z)$ of the form $f(z) = \frac{az+b}{cz+d}$. Beside straightforward iteration of (12.5), a more elegant approach[3] relies on the following property of $f(z)$. For $z = x$, the difference is

$$f(z) - f(x) = \frac{ad - bc}{(d+cz)(d+cx)}(z - x)$$

and, hence, for any two points x_0 and x_1,

$$\frac{f(z) - f(x_0)}{f(z) - f(x_1)} = \left(\frac{cx_1 + d}{cx_0 + d}\right)\left(\frac{z - x_0}{z - x_1}\right)$$

---

[3] The linear fractional transformation $f(z)$ is an automorphism of the extended complex plane and basic in the geometric theory of a complex function for which we refer to the book of Sansone and Gerretsen (1960, vol. 2). Fixed points of an automorphism of the extended plane are solutions of $z = f(z)$, which is a quadratic equation $cz^2 + (d-a)z - b = 0$ and which shows that there are at most two different fixed points.

Let us now confine to the two fixed points, x_0 and x_1, of $f(z)$ that are solution of $f(z) = z$ and let $\xi = \frac{cx_1 + d}{cx_0 + d}$, then

$$\frac{f(z) - x_0}{f(z) - x_1} = \xi \frac{z - x_0}{z - x_1}$$

Now, substitute $z \to f(z)$, then

$$\frac{f(f(z)) - x_0}{f(f(z)) - x_1} = \xi \frac{f(z) - x_0}{f(z) - x_1} = \xi^2 \frac{z - x_0}{z - x_1}$$

Let us denote the iterates of $f(z)$ by $w_n = f_n(z) = f(f_{n-1}(z))$. By iterating, we find that the iterates obey

$$\frac{w_n - x_0}{w_n - x_1} = \xi^n \frac{z - x_0}{z - x_1}$$

or

$$w_n = \frac{x_0(z - x_1) - x_1 \xi^n (z - x_0)}{z - x_1 - \xi^n (z - x_0)} \tag{12.20}$$

Since the probability generating function (3.6) of a geometric random variable Y is of the form $f(z) = \frac{az + b}{cz + d}$, a *geometric branching process* is regarded as a basic reference model in the study of branching processes. The production process in each generation obeys $\Pr[Y = k] = qp^k$ for $k \geq 0$ leading to $\varphi_Y(z) = \frac{q}{1 - pz}$, which is slightly different from (3.6). We know that the equation $\varphi_Y(x) = x$ can have two real zeros in $[0, 1]$, one at $x_1 = 1$ since $\varphi_Y(z)$ is a probability generating function and another at $x_0 = \frac{q}{p} = \frac{1}{E[Y]} = \frac{1}{\mu} = \pi_0$ such that

$$\xi = \frac{-px_1 + 1}{-px_0 + 1} = \frac{q}{p} = \frac{1}{\mu}$$

The functional equation (12.5) associates $w_n = \varphi_{X_n}(z)$ and after substitution in (12.20) we obtain

$$\varphi_{X_n}(z) = \frac{(\mu^{n-1} - 1) z - \mu^{n-1} + \frac{1}{\mu}}{(\mu^n - 1) z - \mu^n + \frac{1}{\mu}} \tag{12.21}$$

In the case that $E[Y] = \mu \to 1$ or $p = q$, using the rule of de l'Hospital gives

$$\varphi_{X_n}(z) = \frac{(n - 1)z - n}{nz - n + 1}$$

From (12.21), the probabilities of extinction at the k-th generation are

$$\Pr[X_k = 0] = \varphi_{X_k}(0) = \frac{1}{\mu} \left(\frac{\mu^k - 1}{\mu^k - \frac{1}{\mu}} \right)$$

If $E[Y] = \mu > 1$, then $\lim_{k \to \infty} \Pr[X_k = 0] = \frac{1}{\mu} = x_0 = \pi_0$ (Theorem 12.3.2) whereas for $E[Y] = \mu \le 1$, we find that $\lim_{k \to \infty} \Pr[X_k = 0] = 1$. If T_0 is the hitting time defined in Section 9.2.2 as the smallest discrete-time k such that $X_k = 0$, then $\Pr[T_0 \le k] = \Pr[X_k = 0]$.

The probability generating function of the scaled random variables $W_k = \frac{X_k}{\mu^k}$ follows from (12.21) and (12.8) as

$$\varphi_{W_k}(z) = \frac{\left(\mu^{k-1} - 1\right) z^{\mu^{-k}} - \mu^{k-1} + \frac{1}{\mu}}{\left(\mu^k - 1\right) z^{\mu^{-k}} - \mu^k + \frac{1}{\mu}}$$

from which $\varphi_{W;Geo}(z) = \lim_{k \to \infty} \varphi_{W_k}(z)$ follows as

$$\varphi_{W;Geo}(z) = \frac{-\frac{1}{\mu} \log z + 1 - \frac{1}{\mu}}{-\log z + 1 - \frac{1}{\mu}} \tag{12.22}$$

and

$$\chi_{W;Geo}(t) = \frac{\frac{t}{\mu} + 1 - \frac{1}{\mu}}{t + 1 - \frac{1}{\mu}} = 1 + \sum_{k=0}^{\infty} \frac{\mu^{k-1}(-1)^k}{(\mu-1)^{k-1}} t^k \tag{12.23}$$

Since $\chi_W(t) = E\left[e^{-Wt}\right]$ and using (2.40), all moments are found as $E\left[W^k\right] = \frac{k!\mu^{k-1}}{(\mu-1)^{k-1}}$. Furthermore, with $\pi_0 = \varphi_W(0) = \frac{1}{\mu}$ and from (2.38), the probability density function follows as

$$f_{W;Geo}(x) = \frac{1}{2\pi i} \int_{c-i\infty}^{c+i\infty} \frac{\frac{t}{\mu} + 1 - \frac{1}{\mu}}{t + 1 - \frac{1}{\mu}} e^{xt}\, dt \qquad (c > 0)$$

By closing the contour for $x > 0$ over the negative $\mathrm{Re}(t)$-plane, we encounter a simple pole at $t = -1 + \frac{1}{\mu} = -(1 - \pi_0) < 0$ (since $\mu > 1$) resulting in

$$f_{W;Geo}(x) = \begin{cases} \left(\frac{1}{\mu} - 1\right)^2 \exp\left(-x\left(1 - \frac{1}{\mu}\right)\right) & x > 0 \\ \frac{1}{\mu}\delta(x) & x = 0 \\ 0 & x < 0 \end{cases} \tag{12.24}$$

From (12.7) the variance is $\mathrm{Var}[W_{Geo}] = \frac{\mu+1}{\mu-1}$. The limit random variable W_{Geo} of a geometric branching process is exponentially distributed with an atom at $x = 0$ equal to the extinction probability $\pi_0 = \frac{1}{\mu}$. From (12.17), the exponent $\beta_{Geo} = 1$ for any value of $\mu \ge 1$. Comparing (12.24) and the general relation (12.19) for small x indicates that the parameter $F_\mu = \left(\frac{1}{\mu} - 1\right)^2$ for a geometric production process.

The limit random variable W for production processes Y of which all moments exist can be computed via Taylor series expansions. In Van Mieghem

(2005), series for both $\chi_{W;Po}(t)$ and $f_{W;Po}(x)$ of a Poisson branching process are presented. Fig. 12.3 illustrates that the probability density function $f_{W;Po}(x)$ of a Poisson branching process is definitely distinct from that of geometric branching process. Since $E[W] = 1$, the variance $\text{Var}[W_{Po}] = \frac{1}{\mu-1}$ of a Poisson limit random variable W_{Po} implies that $f_{W;Po}(x)$ is centered around $x = 1$ more tightly as μ increases.

13

General queueing theory

Queueing theory describes basic phenomena such as the waiting time, the throughput, the losses, the number of queueing items, etc. in queueing systems. Following Kleinrock (1975), any system in which arrivals place demands upon a finite-capacity resource can be broadly termed a *queueing system*.

Queuing theory is a relatively new branch of applied mathematics that is generally considered to have been initiated by A. K. Erlang in 1918 with his paper on the design of automatic telephone exchanges, in which the famous Erlang blocking probability, the Erlang B-formula (14.17), was derived (Brockmeyer *et al.*, 1948, p. 139). It was only after the Second World War, however, that queueing theory was boosted mainly by the introduction of computers and the digitalization of the telecommunications infrastructure. For engineers, the two volumes by Kleinrock (1975, 1976) are perhaps the most well-known, while in applied mathematics, apart from the penetrating influence of Feller (1970, 1971), the *Single Server Queue* of Cohen (1969) is regarded as a landmark. Since Cohen's book, which incorporates most of the important work before 1969, a wealth of books and excellent papers have appeared, an evolution that is still continuing today.

13.1 A queueing system

Examples of queueing abound in daily life: queueing situations at a ticket window in the railway station or post office, at the cash points in the supermarket, the waiting room at the airport, train or hospital, etc. In telecommunications, the packets arriving at the input port of a router or switch are buffered in the output queue before transmission to the next hop towards the destination. In general, a queueing system consists of (a) arriving items

247

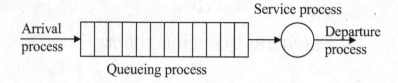

Fig. 13.1. The main processes in a general queueing system.

(packets or customers), (b) a buffer or waiting room, (c) a service center and (d) departures from the system.

The main processes as illustrated in Fig. 13.1 are stochastic in nature. Initially in queueing theory, the main stochastic processes were described in continuous time, while with the introduction of the Asynchronous Transfer Mode (ATM) at the late eighties, many queueing problems were more effectively treated in discrete time, where the basic time unit or time slot was the minimum service time of one ATM cell. In the literature, there is unfortunately no widely adopted standard notation for the main random variables, which often troubles the transparency. Let us start defining the main random variables in continuous time.

13.1.1 *The arrival process*

The arrival process is characterized by the arrival time t_n of the n-th packet (customer) and the interarrival time $\tau_n = t_n - t_{n-1}$ between the n-th and $(n-1)$-th packet. If all interarrival times are i.i.d. random variables with distribution $F_A(t)$, then

$$\Pr\left[\tau_n \leq t\right] = F_A(t)$$

As illustrated in Fig. 8.1, we can associate a counting process $\{N(t), t \geq 0\}$ to the arrival process $\{t_n, t \geq 0\}$ by the equivalence $\{N(t) \geq n\} \iff \{t_n \leq t\}$. In other words, *if all interarrival times are i.i.d.*, the number of arriving packets (customers) is a general renewal process with interarrival time distribution specified by $F_A(t)$. We mention explicitly the condition of independence which was initially considered as a natural assumption. In recent measurements, however, arrivals of IP packets are shown not to obey this simple condition of independence, which has lead to the use of complicated self-similar and long-range dependent arrival processes.

In the sequel, we will use the following notation: $N_A(t)$ is the number of arrivals at time t, while $A(t) = \int_0^t N_A(u)du$ is the total number of arrivals in the interval $[0, t]$.

13.1.2 The service process

The service process is specified in similar way by the service time x_n of the n-th packet (customer). If the random variables x_n are i.i.d. with distribution $F_x(t)$, then

$$\Pr[x_n \leq t] = F_x(t) \tag{13.1}$$

The service process needs additional specifications. First of all, in a single-server queueing system, only one packet (customer) is served at a time. If there is more than one server, more packets can evidently be served simultaneously. Next, we must detail the *service discipline* or *scheduling rule*, which describes the way a packet is treated. There is a large variety of service disciplines. If all packets are of equal priority, the simplest rule is first-in-first-out (FIFO), which serves the packets in the same order in which they arrive. Other types such as last-in-first-out or a random order are possible, though in telecommunication, FIFO occurs most often. If we have packets of different multimedia flows, all with different quality of service requirements, not all packets have equal priority. For instance, a delay-sensitive packet (of e.g. a voice call) must be served as soon as possible preferably before non-delay-sensitive packets (of e.g. a file transfer). In these cases, packets are extracted from the queue by a certain scheduling rule. The simplest case is a two-priority system with a head-of-the line scheduling rule: high-priority packets are always served before low-priority packets. In the sequel, we confine the presentation to a single-server system with one type of packet and a FIFO discipline. Hence, we omit a discussion of scheduling rules. A next assumption is that of *work conservation*: if there is a packet waiting for service, the server will always serve the packet. Thus, the server is only idle if there are no packets waiting in the buffer and immediately starts service when the first packet is placed in the queue or arrives. In a non-work-conservative system, the server may stay idle, even if there are customers waiting (e.g. a situation where patients have to wait during a coffee break in a hospital). Finally, we assume that the arrival process is independent of the service process. Situations where arriving packets of some type (e.g. control) change the way the remaining packets in the buffer are served, or a service discipline that serves at a rate proportional to the number of waiting packets, are not treated.

The service in a router consists in fetching the packet from the buffer, inspecting the header to determine the correct output port and in placing the packet on the output link for transmission.

In this chapter unless the contrary is explicitly mentioned, we consider a single-server queueing system under a work-conservative, FIFO service discipline in which the arrival and service process are independent.

13.1.3 The queueing process

From Fig. 13.1, we observe at least two aspects regarding the queue or buffer: (a) the number of different queues and (b) the number of positions in the queue. In general, a queueing system may have several queues or even a shared queue for different servers. For example, in a router, there is one physical fast memory or buffer in which arriving packets are placed. Depending on the output interfaces, each link driver per output port is a server that extracts the packet destined for its link from the common buffer and transmits the packet on this link. For simplicity, we consider here only one queue with K positions. Often queueing analyses are greatly simplified in the infinite buffer case $K \to \infty$. If the buffer is infinitely long, there is zero loss, as opposed to the finite buffer case in which losses can occur if the queue is full and packets arrive.

So far, the description of the queueing system is complete: we have specified the arrival process, the service process and the physical size of the waiting room or queue. We now turn our attention to desirable quantities that can be deduced from the model specification of the queueing system such as (a) the waiting or queueing time w_n of the n-th packet, (b) the system time $T_n = w_n + x_n$ of the n-th packet, (c) the unfinished work (also called the virtual waiting time or workload) $v(t)$ at time t, (d) the number of packets in the queue $N_Q(t)$ or in the system $N_S(t)$ at time t and (e) the departure time r_n of the n-th packet.

The waiting or queueing time w_n of the n-th packet is only zero if the queue is empty at arrival time t_n. The unfinished work $v(t)$ at time t is the time needed to empty the queueing system or to serve all remaining packets in the system (queue plus server) at time t. Hence, the unfinished work at time t is equal to the sum of the service times of the $N_Q(t)$ buffered packets at time t plus the remaining service time of the packet under service at time t. Precisely at an arrival epoch $t = t_n$ as illustrated in Fig. 13.2, we observe that $v(t_n) = T_n = w_n + x_n$. In addition, $v'(t) = -1$ for all $t \neq t_n$ or $v(t) = \max[T_n - t + t_n, 0]$ for $t \geq t_n$.

The departure times r_n satisfy $r_n = t_n + T_n$. The time during which the server is busy is called a busy period, and likewise, the interval of non-activity is called an idle period.

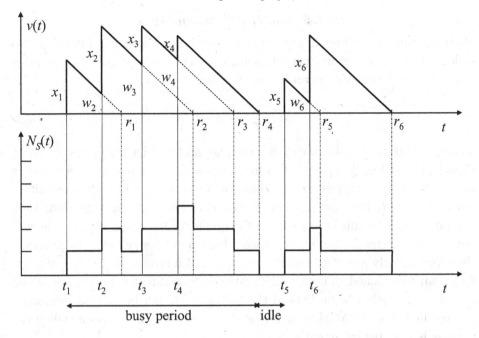

Fig. 13.2. The unfinished work $v(t)$ and the number of packets in the system $N_S(t)$ as function of time. At any new arrival at t_n holds $v(t_n) = w_n + x_n$. The unfinished work $v(t)$ decreases with slope -1 between two arrivals. The waiting times w_n and departure times r_n are also shown. Notice that $w_1 = w_5 = 0$.

13.1.4 The Kendall notation for queueing systems

Kendall introduced a notation that is commonly used to describe or classify the type of a queueing system. The general syntax is $A/B/n/K/m$, where A specifies the interarrival process, B the service process, n the number of servers, K the number of positions in the queue and m restricts the number of allowed arrivals in the queueing system. Examples for both the interarrival distribution A and the service distribution B are M (memoryless or Markovian) for the exponential distribution, G for a general distribution[1] and D for a deterministic distribution. When other letters are used besides these three common assignments, the meaning will be defined. For example, $M/G/1$ stands for a queueing system with exponentially distributed interarrival times, a general service distribution and 1 server. If one of the two last identifiers K and m is not written, they should be interpreted as infinity. Hence, $M/G/1$ has an infinitely long queue and no restriction on the number of allowed arrivals.

---

[1] Often G is written where GI, general independent process, is meant. We interpret G as a general interarrival process, which can be correlated over time.

13.1.5 The traffic intensity ρ

An important parameter in any queueing system is the traffic intensity also called the load or the utilization, defined as the ratio of the mean service time $E[x] = \frac{1}{\mu}$ over the mean interarrival time $E[\tau] = \frac{1}{\lambda}$

$$\rho = \frac{E[x]}{E[\tau]} = \frac{\lambda}{\mu} \tag{13.2}$$

where λ and μ are the mean interarrival and service rate, respectively. Clearly, if $\rho > 1$ or $E[x] > E[\tau]$, which means that the mean service time is longer than the mean interarrival time, then the queue will grow indefinitely long for large t, because packets are arriving faster on average than they can be served. In this case ($\rho > 1$), the queueing system is unstable or will never reach a steady-state. The case where $\rho = 1$ is critical. In practice, therefore, mostly situations where $\rho < 1$ are of interest. If $\rho < 1$, a steady-state can be reached. These considerations are a direct consequence of the law of conservation of packets in the system, but can be proved rigorously by ergodic theory or Markov steady-state theory, which determine when the process is positive recurrent.

13.2 The waiting process: Lindley's approach

From the definition of the waiting time and from Fig. 13.2, a relation between w_{n+1} and w_n is found. Suppose the waiting time for the first packet $w_1 = w$, which is the initialization. If $r_n \leq t_{n+1}$, which means that the n-th packet leaves the queueing system before the $(n+1)$-th packet arrives, the system is idle and $w_{n+1} = 0$. In all other situations, $r_n > t_{n+1}$, the n-th packet is still in the queueing system while the next $(n+1)$-th packet arrives and $w_{n+1} = t_n + w_n + x_n - t_{n+1}$. Indeed, the waiting time of $(n+1)$-th packet equals the system time $T_n = w_n + x_n$ of the n-th packet which started at t_n minus his own arrival time t_{n+1}. During the interval $[t_n, t_{n+1}]$, the queueing system has processed an amount of the unfinished work equal to $t_{n+1} - t_n = \tau_{n+1}$ time units. Hence, we arrive at the general recursion for the waiting time,

$$w_{n+1} = \max(w_n + x_n - \tau_{n+1}, 0) \tag{13.3}$$

Let $\sigma_n = x_n - \tau_{n+1}$, then

$$
\begin{aligned}
w_{n+1} &= \max[0, w_n + \sigma_n] \\
&= \max[0, \max[w_{n-1} + \sigma_{n-1}, 0] + \sigma_n] \\
&= \max[0, \sigma_n, w_{n-1} + \sigma_{n-1} + \sigma_n]
\end{aligned}
$$

and, by iteration,

$$w_{n+1} = \max \left[0, \sigma_n, \sigma_{n-1} + \sigma_n, \sigma_{n-2} + \sigma_{n-1} + \sigma_n, \ldots, \sum_{k=1}^{n} \sigma_k + w_1 \right] \quad (13.4)$$

A number of observations are in order:

First, *if both the interarrival times τ_{n+1} and the service times x_n are i.i.d. random variables* and mutually independent, then the differences σ_n are i.i.d. random variables. In addition, w_n and σ_n are also independent because (13.4) shows that w_n only depends on σ_k with indices $k < n$. Then, the waiting time process $\{w_n\}_{n \geq 1}$ is a discrete time Markov process with a continuous state space (the waiting times w_n are positive real numbers) because the general relation (13.3) reveals that, since the random variable σ_n is independent of w_n, the waiting time for the $(n+1)$-th packet is only dependent on the waiting time of the previous n-th packet. This is the Markov property. Since the state space is a continuum, it is not a Markov chain, merely a Markov process.

Second, if there exists a packet m for which $w_m = 0$ (e.g. packet $m = 5$ in Fig. 13.2), which means that the m-th packet finds the system empty, then all packets after the m-th packet are isolated from the effects of those before the m-th. Mathematically, this separation between two busy periods directly follows from (13.3) because $w_{m+1} = \max [0, \sigma_m]$ leading via iteration for $n \geq m$ to

$$w_{n+1} = \max \left[0, \sigma_n, \sigma_{n-1} + \sigma_n, \ldots, \sum_{k=m+1}^{n} \sigma_k, \sum_{k=m}^{n} \sigma_k \right]$$

In other words, this relation is similar to (13.4) as if the system were started from $k = m$ and $w_m = 0$ instead of $k = 1$ with $w_1 = w$. Any busy period can be regarded as a renewal of the waiting process, independent of the previous busy periods.

Third, again invoking the assumption that σ_n are i.i.d. random variables, then the order in the sequence $\{\sigma_n\}_{n \geq 1}$ is of no importance in (13.4) and we may relabel the random variables in (13.4) as $\sigma_k \to \sigma_{n-k+1}$ to obtain a new random variable

$$\omega_{n+1} = \max \left[0, \sigma_1, \sigma_1 + \sigma_2, \ldots, \sum_{k=1}^{n-1} \sigma_k, \sum_{k=1}^{n} \sigma_k + w_1 \right]$$

which is identically distributed as w_{n+1}. The interest of this observation is that, provided $w_1 = 0$ and, hence, $\omega_n = \max_{1 \leq j \leq n} \sum_{k=0}^{n-j} \sigma_k$ where $\sigma_0 = 0$, the sequence $\{\omega_n\}_{n \geq 1}$ can only increase with n, because the maximum cannot decrease[2] if an additional term $\sum_{k=0}^{n+1} \sigma_k$ is added. Thus, if $w_1 = 0$, the event $\{\omega_{n+1} < x\}$ is always contained in $\{\omega_n < x\}$. In steady-state, which is reached if $n \to \infty$,

$$\lim_{n \to \infty} \{\omega_n < x\} = \cap_{n=1}^{\infty} \{\omega_n < x\} = \{\sup_{j \geq 0} \sum_{k=0}^{j} \sigma_k < x\}$$

which means that the random variable ω_n with same distribution as the waiting time w_n converges to a limit random variable that is the supremum of the terms $\sum_{k=0}^{j} \sigma_k$ in the series. From this relation, it follows that the steady-state distribution $W(x)$ of the waiting time is

$$W(x) = \lim_{n \to \infty} \Pr[w_n < x] = \lim_{n \to \infty} \Pr[\omega_n < x] = \Pr\left[\sup_{j \geq 0} \sum_{k=0}^{j} \sigma_k < x\right]$$

if the latter probability exists, i.e. not zero for all x. Lindley has proved that, if $\rho < 1$, the latter corresponds to a proper probability distribution. In other words, the steady-state distribution of the waiting time in a GI/G/1 system[3] exists. Alternatively, the Markov process $\{w_n\}_{n \geq 1}$ is ergodic if $\rho < 1$.

Lindley's proof is as follows. Due to the assumption that σ_n are i.i.d. random variables, the Strong Law of Large Numbers (6.3) is applicable: $\Pr\left[\lim_{n \to \infty} \frac{1}{n} \sum_{k=0}^{n} \sigma_k = E[\sigma]\right] = 1$ where $E[\sigma] = E[x] - E[\tau] < 0$ (the mean service time is smaller than the mean interarrival time) if $\rho < 1$ while $E[\sigma] > 0$ if $\rho > 1$. In case $\rho > 1$, there exists a number $\nu > 0$ and $\beta > 1$ such that, for all $n > \nu$, holds $\sum_{k=0}^{n} \sigma_k \geq \frac{1}{\beta} E[\sigma] n$ with probability 1. For large ν, $\sum_{k=0}^{n} \sigma_k$ can be made larger than any fixed x such that $\Pr\left[\sup_{j \geq 0} \sum_{k=0}^{j} \sigma_k < x\right] = 0$. In case $\rho < 1$, we have for sufficiently large n that $\sum_{k=0}^{n} \sigma_k < 0$. Thus, for any $x > 0$ and $\epsilon > 0$, there exists a number ν (independent of x) such that, for all $n > \nu$,

$$\Pr\left[\sum_{k=0}^{n} \sigma_k < x\right] \geq \Pr\left[\sum_{k=0}^{n} \sigma_k < 0\right] > 1 - \epsilon$$

while, for $n < \nu$, we can always find a number $\xi > 0$ such that for all $x > \xi$,

$$\Pr\left[\sum_{k=0}^{n} \sigma_k < x\right] > 1 - \epsilon$$

Since $\sup_{j \geq 0} \sum_{k=0}^{j} \sigma_k$ is attained for $j < \nu$ or $j > \nu$ and because both regimes can be bounded by the same lower bound,

$$\Pr\left[\sup_{j \geq 0} \sum_{k=0}^{j} \sigma_k < x\right] > \Pr\left[\sum_{k=0}^{n < \nu} \sigma_k < x\right] 1_{j < \nu} + \Pr\left[\sum_{k=0}^{n > \nu} \sigma_k < x\right] 1_{j > \nu}$$
$$> 1 - \epsilon$$

Clearly, $\lim_{x \to \infty} \Pr\left[\sup_{j \geq 0} \sum_{k=0}^{j} \sigma_k < x\right] = 1$ and $\Pr[w_n < 0] = 0$, thus $\Pr\left[\sup_{j \geq 0} \sum_{k=0}^{j} \sigma_k < x\right]$

---

[2] This observation cannot be made from (13.4) because σ_n, which affects all but the first term in the maximum, can be negative.

[3] Notice that the analysis crucially relies on the independence of the interarrival and service process.

is non-decreasing and a proper probability distribution. We omit the considerations for the case $\rho = 1$. □

We now concentrate on the computation of the steady-state distribution for the waiting time (in the queue) in the case that the load $\rho < 1$ and under the confining assumption that *both the interarrival times τ_{n+1} and the service times x_n are i.i.d. random variables*. We find from (13.3) that

$$\Pr\left[w_{n+1} < x\right] = \Pr\left[w_n + \sigma_n < x\right] \qquad \text{if } x \geq 0$$
$$\Pr\left[w_{n+1} < x\right] = 0 \qquad \text{if } x < 0$$

With the law of total probability (2.46) and since σ_n can be negative, the right-hand side is

$$\Pr\left[w_n + \sigma_n < x\right] = \int_{-\infty}^{\infty} \Pr\left[w_n < x - s | \sigma_n = s\right] \frac{d}{ds} \Pr\left[\sigma_n < s\right] ds$$

Using the independence of w_n and σ_n, and that $w_n \geq 0$, we obtain for $x \geq 0$,

$$\Pr\left[w_n + \sigma_n < x\right] = \int_{-\infty}^{x} \Pr\left[w_n < x - s\right] d\Pr\left[\sigma_n < s\right]$$

The distribution $\Pr\left[\sigma_n < s\right] = \Pr\left[x_n - \tau_{n+1} < s\right] = C_n(s)$ can be computed (see Problem (v) in Chapter 3) provided the interarrival and service process are known. Proceeding to the steady-state by letting $n \to \infty$ amounts to Lindley's integral equation in $W(x) = \lim_{n\to\infty} \Pr\left[w_n < x\right]$ with $C(x) = \lim_{n\to\infty} C_n(x)$,

$$W(x) = \int_{-\infty}^{x} W(x - s)dC(s) \qquad \text{if } x \geq 0 \qquad (13.5)$$
$$= 0 \qquad \text{if } x < 0$$

The integral equation (13.5) is of the Wiener-Hopf type and treated in general by Titchmarsh (1948, Section 11.17) and specifically by Kleinrock (1975, Section 8.2) and Cohen (1969, p. 337). Apart from Lindley's approach, Pollaczek has used variants of the complex integral expression for

$$\max(x, 0) = \frac{xe^{-ax}}{2\pi i} \int_{c-i\infty}^{c+i\infty} \frac{e^{xz}}{z - a} dz \qquad (c > \text{Re}(a) > 0)$$

to treat the complicating non-linear function $\max(x, 0)$ in (13.3). Several other approaches (Kleinrock, 1975, Chapter 8) have been proposed to solve (13.3). We will only discuss the approach due to Beneš, because his approach does not make the confining assumption that both the interarrival times τ_{n+1} and the service times x_n are i.i.d. random variables. As mentioned before, in Internet traffic, which has been shown to be long-range dependent (i.e. correlated over many time units mainly due to TCP's control loop), the interarrival times can be far from independent.

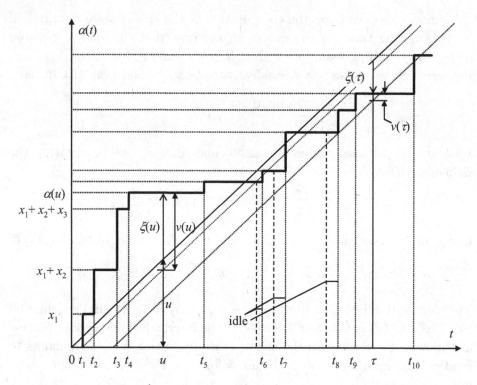

Fig. 13.3. The amount of work arriving to the queueing system $\alpha(t)$ versus time t. At $t = u$, we observe that $\xi(t) = \alpha(t) - t + v(0^-) > 0$. The largest value of $\xi(t) - \xi(u)$ is found for $u = t_1$ because $\xi(t_1) = -t_1$, the only negative value of $\xi(t)$ in $[0, u)$. Graphically, we shift the line at 45° so to intersect the point $(t_1, \alpha(t_1))$ to determine $v(u)$. At $t = \tau$, $\xi(\tau) < 0$ and the largest negative value of $\xi(t)$ in $[0, \tau)$ is attained at $t = t_8$. Three of the five idle periods have also been shown.

13.3 The Beneš approach to the unfinished work

Instead of observing the queueing system at a time t, the Beneš approach considers the behavior over a time interval $[0, t)$. Let $\alpha(t)$, $\eta(t)$ and $b(t)$ denote the amount of work arriving to the queueing system in the interval $[0, t)$, the total idle time of the server and the total busy time of the server in the interval $[0, t)$, respectively. The amount of work arriving to the system is expressed in units of time and must be regarded as the time needed to process this work, similarly to the definition of the unfinished work. If the work to process arrives at discrete times, then $\alpha(t)$ increases in jumps, as illustrated in Fig. 13.3,

$$\alpha(t) = \sum_{j=0}^{A(t)} x_j$$

In general, however, the work may arrive continuously over time with possibly jumps at certain times. The purpose is to determine the unfinished work or virtual waiting time $v(t)$ at time instant t, and not over a time interval $[0, t)$ as the previously defined quantities and $A(t) = \int_0^t N_A(u)du$. Clearly, for $t > 0$, the unfinished work at time t consists of the total amount of work brought in by arrivals during $[0, t)$ plus the amount of work present just before $t = 0$ minus the total time the server has been active,

$$v(t) = v(0^-) + \alpha(t) - b(t) \tag{13.6}$$

From the definitions above,

$$\eta(t) + b(t) = t \tag{13.7}$$

Moreover, $\alpha(t)$, $\eta(t)$ and $b(t)$ are non-decreasing and right continuous (jumps may occur) functions of time t. Since $\eta(t)$ and $b(t)$ are complementary, it is convenient to eliminate $b(t)$ from (13.6) and (13.7) and further concentrate on the total idle time $\eta(t)$ given as

$$\eta(t) = v(t) + t - v(0^-) - \alpha(t) \tag{13.8}$$

If $v(u) > 0$ at any time $u \in [0, t)$, then $\eta(t) = 0$. On the other hand, if $v(u) = 0$ at some time $u \in [0, t)$, then it follows from (13.8) that

$$\eta(u) = u - v(0^-) - \alpha(u) \tag{13.9}$$

Since $\eta(t)$ is non-decreasing in t, the total idle time in the interval $[0, t)$ at moments u when the buffer is empty ($v(u) = 0$) is the largest value for η that can be reached in $[0, t)$,

$$\eta(t) = \sup_{0 < u < t} \left(u - v(0^-) - \alpha(u) \right)$$

and the supremum is needed because $\alpha(t)$ can increase discontinuously (in jumps). Combining the two regimes, we obtain in general that

$$\eta(t) = \max \left[0, \sup_{0 < u < t} \left(u - v(0^-) - \alpha(u) \right) \right] \tag{13.10}$$

Equating the two general expressions (13.8) and (13.10) for the total idle time of the server leads to an new relation for the unfinished work,

$$v(t) = v(0^-) + \alpha(t) - t + \max\left[0, \sup_{0<u<t} (u - v(0^-) - \alpha(u))\right]$$

$$= \max\left[v(0^-) + \alpha(t) - t, \sup_{0<u<t} \{\alpha(t) - t - (\alpha(u) - u)\}\right]$$

The quantity $\xi(t) = \alpha(t) - t + v(0^-)$ is recognized as the server overload during $[0, t)$, while $\alpha(t) - t$ is the amount of excess work arriving during the interval $[0, t)$. Thus, $\xi(t) - \xi(u)$ is the amount of excess work during $[u, t)$ or the overload of the server during $[u, t)$ provided $u > 0$ and $\xi(0) = v(0^-)$. Then,   .

$$v(t) = \max\left[\xi(t), \sup_{0<u<t} \{\xi(t) - \xi(u)\}\right]$$

and, with the convention that $v(0^-) = \sup_{0<u<0^-} \{\xi(0) - \xi(u)\} = \xi(0)$,

$$v(t) = \sup_{0<u<t} \{\xi(t) - \xi(u)\} \tag{13.11}$$

The unfinished work $v(t)$ at time t is equal to the largest value of the overload or excess work during any interval $[u, t) \subset [0, t)$. The relation (13.11) is illustrated and further explained in Fig. 13.3. This general relation (13.11) shows that the unfinished work is the maximum of a stochastic process. Furthermore, if $v(t) = 0$, (13.11) indicates that $\sup_{0<u<t} \{\xi(t) - \xi(u)\} = 0$. Let u^* denote the value at which $\sup_{0<u<t} \{\xi(t) - \xi(u)\} = \xi(t) - \xi(u^*) = 0$. But $\xi(u^*)$ is the lowest value of $\xi(t)$ in $[0, t)$ and, unless an arrival occurs during the interval $[t, t+ \Delta t]$, $\xi(t + \Delta t) = \xi(t) - \Delta t < \xi(t)$. This argument shows that, as soon as a new idle period begins, $\xi(t)$ attains the minimum value so far.

During the idle period as shown in Fig. 13.4, $\xi(t)$ further decreases linearly with slope -1 towards a new minimum $\xi(b_j)$ in $[0, b_j]$ until the beginning of a new busy period, say the j-th at $t = b_j$. Then, for all $b_j < t < b_{j+1}$,

$$v(t) = \xi(t) - \xi(b_j) = \sup_{b_j<u<t} \{\xi(t) - \xi(u)\}$$

In other words, we observe that idle periods decouple the past behavior from future behavior, as deduced earlier from the waiting time analysis in Section 13.2. As illustrated in Fig. 13.4, the series $\{\xi(b_j)\}$, where b_j denotes the start of the j-th busy period, is monotonously decreasing in b_j, i.e. $\xi(b_j) > \xi(b_{j+1})$ for any j.

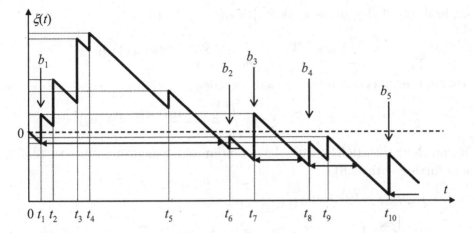

Fig. 13.4. The excess work $\xi(t)$ for the same process as in previous plot. The arrows with b_j denote the start of the j-th busy period. Observe that $\xi(b_j)$ is the minimum so far and that a busy period ends at $t > b_j$ for which $\xi(t) = \xi(b_j)$. The length of a busy period has been represented by a double arrow.

Let us proceed to compute the distribution of the unfinished work following an idea due to Beneš. Beneš applies the identity[4], valid for all z,

$$e^{-zt} = 1 - z \int_0^t e^{-z\tau} d\tau$$

to the total idle time of the server by putting $\tau = \eta(u)$

$$e^{-zt} = 1 - z \int_{\eta^{-1}(0)}^{\eta^{-1}(t)} e^{-z\eta(u)} d\eta(u)$$

where $\eta^{-1}(t)$ is the inverse function. Note that $\eta(0) = 0$ and that $d\eta(u) = 1_{\{v(u)=0\}} du = \delta\left(v(u)\right) du$ where $\delta\left(x\right)$ is the Dirac impulse. Let $t \to \eta(t)$, then

$$e^{-z\eta(t)} = 1 - z \int_0^t e^{-z\eta(u)} \delta\left(v(u)\right) du$$

Substituting (13.9) in the integral, which is only valid if $v(u) = 0$, and (13.8) at the left-hand side, which is generally valid, gives

$$e^{-z\left(v(t)+t-v(0^-)-\alpha(t)\right)} = 1 - z \int_0^t e^{-z\left(u-v(0^-)-\alpha(u)\right)} \delta\left(v(u)\right) du$$

---

[4] Borovkov (1976, p. 30) proposes another but less simple approach by avoiding the use of the identity ingeniously introduced by Beneš.

or, in terms of the excess work $\xi(t) = \alpha(t) - t + v(0^-)$,

$$e^{-zv(t)} = e^{-z\xi(t)} - z \int_0^t e^{-z(\xi(t)-\xi(u))} \delta\left(v(u)\right) du$$

Taking the expectation of both sides yields,

$$E\left[e^{-zv(t)}\right] = E\left[e^{-z\xi(t)}\right] - z \int_0^t E\left[e^{-z(\xi(t)-\xi(u))}\delta\left(v(u)\right)\right] du$$

Recall that, with (2.34), with the definition of a generating function (2.37), and further with (2.61),

$$Q = E\left[e^{-z(\xi(t)-\xi(u))}\delta\left(v(u)\right)\right]$$

$$= \int_{-\infty}^{\infty}\int_{-\infty}^{\infty} e^{-zx}\delta(y)\frac{\partial^2}{\partial x\partial y}\Pr\left[\xi(t)-\xi(u)\le x, v(u)\le y\right] dxdy$$

and with (2.45), we have

$$Q = \int_{-\infty}^{\infty}\int_{-\infty}^{\infty} e^{-zx}\delta(y)\frac{\partial^2}{\partial x\partial y}\Pr\left[\xi(t)-\xi(u)\le x|v(u)\le y\right]\Pr\left[v(u)\le y\right] dxdy$$

$$= \int_{-\infty}^{\infty} dx e^{-zx}\frac{\partial}{\partial x}\int_{-\infty}^{\infty}\delta(y)\Pr\left[\xi(t)-\xi(u)\le x|v(u)\le y\right]\frac{\partial \Pr\left[v(u)\le y\right]}{\partial y} dy$$

$$= \int_{-\infty}^{\infty} e^{-zx}\frac{d}{dx}\Pr\left[\xi(t)-\xi(u)\le x|v(u)=0\right]\Pr\left[v(u)=0\right] dx$$

Combining all leads to

$$\int_{-\infty}^{\infty} e^{-zx} f_{v(t)}(x)dx = \int_{-\infty}^{\infty} e^{-zx} f_{\xi(t)}(x)dx$$

$$- z\int_{-\infty}^{\infty} e^{-zx}\frac{d}{dx}\left(\int_0^t \Pr\left[\xi(t)-\xi(u)\le x|v(u)=0\right]\Pr\left[v(u)=0\right] du\right) dx$$

By partial integration, we can remove the factor z at both sides. Indeed, since $\Pr\left[v(t)\le x\right] = 0$ for $x < 0$,

$$\int_{-\infty}^{\infty} e^{-zx} f_{v(t)}(x)dx = z\int_{-\infty}^{\infty} e^{-zx}\Pr\left[v(t)\le x\right] dx$$

Hence, we arrive at

$$\int_{-\infty}^{\infty} e^{-zx}\Pr\left[v(t)\le x\right] dx = \int_{-\infty}^{\infty} e^{-zx}\bigg[\Pr\left[\xi(t)\le x\right]$$

$$- \frac{d}{dx}\left(\int_0^t \Pr\left[\xi(t)-\xi(u)\le x|v(u)=0\right]\Pr\left[v(u)=0\right] du\right)\bigg] dx$$

which is equivalent to

$$\Pr\left[v(t)\le x\right] = \Pr\left[\xi(t)\le x\right] - \int_0^t \frac{d\Pr\left[\xi(t)-\xi(u)\le x|v(u)=0\right]}{dx}\Pr\left[v(u)=0\right] du$$

$$(13.12)$$

This general relation for the distribution of the unfinished work in terms of the excess work is the Beneš equation. If $v(u) = 0$ for all $u \in [0, t)$, this means that during that interval no work arrives and that $\xi(t) - \xi(u) = u - t$ or that $-t \le \xi(t) - \xi(u) \le 0$ for any $u \in [0, t)$. Thus, if we choose $x \in [-t, 0)$ such that the event $\{\xi(t) - \xi(u) = u - t \le x\}$ is possible, the probabilities appearing in the right-hand side are not identically zero while $\Pr[v(t) \le x] = 0$. Hence, for $x \in [-t, 0)$, the Beneš equation reduces to

$$\Pr[\xi(t) \le x] = \int_0^t \frac{d\Pr[\xi(t) - \xi(u) \le x | v(u) = 0]}{dx} \Pr[v(u) = 0]\, du$$

from which the unknown probability of an empty system $\Pr[v(u) = 0]$ can be found[5] for $t + x \le u \le t$. The Beneš equation translates the problem of finding the time-dependent virtual waiting time or unfinished work in an integral equation that, in principle, can be solved. We further note that in the derivation hardly any assumptions about the queueing system nor the arrival process are made such that the Beneš equation provides the most general description of the unfinished work in any queueing system. Of course, the price for generality is a considerable complexity in the integral equation to be solved. However, we will see examples[6] of its use in ATM.

13.3.1 A constant service rate

If the server operates deterministically as in ATM, for example, the amount of work arriving to the queueing system in the interval $[0, t)$ simplifies to $\alpha(t) = A(t)$, the number of ATM cells in the interval $[0, t)$, because $x_j = x$ is the time to process one ATM cell, which we take as time unit $x = 1$. With this convention, we have that $\xi(t) = A(t) - t$. After substitution of $u = t - y$, the integral I in (13.12) is

$$I = \int_0^t \frac{d\Pr[\xi(t) - \xi(t-y) \le x | v(t-y) = 0]}{dx} \Pr[v(t-y) = 0]\, dy$$

and the event

$$\{\xi(t) - \xi(t-y) \le x\} = \{A(t) - A(t-y) \le x + y\}$$

---

[5] This relation in the unknown function $f(u) = \Pr[v(u) = 0]$ is a Volterra equation of the first kind (see e.g. Morse and Feshbach (1978, Chapter 8))

$$g(z) = \int_a^z K(z|u)f(u)du$$

These integral equations frequently appear in physics in boundary problems, potential and Green's function theory.

[6] Borovkov (1976) investigates the Beneš method in more detail. He further derives from (13.12) formulae for light and heavy traffic, and the discrete time process.

Since $A(t) - A(t-y)$ is a non-negative integer k and thus a discrete random variable, the probability density function is

$$\frac{d\Pr[\xi(t)-\xi(t-y)\le x|v(t-y)=0]}{dx} = \Pr[A(t)-A(t-y)= k|v(t-y)=0]\,1_{x+y=k}$$

which implies that only values at $y = k - x$ contribute to the integral I. Hence, $0 \le y \le t$ implies that $\lceil x \rceil \le k \le \lfloor x + t \rfloor$ where $\lfloor z \rfloor$ (respectively $\lceil z \rceil$) are the integer equal to or smaller (respectively larger) than z,

$$I = \sum_{k=\lceil x \rceil}^{\lfloor x+t \rfloor} \Pr\left[A(t) - A(t + x - k) = k|v(t+x-k) = 0\right] \Pr\left[v(t + x - k) = 0\right]$$

Hence, for a discrete queue with time slots equal to the constant service time, the Beneš equation reduces to

$$\Pr\left[v(t) \le x\right] = \Pr\left[A(t) \le \lfloor x + t \rfloor\right] \tag{13.13}$$
$$-\sum_{k=\lceil x \rceil}^{\lfloor x+t \rfloor}\Pr\left[A(t)-A(t+x-k)= k|v(t+x-k)=0\right]\Pr[v(t+x-k)=0]$$

13.3.2 The steady-state distribution of the virtual waiting time

Let us turn to the steady-state distribution $V(x) = \lim_{t\to\infty} \Pr\left[v(t) \le x\right]$. Since $\alpha(t)$ is the amount of work arriving to the queueing system in the interval $[0,t)$, $\frac{\alpha(t)}{t}$ is the mean amount of work arriving in that interval $[0,t)$. The steady-state stability condition, equivalent to $\rho < 1$, requires that

$$\lim_{t\to\infty} \frac{\alpha(t)}{t} = \rho < 1$$

because the server capacity is 1 unit of work per unit of time. Since $\alpha(t)$ is not decreasing, $\xi(t)$ decreases continuously with slope -1 between two arrivals and increases (possibly discontinuously with jumps) during arrival epochs, as illustrated in Fig. 13.3. In the stable steady-state regime where $\rho < 1$ and $\lim_{t\to\infty} \frac{\alpha(t)}{t} < 1$, we find that

$$\lim_{t\to\infty} \xi(t) = \lim_{t\to\infty} t\left(\frac{\alpha(t)}{t} - 1\right) = -\infty$$

and thus $\lim_{t\to\infty} \Pr\left[\xi(t) \le x\right] = 1$ and $\lim_{t\to\infty} \frac{\xi(t)}{t} = \rho - 1 < 0$. From (13.7), we see that

$$\lim_{t\to\infty} \frac{\eta(t)}{t} = 1 - \lim_{t\to\infty} \frac{b(t)}{t} = 1 - \rho$$

which, with (13.9), suggests that

$$V(0) = \lim_{t \to \infty} \Pr[v(t) = 0] = 1 - \rho \qquad (13.14)$$

If the Strong Law of Large Numbers is applicable, which implies that the lengths of the idle periods are independent and identically distributed, this relation $V(0) = 1 - \rho$ is proved to be true by Borovkov (1976, pp. 33–34). Hence, for any stationary single-server system with traffic intensity ρ, the probability of an empty system at an arbitrary time is $1 - \rho$.

Taking the limit $t \to \infty$ in (13.12) then yields

$$V(x) = 1 - \lim_{t \to \infty} \int_0^t \frac{d\Pr[\xi(t) - \xi(t - y) \le x | v(t - y) = 0]}{dx} \Pr[v(t - y) = 0] dy$$

The tail probability $1 - V(x) = \lim_{t \to \infty} \Pr[v(t) > x] = \Pr[v(t_\infty) > x]$ is

$$\Pr[v(t_\infty) > x] = \int_0^\infty \frac{d\Pr[\xi(t_\infty) - \xi(t_\infty - y) \le x | v(t_\infty - y) = 0]}{dx} \Pr[v(t_\infty - y) = 0] dy$$

This relation shows that, at a point in the steady-state $t_\infty \to \infty$, the contributions to $V(x)$ are due to arrivals and idle periods in the past. The corresponding steady-state equation for (13.13) is

$$\Pr[v(t_\infty) > x] = \sum_{k = \lceil x \rceil}^{\infty} \Pr\left[v(t_\infty + x - k) = 0 \,\bigg|\, \int_{t_\infty + x - k}^{t_\infty} N_A(u)\, du = k\right]$$

$$\times \Pr\left[\int_{t_\infty + x - k}^{t_\infty} N_A(u)\, du = k\right] \qquad (13.15)$$

13.4 The counting process

A similar general conservation relation to (13.3) can be deduced for the counting process,

$$N_S(r_{k+1}) = \max(N_S(r_k) - 1, 0) + N_A(r_{k+1}) - N_A(r_k)$$

The number of packets in the system at the departure time of the $(k + 1)$-th packet equals the number of packets in the system at the departure time of the previous packet k minus that packet, but increased by the number of arrivals in the time interval $[r_k, r_{k+1}]$. Similarly, for the queue (which is system minus the packet currently under service),

$$N_Q(r_{k+1}) = \max(N_Q(r_k) + N_A(r_{k+1}) - N_A(r_k) - 1, 0)$$

which is the direct analog of (13.3) for the waiting time.

Whereas the waiting process is more natural to consider in problems where interarrival times are specified, the counting process has more advantages in a discrete time analysis. In the latter, the queueing system is observed at certain moments in time, for instance, at the beginning of a timeslot k that starts immediately[7] after the departure of the k-th packet and is equal to the interval $[r_k, r_{k+1}]$, for all $r_k > 0$ and $r_0 = 0$. It will be convenient to simplify the notation: $S_k = N_S(r_k^+)$ denotes the system content (i.e. the number of occupied queue positions including the packets currently being served) at the beginning of timeslot k, $Q_k = N_Q(r_k^+)$ is the queue content at the beginning of timeslot k, and X_k and A_k are the number of served packets and of arriving packets during timeslot k respectively. The system content satisfies the continuity (or balance) equation

$$S_{k+1} = (S_k - X_k)^+ + A_k \qquad (13.16)$$

whereas the queue content obeys

$$Q_{k+1} = (Q_k - X_k + A_k)^+ \qquad (13.17)$$

where $(x)^+ \equiv \max(x, 0)$. On the other hand, the relation between system and queue content implies that $Q_k = (S_k - X_k)^+$ such that (13.16) is rewritten as

$$S_{k+1} = Q_k + A_k \qquad (13.18)$$

The number of cells at the beginning of a timeslot $k + 1$ in the system is the sum of the number of queued packets at the beginning of the previous timeslot k and the newly arrived packets during timeslot k.

13.4.1 *Queue observations*

It is worthwhile to investigate the relation between observations at various instances of time of the queueing process $\{N_S(t), t \geq 0\}$ which represents the number of packets in the system at time t. As seen before and as illustrated in Fig. 13.5, two time instances seem natural: an inspection at departure times where $N_S(r_n^+)$ describes the number of packets in the system just after departure of the n-th packet and an observation at arrival times where $N_S(t_n^-)$ describes the number of packets in the system just before the n-th packet enters.

Suppose that the n-th packet leaves $N_S(r_n^+) = k \leq j$ packets behind in the system. This implies that precisely k arrivals after the n-th packet have entered the system. Hence, the $(n + k + 1)$-th packet sees, just before

---

[7] We write $x^+ = x + \epsilon$ and $x^- = x - \epsilon$ where $\epsilon > 0$ is an arbitrary, positive real number. The notation x^+ should not be confused with $(x)^+ = \max(x, 0)$.

entering the system, at most k packets, because during the period $t = r_n$ and $t = t_{n+k+1} \geq r_n$, only departures are possible. Thus, $N_S(t_{n+k+1}^-) \leq k$ and, clearly, for $j \geq k$, the $(n+j+1)$-th packet observes no more than $N_S(t_{n+j+1}^-) \leq j$. Hence, the following implication holds

$$\left\{ N_S(r_n^+) \leq j \right\} \Longrightarrow \left\{ N_S(t_{n+j+1}^-) \leq j \right\}$$

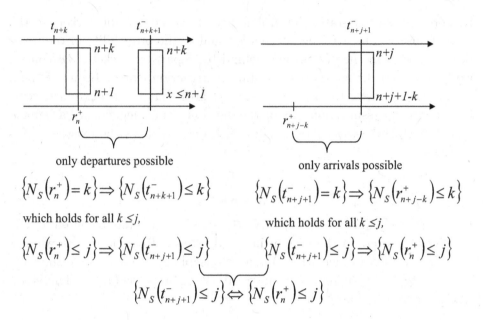

Fig. 13.5. Relation between queue observations at arrival and at departure epochs.

Consider now the converse. Suppose that the $(n+j+1)$-th packet sees precisely $k \leq j$ packets in front of it upon arrival: $N_S(t_{n+j+1}^-) = k \leq j$. This implies that the $(n+j+1-k)$-th packet is the first packet that will leave the system after $t = t_{n+j+1}$ and that the $(n+j-k)$-th packet has already left the system. At its departure at $t = r_{n+j-k} \leq t_{n+j+1}$, it has observed at most k packets behind it, because only arrivals are possible in the interval $[r_{n+j-k}, t_{n+j+1})$. Hence, $N_S(r_{n+j-k}^+) \leq k$ and, set $k = j$, then $N_S(r_n^+) \leq j$ leading to the implication

$$\left\{ N_S(t_{n+j+1}^-) \leq j \right\} \Longrightarrow \left\{ N_S(r_n^+) \leq j \right\}$$

Combining both implications leads to the equivalence,

$$\left\{ N_S(t_{n+j+1}^-) \leq j \right\} \Longleftrightarrow \left\{ N_S(r_n^+) \leq j \right\}$$

or, for any sample path (or realization), it holds, for any non-zero integer j, that

$$\Pr\left[N_S(t^-_{n+j+1}) \leq j\right] = \Pr\left[N_S(r^+_n) \leq j\right]$$

In steady-state for $n \to \infty$, with $\lim_{n\to\infty} N_S(t^-_n) = N_{S;A}$ and $\lim_{n\to\infty} N_S(r^+_n) = N_{S;D}$, we find that

$$\Pr\left[N_{S;A} = j\right] = \Pr\left[N_{S;D} = j\right] \tag{13.19}$$

In words, in steady-state, the probability of the number of packets in the system observed by arriving packets is equal to the probability of the number of packets in the system left behind by departing packets. Of course, we have assumed that the steady-state distribution exists. If one of these distributions exists, the analysis demonstrates that the other must exist. Notice that no assumptions about the distribution or dependence are made and that (13.19) is a general result which only assumes the existence of a steady-state.

13.5 PASTA

Let us denote by $\lim_{t\to\infty} N_S(t) = N_S$ the steady-state system content or the number of packets in the system in steady-state. To compute the waiting time distribution (under a FIFO service discipline), we must take the view of how a typical arriving packet in steady-state finds the queue. Therefore, it is of interest to know when

$$\Pr\left[N_{S;A} = j\right] \overset{?}{=} \Pr\left[N_S = j\right] \tag{13.20}$$

The equality would imply that, in steady-state, the probability that an arriving packet finds the system in state j equals the probability that the system *is* in state j. Recall with (6.1) that the existence of the probabilities means that $\Pr\left[N_S = j\right]$ also equals the long-run fraction of the time the system contains j packets or is in state j. Similarly, $\Pr\left[N_{S;A} = j\right]$ also equals the long-run fraction of arriving packets that see the system in state j. In general, relation (13.20) is unfortunately not true. For example, consider a D/D/1 queue with a constant interarrival time τ_c and a constant service time $x_c < \tau_c$. Clearly, the D/D/1 system has a periodic service cycle: a busy period takes x_c time units and the idle period equals $\tau_c - x_c$ time units. Thus, every arriving packet always finds the system empty and concludes $\Pr\left[N_{S;A} = 0\right] = 1$, while $\Pr\left[N_S = 1\right] = \frac{x_c}{\tau_c}$ and $\Pr\left[N_S = 0\right] = \frac{\tau_c - x_c}{\tau_c}$. The waiting time computation of the GI/D/c system in Section 14.4.2 is another counter example. Since the arrival process $\{N_A(t), t \geq 0\}$ interacts with the system process $\{N_S(t), t \geq 0\}$ because every arrival increases the system content with one, they are dependent processes. Relation (13.20) is true for Poisson arrivals and this property is called *"Poisson arrivals see time averages"* (PASTA).

Theorem 13.5.1 (PASTA) *The long-run fraction of time that a process spends in state j is equal to the long-run fraction of Poisson arrivals that find the process in state j,*

$$\Pr[N_{S;A} = j] = \Pr[N_S = j]$$

Proof: See[8] e.g. Wolff (1982). □

The Poisson process has the typical property that future increments are independent of the past and, thus also of the past system history. In certain sense, Poisson arrivals perform a random sampling which is sufficient to characterize the steady-state of the system exactly. The PASTA property also applies to Markov chains. The transitions in continuous time Markov chains are Poisson processes if self-transitions are allowed (see Section 10.4.1). For any state j, the fraction of Poisson events that see the chain in state j is π_j, which (see Lemma 6.1.2) also equals the fraction of time the chain is in state j.

13.6 Little's Law

Little's Law is perhaps the simplest of the general queueing formulae.

Theorem 13.6.1 (Little's Law) *The average number of packets (customers) in the system $E[N_S]$ (or in the queue $E[N_Q]$) equals the average arrival rate λ times the average time spent in the system $E[T]$ (or in the queue $E[w]$),*

$$E[N_S] = \lambda E[T] \tag{13.21}$$
$$E[N_Q] = \lambda E[w]$$

---

[8] Although Wolff's general proof (Wolff, 1982) only contains two pages, it is based on martingales and on axiomatic probability theory.

Little's Law holds if two of the three limits

$$\lim_{t\to\infty} \frac{A(t)}{t} = \lambda \tag{13.22}$$

$$\lim_{t\to\infty} \frac{1}{t}\int_0^t N_S(u)du = E[N_S] \tag{13.23}$$

$$\lim_{n\to\infty} \frac{1}{n}\sum_{k=1}^n T_k = E[T] \tag{13.24}$$

exist.

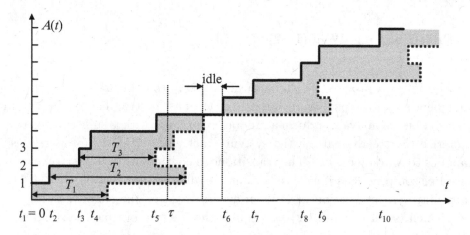

Fig. 13.6. The arrival (bold) and departure (dotted) process, together with the system time T_k for each packet in the queueing system.

Proof: Recall that $A(t)$ represents the total number of arrivals in time interval $[0,t]$. If $N_S(t) = 0$ or the system is idle at time t, then

$$\int_0^t N_S(u)du = \int_0^t \sum_{j=1}^\infty 1_{\{u\in[t_j,t_j+T_j)}du = \sum_{j=1}^{A(t)} \int_0^t 1_{\{u\in[t_j,t_j+T_j)}du = \sum_{k=1}^{A(t)} T_k$$

The general case where $N_S(t) \geq 0$ is more complicated as Fig. 13.6 shows for $t = \tau$ because not all intervals $[t_j, t_j + T_j)$ for $1 \leq j \leq A(\tau)$ are contained in $[0,\tau)$. Hence, $\sum_{k=1}^{A(\tau)} T_k$ counts too much and is an upper bound for $\int_0^\tau N_S(u)du$. If $D(t)$ denotes the number of departures in $[0,t]$, Fig. 13.6 illustrates that the area (in grey) in an interval $[0,t]$, which equals the total number of packets in the system in that interval $\int_0^t (A(u) - D(u))du = \int_0^t N_S(u)du$, can be bounded for any realization (sample path) and any $t \geq 0$

by

$$\sum_{k:T_k+t_k\leq t} T_k \leq \int_0^t N_S(u)du \leq \sum_{k=1}^{A(t)} T_k$$

where the lower bound only counts the packets that have left the system by time t. By dividing by t, we have

$$\frac{A(t)}{t}\sum_{k:T_k+t_k\leq t}\frac{T_k}{A(t)} \leq \frac{1}{t}\int_0^t N_S(u)du \leq \frac{A(t)}{t}\sum_{k=1}^{A(t)}\frac{T_k}{A(t)} \qquad (13.25)$$

Since we assume that the limit (13.22) exists, we have that $A(t) = O(t)$ for $t \to \infty$. From the existence of the limit (13.24), we can thus write

$$\lim_{t\to\infty}\sum_{k=1}^{A(t)}\frac{T_k}{A(t)} = E\,[T]$$

When $t \to \infty$ in (13.25) and using the limits defined above, we find that the upper bound converges to $\lambda E\,[T]$. In order to proof (13.21), it remains to show that also the lower bound in (13.25) converges to the same limit $\lambda E\,[T]$. Since $A(t) = \lambda t + o(t)$ for $t \to \infty$, it follows for the sequence of arrival times t_n that $t_n \to \infty$ if $n \to \infty$ and that

$$\frac{n}{t_n} = \frac{A(t_n)}{t_n} \to \lambda \qquad\qquad \text{as } n \to \infty$$

The convergence of the series (13.24) implies for $n \to \infty$ that

$$\frac{T_n}{n} = \sum_{k=1}^{n}\frac{T_k}{n} - \frac{n-1}{n}\sum_{k=1}^{n-1}\frac{T_k}{n-1} \to 0$$

Combining both relations leads to

$$\frac{T_n}{t_n} = \frac{T_n}{n}\frac{n}{t_n} \to 0 \qquad\qquad \text{as } n \to \infty$$

which implies that, for any $\varepsilon > 0$, there exists a fixed m such that, for all $k > m$, we have that $\frac{T_k}{t_k} < \varepsilon$ or $t_k + T_k < (1+\varepsilon)t_k$. For $t > t_m$, the lower bound in (13.25) is

$$\sum_{k>m:T_k+t_k\leq t} T_k + \sum_{k=1}^{m} T_k = \sum_{k=1}^{A(t/(1+\varepsilon))} T_k$$

or

$$\frac{1}{t} \sum_{k \geq 1: T_k + t_k \leq t} T_k = \frac{1}{1+\varepsilon} \frac{A(t/(1+\varepsilon))}{t/(1+\varepsilon)} \sum_{k=1}^{A(t/(1+\varepsilon))} \frac{T_k}{A(t/(1+\varepsilon))}$$

In the limit $t \to \infty$, we obtain

$$\frac{1}{t} \sum_{k \geq 1: T_k + t_k \leq t} T_k \to \frac{\lambda}{1+\varepsilon} E[T]$$

Since ε can be made arbitrarily small, this finally proves (13.21).   □

Although the proof may seem rather technical[9] for, after all, an intuitive result, it reveals that no assumptions about the distributions of arrival and service process apart from steady-state convergence are made. There are no probabilistic arguments used. In essence Little's Law is proved by showing that two limits exist for any sample path or realization of the process, which guarantees a very general theorem. Moreover, no assumptions about the service discipline, nor about the dependence between arrival and service process or about the number of servers are made which means that Little's Law also holds for non-FIFO scheduling disciplines, in fact for *any* scheduling discipline! Little's Law connects three essential quantities: once two of them are known the third is determined by (13.21). Little's Law is very important in operations where it relates the average inventory (similar to $E[N_S]$), the average flow rate or throughput λ and the average flow time $E[T]$ in a process flow of products or services. Several examples can be found in Chapter 14, in Anupindi *et al.* (2006) and Bertsekas and Gallager (1992, pp. 157–162).

---

[9] We have chosen for a very general proof. Other proofs (e.g. in Ross (1996) and Gallager (1996)) use arguments from renewal reward theory (Section 8.4) which makes their proofs less general because they require that the system has renewals.

14

Queueing models

This chapter presents some of the simplest and most basic queueing models. Unfortunately, most queueing problems are not available in analytic form and many queueing problems require a specific and sometimes tailor-made solution.

Beside the simple and classical queueing models, we also present two other exact solvable models that have played a key role in the development of Asynchronous Transfer Mode (ATM). In these ATM queueing systems the service discipline is deterministic and only the arrival process is the distinguishing element. The first is the N*D/D/1 queue (Roberts, 1991, Section 6.2) whose solution relies on the Beneš approach. The arrivals consist of N periodic sources each with period of D time slots, but randomly phased with respect to each other. The second model is the fluid flow model of Anick *et al.* (1982), known as the AMS-queue, which considers N on-off sources as input. The solution uses Markov theory. Since the Markov transition probability matrix has a special tri-band diagonal structure, the eigenvector and eigenvalue decomposition can be computed analytically.

We would like to refer to a few other models. Norros (1994) succeeded in deriving the asymptotic probability distribution of the unfinished work for a queue with self-similar input, modeled via a fractal Brownian motion. The resulting asymptotic probability distribution turns out to be a Weibull distribution (3.40). Finally, Neuts (1989) has established a matrix analytic framework and was the founder of the class of Markov Modulated arrival processes and derivatives as the Batch Markovian Arrival process (BMAP).

14.1 The M/M/1 queue

The M/M/1 queue consists of a Poisson arrival process of packets with exponentially distributed interarrival times, a service process with exponentially distributed service time, one server and an infinitely long queue. The M/M/1 queue is a basic model in queueing theory for several reasons. First, as shown below, the M/M/1 queue can be computed in analytic form, even the transient time behavior. Apart from the computational advantage, the M/M/1 queue possesses the basic feature of queuing systems: the quantities of interest (waiting time, number of packets, etc.) increase monotonously with the traffic intensity ρ.

Packets arrive in the M/M/1 queue with interarrival rate λ and are served with service rate μ. The M/M/1 queue is precisely described by a constant rate birth and dead process. Any arrival of a packet to the queueing system can be regarded as a birth. The current state k that reflects the number of packets in the M/M/1 system jumps to state $k + 1$ at the arrival of a new packet and the transition rate equals the interarrival rate λ: on average every $\frac{1}{\lambda}$ time units a packet arrives to the system. A packet leaves the M/M/1 system after service, which corresponds to a death: at each departure from the system the current state is decreased by one, with death rate μ equal to the service rate μ: on average every $\frac{1}{\mu}$ time units, a packet is served. In the sequel, we concentrate on the steady-state behavior and refer for the transient behavior to the discussion of the birth and death process in Section 11.3.3.

14.1.1 The system content in steady-state

From the analogy with the constant rate birth and death process as studied in Section 11.3.3, we obtain immediately the steady-state queueing distribution (11.23) as

$$\Pr[N_S = j] = (1 - \rho)\rho^j \qquad\qquad j \geq 0 \qquad\qquad (14.1)$$

where $N_S = \lim_{t\to\infty} N_S(t)$ is the number of packets in the system in the stationary regime. In other words, $\Pr[N_S = j]$ is the probability that the M/M/1 system (queue plus server) contains j packets. It has been shown in Section 11.3 that the M/M/1 queue is ergodic (i.e. that an equilibrium or steady-state exists) if $\rho = \frac{\lambda}{\mu} < 1$, which is a characteristic of a general queueing system. The probability density function of the system content (14.1) is a geometric distribution reflecting the memoryless property. We observe that the M/M/1 system is empty with probability $\Pr[N_S = 0] = 1 - \rho$ and of all states, the empty state has the highest probability. Immediately, the chance that there is a packet in the M/M/1 system is precisely equal to the traffic intensity ρ, namely $\Pr[N_S > 0] = 1 - \Pr[N_S = 0] = \rho$.

The corresponding probability generating function (2.18) is

$$\varphi_{N_S}(z) = \sum_{k=0}^{\infty} \Pr\left[N_S = k\right] z^k = \frac{1-\rho}{1-\rho z}$$

The average number of packets in the M/M/1 system $E[N_S] = \varphi'_S(1)$ equals

$$E\left[N_{S;M/M/1}\right] = \frac{\rho}{1-\rho}$$

while the variance $\mathrm{Var}[N_S]$ follows from (2.27) as

$$\mathrm{Var}\left[N_{S;M/M/1}\right] = \frac{\rho}{(1-\rho)^2}$$

Both the mean and variance of the number of packets in the system diverges as $\rho \to 1$. When the interarrival rate tends to the service rate, the queue grows indefinitely long with indefinitely large variation. From Little's law (13.21), the average time spent in the M/M/1 system equals

$$E\left[T_{M/M/1}\right] = \frac{E[N_S]}{\lambda} = \frac{1}{\mu(1-\rho)} = \frac{1}{\mu-\lambda} \tag{14.2}$$

where $\rho < 1$ or, equivalently, $\lambda < \mu$. If $\rho = 0$, there is no load in the system and the average waiting time attains its minimum equal to the average service time $\frac{1}{\mu}$. In the other limit $\rho \to 1$, the average waiting time grows unboundedly, just as the queue length or number of packets in the system. The behavior of the M/M/1 system in the limit $\rho \to 1$ is characteristic for the average of quantities $(N_S, w, T, \ldots)$ in many queueing systems: a simple pole at $\rho = 1$.

As a remark, the average waiting time in the M/M/1 queue follows, after taking expectations from the general relation $T_n - x_n = w_n$ or $E[w] = E[T] - \frac{1}{\mu}$, as

$$E\left[w_{M/M/1}\right] = \frac{1}{\mu(1-\rho)} - \frac{1}{\mu} = \frac{\rho}{\mu(1-\rho)} \tag{14.3}$$

14.1.2 The virtual waiting time

For the M/M/1 queue, the virtual waiting time $v(t)$ at some time t consists of (a) the residual service time of the packet currently under service, (b) the time needed to serve the $N_Q(t)$ packets in the queue. As mentioned in Section 13.1.3, the virtual waiting time at arrival epochs equals the system time T_n. In other words, if a new packet, say the n-th packet, enters the M/M/1 system at $t = t_n$, the total time (system time) that the packet spends in the M/M/1 system equals $v(t_n) = T_n$. At $t = t_n^-$, the number $N_S(t_n^-)$ does not include the new packet at the last position and the packet "sees" $N_S(t_n^-)$ other packets in the system (queue plus the packet in the server) in front of it. We assume further that the server operates in FIFO (first in, first out) order. Since the service time is exponentially distributed and possesses the memoryless property, the residual or remaining service time of the packet currently under service has the same distribution. In other words, it does not matter how long the packet has already been under service. The more general argument is that the PASTA property applies. The system time of the n-th packet is thus the sum of $N_S(t_n^-)+1$ exponential i.i.d. random variables. As shown in Section 3.3.1, if $N_S(t_n^-) = k$, the system time has an Erlang distribution given by (3.24) with $n = k + 1$,

$$f_{T_n}(t|N_S(t_n^-) = k) = \frac{\mu(\mu t)^k}{k!}e^{-\mu t}$$

Using the law of total probability (2.46), the system time T_n of the n-th packet or the virtual waiting time at time $t = t_n$ becomes

$$f_{T_n}(t) = \frac{d}{dt}\Pr[T_n \leq t]$$

$$= \sum_{k=0}^{\infty} f_{T_n}(t|N_S(t_n^-) = k)\Pr[N_S(t_n^-) = k]$$

$$= \mu e^{-\mu t}\sum_{k=0}^{\infty}\frac{(\mu t)^k}{k!}\Pr[N_S(t_n^-) = k]$$

In Section 11.3.3, $s_k(t_n^-) = \Pr[N_S(t_n^-) = k]$ is computed in (11.27) assuming that the system starts with j packets, i.e. $s_k(0) = \delta_{kj}$. In steady-state, where $t_n \rightarrow \infty$, it is shown that $s_k(t_n^-) \rightarrow (1-\rho)\rho^k$. In most cases, however, a time-dependent solution is not available in closed form. Fortunately, for Poisson arrivals, the PASTA property helps to circumvent this inconvenience. Based on the PASTA property, in steady-state, $\lim_{n\to\infty}\Pr[N_S(t_n^-) = k] = \Pr[N_S = k]$ given by (14.1). The probability density function $f_T(t) = \lim_{n\to\infty} f_{T_n}(t)$ of the steady-state system time T (or the total waiting time of a packet) is

$$f_T(t) = \mu e^{-\mu t}\sum_{k=0}^{\infty}\frac{(\mu t)^k}{k!}(1-\rho)\rho^k$$

or

$$f_T(t) = (1-\rho)\mu e^{-(1-\rho)\mu t} \tag{14.4}$$

In summary, the total time spent in the M/M/1 system in steady-state ($\rho < 1$) has an exponential distribution with mean $\frac{1}{(1-\rho)\mu} = \frac{1}{\mu-\lambda}$, which has been found above in (14.2) by Little's law. Similarly[1], the waiting time in the M/M/1 queue is

$$f_w(t) = (1 - \rho)\,\delta(t) + (1 - \rho)\,\lambda e^{-(1-\rho)\mu t} \tag{14.5}$$

where the first term with Dirac function reflects a zero queueing time provided the system is empty, which has probability $\Pr[N_S = 0] = 1 - \rho$.

14.1.3 The departure process from the M/M/1 queue

There is a remarkable theorem due to Burke which has far-reaching consequences for networks of M/M/1 queues.

Theorem 14.1.1 (Burke) *In a steady-state M/M/1 queue, the departure process is a Poisson process with rate λ*

Burke's Theorem is equivalent to the statement that the interdeparture times $r_n - r_{n-1}$ in steady-state are i.i.d. exponential random variables with mean $\frac{1}{\lambda}$.

Proof: Let us denote the probability density function of the interdeparture time r by

$$f_r(t) = \frac{d}{dt}\Pr[r \le t]$$

In steady-state, it holds in general that $\Pr[N_{S;A} = j] = \Pr[N_{S;D} = j]$, as shown in Section 13.4.1, while the PASTA property (Theorem 13.5.1) states that $\Pr[N_{S;A} = j] = \Pr[N_S = j]$. Hence, in steady-state in the M/M/1 queue, departing packets see the steady-state system content, i.e. $\Pr[N_{S;D} = j] = \Pr[N_S = j]$. Moreover, in steady-state, the departure process can be decomposed into two different situations after the departure of a packet: (a) the departing packet sees an empty system (which is equivalent to "the system is empty") or (b) the departing packet sees a next packet in the queue (which is equivalent to "the system serves immediately the next packet in

---

[1] The Laplace transform of the waiting time in the queue follows from $T_n = w_n + x_n$ as

$$\varphi_w(z) = \frac{\varphi_T(z)}{\varphi_x(z)} = \frac{(1 - \rho)\,\mu}{z + (1 - \rho)\,\mu}\,\frac{z + \mu}{\mu}$$
$$= (1 - \rho) + \rho\frac{(1 - \rho)\,\mu}{z + (1 - \rho)\,\mu}$$

which, after inverse Laplace transformation, gives (14.5).

the queue"),

$$\Pr\left[r \le t\right] = \Pr\left[r \le t | N_S = 0\right]\Pr\left[N_S = 0\right] + \Pr\left[r \le t | N_S > 0\right]\Pr\left[N_S > 0\right]$$

In case (a), we must await for the next packet to arrive and to be served. This total time is the sum of an exponential random variable with rate λ and an exponential random variable with rate μ. It is more convenient to compute the Laplace transform as shown in Section 3.3.1,

$$\varphi_{r|N_S=0}(z) = \int_0^\infty e^{-zt}d\left(\Pr\left[r \le t | N_S = 0\right]\right) = \frac{\lambda}{z+\lambda}\frac{\mu}{z+\mu}$$

In case (b), the next packet leaves the M/M/1 queue after an exponential service time with rate μ,

$$\varphi_{r|N_S>0}(z) = \int_0^\infty e^{-zt}d\left(\Pr\left[r \le t | N_S > 0\right]\right) = \frac{\mu}{z+\mu}$$

Hence,

$$\varphi_r(z) = \varphi_{r|N_S=0}(z)\Pr\left[N_S = 0\right] + \varphi_{r|N_S>0}(z)\Pr\left[N_S > 0\right]$$
$$= \frac{\lambda}{z+\lambda}\frac{\mu}{z+\mu}(1-\rho) + \frac{\mu}{z+\mu}\rho = \frac{\lambda}{z+\lambda}$$

which proves the theorem.   □

Burke's Theorem states that the steady-state arrival and departure process of the M/M/1 queue are the same! Consequently, the steady-state departure rate equals the steady-state arrival rate λ.

14.2 Variants of the M/M/1 queue

A number of readily obtained variants from the birth-death analogy are worth considering here. Mainly a steady-state analysis is presented.

14.2.1 The M/M/m queue

Instead of one server, we consider the case with m servers. The buffer is still infinitely long and the interarrival process is exponential with interarrival rate λ. The M/M/m queue can model a router with m physically different interfaces (or output ports) with same transmission rate μ towards the same next hop. All packets destined to that next hop can be transmitted over any of the m interfaces. This type of load balancing frequently occurs in the Internet.

As shown in Fig. 14.1, the M/M/m system can still be described by a birth and death process with birth rate $\lambda_k = \lambda$, but with death rate $\mu_k = k\mu$ for $0 \leq k \leq m$ and $\mu_k = m\mu$ if $k \geq m$. Indeed, if there are $k \leq m$ packets in system, they can all be served and the departure (or death) rate from the system is $k\mu$. Only if there are more packets $k > m$, only m of them can be served such that the death rate is limited to the maximum service rate $m\mu$.

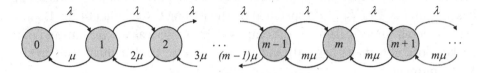

Fig. 14.1. The birth–death process corresponding to the M/M/m queue.

14.2.1.1 System content

From the basic steady-state relations for the birth and death process (11.15) and (11.16), we find

$$\Pr[N_S = 0] = \frac{1}{1 + \sum_{j=1}^{m-1} \frac{\lambda^j}{j!\mu^j} + \sum_{j=m}^{\infty} \frac{\lambda^j}{m^{j-m}m!\mu^j}}$$

$$= \frac{1}{\sum_{j=0}^{m-1} \frac{\lambda^j}{j!\mu^j} + \frac{\lambda^m}{m!\mu^m} \frac{1}{1 - \frac{\lambda}{m\mu}}} \qquad (14.6)$$

$$\Pr[N_S = j] = \frac{\lambda^j}{j!\mu^j} \Pr[N_S = 0] \qquad\qquad j \leq m \qquad (14.7)$$

$$= \frac{m^m}{m!} \left(\frac{\lambda}{m\mu}\right)^j \Pr[N_S = 0] \qquad j \geq m \qquad (14.8)$$

The traffic intensity is $\rho = \frac{\lambda}{m\mu}$, the ratio between average interarrival rate and average (maximum) service rate. Again, $\rho < 1$ corresponds to the stable (ergodic) regime.

For the M/M/m system it is of interest to know what the probability of queueing is. Queueing occurs when an arriving packet finds all servers busy, which happens with probability $\Pr[N_S \geq m]$, or explicitly,

$$\Pr[N_S \geq m] = \frac{\Pr[N_S = 0]}{m!\left(1 - \frac{\lambda}{m\mu}\right)} \frac{\lambda^m}{\mu^m} \qquad (14.9)$$

This probability also corresponds to a situation in classical telephony where no trunk is available for an arriving call. Relation (14.9) is known as the *Erlang C formula*.

14.2.1.2 Waiting (or queueing) time

Instead of computing the virtual waiting time (or system time, unfinished work), we will now concentrate on the waiting time of a packet in the M/M/m queue. The system time can be deduced from the basic relation $T_n = w_n + x_n$, where x_n is an exponential random variable with rate $m\mu$.

A packet only experiences queueing if all servers are occupied. This event has probability $\Pr[N_S \geq m]$ specified by the Erlang C formula (14.9). Hence, the queueing time w can be decomposed into two cases: (a) an arriving packet does not queue ($w = 0$) and (b) an arriving packet must wait in the M/M/m queue,

$$
\begin{aligned}
\Pr[w \leq t] &= \Pr[w \leq t | N_S < m] \Pr[N_S < m] + \Pr[w \leq t | N_S \geq m] \Pr[N_S \geq m] \\
&= \Pr[N_S < m] + \Pr[w \leq t | N_S \geq m] \Pr[N_S \geq m] \\
&= 1 - \Pr[N_S \geq m] + \Pr[w \leq t | N_S \geq m] \Pr[N_S \geq m] \qquad (14.10)
\end{aligned}
$$

It remains to compute $\Pr[w \leq t | N_S \geq m]$. The reasoning is analogous to that in the M/M/1 queue. An arriving packet must wait for the packet currently under service and for the j packets already in the queue before it. Thus, w equals the sum of $j + 1$ exponentially random variables with rate $m\mu$ because for the M/M/m queue, the service rate is $m\mu$. Hence (see Section 3.3.1), the distribution for the waiting time w in the queue, provided j packets are in the queue, is an Erlang distribution,

$$
f_w(t | N_Q = j) = \frac{m\mu(m\mu t)^j}{j!} e^{-m\mu t}
$$

Furthermore, the number of packets in the queue N_Q in steady-state is related to the system content as $N_S = m + N_Q$. Using the law of total probability (2.46), the waiting time in the queue in steady-state is

$$
\begin{aligned}
f_w(t | N_S \geq m) &= \frac{d}{dt} \Pr[w \leq t | N_S \geq m] \\
&= \sum_{j=0}^{\infty} f_w(t | N_S = m + j) \Pr[N_S = m + j | N_S \geq m]
\end{aligned}
$$

The conditional probability $\Pr[N_S = m + j | N_S \geq m]$ follows from (2.44) and (14.8) with $\rho = \frac{\lambda}{m\mu}$ as

$$
\begin{aligned}
\Pr[N_S = m + j | N_S \geq m] &= \frac{\Pr[N_S = m + j]}{\Pr[N_S \geq m]} = \frac{\Pr[N_S = 0]}{\Pr[N_S \geq m]} \frac{m^m}{m!} \left(\frac{\lambda}{m\mu}\right)^{j+m} \\
&= \left(1 - \frac{\lambda}{m\mu}\right) \left(\frac{\lambda}{m\mu}\right)^j = (1 - \rho) \rho^j
\end{aligned}
$$

We observe from (14.1) that, if all m servers are busy, the system content of an M/M/m system behaves as that in a M/M/1 system. Thus, the conditional probability density function for the waiting time in the M/M/m queue is also an exponential distribution,

$$f_w(t|N_S \geq m) = (1-\rho)\, m\mu e^{-m\mu t} \sum_{j=0}^{\infty} \frac{(m\mu t \rho)^j}{j!}$$

$$= (1-\rho)\, m\mu e^{-(1-\rho)m\mu t}$$

or

$$\Pr\left[w \leq t | N_S \geq m\right] = 1 - e^{-(1-\rho)m\mu t}$$

Substitution in (14.10) finally results in the distribution of the waiting time in the queue of the M/M/m system,

$$F_w(t) = \Pr\left[w \leq t\right] = 1 - \Pr\left[N_S \geq m\right] e^{-(1-\rho)m\mu t} \tag{14.11}$$

Since $F_w(0) = 1 - \Pr\left[N_S \geq m\right] > 0$ while obviously $F_w(0^-) = 0$, there is probability mass at $t = 0$, which is reflected by a Dirac impulse in the probability density function,

$$f_w(t) = (1 - \Pr\left[N_S \geq m\right])\, \delta(t) + (1-\rho)\, m\mu \Pr\left[N_S \geq m\right] e^{-(1-\rho)m\mu t} \tag{14.12}$$

The pdf of the system time $T = w + x$ follows after convolution of (14.12) and $f_x(t) = \mu e^{-\mu t}$ as

$$f_T(t) = (1 - \Pr\left[N_S \geq m\right])\, \mu e^{-\mu t} + \frac{\Pr\left[N_S \geq m\right]}{1 - m(1-\rho)}\, (1-\rho)\, m\mu \left(e^{-(1-\rho)m\mu t} - e^{-\mu t}\right) \tag{14.13}$$

and the average system time can be computed from (14.13) with (2.33) or directly from $E\left[T\right] = E\left[w\right] + E\left[x\right]$ as

$$E\left[T\right] = \frac{1}{\mu} + \frac{\Pr\left[N_S \geq m\right]}{m\mu(1-\rho)} \tag{14.14}$$

Also, in the single-server case ($m = 1$), (14.12) reduces to the pdf (14.5) of the M/M/1 queue. Furthermore, Burke's Theorem 14.1.1 can be extended to the M/M/m queue: the arrival and departure process of the M/M/m queue are both Poisson processes with rate λ.

14.2.2 The M/M/m/m queue

The difference with the M/M/m queue is that the number of packets (calls) in the M/M/m/m queue is limited to m. Hence, when more than m packets (calls) arrive, they are lost. This situations corresponds with classical telephony where a conversation is possible if no more than m trunks are occupied, otherwise you hear a busy tone and the connection cannot be set-up. The limitation to m arrivals is modeled in the birth and death process by limiting the interarrival rates, $\lambda_k = \lambda$ if $k < m$ and $\lambda_k = 0$ if $k \geq m$. The death rates are the same as in the M/M/m queue, $\mu_k = k\mu$ for $0 \leq k \leq m$ and $\mu_k = m\mu$ if $k \geq m$.

From the basic steady-state relations for the birth and death process (11.15) and (11.16), we find

$$\Pr[N_S = 0] = \frac{1}{\sum_{j=0}^{m} \frac{\lambda^j}{j!\mu^j}} \tag{14.15}$$

$$\Pr[N_S = j] = \frac{\lambda^j}{j!\mu^j} \Pr[N_S = 0] \qquad j \leq m$$

$$= 0 \qquad\qquad\qquad\qquad j > m \tag{14.16}$$

The quantity of interest in the M/M/m/m system is the probability that all trunks (servers) are busy, which is known as the *Erlang B formula*,

$$\Pr[N_S = m] = \frac{\frac{\lambda^m}{m!\mu^m}}{\sum_{j=0}^{m} \frac{\lambda^j}{j!\mu^j}} \tag{14.17}$$

In practice, a telephony exchange is dimensioned (i.e. the number of lines m is determined) such that the blocking probability $\Pr[N_S = m]$ is below a certain level, say below 10^{-4}. In summary, the Erlang B formula (14.17) determines the blocking probability or loss probability (because only m calls or packets are allowed to the system), while the Erlang C formula (14.9) is the probability that a packet must wait in the (infinitely long) queue because all servers are busy.

Although the Erlang B formula (14.17) has been derived in the context of the M/M/m/m queue, it holds under much weaker assumptions, a fact already known to Erlang, as mentioned by Kelly (1991). Kelly starts his memoir on Loss Networks with the Erlang B formula, which Erlang obtained from his powerful method of statistical equilibrium. The latter concept is now identified as the steady-state of Markov processes. Kelly further relates the impact of the Erlang B formula from telephony to interacting particle system and phase transitions in nature (e.g. the famous Ising model). Much effort has been devoted over time to generalize Erlang's results as far as possible. The Erlang B formula (14.17) holds for the M/G/m/m queue as well, thus for an arbitrary service process provided the mean service rate (per server) equals μ. The proof by Gnedenko and Kovalenko (1989, pp. 237–240) is long and complicated, whereas the proof of Ross (1996, Section 5.7.2) is more elegant and is based on the time-reversed Markov chain. As a corollary, Ross demonstrates that the departure process (including both lost and served packets) of the M/G/m/m system is a Poisson process with rate λ.

Example 1 In case $m \to \infty$, the expression (14.15) and (14.6) tend to $\lim_{m \to \infty} \Pr[N_S = 0] = \exp\left(-\frac{\lambda}{\mu}\right)$ while (14.16) and (14.7) become

$$\Pr[N_S = j] = \frac{\lambda^j}{j!\mu^j} \exp\left(-\frac{\lambda}{\mu}\right)$$

This queueing system is denoted as M/M/∞. Thus, the number in the M/M/∞ system (in steady-state) is Poisson distributed with parameter $\frac{\lambda}{\mu}$. Hence, in case $m \to \infty$, the average number in the system $E[N_S] = \rho$ (as follows from (3.11)) and the average time in the system follows from Little's theorem (13.21) as $E[T] = \frac{1}{\mu}$. The fact that, if $m \to \infty$, the mean time in the system M/M/∞ equals the average service time has a consistent explanation: if the number of servers $m \to \infty$, implying that there is an infinite service capacity, it means that there is no waiting room and the only time a packet is in the system is his service time $\frac{1}{\mu}$.

Example 2 Consider two voice over IP (VoIP) gateways connected by a link with capacity C. Denote the capacity of a voice call by C_{voice} (in bit/s). For example in ISDN, $C_{\text{voice}} = 64$ kb/s. In general, C_{voice} in VoIP depends on the specifics of the codecs used. The arrival rate λ of voice traffic can be expressed in terms of the number a of call attempts per hour and the mean call duration d (in seconds) as

$$\lambda = \frac{a \times d}{3600}$$

The number m of calls that the link can carry simultaneously is

$$m = \frac{C}{C_{\text{voice}}}$$

Since the arrival process of voice calls is well modeled by a Poisson process with exponential holding time, the Erlang B formula (14.17) is applicable to compute the blocking probability ϵ or grade of service (GoS) as

$$\frac{\frac{r^m}{m!}}{\sum_{j=0}^m \frac{r^j}{j!}} \leq \epsilon \tag{14.18}$$

where $r = \frac{\lambda}{\mu} = m\rho$ and $\rho = \frac{\lambda}{m\mu}$ is the traffic intensity. This relation (14.18) specifies the probability that admission control will have to refuse a call request between the two VoIP gateways because the link is already transporting m calls. An Internet service provider can make a trade-off between the link capacity C (by hiring more links or a higher capacity link from a network provider) and the blocking probability ϵ or GoS. The latter must be small enough to keep its subscribed customers, but large enough to make profit. A reasonable value for GoS seems $\epsilon = 10^{-4}$. If the Internet service provider hires a 2 Mb/s link and offers its customers VoIP software with codec rate 40 kb/s (G.726 standard), then $m = 50$. Since the left-hand side of (14.18) is strictly increasing in ρ, solving the equation (14.18) for r yields $r \leq 28.87$ or the traffic intensity equals $\rho = 0.5775$. Furthermore, since $\mu = \frac{C}{m} = 40$ kb/s, we obtain $\lambda = 1.155$ Mb/s. If the mean call duration d (in seconds) is known, the number of call attempts per hour then follow as $a = \frac{4158129}{d}$. If we assume that a telephone call lasts on average 2 minutes or $d = 120$ s, the number of call attempts per hour that the Internet service provider can handle with a GoS of 10^{-4} equals $a = 34651$.

14.2.3 The M/M/1/K queue

In contrast to the basic M/M/1 queue, the M/M/1/K system cannot contain more than K packets (including the packet in the server). Arriving packets that find the system completely occupied (with K packets), are refused service and are to be considered as lost (or marked).

In the basic steady-state relations for the birth and death process (11.15) the appearing summation is limited to K instead of infinity or $\lambda_k = \lambda$ if $k < K$ and $\lambda_k = 0$ if $k \geq K$. Thus, with $\rho = \frac{\lambda}{\mu}$ and

$$\Pr\left[N_S = 0\right] = \frac{1}{\sum_{j=0}^{K} \rho^j} = \frac{1 - \rho}{1 - \rho^{K+1}}$$

the pdf of the system content for the M/M/1/K system becomes,

$$\Pr\left[N_S = j\right] = \frac{(1 - \rho)\rho^j}{1 - \rho^{K+1}} \qquad 0 \leq j \leq K \qquad (14.19)$$
$$= 0 \qquad j > K$$

The probability that j positions in the M/M/1/K system are occupied is proportional to that in the infinite system (14.1) with proportionality factor $\left(1 - \rho^{K+1}\right)^{-1}$.

The probability that the system is completely filled with K packets equals

$$\Pr\left[N_S = K\right] = \frac{(1 - \rho)\rho^K}{1 - \rho^{K+1}} \qquad (14.20)$$

This probability also equals the loss probability for packets in the M/M/1/K system. Regarding the QoS problem in multimedia based on IP-networks, a first crude estimate of the packet loss in a router with K positions can be derived from (14.20). The estimate is rather crude because the arrival process of packets in the Internet is likely not a Poisson process and the variable length of the packets does not necessarily lead to an exponential service rate.

14.3 The M/G/1 queue

The most general single-server queueing system with Poisson arrivals is the M/G/1 queue. The service time distribution $F_x(t)$ can be any arbitrary distribution. Due to its importance, we will derive the system content and the waiting time distribution in steady-state.

In order to describe the M/G/1 queueing system, special observation points in time must be chosen such that the equations for the evolution of the number of packets in the system are most conveniently deduced. The set of departure times $\{r_n\}$ appears to be a suitable set. Any other set of observation points is likely to lead to a more complex mathematical treatment, mainly because the remaining service time of the packet just under service is a stochastic variable. If the M/G/1 queue is observed at departure times, the evolution of the number of packets in the system $N_S(r_n^+)$ that the departing packets leave behind is a discrete Markov chain, namely the embedded Markov chain of the M/G/1 queue continuous-time process. Section 13.4.1 has shown that, in steady-state, relation (13.19) tells that the distribution of the number of packets in the system observed by arrivals equals that left behind by departures. In addition, the PASTA Theorem 13.5.1 states that, in steady-state, Poisson arrivals observe the actual distribution of the number of packets in the system. This PASTA property makes that the embedded Markov chain observing the system at departure epoch only, nevertheless provides the steady-state solution since the arrival process is Poisson.

Let us concentrate in deriving the transition probabilities that specify this embedded Markov chain entirely apart from an initial state distribution. With the notation of Section 13.4, $S_k = N_S(r_k^+)$ and $\mathcal{A}_k$ denote the number of packets in the system at the discrete-time r_k and the number of arrivals during a time interval $[r_k, r_{k+1}]$, respectively. The transition probability of the embedded Markov chain is

$$V_{ij} = \Pr\left[S_{k+1} = j | S_k = i\right]$$

and the evolution over time follows from (13.16) as

$$\mathcal{S}_{k+1} = (\mathcal{S}_k - 1)^+ + \mathcal{A}_k \qquad (14.21)$$

Hence, since $\mathcal{A}_k \geq 0$, we see that $\mathcal{S}_{k+1} \geq (\mathcal{S}_k - 1)^+$ or that $\mathcal{S}_{k+1} < (\mathcal{S}_k - 1)^+$ is impossible. Hence, $V_{ij} = 0$ for $j < i$ and $i > 1$ while, for $i > 0$, $V_{ij} = \Pr[i - 1 + \mathcal{A}_k = j | \mathcal{S}_k = i] = \Pr[\mathcal{A}_k = j - i + 1]$. The case for $i = 0$ results in $V_{0j} = \Pr[\mathcal{A}_k = j] = V_{1j}$. Denoting $a_j = \Pr[\mathcal{A}_k = j]$, the transition probability matrix becomes[2]

$$V = \begin{bmatrix} a_0 & a_1 & a_2 & a_3 & \cdots \\ a_0 & a_1 & a_2 & a_3 & \cdots \\ 0 & a_0 & a_1 & a_2 & \cdots \\ 0 & 0 & a_0 & a_1 & \cdots \\ \vdots & \vdots & \vdots & \vdots & \ddots \end{bmatrix}$$

and the corresponding transition graph is sketched in Fig. 14.2.

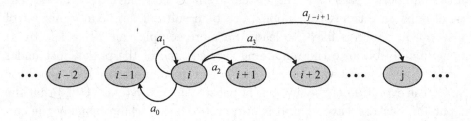

Fig. 14.2. State transition graph for the M/G/1 embedded Markov chain.

The number of Poisson arrivals during a time slot $[r_k, r_{k+1}]$ clearly depends on the length of the service time $x_{k+1} = r_{k+1} - r_k$ that is distributed according to $F_x(t)$, which is independent of a specific packet k. Furthermore, the arrival process is a Poisson process with rate λ and independent of the state of the queueing process, thus $\Pr[\mathcal{A}_k = j] = \Pr[\mathcal{A} = j]$. Hence, using the law of total probability (2.46),

$$\Pr[\mathcal{A} = j] = \int_0^\infty \Pr[\mathcal{A} = j | x = t] \, dF_x(t)$$

$$= \int_0^\infty e^{-\lambda t} \frac{(\lambda t)^j}{j!} dF_x(t)$$

---

[2] The structure of this transition probability matrix V has been investigated in great depth by Neuts (1989). Moreover, V belongs to the class of matrices whose eigenstructure is explicitly given in Appendix A.5.3.

If we denote the Laplace transform of the service time by

$$\varphi_x(z) = \int_0^\infty e^{-zt} dF_x(t)$$

then we observe that

$$a_j = \Pr\left[\mathcal{A} = j\right] = \frac{\lambda^j}{j!} \int_0^\infty e^{-\lambda t} t^j dF_x(t) = \frac{(-\lambda)^j}{j!} \left. \frac{d^j \varphi_x(z)}{dz^j} \right|_{z=\lambda} \qquad (14.22)$$

so that the transition probability matrix V is specified. Since all $a_j > 0$ for all $j > 0$, Fig. 14.2 indicates that all states are reachable from an arbitrary state i as the Markov process evolves over time in at least i steps. These i steps occur in the transition from state i to state 0. This implies that the Markov process is irreducible and the steady-state stability requirement $\rho < 1$ makes the Markov process ergodic. The steady-state vector with components $\pi_j = \Pr[N_{S;D} = j]$ where $\lim_{n\to\infty} N_S\left(r_n^+\right) = N_{S;D}$ follows from (9.22) as solution of $\pi = \pi V$, where V is a matrix of infinite dimensions.

14.3.1 *The system content in steady-state*

Rather than pursuing with the matrix analysis that is explored by Neuts (1989), we present an alternative method to determine the steady-state distribution π_j using generating functions. The generating function approach will lead in an elegant way to the celebrated Pollaczek-Khinchin equation. The probability generating function (pgf) $G_k(z)$ of a discrete random variable $\mathcal{G}_k$ is defined in (2.18) as

$$G_k(z) \triangleq E\left[z^{\mathcal{G}_k}\right] = \sum_{j=0}^\infty g_k[j] z^j \qquad (14.23)$$

where $g_k[j] = \Pr[\mathcal{G}_k = j]$. From (14.21), we have

$$S_{k+1}(z) = E\left[z^{(\mathcal{S}_k - 1)^+ + \mathcal{A}_k}\right] \qquad (14.24)$$

Anticipating the corresponding result (14.31) derived in Section 14.4 for the GI/D/m system in discrete-time, we observe that the generating function $S_k(z)$ satisfies a formally similar equation when $m = 1$ in (14.31). This correspondence points to a more general framework because, by choosing appropriate observation points, the M/G/1 and GI/D/1 systems (in discrete-time) obey a same equation formally. Since the results deduced for the GI/D/m system are more general because of the m servers instead of 1 here, we content ourselves here to copy the result (14.35) derived below[3] in

---

[3] Notice that the notation $A(z)$ here is different from $A(t) = \int_0^t N_A(u)du$ used before.

Section 14.4,

$$S(z) = \left(1 - A'(1)\right) \frac{(z-1)\,A(z)}{z - A(z)}$$

We further continue to introduce in this general equation the details of the M/G/1 queueing system by specifying $A(z) = \sum_{j=0}^{\infty} \Pr\left[\mathcal{A} = j\right] z^j$. With (14.22), we find the Taylor expansion,

$$A(z) = \sum_{j=0}^{\infty} \frac{(-\lambda z)^j}{j!} \left. \frac{d^j \varphi_x(z)}{dz^j} \right|_{z=\lambda} = \varphi_x(\lambda - \lambda z) \tag{14.25}$$

and the probability generating function of the system content of the steady-state M/G/1 queueing system,

$$S(z) = \left(1 + \lambda \varphi_x'(0)\right) \frac{(z-1)\,\varphi_x(\lambda - \lambda z)}{z - \varphi_x(\lambda - \lambda z)}$$

But, since $\varphi_x'(0) = -E\left[x\right] = -\frac{1}{\mu}$, which is the average service time, we finally arrive using (13.2) at the famous *Pollaczek-Khinchin equation*,

$$S(z) = (1 - \rho) \frac{(z-1)\,\varphi_x(\lambda - \lambda z)}{z - \varphi_x(\lambda - \lambda z)} \tag{14.26}$$

Let us further investigate what can be concluded from the Pollaczek-Khinchin equation (14.26). First, we can verify by executing the limit $z \to 1$ that $S(1) = 1$, the normalization condition for any probability generating function. More interestingly, the average number of packets in the M/G/1 system follows after some tedious manipulations using de l'Hospital's rule as

$$E\left[N_S\right] = S'(1)$$

$$= \rho + \frac{\lambda^2 \varphi_x''(0)}{2(1 - \rho)}$$

From (2.42), $\varphi_x''(0) = E\left[x^2\right]$,

$$E\left[N_S\right] = \rho + \frac{\lambda^2 E\left[x^2\right]}{2(1 - \rho)} \tag{14.27}$$

Hence, the average number of packets in the M/G/1 system (in steady-state) is proportional to the second moment of the service time distribution. Since $E\left[x^2\right] = \text{Var}[x] + (E\left[x\right])^2$, the relation[4] (14.27) shows that, for equal average

---

[4] Sometimes the coefficient of variation for the service time, $C_X = \frac{\sqrt{\text{Var}[X]}}{E[X]}$, is used such that

$$E\left[N_S\right] = \rho + \frac{\rho^2 \left(1 + C_X^2\right)}{2(1 - \rho)}$$

service rates, the service process with highest variability leads to the largest average number of packets in the system. One of the early successes of the Japanese industry was the "just in time" (JIT) principle, which essentially tries to minimize the variability in a manufacturing process. Minimization of variability is also very important in the design of scheduling rules: the less variability, the more efficiently buffer places in a router are used. Since a deterministic server has the lowest variance (namely zero), the M/D/1 queue will occupy on average the lowest number of packets. This design principle was used in ATM, where all service times precisely equal the time needed to serve one ATM cell. The average time spent in the system follows directly from Little's law (13.21),

$$E[T] = \frac{E[N_S]}{\lambda} = E[x] + \frac{\lambda E[x^2]}{2(1-\rho)}$$

and, since $E[T] = E[x] + E[w]$, the average waiting time in the queue is

$$E[w] = \frac{\lambda E[x^2]}{2(1-\rho)} \tag{14.28}$$

Observe a general property of "averages" in queueing systems: there is a simple pole at $\rho = 1$. Both the average number of packets in the system (and in the queue) and the average waiting time grows unboundedly as $\rho \to 1$.

14.3.2 The waiting time in steady-state

The derivation of the pgf (14.26) of the steady-state system content $S(z)$ has not made any assumption about the service discipline that determines the order in which packets are served. The waiting time (in the queue) and the system time (total time spent in the M/G/1 queueing system) will, of course, dependent on the order. As mentioned earlier, a FIFO service discipline is assumed. At each departure time r_k, the number of packets left behind by that k-th packet is precisely $N_S(r_k)$. With FIFO, this implies that during the total time T_k that the k-th packet has spent in the M/G/1 queueing system, precisely $N_S(r_k)$ packets have arrived. Similarly, as above in (14.22), we compute the number of Poisson $\mathcal{A}^*$ arrivals during T_k (instead of x_k) and directly find, in steady-state,

$$\Pr[\mathcal{A}^* = j] = \frac{\lambda^j}{j!} \int_0^\infty e^{-\lambda t} t^j \, dF_T(t) = \frac{(-\lambda)^j}{j!} \left. \frac{d^j \varphi_T(z)}{dz^j} \right|_{z=\lambda}$$

and the corresponding pgf,

$$A^*(z) = \sum_{j=0}^{\infty} \Pr\left[\mathcal{A}^* = j\right] z^j = \varphi_T(\lambda - \lambda z)$$

where $\varphi_T(z)$ is the Laplace transform of the system time T. Since the number of Poisson arrivals $\mathcal{A}^*$ during the system time T of a packet in steady-state equals the number of packets left behind by that packet, $\Pr\left[\mathcal{A}^* = j\right] = \Pr\left[N_{S;D} = j\right]$. The PASTA property (Theorem 13.5.1) states that, in steady-state, the observed number of packets in the queue at departure or arrival times is equal in distribution to the actual number of packets in the queue or that $\Pr\left[N_{S;D} = j\right] = \Pr\left[N_S = j\right]$. By considering the pgfs of both sides, $A^*(z) = S(z)$, such that with (14.26),

$$\varphi_T(\lambda - \lambda z) = (1 - \rho)\frac{(z - 1)\,\varphi_x(\lambda - \lambda z)}{z - \varphi_x(\lambda - \lambda z)}$$

After a change of variable $s = \lambda - \lambda z$, we end up with the result that the Laplace transform of the total system time in steady-state is a function of the Laplace transform of the service time

$$\varphi_T(s) = (1 - \rho)\frac{s\varphi_x(s)}{s - \lambda + \lambda\varphi_x(s)} \tag{14.29}$$

Since $T_k = w_k + x_k$ and, in steady-state $T = w + x$, we have that $E\left[e^{-sT}\right] = E\left[e^{-sw}e^{-sx}\right] = E\left[e^{-sw}\right]E\left[e^{-sx}\right]$, where the latter follows from independence between x and w. Hence, $\varphi_T(s) = \varphi_x(s)\varphi_w(s)$, from which the Laplace transform of the waiting time in the queue follows as

$$\varphi_w(s) = (1 - \rho)\frac{s}{s - \lambda + \lambda\varphi_x(s)} \tag{14.30}$$

Due to the correspondence with (14.26), these two relations (14.29) and (14.30) are also called Pollaczek-Khinchin equations for the system time and waiting time respectively. For example, for an exponential service time with average $\frac{1}{\mu}$, $\varphi_x(s) = \frac{\mu}{s+\mu}$ and (14.29) becomes $\varphi_T(s) = \frac{(1-\rho)\mu}{s+(\mu-\lambda)}$, which is indeed the Laplace transform of the pdf of total system time (14.4) in the M/M/1 queue. The relation (14.30) can be written in terms of the residual service time (8.17), which after Laplace transform becomes $\varphi_{rw}(s) = \frac{1-\varphi_x(s)}{sE[x]}$, as

$$\varphi_w(s) = \frac{1 - \rho}{1 - \rho\varphi_{rw}(s)}$$

It shows that the dominant tail behavior (see Section 5.7) arises from the pole at $\varphi_{rw}(s) = \frac{1}{\rho}$. By formal expansion into a Taylor series (only valid for

$|\rho\varphi_{rw}(s)| < 1)$, we find

$$\varphi_w(s) = (1 - \rho) \sum_{k=0}^{\infty} \rho^k \varphi_{rw}^k(s)$$

or, after taking the inverse Laplace transform,

$$f_w(t) = \frac{d}{dt} \Pr[w \le t] = \sum_{k=0}^{\infty} (1 - \rho) \rho^k f_{rw}^{(k*)}(t)$$

The pdf $f_w(t)$ of the waiting time in the queue can be interpreted as a sum of convolved residual service time pdfs $f_{rw}^{(k*)}(t)$ weighted by $(1 - \rho)\rho^k = \Pr[N_S = k]$, the steady-state probability of the system content in the M/M/1 system (14.1).

14.4 The GI/D/m queue

The analysis of the GI/D/m queue illustrates a discrete-time approach to queueing. Since each of the m servers operate deterministically which means that per unit time precisely one packet (or an ATM cell or customer) is served, the basic time unit in the analysis, also called a time slot, is equal to that service time. Hence, the arrival process is expressed as a counting process: instead of specifying the interarrival rate, the number of arrivals at each time slot is used.

In the sequel, we confine ourselves to a deterministic server discipline that removes during timeslot k precisely m cells from the queue. Hence, we have $\mathcal{X}_k = m$ and $X_k(z) = E\left[z^{\mathcal{X}_k}\right] = z^m$. Substituting (13.16) in (14.23) leads to

$$S_{k+1}(z) = E\left[z^{(S_k-m)^+ + A_k}\right] \tag{14.31}$$

At this point, a further general evaluation of the expression (14.31) is only possible by assuming *independence* between the random variables A_k and Q_k. From (13.18), it then follows that $S_{k+1}(z) = Q_k(z)A_k(z)$. This crucial assumption facilitates the analysis considerably. For,

$$
\begin{aligned}
S_{k+1}(z) &= E\left[z^{(S_k-m)^+} z^{A_k}\right] \\
&= E\left[z^{(S_k-m)^+}\right] E\left[z^{A_k}\right] && \text{(by independence)} \\
&= A_k(z) \sum_{j=0}^{\infty} \Pr[(S_k - m)^+ = j]\, z^j && \text{(by definition (14.23))}
\end{aligned}
$$

The summation can be worked out as

$$B = \sum_{j=0}^{\infty} \Pr[(\mathcal{S}_k - m)^+ = j]\, z^j$$

$$= \Pr[(\mathcal{S}_k - m)^+ = 0] + \sum_{j=1}^{\infty} \Pr[\mathcal{S}_k = m + j]\, z^j$$

$$= \sum_{j=0}^{m} \Pr[\mathcal{S}_k = j] + \sum_{j=1+m}^{\infty} \Pr[\mathcal{S}_k = j]\, z^{j-m}$$

Setting in terms of the generating function of $S_k(z)$ yields

$$B = \sum_{j=0}^{m} \Pr[\mathcal{S}_k = j] + z^{-m}\left(\sum_{j=0}^{\infty} \Pr[\mathcal{S}_k = j]\, z^j - \sum_{j=0}^{m} \Pr[\mathcal{S}_k = j]\, z^j\right)$$

$$= z^{-m}\left(S_k(z) - \sum_{j=0}^{m} \Pr[\mathcal{S}_k = j]\,(z^j - z^m)\right)$$

Finally, we obtain a recursion relation for the generating function of the system content in the GI/D/m system at discrete-time k,

$$S_{k+1}(z) = A_k(z)z^{-m}\left(S_k(z) - \sum_{j=0}^{m-1} \Pr[\mathcal{S}_k = j]\,(z^j - z^m)\right) \qquad (14.32)$$

In the single-server case $m = 1$, where precisely one cell is served per time slot (provided the queue is not empty), equation (14.32) simplifies with $S_k(0) = \Pr[\mathcal{S}_k = 0]$ to

$$S_{k+1}(z) = A_k(z)z^{-1}\left\{S_k(z) - \Pr[\mathcal{S}_k = 0]\,(1 - z)\right\}$$

$$= A_k(z)\left[\frac{S_k(z) - S_k(0)}{z} + S_k(0)\right] \qquad (14.33)$$

Notice that $S_{k+1}(0) = A_k(0)(S_k'(0) + S_k(0))$.

14.4.1 The steady-state of the GI/D/m queue

The steady-state behavior is reached if the system's distributions do not change anymore in time. With $\lim_{k\to\infty} S_k(z) = S(z)$ and $\lim_{k\to\infty} A_k(z) = A(z)$, (14.32) reduces in steady-state to

$$S(z) = \frac{A(z)\sum_{j=0}^{m-1} \Pr[\mathcal{S} = j]\,(z^m - z^j)}{z^m - A(z)}$$

Recall that $X_k(z) = z^m$ is the generating function of the service process. At this point, we use the same powerful argument from complex analysis as in Section 11.3.3. Since a generating function of a probability distribution is analytic inside and on the unit circle, the possible zeros of $z^m - A(z)$ inside that unit circle must be precisely cancelled by zeros in the numerator. Clearly, $z = 1$ is a zero of $z^m - A(z)$. On the unit circle, excluding the point $z = 1$ where $A(1) = 1$, we know[5] that $|A(z)| < 1$ for all points on the unit circle $|z| = 1$ (except for $z = 1$).

The region around $z = 1$ deserves some closer investigation. From the Taylor expansions

$$A(z) = 1 + \lambda(z - 1) + o(z - 1)$$
$$z^m = 1 + m(z - 1) + o(z - 1)$$

and the fact that $\lambda < m$ because a steady-state requires $\rho < 1$, we substitute $z - 1 = \varepsilon e^{i\theta}$, which describes a circle with radius ε around $z = 1$. Along this circle with arbitrarily small radius ε, we find that

$$|A(z)| = \left|1 + \lambda \varepsilon e^{i\theta} + o(\varepsilon)\right| = \sqrt{(1 + \lambda \varepsilon \cos \theta + o(\varepsilon))^2 + (\lambda \varepsilon \sin \theta + o(\varepsilon))^2}$$

$$|z^m| = \left|1 + m \varepsilon e^{i\theta} + o(\varepsilon)\right| = \sqrt{(1 + m \varepsilon \cos \theta + o(\varepsilon))^2 + (m \varepsilon \sin \theta + o(\varepsilon))^2}$$

which demonstrates that $|z^m| > |A(z)|$ on this arbitrary small circle if $\cos \theta > 0$ or $-\frac{\pi}{2} < \theta < \frac{\pi}{2}$. Invoking Rouché's Theorem 11.3.1 with $f(z) = z^m$ and $g(z) = -A(z)$ on the contour C, the unit circle including the point $z = 1$ by an arbitrarily small arc ε on the right of $z = 1$ as illustrated in Fig. 14.3, such that $|f(z)| > |g(z)|$ on the contour C, shows that $z^m - A(z)$ has precisely m zeros $\zeta_1, \zeta_2, \ldots, \zeta_m = 1$ inside that contour C.

Since $Q(z) = \frac{S(z)}{A(z)}$ is the generating function of the number of occupied buffer positions, it is also analytic inside the unit circle. Therefore, the zeros $\{\zeta_n\}_{1 \le n \le m-1}$ and $\zeta_m = 1$ are also zeros of $p(z) = \sum_{j=0}^{m-1} \Pr[\mathcal{S} = j] \, (z^m - z^j)$. This leads to a set of m equations for each $\zeta_n \ne 0$,

$$\sum_{j=0}^{m-1} \Pr[\mathcal{S} = j] \, (\zeta_n^m - \zeta_n^j) = 0$$

which determine the unknown probabilities $\Pr[\mathcal{S} = j]$. Since $p(z)$ is a poly-

---

[5] For any probability generating function $\varphi_G(z)$ it holds for $|z| \le 1$ that

$$|\varphi_G(z)| = \left|\sum_{j=0}^{\infty} \Pr[G = j] \, z^j\right| \le \sum_{j=0}^{\infty} \Pr[G = j] \, |z^j| \le \sum_{j=0}^{\infty} \Pr[G = j] = 1$$

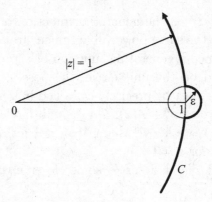

Fig. 14.3. Details of the contour C in the neighborhood of $z = 1$.

nomial of degree m, $p(z)$ is entirely determined by its zeros as

$$p(z) = \alpha(z - 1) \prod_{n=1}^{m-1} (z - \zeta_n)$$

The unknown α is determined from the normalization condition $S(1) = Q(1) = 1$, which is explicitly

$$\lim_{z \to 1} \frac{p(z)}{z^m - A(z)} = 1$$

With de l'Hospital's rule,

$$\alpha^{-1} = \prod_{n=1}^{m-1} (1 - \zeta_n) \lim_{z \to 1} \frac{1}{mz^{m-1} - A'(z)} = \frac{\prod_{n=1}^{m-1}(1 - \zeta_n)}{m - \lambda}$$

Finally, we arrive at the generating function of the buffer content via that of the system content $S(z) = A(z)Q(z)$,

$$Q(z) = \frac{(m - \lambda)(z - 1)}{z^m - A(z)} \prod_{n=1}^{m-1} \frac{z - \zeta_n}{1 - \zeta_n} \tag{14.34}$$

With $A'(1) = \lambda$ the single-server case ($m = 1$) in (14.34) reduces to the well-known result for the pgf of the system and buffer content of a GI/D/1 system respectively as

$$S(z) = (1 - A'(1)) \frac{(z - 1) A(z)}{z - A(z)} \tag{14.35}$$

$$Q(z) = (1 - \lambda) \frac{z - 1}{z - A(z)} \tag{14.36}$$

The probability of an empty buffer, $Q(0) = \Pr[\mathcal{Q} = 0]$, immediately follows from (14.34). The average queue length for the single-server $(m = 1)$ is obtained as $Q'(1)$, or

$$E[N_Q] = E[\mathcal{Q}] = \frac{A''(1)}{2(1 - \lambda)} \tag{14.37}$$

14.4.2 The waiting time in a GI/D/m system

Let $T = w + 1$ denote the steady-state system time of an arbitrary packet, a "test" packet, in units of a timeslot. In addition to the system content $\mathcal{S}$ that describes the number of packets in the system at the beginning of a time slot, an additional random variable must be introduced: $\mathcal{F}$ denotes the number of packets that arrive in the same timeslot just before the "test" packet. In the D-server and assuming a FIFO discipline, these $\mathcal{F}$ packets will be served before the "test" packet, possibly in the same time slot. The system time of the "test" packet equals

$$T = \left\lfloor \frac{(\mathcal{S} - m)^+ + \mathcal{F}}{m} \right\rfloor + 1$$

where $\lfloor x \rfloor$ denotes the largest integer smaller than or equal to x. Indeed, $(\mathcal{S} - m)^+ + \mathcal{F}$ are the number of packets in the system just before the arrival of the "test" packet. At the beginning of a timeslot, at most m packets are served, which explains the integer division. The service time takes precisely one additional time slot. Let us simplify the notation by defining $\mathcal{R} = (\mathcal{S} - m)^+ + \mathcal{F}$. From this expression for the system time, we deduce, for each integer $k \geq 1$ (the minimal waiting time in the system equals 1 timeslot), that

$$\Pr[T = k] = \sum_{j=0}^{m-1} \Pr[\mathcal{R} = (k - 1)m + j]$$

such that the generating function of the waiting time $T(z)$ is

$$T(z) = \sum_{k=1}^{\infty} \Pr[T = k] z^k = \sum_{k=1}^{\infty} \sum_{j=0}^{m-1} \Pr[\mathcal{R} = (k - 1)m + j] z^k$$

$$= \sum_{j=0}^{m-1} \sum_{k=0}^{\infty} \Pr[\mathcal{R} = km + j] z^{k+1}$$

Also,

$$T\left(z^m\right) = z^m \sum_{j=0}^{m-1} z^{-j} \sum_{k=0}^{\infty} \Pr\left[\mathcal{R} = km + j\right] z^{mk+j}$$

$$= z^m \sum_{j=0}^{m-1} z^{-j} \sum_{k=0}^{\infty} \sum_{n=0}^{\infty} \Pr\left[\mathcal{R} = n\right] z^n \delta_{n,mk+j}$$

$$= z^m \sum_{j=0}^{m-1} z^{-j} \sum_{n=0}^{\infty} \Pr\left[\mathcal{R} = n\right] z^n \sum_{k=0}^{\infty} \delta_{n,mk+j}$$

where the Kronecker delta $\delta_{k,m} = 1$ if $k = m$ else $\delta_{k,m} = 0$. The sum

$$\sum_{k=0}^{\infty} \delta_{n,mk+j} = \sum_{k=-\infty}^{\infty} \delta_{n-j,mk} = 1_{m|n-j}$$

is one if m divides $n - j$ else it is zero. Such expression can be written as

$$\sum_{k=-\infty}^{\infty} \delta_{n-j,mk} = \frac{1 - e^{2\pi i(n-j)}}{m\left(1 - e^{2\pi i(n-j)/m}\right)} = \sum_{k=0}^{m-1} e^{2\pi ik(n-j)/m}$$

Using the latter summation yields

$$T\left(z^m\right) = \frac{z^m}{m} \sum_{j=0}^{m-1} z^{-j} \sum_{n=0}^{\infty} \Pr\left[\mathcal{R} = n\right] z^n \sum_{k=0}^{m-1} e^{2\pi ik(n-j)/m}$$

$$= \frac{z^m}{m} \sum_{k=0}^{m-1} \sum_{j=0}^{m-1} \left(ze^{2\pi ik/m}\right)^{-j} \sum_{n=0}^{\infty} \Pr\left[\mathcal{R} = n\right] \left(ze^{2\pi ik/m}\right)^n$$

$$= \frac{z^m}{m} \sum_{k=0}^{m-1} \frac{1 - z^{-m}}{1 - \left(ze^{2\pi ik/m}\right)^{-1}} R\left(ze^{2\pi ik/m}\right)$$

where we have introduced the generating function $R(z) = \sum_{n=0}^{\infty} \Pr\left[\mathcal{R} = n\right] z^n$. Thus, we arrive at

$$T\left(z^m\right) = \frac{1}{m} \sum_{k=0}^{m-1} \frac{z^m - 1}{1 - \left(ze^{2\pi ik/m}\right)^{-1}} R\left(ze^{2\pi ik/m}\right)$$

The generating function $R(z)$ can be specified further since

$$R(z) = E\left[z^{(\mathcal{S}-m)^+ + \mathcal{F}}\right] = E\left[z^{\mathcal{F}}\right] E\left[z^{(\mathcal{S}-m)^+}\right]$$

where independence of the arrival process and queueing process (GI) has been used. From (14.31), the corresponding steady-state relation is

$$S(z) = E\left[z^{(\mathcal{S}-m)^+ + \mathcal{A}}\right] = E\left[z^{\mathcal{A}}\right] E\left[z^{(\mathcal{S}-m)^+}\right] = A(z)E\left[z^{(\mathcal{S}-m)^+}\right]$$

while from $S(z) = A(z)Q(z)$, we observe that $Q(z) = E\left[z^{(\mathcal{S}-m)^+}\right]$. Hence,

$$R(z) = F(z)Q(z)$$

where $Q(z)$ is given by (14.34).

We now turn our attention to the determination of $F(z)$, the generating function of the number of packets in front of the "test" packet. The "test" packet has been uniformly chosen out of the total flow of arriving packets at the system. Let us denote by $\mathcal{A}^*$ the number of arriving packets in the same time slot as the "test" packet. The random variable $\mathcal{A}^*$ is not the same as the number of arriving cells $\mathcal{A}$ per time slot. For example, we know that there is at least one arrival in the time slot of the "test" packet, namely, the "test" packet itself, hence, $\Pr[\mathcal{A}^* = 0] = 0$. Furthermore, the larger the number of arriving packets in a time slot, the higher the probability that the "test" packet is chosen out of those packets in this time slot. Hence, $\Pr[\mathcal{A}^* = j]$ is proportional to the number j of arriving packets in a time slot. In addition, $\Pr[\mathcal{A}^* = j]$ is also proportional to $\Pr[\mathcal{A} = j]$, which describes how likely a number j of arriving packets is. Combining both shows that

$$\Pr[\mathcal{A}^* = j] = \alpha j \Pr[\mathcal{A} = j]$$

with the proportionality factor equal to $\alpha = \frac{1}{E[\mathcal{A}]}$ because $\sum_{j=0}^{\infty} \Pr[\mathcal{A}^* = j] = 1$. The "test" packet is uniformly distributed among the arriving packets $\mathcal{A}^*$ in the time slot of the "test" packet (in steady-state). The probability of having precisely k packets in front of the "test" packet given $\mathcal{A}^* = j \geq 1$ equals

$$\Pr[\mathcal{F} = k | \mathcal{A}^* = j] = \frac{1_{k<j}}{j}$$

Indeed, the "test" packet has equal probability $\frac{1}{j}$ of occupying any of the j possible positions. The occupation of a position $k + 1$ implies precisely k cells in front of the "test" packet in a FIFO discipline. Using the law of total probability (2.46),

$$\Pr[\mathcal{F} = k] = \sum_{j=1}^{\infty} \Pr[\mathcal{F} = k | \mathcal{A}^* = j] \Pr[\mathcal{A}^* = j]$$

$$= \sum_{j=k+1}^{\infty} \frac{1}{j} \frac{j \Pr[\mathcal{A} = j]}{E[\mathcal{A}]} = \frac{1}{E[\mathcal{A}]} \sum_{j=k+1}^{\infty} \Pr[\mathcal{A} = j]$$

The generating function $F(z)$ becomes

$$F(z) = \sum_{k=0}^{\infty} \Pr[\mathcal{F} = k] z^k = \frac{1}{E[\mathcal{A}]} \sum_{k=0}^{\infty} \sum_{j=k+1}^{\infty} \Pr[\mathcal{A} = j] z^k$$

$$= \frac{1}{E[\mathcal{A}]} \sum_{j=1}^{\infty} \Pr[\mathcal{A} = j] \sum_{k=1}^{j} z^{k-1} = \frac{1}{E[\mathcal{A}]} \sum_{j=1}^{\infty} \Pr[\mathcal{A} = j] \frac{1 - z^j}{1 - z}$$

$$= \frac{1}{E[\mathcal{A}](z-1)} \left(\sum_{j=1}^{\infty} \Pr[\mathcal{A} = j] z^j - \sum_{j=1}^{\infty} \Pr[\mathcal{A} = j] \right)$$

$$= \frac{z(A(z) - \Pr[\mathcal{A} = 0] - 1 + \Pr[\mathcal{A} = 0])}{E[\mathcal{A}](z - 1)}$$

Since $E[\mathcal{A}] = A'(1) = \lambda$, we finally arrive at

$$F(z) = \frac{A(z) - 1}{(z - 1) A'(1)}$$

Combining all involved expressions leads to the expression of the generating function of the total time spent in the GI/D/m queue

$$T(z^m) = \frac{m - \lambda}{m\lambda} \sum_{k=0}^{m-1} \frac{z^m - 1}{1 - (ze^{2\pi i k/m})^{-1}} \frac{A(ze^{2\pi i k/m}) - 1}{z^m - A(ze^{2\pi i k/m})} \prod_{n=1}^{m-1} \frac{ze^{2\pi i k/m} - \zeta_n}{1 - \zeta_n}$$

For the single-server case ($m = 1$), the generating function of the system time (queueing time plus service time) considerably simplifies to

$$T(z) = \left(\frac{1 - \lambda}{\lambda} \right) z \frac{A(z) - 1}{z - A(z)}$$

from which $E[T]$ and $\text{Var}[T]$ readily follow. The computation of the pdf given the arrival process $A(z)$ is more complex, as illustrated for the M/D/1/K queue in the next section.

14.5 The M/D/1/K queue

Suppose we have a buffer of K cells and an aggregate arrival stream consisting of a large number of sources with none of them dominating the others. This input process is well modeled by a Poisson process with arrival rate λ. Both the input process as well as the buffer content and the output process have been simulated in Fig. 14.4. Observe the effect of variations and maximum number of cells in the queue and input process!

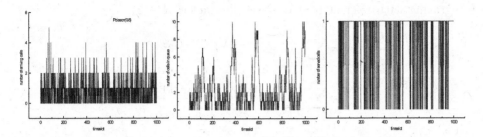

Fig. 14.4. On the left, the Poisson input process with $\lambda = 0.8$ in terms of the number of cells versus the timeslot. In the middle, the buffer occupancy for a buffer with $K = 20$ as function of time. On the right, the M/D/1/K output process in cells served per timeslot.

14.5.1 The pdf of the buffer occupancy in the M/D/1 queue

For a Poisson process, $\Pr[\mathcal{A} = k] = \frac{\lambda^k}{k!}e^{-\lambda}$ and $A(z) = e^{\lambda(z-1)}$. The pgf $Q(z)$ of the buffer content immediately follows from (14.36) as

$$Q(z) = (1 - \lambda)\frac{(1 - z)\,e^{\lambda(1-z)}}{1 - z\,e^{\lambda(1-z)}}$$

The average queue length is obtained from (14.37) as

$$E\left[N_{Q;M/D/1}\right] = E\left[\mathcal{Q}\right] = \frac{\lambda^2}{2\,(1 - \lambda)}$$

and Little's law (13.21) provides the average waiting time in the queue

$$E\left[w_{M/D/1}\right] = \frac{E\left[N_{Q;M/D/1}\right]}{\lambda} = \frac{\lambda}{2\,(1 - \lambda)} \tag{14.38}$$

Since $z\,e^{\lambda(1-z)}$ is analytic everywhere, there always exists a neighborhood (depending on λ) around $z = 0$ for which $|z\,e^{\lambda(1-z)}| \leq 1$. Hence, we can use the series expansion for the geometric series to obtain

$$\frac{Q(z)}{1 - \lambda} = (1 - z)\,e^{\lambda(1-z)}\sum_{k=0}^{\infty} z^k\,e^{k\lambda(1-z)} = \sum_{k=0}^{\infty}(1 - z)\,z^k\,e^{(k+1)\lambda(1-z)}$$

Integrating with respect to λ removes the factor $(1 - z)$,

$$\int \frac{Q(z)}{1 - \lambda} d\lambda = \sum_{k=0}^{\infty} \frac{z^k}{k+1} e^{(k+1)\lambda} e^{-(k+1)\lambda z}$$

$$= \sum_{k=0}^{\infty} \frac{z^k}{k+1} e^{(k+1)\lambda} \sum_{n=0}^{\infty} \frac{(-1)^n (k+1)^n \lambda^n}{n!} z^n$$

$$= \sum_{k=0}^{\infty} \sum_{n=0}^{\infty} \frac{(-1)^n}{n!} e^{(k+1)\lambda} (k+1)^{n-1} \lambda^n z^{n+k}$$

After a change in the variable $m = n + k$ with $m \geq 0$, which implies that $k \leq m$ since $n = m - k \geq 0$, we have

$$\int \frac{Q(z)}{1 - \lambda} d\lambda = \sum_{m=0}^{\infty} \left(\sum_{k=0}^{m} \frac{(-1)^{m-k}}{(m-k)!} e^{(k+1)\lambda} (k+1)^{m-k-1} \lambda^{m-k} \right) z^m$$

$$= \sum_{m=0}^{\infty} \left(\sum_{k=1}^{m+1} \frac{(-1)^{m+1-k}}{(m+1-k)!} e^{k\lambda} k^{m-k} \lambda^{m+1-k} \right) z^m$$

Differentiation with respect to λ, gives

$$Q(z) = (1 - \lambda) \sum_{m=0}^{\infty} \left(\sum_{k=1}^{m+1} (-1)^{m+1-k} e^{k\lambda} \left[\frac{(k\lambda)^{m+1-k}}{(m+1-k)!} + \frac{(k\lambda)^{m-k}}{(m-k)!} \right] \right) z^m$$

from which we finally deduce the probability $q[m] = \Pr[\mathcal{Q} = m]$ that the m-th position in the buffer is occupied[6]

$$q[m] = (1 - \lambda) \sum_{k=1}^{m+1} (-1)^{m+1-k} e^{k\lambda} \left[\frac{(k\lambda)^{m+1-k}}{(m+1-k)!} + \frac{(k\lambda)^{m-k}}{(m-k)!} \right] \qquad (14.39)$$

One readily observes from the derivation above that the probability $s[m] = \Pr[\mathcal{S} = m]$ that m positions in the system are occupied, is $s[m] = q[m-1]$

---

[6] Explicitly, we have for the queue content probabilities

$$q[0] = e^{\lambda} (1 - \lambda)$$

$$q[1] = e^{\lambda} (1 - \lambda) (e^{\lambda} - \lambda - 1)$$

$$q[2] = e^{\lambda} (1 - \lambda) (\frac{\lambda^2}{2} - e^{2\lambda} + \lambda - e^{\lambda} - 2\lambda e^{\lambda})$$

while the system content probabilities are

$$s[0] = (1 - \lambda)$$

$$s[1] = (e^{\lambda} - 1) (1 - \lambda)$$

$$s[m] = q[m-1] \qquad (m \geq 2)$$

for $m \geq 2$ because the z-transform is $S(z) = (1 - \lambda)\frac{(1-z)}{1 - z\,e^{\lambda(1-z)}}$. This result is a characteristic property of a deterministic server. Next, we rewrite (14.39) as

$$q[m] = (1 - \lambda)\left[\sum_{k=1}^{m+1} e^{k\lambda}\frac{(-k\lambda)^{m+1-k}}{(m+1-k)!} - \sum_{k=1}^{m} e^{k\lambda}\frac{(-k\lambda)^{m-k}}{(m-k)!}\right]$$

$$= (1 - \lambda)\left[e^{(m+1)\lambda}g(-\lambda e^{-\lambda}; m+1) - e^{m\lambda}g(-\lambda e^{-\lambda}; m)\right] \qquad (14.40)$$

where

$$g(x; m) = \sum_{k=0}^{m}(m-k)^k\frac{x^k}{k!} \qquad (14.41)$$

Due to the nature of the differences, we immediately find the cumulative distribution,

$$\sum_{m=1}^{K} q[m] = (1 - \lambda)\left[e^{(K+1)\lambda}g(-\lambda e^\lambda; K+1) - e^\lambda\right]$$

and since $q[0] = (1 - \lambda)\,e^\lambda$, we arrive at

$$\Pr[\mathcal{Q} \leq K] = \sum_{m=0}^{K} q[m] = (1 - \lambda)\,e^{(K+1)\lambda}g(-\lambda e^\lambda; K+1) \qquad (14.42)$$

The expressions in (14.40) are numerically only useful for small m because the series is alternating. This problem may be solved by considering a famous result due to Lagrange (Markushevich, 1985, Vol. 2, Chapter 3, Section 14)

$$e^{bz} = 1 + b\sum_{n=1}^{\infty}\frac{(b+na)^{n-1}}{n!}\left(ze^{-az}\right)^n \qquad (14.43)$$

that converges for $|z| \leq a^{-1}$. Differentiation of (14.43) with respect to $w = ze^{-az}$ leads to

$$\frac{e^{(b+a)z}}{1 - az} = \sum_{n=0}^{\infty}\frac{(b+a+na)^n}{n!}z^n e^{-naz}$$

because $\frac{d}{dw}e^{bz} = \frac{d}{dz}e^{bz}\frac{dz}{dw} = \frac{be^{bz}}{(1-az)e^{-az}}$. Choosing $a = -1$, $b + a = m$ and $z = -\lambda$, we obtain

$$\frac{e^{-m\lambda}}{1 - \lambda} = \sum_{n=0}^{\infty}\frac{(m-n)^n}{n!}(-\lambda e^{-\lambda})^n = g(-\lambda e^{-\lambda}; m) + \sum_{n=m+1}^{\infty}\frac{(n-m)^n}{n!}(\lambda e^{-\lambda})^n$$

and thus

$$g(-\lambda e^{-\lambda}; m) = \frac{e^{-m\lambda}}{1-\lambda} - \sum_{n=m+1}^{\infty} \frac{(n-m)^n}{n!}(\lambda e^{-\lambda})^n$$

where the infinite series consists of merely positive terms. Substituted in (14.42), this finally yields

$$\Pr[\mathcal{Q} > K] = (1-\lambda)\,\lambda^{K+1} \sum_{n=1}^{\infty} \frac{n^{n+K+1}}{(n+K+1)!}(\lambda e^{-\lambda})^n \qquad (14.44)$$

In the heavy traffic limit $\rho = \lambda \to 1$, the dominant zero (5.35) of $1 - z\,e^{\lambda(1-z)}$ is approximately equal to $\zeta \approx 1 + \frac{2(1-\lambda)}{\lambda}$ and the resulting tail asymptotic (5.31) for the buffer occupancy pdf is

$$\Pr[\mathcal{Q} > K] \approx -\frac{1-\rho}{1-\rho\zeta}\zeta^{-K-1} \approx e^{-K\log\zeta} \sim e^{-2K\frac{(1-\rho)}{\rho}} \qquad (14.45)$$

14.6 The N*D/D/1 queue

The N*D/D/1 queue is a basic model for constant bit rate sources in ATM, as shown in Fig. 14.5. The input process consists of a superposition of N independent periodic sources with the same period D but randomly phased, i.e. arbitrarily shifted in time with respect to each other. The server operates deterministically and serves one ATM cell per timeslot. The buffer size is assumed to be infinitely long mainly to enable an exact analytic solution. During a time period D measured in time slots or server time units, precisely N cells arrive such that the traffic intensity or load (13.2) equals $\rho = \frac{N}{D}$.

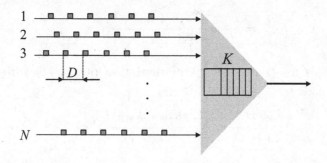

Fig. 14.5. Sketch of an ATM concentrator where N input lines are multiplexed onto a single output line. The N*D/D/1 queue models this ATM basic switching unit accurately.

Whereas the arrivals in the M/D/1 queue are uncorrelated, the successive interarrival times in the N*D/D/1 queue are negatively correlated. For the same average arrival rate, this more regular arrival process results in shorter queues than in the M/D/1 queue, where the higher variability in the arrival process causes longer queues.

Due to the dependence of the arrivals over many timeslots, the solution method is based on the Beneš approach and starts from the complementary distribution (13.15) for the virtual waiting time or unfinished work in steady-state, $\Pr[v(t_\infty) > x] = \lim_{t\to\infty} \Pr[v(t) > x]$. Applied to the N*D/D/1 queue the unfinished work equals the number of ATM cells in the system, thus $\Pr[v(t_\infty) > x] = \Pr[N_S > x]$. Hence, in the steady-state for $\rho < 1$ or $N < D$,

$$\Pr[N_S > x] = \sum_{k=\lceil x \rceil}^{\infty} \Pr\left[v(t_\infty + x - k) = 0 \left| \int_{t_\infty + x - k}^{t_\infty} N_A(u)\, du = k\right.\right]$$

$$\times \Pr\left[\int_{t_\infty + x - k}^{t_\infty} N_A(u)\, du = k\right]$$

The periodic cell trains with period equal to D timeslots at each input line lead to a periodic aggregated arrival stream of the N input lines also with period D. Each cell train transports precisely one cell per period D, which allows us to observe the characteristics of the aggregated arrival process during the time interval $[0, D)$. The computations are most conveniently performed if we choose the steady-state observation point $t_\infty = D$. Each of the N ATM cells arrives uniformly in $[0, D]$ due to the random phasing of each cell train and the probability that it arrives in $[D + x - k, D]$ is $p = \frac{k-x}{D}$. Hence, the number of arrivals in $[D + x - k, D]$ is a sum of Bernoulli random variables, which is binomially distributed,

$$\Pr\left[\int_{D+x-k}^{D} N_A(u)\, du = k\right] = \binom{N}{k}\left(\frac{k-x}{D}\right)^k \left(1 - \frac{k-x}{D}\right)^{N-k}$$

The conditional probability is obtained as follows. The unfinished work at time $D + x - k$ only depends on past arrivals in the interval $[0, D + x - k]$. Given that the number of arrivals in $[D + x - k, D]$ equals k while there are always precisely N in $[0, D]$, the number of arrivals in $[0, D + x - k]$ equals $N - k$ and the corresponding traffic intensity $\rho' = \frac{N-k}{D+x-k} < 1$ since $N < D$ and, thus, $N - k < D - k$ for any k. From Section 13.3.2, we use the *local stationary result*: for any stationary single-server queueing system with traffic intensity ρ, the probability of an empty system at an arbitrary time is $1 - \rho$. If we take a random point in $t \in [0, D + x - k]$, then stationarity implies that $\Pr[v(t) = 0] = 1 - \rho' = \frac{D+x-N}{D+x-k}$. Since $\rho' < 1$, the system is necessarily empty at some instant t^* in $[0, D + x - k)$. As explained in Section 13.2 and Section 13.3, we may consider that the system restarts from t^* on ignoring the past. But the probability $\Pr[v(t) = 0]$ at a random time in the interval $[0, t^*]$ and in $[t^*, D + x - k]$ is the same, which means that $\Pr[v(D + x - k) = 0] = \frac{D+x-N}{D+x-k}$ because we can periodically repeat the system's process in $[0, t^*]$ while omitting any activity in $[t^*, D + x - k]$. With respect to this newly constructed periodic arrival pattern, the point $t = D + x - k$ is arbitrary such that the *local stationary result* is applicable. In summary, we arrive at the overflow probability in a N*D/D/1 system,

$$\Pr[N_S > x] = \sum_{k=\lceil x \rceil}^{N} \frac{D+x-N}{D+x-k} \binom{N}{k} \left(\frac{k-x}{D}\right)^k \left(1 - \frac{k-x}{D}\right)^{N-k} \quad (14.46)$$

Observe that $\Pr[N_S > N] = 0$. Rewriting (14.46) yields

$$\Pr[N_S > x] = \frac{D+x-N}{D^N} \sum_{k=\lceil x \rceil}^{N} \binom{N}{k} (k-x)^k (D+x-k)^{N-k-1}$$

$$= \frac{D+x-N}{D^N} \sum_{j=0}^{N-\lceil x \rceil} \binom{N}{j} (N-j-x)^{N-j} (D+x-N+j)^{j-1}$$

$$= \frac{D+x-N}{D^N} \sum_{j=0}^{N} \binom{N}{j} (D+x-N+j)^{j-1} (N-x-j)^{N-j}$$

$$- \frac{D+x-N}{D^N} \sum_{j=N-\lceil x \rceil+1}^{N} \binom{N}{j} (D+x-N+j)^{j-1} (N-j-x)^{N-j}$$

Applying Abel's identity (Comtet, 1974, p. 128), valid for all u, y, z,

$$(u+y)^n = u \sum_{k=0}^{n} \binom{n}{k} (u-kz)^{k-1} (y+kz)^{n-k} \quad (14.47)$$

with $n = N$, $u = D + x - N$, $y = N - x$ and $z = -1$ gives

$$\Pr[N_S \leq x] = \frac{D+x-N}{D^N} \sum_{j=N-\lceil x \rceil+1}^{N} \binom{N}{j} (D+x-N+j)^{j-1} (N-j-x)^{N-j}$$

$$(14.48)$$

demonstrating that, indeed, $\Pr[N_S \leq 0] = 0$. For small x, relation (14.48) is convenient, while (14.46) is more suited for large $x \to N$. For example, $\Pr[N_S \leq 1] = \frac{D+1-N}{D+1} \left(1 + \frac{1}{D}\right)^N$, while $\Pr[N_S > N - 1] = \left(\frac{1}{D}\right)^N$.

In the heavy traffic regime for $\rho = \frac{N}{D} \to 1$, a Brownian motion approximation (Roberts, 1991) is

$$\Pr\left[N_S > x\right] \simeq e^{-2x\left(\frac{x}{N} + \frac{1-\rho}{\rho}\right)} \tag{14.49}$$

Figure 14.6 compares the exact (14.46) overflow probability and the Brownian approximation (14.49) for $\rho = 0.95$. Observe from (14.45) that

$$\Pr\left[N_S > x\right] \simeq e^{-\frac{2x^2}{N}} \Pr\left[\mathcal{Q}_{\mathrm{M/D/1}} > x\right]$$

which shows that, for sufficiently high N, the overflow probability of the N*D/D/1 queue tends to that of the M/D/1 queue. Thus, an arrival process consisting of a superposition of a large number of periodic processes tends to a Poisson arrival process. The decaying factor $e^{-\frac{2x^2}{N}}$ reflects the effect of the negative correlations in the arrival process and shows that a Poisson process overestimates the tail probability in heavy traffic. Comparison of (14.46) and (14.44) for lower loads $\rho = \frac{N}{D}$ illustrates that the Poisson approximation becomes more accurate.

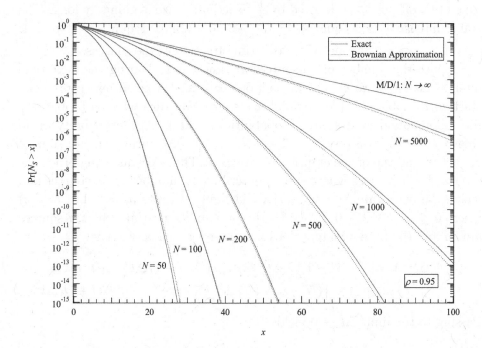

Fig. 14.6. The overflow probability $\Pr\left[N_S > x\right]$ in the N*D/D/1 queue for $\rho = 0.95$ and various number of sources N.

14.7 The AMS queue

Many arrival patterns in telecommunication exhibit active or "on" periods succeeded by silent inactive or "off" periods. At the burst or flow level phenomena of the order of time of an on-off period are dominant and the finer details of the individual packet arrivals within an on-period can be ignored. The stream of packets can be regarded as a continuous fluid characterized by the flow arrival rate.

The AMS queue is perhaps the simplest exact solvable queueing model that describes the queueing behavior at the burst or flow level. The AMS queue named after Anick, Mitra and Sondhi (Anick *et al.*, 1982; Mitra, 1988) considers N homogeneous, independent on-off sources in a continuous fluid flow approach. For each source, both the on- and off-period are exponentially distributed, which makes the model Markovian by Theorem 10.2.3. In the on-period each source emits a unit amount of information. Hence, at each moment in time when r sources are in the on-period, r packets (units of information) arrive at the buffer. The service time is constant and equal to $c < N$ packets per unit time. If $c > N$, then the buffer is always empty. The time unit is chosen as the average time of an on-period while the average time of an off-period is denoted by $\frac{1}{\lambda}$. The buffer size is infinitely long. The traffic intensity then equals $\rho = \frac{N\lambda}{c(1+\lambda)}$ and stability requires that $\rho < 1$. The ratio $\frac{\lambda}{1+\lambda}$ is the long term "on" time fraction of the sources.

Suppose the number of on-sources at time t is i. During the next time interval Δt only two elementary actions can take place: a new source can start with probability $(N-i)\lambda\Delta t$ or a source can turn off with probability $i\Delta t$. Compound events have probabilities $O(\Delta t^2)$. The probability of no change in the arrival process is $1 - [(N-i)\lambda + i]\Delta t$ during which i sources are active and the queue empties at rate $c-i$. The AMS queueing process is a birth-death process where state i describes the number of on-sources and where the birth rate $\lambda_i = (N-i)\lambda$ and the death rate $\mu_i = i$. Let $P_i(t,x)$ where $0 \le i \le N$, $t \ge 0$, $x \ge 0$ be the probability that at time t, i sources are on and the buffer content does not exceed x. Then, we have

$$P_i(t+\Delta t, x) = [N-(i-1)]\lambda\,\Delta t\, P_{i-1}(t,x) + (i+1)\,\Delta t\, P_{i+1}(t,x)$$
$$+ [1 - \{(N-i)\lambda + i\}\,\Delta t]\, P_i(t, x-(i-c)\,\Delta t) + O(\Delta t^2)$$

Passing to the limit $\Delta t \to 0$ yields

$$\frac{\partial P_i(t,x)}{\partial t} + (i-c)\frac{\partial P_i(t,x)}{\partial x} = (N-i+1)\lambda P_{i-1}(t,x) - [(N-i)\lambda + i]P_i(t,x)$$
$$+ (i+1)P_{i+1}(t,x)$$

The time-independent equilibrium probabilities, $\pi_i(x) = \lim_{t \to \infty} P_i(t, x)$, reflect the steady-state where i sources are on and the buffer content does not exceed x. Setting $\frac{\partial P_i(t,x)}{\partial t} = 0$, the steady-state equations become for $0 \le i \le N$

$$(i - c)\frac{d\pi_i(x)}{dx} = (N - i + 1)\,\lambda\,\pi_{i-1}(x) - [(N - i)\,\lambda + i]\,\pi_i(x) + (i + 1)\,\pi_{i+1}(x)$$
(14.50)

In matrix notation, where $\pi(x)$ is a column vector as opposed to Markov theory where $\pi(x)$ is a row vector,

$$D\frac{d\pi(x)}{dx} = Q\pi(x)$$
(14.51)

and $D = \text{diag}[-c, 1 - c, 2 - c, \ldots, N - c]$ and Q is a tri-diagonal $(N + 1) \times (N + 1)$ matrix,

$$Q = \begin{bmatrix} -N\lambda & 1 & 0 & 0 & \cdots & 0 \\ N\lambda & -[(N-1)\lambda + 1] & 2 & 0 & \cdots & 0 \\ 0 & (N-1)\lambda & -[(N-2)\lambda + 2] & 3 & \cdots & 0 \\ \vdots & \vdots & \vdots & & \ddots & \vdots & \vdots \\ 0 & 0 & 0 & \cdots & -[\lambda + (N-1)] & N \\ 0 & 0 & 0 & \cdots & \lambda & -N \end{bmatrix}$$

The buffer overflow probability is $\Pr[N_S > x] = 1 - \|\pi(x)\|_1 = 1 - \sum_{j=0}^{N}\pi_j(x)$, which implies that $\|\pi(\infty)\|_1 = 1$. Moreover, $\lim_{x \to \infty}\frac{d\pi(x)}{dx} = 0$ and $Q\pi(\infty) = 0$ corresponds to the steady-state of the continuous Markov chain (arrival process and service process as a whole). Furthermore,

$$\pi_i(\infty) = \frac{1}{(1 + \lambda)^N}\binom{N}{i}\lambda^i$$
(14.52)

is the probability that i out of N sources are on simultaneously irrespective of what the buffer level in the system is.

As shown in Section 10.2.2, besides $\pi(x) = e^{D^{-1}Qx}\,\pi(0)$, the solution of (14.51) can be expressed in terms of the eigenvalues ζ_j, the corresponding right-eigenvector x_j and left-eigenvector y_j of $D^{-1}Q$,

$$\pi(x) = \sum_{j=0}^{N} e^{\zeta_j x}x_j\left(y_j^T\pi(0)\right)$$

where, as shown in Appendix A.5.2.2, the eigenvalues are labeled in increasing order $\zeta_{N-[c]-1} < \cdots < \zeta_1 < \zeta_0 < \zeta_N = 0 < \zeta_{N-1} < \cdots < \zeta_{N-[c]}$. This way of writing distinguishes between underload and overload eigenvalues. Only bounded solutions are allowed. As shown in Appendix A.5.2.2, there are precisely $N - [c] - 1$ negative real eigenvalues such that $j \in [0, N - [c] - 1]$. In addition, $j = N$ that corresponds to the eigenvalue $\zeta_N = 0$ and the $\pi(\infty)$ eigenvector. The general bounded solution of (14.51) is

$$\pi(x) = \pi(\infty) + \sum_{j=0}^{N-[c]-1} a_j e^{\zeta_j x} x_j \qquad (14.53)$$

where the scalar coefficients $a_j = y_j^T \pi(0)$ still need to be determined. Rather than determining $y_j^T \pi(0)$ as in Appendix A.5.2.2, a more elegant and physical method is used. The eigenvalue solution in Appendix A.5.2.2 has scaled the eigenvectors by setting the N component equal to 1, hence, $(x_j)_N = 1$. Writing the N-th component in (14.53) gives with (14.52)

$$\pi_N(x) = \frac{\lambda^N}{(1+\lambda)^N} + \sum_{j=0}^{N-[c]-1} a_j e^{\zeta_j x} \qquad (14.54)$$

The most convenient choice of x is $x = 0$. If the number of on-source j at any time exceeds the service rate c, then the buffer builds-up and cannot be empty,

$$\pi_j(0) = 0 \qquad \text{for } [c] + 1 \le j \le N$$

This observation provides one equation in (14.54) for the coefficients a_k,

$$\sum_{j=0}^{N-[c]-1} a_j = -\frac{\lambda^N}{(1+\lambda)^N}$$

and shows that $N - [c] - 1$ additional equations are needed to determine all coefficients a_j. By differentiating (14.54) m-times and evaluating at $x = 0$, we find these additional equations

$$\left. \frac{d^m \pi_N(x)}{dx^m} \right|_{x=0} = \sum_{j=0}^{N-[c]-1} a_j \zeta_j^m$$

which will be determined with the help of the differential equation (14.51). Indeed, for $m = 1$, the differential equation (14.51) gives

$$\left. \frac{d\pi_N(x)}{dx} \right|_{x=0} = D^{-1} Q \pi(0)$$

The important observation is that the effect of multiplication by $D^{-1}Q$ decreases the number of zero components in $\pi(0)$ by 1, i.e. $\left. \frac{d\pi_j(x)}{dx} \right|_{x=0} = 0$ for $[c] + 2 \le j \le N$. Any additional multiplication by $D^{-1}Q$ has the same effect. Since $\frac{d^m \pi(x)}{dx^m} = \left(D^{-1}Q\right)^m \pi(x)$, we thus find, for $0 \le m \le N - [c] - 1$ that

$$\left. \frac{d^m \pi_N(x)}{dx^m} \right|_{x=0} = 0$$

We write these $N - [c]$ equation in the unknown a_k in matrix form,

$$
\begin{bmatrix}
1 & 1 & 1 & \cdots & 1 \\
\zeta_1 & \zeta_2 & \zeta_3 & \cdots & \zeta_{N-[c]-1} \\
\zeta_1^2 & \zeta_2^2 & \zeta_3^2 & \cdots & \zeta_{N-[c]-1}^2 \\
\vdots & \vdots & \vdots & \ddots & \vdots \\
\zeta_1^{N-[c]-2} & \zeta_2^{N-[c]-2} & \zeta_3^{N-[c]-2} & \cdots & \zeta_{N-[c]-1}^{N-[c]-2} \\
\zeta_1^{N-[c]-1} & \zeta_2^{N-[c]-1} & \zeta_3^{N-[c]-1} & \cdots & \zeta_{N-[c]-1}^{N-[c]-1}
\end{bmatrix}
\cdot
\begin{bmatrix}
a_0 \\
a_1 \\
a_2 \\
\vdots \\
a_{N-[c]-1}
\end{bmatrix}
=
\begin{bmatrix}
\frac{-\lambda^N}{(1+\lambda)^N} \\
0 \\
0 \\
\vdots \\
0 \\
0
\end{bmatrix}
$$

and recognize the matrix, denoted by V, as a Vandermonde matrix (Section A.1, art. **5**) with

$$\det(V) = \prod_{i=0}^{N-[c]-1} \prod_{j=i+1}^{N-[c]-1} (\zeta_j - \zeta_i)$$

Since all eigenvalues appearing in the Vandermonde matrix are distinct (Appendix A.5.2.2) $\det(V) \ne 0$ and a unique solution follows for all $0 \le j \le N - [c] - 1$ from Cramer's theorem as

$$a_j = -\left(\frac{\lambda}{1+\lambda}\right)^N \prod_{i=0; i \ne j}^{N-[c]-1} \frac{\zeta_j}{\zeta_i - \zeta_j} \tag{14.55}$$

Together with the exact determination of the eigenvalues ζ_j and corresponding right-eigenvector x_j explicitly given in Appendix A.5.2.2, the coefficients a_j completely solve the AMS queue.

The buffer overflow probability $\Pr[N_S > x] = 1 - \sum_{j=0}^{N} \pi_j(x)$ becomes with $\sum_{j=0}^{N} \pi_j(\infty) = 1$,

$$\Pr[N_S > x] = - \sum_{j=0}^{N-[c]-1} a_j e^{\zeta_j x} \sum_{l=0}^{N} (x_j)_l$$

Using the explicit form of the generating function (A.44) where the roots r_1 and r_2 belonging to eigenvalue ζ_k are specified in (A.42) and the residue $k = k_j = c_1$ in (A.43), the buffer overflow probability is

$$\Pr[N_S > x] = - \sum_{j=0}^{N-[c]-1} a_j e^{\zeta_j x} (1 - r_1)^{k_j} (1 - r_2)^{N-k_j} \tag{14.56}$$

For large x, $\Pr[N_S > x]$ will be dominated by the exponential with the largest negative eigenvalue ζ_0 (for which $k_0 = N$),

$$\Pr[N_S > x] \sim -a_0 e^{\zeta_0 x} \sum_{j=0}^{N} (x_N)_j$$

Writing that largest negative eigenvalue (A.47) in terms of the traffic intensity ρ, gives

$$\zeta_0 = -\frac{(1+\lambda)(1-\rho)}{1-\frac{c}{N}}$$

From (A.49), we have $\sum_{j=0}^{N} (x_N)_j = \left(\frac{N}{c}\right)^N$. Combined with (14.55), the asymptotic formula for the buffer overflow probability becomes

$$\Pr[N_S > x] \sim \rho^N e^{\zeta_0 x} \prod_{i=1}^{N-[c]-1} \frac{\zeta_i}{\zeta_i - \zeta_0} \qquad (14.57)$$

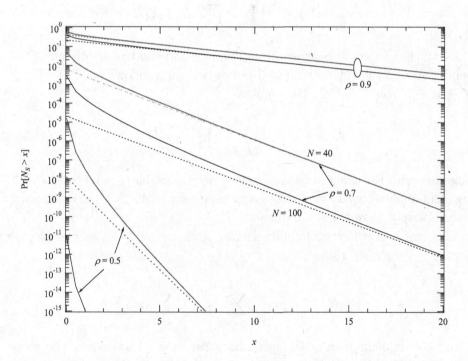

Fig. 14.7. The overflow probability (14.56) in the AMS queue versus the buffer level x for fixed $\lambda = \frac{1}{2}$. For each traffic intensity $\rho = 0.5$, 0.7 and 0.9, the upper curve corresponds to $N = 40$ and the lower to $N = 100$. The asymptotic formula (14.57) is shown in dotted line.

Figure 14.7 shows both the exact (14.56) and asymptotic (14.57) overflow probability as function of x for various traffic intensities ρ and two size of N. The average off-period in Fig. 14.7 is two times the average on-period. For large values of x and large traffic intensities ρ, the asymptotic formula is adequate. For smaller values, clear differences are observed. As mentioned in Section 5.7, the asymptotic regime that nearly coincides with (14.57) refers to the burst scale phenomena while the non-asymptotic regime reflects the smaller scale variations. The AMS queue allows us to analyze the effect of the burstiness of a source by varying λ.

14.8 The cell loss ratio

Due to its importance in ATM and in future time-critical communication services, the QoS loss-performance measure, the cell loss ratio, deserves some attention. In designing a switch for time-critical services with strict delay requirements smaller than D^*, the buffer size K is dimensioned as follows. The order of magnitude of D^* is about 10 ms, the maximum end-to-end delay for high-quality telephony (world wide) advised in ITU-T standards. Let $E[H]$ be the average number of hops of a path in an ATM network that rarely exceeds 10 hops. The buffer size K is determined such that the maximum waiting time of a cell never exceeds $\frac{D^*}{E[H]}$, thus, $\frac{K}{\mu} \le \frac{D^*}{10}$. For example, for STM-1 links where $C = 155$ Mb/s, we have that $\mu \simeq 366\,800$ ATM cell/s such that $K \le \frac{\mu D^*}{10} \simeq 366$ ATM cell buffer positions. This first-order estimate shows that the ATM buffer for time-critical traffic consists of a few hundreds of ATM cell positions. That small number for K indeed assures that the delay constraints are met, but introduces the probability to loose cells. Hence, the QoS parameter to be controlled for time-critical services is the cell loss ratio.

The cell loss ratio clr is defined as the ratio of the long-time average number of lost cells because of buffer overflow to the long-time average number of cells that arrive in steady-state. There are typically two different views to describe the cell loss ratio: a conservation-based and a combinatorial one. The conservation law simply states that cells entering the system also must leave it. The average number of entering cells are all those offered per time slot minus the ones that have been rejected, thus $(1 - clr)\lambda$. On the other side, the average number of cells that leave the system are related to the server activity as $(1 - q[0])\mu$, where μ is the service rate and $q[j] = \Pr[N_Q = j]$. Hence, we have

$$(1 - clr)\,\lambda = (1 - q[0])\,\mu \qquad (14.58)$$

In the combinatorial view, only the arrival process is viewed from a position in the buffer and the number of ways in which cells are lost are counted leading to

$$clr = \frac{1}{A'(1)} \sum_{n=0}^{\infty} n \sum_{j=0}^{K} q[K-j] a[j+n] \tag{14.59}$$

with $A'(1) = \lambda$ and $a[j] = \Pr[\mathcal{A} = j]$. Although equation (14.58) is simple, its practical use is limited since the quantities involved are to be known with extremely high accuracy if clr is of the order of 10^{-10}, which in practice means a virtually loss-free service. Therefore, we confine ourselves to the combinatorial result and express (14.59) in terms of a generating function as

$$clr\, A'(1) = \frac{dV(z)}{dz}\bigg|_{z=1} \tag{14.60}$$

where

$$V(z) = \sum_{n=0}^{\infty} z^n \sum_{j=0}^{K} q[K-j]a[j+n] = \sum_{n=0}^{\infty} z^n \sum_{j=0}^{K} q[K-j] z^{-j} a[j+n] z^j$$

$$= \sum_{j=0}^{K} q[K-j] z^{-j} \left(\sum_{n=0}^{\infty} a[j+n] z^{j+n} \right)$$

Rearranging in terms of the generating function for the arrivals $A(z)$ and for the buffer occupancy $Q(z) = \sum_{j=0}^{K} q[j] z^j$, where $q[j] = 0$ for $j > K$, yields

$$V(z) = \sum_{j=0}^{K} q[K-j] z^{-j} \left(A(z) - \sum_{n=0}^{j-1} a[n] z^n \right)$$

$$= A(z) z^{-K} \sum_{j=0}^{K} q[K-j] z^{K-j} - \sum_{j=0}^{K} q[K-j] z^{-j} \sum_{n=0}^{j-1} a[n] z^n$$

$$= z^{-K} A(z) Q(z) - z^{-K} \sum_{j=0}^{K} q[j] z^j \sum_{n=0}^{K-j-1} a[n] z^n \tag{14.61}$$

In order to express the cell loss ratio entirely in terms of the generating functions $A(z)$ and $Q(z)$, we employ (2.20),

$$\sum_{j=0}^{n} y[j] z^j = \sum_{j=0}^{n} z^j \left(\frac{1}{2\pi i} \oint_{C(0)} \frac{Y(\omega)}{\omega^{j+1}} d\omega \right) = \frac{1}{2\pi i} \oint_C \frac{Y(\omega)}{\omega - z} \left[1 - \left(\frac{z}{\omega}\right)^{n+1} \right] d\omega$$

$$= Y(z) - \frac{1}{2\pi i} \oint_C \frac{Y(\omega)}{\omega - z} \left(\frac{z}{\omega}\right)^{n+1} d\omega \tag{14.62}$$

where C is a contour enclosing the origin *and* the point z and lying within the convergence region of $Y(z)$. Combining (14.61) and (14.62), we rewrite $V(z)$ as

$$V(z) = z^{-K}A(z)Q(z) - z^{-K}Q(z)A(z) + \frac{1}{2\pi i}\int_C \frac{A(\omega)Q(\omega)}{(\omega - z)\,\omega^K}d\omega$$

$$= \frac{1}{2\pi i}\int_C \frac{A(\omega)Q(\omega)}{(\omega - z)\,\omega^K}d\omega$$

Finally, our expression for the cell loss ratio in a GI/G/1/K system reads

$$clr = \frac{1}{2\pi i A'(1)}\int_C \frac{A(\omega)Q(\omega)}{(\omega - 1)^2\,\omega^K}d\omega \tag{14.63}$$

where the contour C encloses both the origin and the point $z = 1$ and lies in the convergence region of $A(z)$. Usually, $A(z)$ is known while $Q(z)$ proves to be more complicated to obtain. The product $Q(z)A(z) = S(z)$ is the pgf of the system content.

If $Q(z)$ and $A(z)$ are meromorphic functions[7] and if

$$\lim_{z\to\infty}\left|\frac{A(z)\,Q(z)}{(z-1)^2\,z^{K-1}}\right| = 0,$$

the contour C in (14.63) can be closed over $|\omega| > 1$-plane to get

$$clr = -\frac{1}{A'(1)}\sum_p \text{Res}_{\omega\to p}\frac{A(\omega)Q(\omega)}{\omega^K\,(\omega - 1)^2} \tag{14.64}$$

where p are the poles of $A(z)Q(z)$ outside the unit circle. If these conditions are met, a non-trivial evaluation of the cell loss ratio can be obtained. In case the buffer pgf of the finite system is known, then $Q(z)$ is a polynomial of degree at most K so that the only pole of $\frac{Q(z)}{z^K}$ is zero and $\lim_{z\to\infty}\frac{Q(z)}{z^K} = q(K) \leq 1$ and the above conditions simplify to $\lim_{z\to\infty}\left|\frac{z\,A(z)}{(z-1)^2}\right| = 0$. Executing (14.63) then leads to

$$clr = -\frac{1}{A'(1)}\sum_p \frac{Q(p)}{p^K\,(p-1)^2}\,\text{Res}_{\omega\to p}A(\omega) \tag{14.65}$$

where only the poles p of the arrival process $A(z)$ play a role. For example, if the number of arrivals has a geometric distribution $a[k] = (1-\alpha)\alpha^k$ with $0 \leq \alpha \leq 1$ with generating function (3.6), $A_{\text{geo}}(z) = \frac{1-\alpha}{1-\alpha z}$, then the conditions for (14.65) are satisfied and we obtain,

$$clr_{\text{geo}} = \alpha^K Q\left(\frac{1}{\alpha}\right)$$

---

[7] Functions that only have poles in the complex plane.

An important class excluded from (14.64) consists of entire functions $A(z)$ that possess a Taylor series expansion converging for all complex variables z. The pgf of a Poisson process with parameter λ, $A_{\text{Poisson}}(z) = e^{\lambda(z-1)}$, is an important representative of that class. For a Poisson arrival process, (14.63) is

$$clr_{\text{Poisson}} = \frac{e^{-\lambda}}{2\pi i \lambda} \int_C \frac{e^{\lambda\omega} Q(\omega)}{(\omega - 1)^2 \, \omega^K} d\omega$$

Deforming the contour to enclose the negative half ω-plane $(\text{Re}(\omega) < c)$ yields

$$clr_{\text{Poisson}} = \frac{e^{-\lambda}}{2\pi i \lambda} \int_{c-i\infty}^{c+i\infty} \frac{e^{\lambda\omega} Q(\omega)}{(\omega - 1)^2 \, \omega^K} d\omega \cdot$$

where the real number c exceeds unity. This expression is recognized as an inverse Laplace transform and since the argument of the Laplace transform is a rational function, an exact evaluation is possible leading, however, again to (14.59). Hence, the combinatorial view does not offer much insight immediately which suggests to consider a conservation-based approach. Indeed, it is well known that, owing to the PASTA property (Theorem 13.5.1), an exact expression (Syski, 1986; Bisdikian *et al.*, 1992; Steyaert and Bruneel, 1994) in continuous-time for the cell loss ratio in a M/G/1/K system can be derived, with the result

$$clr_{\text{M/G/1/K;cont}} = (1 - \rho)\frac{\Pr[Q > K - 1]}{1 - \rho \Pr[Q > K - 1]} \tag{14.66}$$

where, as usual, the traffic intensity $\rho = \frac{\lambda}{\mu}$ and $\Pr[Q > K - 1]$ is the overflow probability in the corresponding infinite system M/G/1. Transforming (14.66) to discrete-time using $clr_{\text{discr}} = \frac{clr_{\text{cont}}}{\rho(1-clr_{\text{cont}})}$ yields

$$clr_{\text{M/G/1/K;discr}} = \frac{1 - \rho}{\rho} \frac{\Pr[Q > K - 1]}{1 - \Pr[Q > K - 1]} \tag{14.67}$$

14.9 Problems

(i) A router processes 80% of the time data packets. On average 3.2 packets are waiting for service. What is the mean waiting time of a packet given that the mean processing time equals $\frac{1}{\mu}$?

(ii) Compute in a M/M/m/m queue the average number of busy servers.

(iii) Let us model a router by a M/M/1 system with average service time equal to 0.5 s.

(a) What is the relation between the average response time (average system time) and the arrival rate λ?

(b) How many jobs/s can be processed for a given average response time of 2.5 s?

(c) What is the increase in average response time if the arrival rate increases by 10%?

(iv) Assume that company has a call center with two phone lines for service. During some measurements it was observed that both the lines are busy 10% of the time. On the other hand, the average call holding time was 10 minutes. Calculate the call blocking probability in the case that the average call holding time increases from 10 minutes to 15 minutes. Call arrivals are Poissonean with constant rate.

(v) Consider a queueing network with Poisson arrivals consisting of two infinitely long single-server queues in tandem with exponential service times. We assume that the service times of a customer at the first and second queue are mutually independent as well as independent of the arrival process. Let the rate of the Poisson arrival process be λ, and let the mean service rates at queues 1 and 2 be μ_1 and μ_2, respectively. Moreover, assume that $\lambda < \mu_1$ and $\lambda < \mu_2$. Give the probability that in steady-state there are n customers at queue 1 and m customers at queue 2.

(vi) Let us consider the following simple design question: which queue of the M/M/m family is most suitable if the arrival rate is λ and the required service rate is $k\mu$, with $k > 1$. We have the three options illustrated in Fig. 14.8 at our disposal. Since all queues have infinite

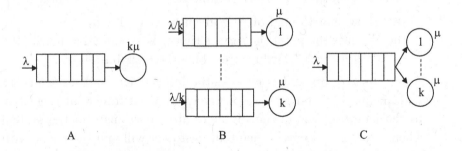

Fig. 14.8. Three different options: (A) one M/M/1 queue with service rate $k\mu$, (B) k M/M/1 queue with service rate μ and (C) one M/M/k queue with service rate $k\mu$.

buffers and the same traffic intensity $\rho = \frac{\lambda}{k\mu}$, and thus the same throughput. The QoS qualifier of interest here is the delay, more precisely, the system time T of a packet. Compare the average system time and draw conclusions.

(vii) An aeroplane takes exactly 5 minutes to land after the airport's traffic control has sent the signal to land. Aeroplanes arrive at random with an average rate of 6/hour. How long can an aeroplane expect to circle before getting the signal to land? (Only one aeroplane can land at a time)

(viii) There are two kinds of connection requests arriving at a base station of a mobile telephone network: connection requests generated by new calls (that originate from the same cell as the base station) or handovers (that originate from a different cell, but are transferred to the cell of the base station). The handovers are supposed not to experience blocking. Therefore, the base station has to reject some of the new call connection requests. Every accepted connection request occupies one of the M available channels. During a busy hour, the average measured channel occupation time of a call is 1.64 minutes irrespective of the type of call. Furthermore, the average number of active calls is 52 and the measured blocking is 2% of the number of all the connection requests. The average interarrival time between two consecutive new call connection requests in the cell is 3 seconds.

(a) Calculate the arrival rate (in calls/minute) for the handover calls.

(b) What is the percentage of new calls that are blocked?

(ix) Let N denote the number of Poisson arrivals with rate λ during the service time x (random variable) of a packet. Assume that the Laplace transform of the service time $\varphi_x(s) = E[e^{-sx}]$ is known.

(a) Show that the pgf of N is given by $\varphi_x(\lambda(1 - z))$.

(b) What is the pgf if the service time x is exponential distributed with mean $\frac{1}{\mu}$? Deduce from this the distribution of N.

(x) A single-server queue has exponential inter-arrival and service times with means λ^{-1} and μ^{-1}, respectively. New customers are sensitive to the length of the queue. If there are i customers in the system when a customer arrives, then that customer will join the queue with a probability $(i+1)^{-1}$, otherwise he/she departs and does not return. Find the steady-state probability distribution of this queuing system.

(xi) *The M/M/m/m/s queue (The Engset-formula)*. Consider a system with m connections and s customers who all desire to telephone and, hence, need to obtain a connection or line. Each customer can occupy at most one line. The group of s customers consists of two subgroups.

When a line has been assigned to a customer, this customer is transferred from the "still demanding subgroup" to the "served group". The number of call attempts decreases with the size of the "served group" whose members all occupy one line. More precisely, the arrival rate in the Engset model is proportional to the size of the "still demanding subgroup" and the number of arrivals is exponential. The holding time of a line is also exponentially distributed with mean $\frac{1}{\mu}$.

(a) Describe the M/M/m/m/s queue as a birth-death process.

(b) Compute the steady-state.

(c) Compute the blocking probability (similar to the blocking in the Erlang model).

(xii) Compare the cell loss ratio of the M/M/1/K and of the discrete M/1/D/K using the dominant pole approximation in Section 5.7. *Hint:* approximate the cell loss ratio by the overflow probability.

Part III

Physics of networks

15

General characteristics of graphs

The structure or interconnection pattern of a network can be represented by a graph. Properties of the graph of a network often relate to performance measures or specific characteristics of that network. For example, routing is an essential functionality in many networks. The computational complexity of shortest path routing depends on the hopcount in the underlying graph. This chapter mainly focuses on general properties of graphs that are of interest to Internet modeling.

Mainly driven by the Internet, a large impetus from different fields in science makes the understanding of the growth and the structure of graphs one of the currently most studied and exciting research areas. The recent books by Barabasi (2002) and Dorogovtsev and Mendes (2003) nicely reflect the current state of the art in stochastic graph theory and its applications to, for example, the Internet, the World Wide Web, and social and biological networks.

15.1 Introduction

Network topologies as drawn in Fig. 15.1 are examples of graphs. A graph G is a data structure consisting of a set of V vertices connected by a set of E edges. In stochastic graph theory and communications networking, the vertices and edges are called nodes and links, respectively. In order to differentiate between the expectation operator $E[.]$, the set of links is denoted by $\mathcal{L}$ and the number of links by L and similarly, the set of nodes by $\mathcal{N}$ and number of nodes by N. Thus, the usual notation of a graph $G(V, E)$ in graph theory is here denoted by $G(N, L)$.

The full mesh or complete graph K_N consists of N nodes and $L = L_{\max} = \frac{N(N-1)}{2}$ links, where every node has a link to every other node. The graph that is generated by the statement "any i is directly connected to any j" in

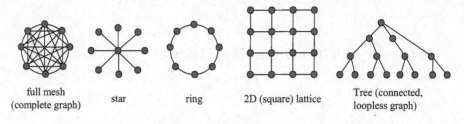

full mesh
(complete graph)   star   ring   2D (square) lattice   Tree (connected, loopless graph)

Fig. 15.1. Several types of network topologies or graphs.

a population of N members, is a complete graph K_N. Since in K_N the number of links $L_{\max} = O\left(N^2\right)$ for large N, it demonstrates "Metcalfe's law": the value of networking increases quadratically in the number of connected members.

The interconnection pattern of a network with N nodes can be represented by an *adjacency matrix* A consisting of elements a_{ij} that are either one or zero depending on whether there is a link between node i and j or not. The adjacency matrix is a real symmetric $N \times N$ matrix when we assume bi-directional transport over links. If there is a link from i to j ($a_{ij} = 1$) then there is a link from j to i ($a_{ji} = 1$) for any $j \neq i$. Moreover, we exclude self-loops ($a_{jj} = 0$) or multiple links between two nodes i and j. More properties of the adjacency matrix of a graph are found in Appendix B.

A *walk* from node A to node B with $k - 1$ hops or links is the node list $\mathcal{W}_{A \to B} = n_1 \to n_2 \to \cdots n_{k-1} \to n_k$ where $n_1 = A$ and $n_k = B$. A *path* from node A to node B with $k - 1$ hops or links is the node list $\mathcal{P}_{A \to B} = n_1 \to n_2 \to \cdots n_{k-1} \to n_k$ where $n_1 = A$ and $n_k = B$ and where $n_j \neq n_i$ for each index i and j. Sometimes the shorter notation $\mathcal{P}_{A \to B} = n_1 n_2 \cdots n_{k-1} n_k$ is used. All links $n_i \to n_j$ and the nodes n_j in the path $\mathcal{P}_{A \to B}$ are different, whereas in a walk $\mathcal{W}_{A \to B}$ no restrictions on the node list is put. If the starting node A equals the destination node B, that path $\mathcal{P}_{A \to A}$ is called a *cycle* or *loop*. In telecommunications networks, paths and not walks are basic entities in connecting two communicating parties. Two paths between A and B are node(link)-disjoint if they have no nodes(links) in common.

Apart from the topological structure specified via the adjacency matrix A, the link between node i and j is further characterized by a link weight $w(i \to j)$, most often a positive real number[1] that reflects the

---

[1] In quality of service routing, a link is specified by a vector $\vec{w}(i \to j)$ with positive components, each reflecting a metric (such as delay, jitter, loss, monetary cost, administrative weight, physical distance, available capacity, priority, etc.).

importance of that particular link. Often symmetry in both directions, $w(i \to j) = w(j \to i)$, is assumed leading to undirected graphs. Although this assumption seems rather trivial, we point out that in telecommunications, transport of information in up-link and down-link is, in general, not symmetrical. Via measurements in the Internet, Paxson (1997) found in 1995 that about 50% of the paths from $A \to B$ were different from those from $B \to A$. Furthermore, it is often assumed that the link metric $w(i \to j)$ is independent from $w(k \to l)$ for all links $(i \to j)$ different from $(k \to l)$. In the Internet's intra-domain routing protocol, the Open Shortest Path First (OSPF) protocol, network operators have the freedom[2] to specify the link weight $w(i \to j) > 0$ on the interfaces of their routers.

15.2 The number of paths with j hops

Let $X_j(A \to B; N)$ denote the number of paths with j hops between a source node A and a destination node B. The most general expression for the number of paths with j hops between node A and node B is

$$X_j(A \to B; N) = \sum_{k_1 \notin \{A,B\}} \sum_{k_2 \notin \{A,k_1,B\}} \cdots \sum_{k_{j-1} \notin \{A,k_1,\cdots,k_{j-2},B\}} 1_{A \to k_1} \cdot 1_{k_1 \to k_2} \cdots 1_{k_{j-1} \to B}$$

(15.1)

where 1_x is the indicator function. The number of paths with one hop equals $X_1(A \to B; N) = 1_{A \to B}$. The maximum number of j hop paths is attained in the complete graph K_N where $1_{k_1 \to k_2} = 1$ for each link $k_1 \to k_2$ and equals

$$\max(X_j(A \to B; N)) = \frac{(N-2)!}{(N-j-1)!}$$

(15.2)

The maximum number of hops in any path is $N-1$. This maximum occurs, for example, in a line graph where the path runs from the one extreme node to the other or in a ring (see Fig. 15.1) between neighboring nodes where there is a one hop and a $(N-1)$-hops path.

The total number of paths M_N between two nodes in the complete graph is

$$M_N = \sum_{j=1}^{N-1} \max(X_j(A \to B; N)) = \sum_{j=1}^{N-1} \frac{(N-2)!}{(N-j-1)!} = (N-2)! \sum_{k=0}^{N-2} \frac{1}{k!}$$

$$= (N-2)!e - R$$

---

[2] In Cisco's OSPF implementation, it is suggested to use $w(i \to j) = \frac{10^8}{B(i \to j)}$ where $B(i \to j)$ denotes the capacity (in bit/s) of the link between nodes i and j. An approach to optimize the OSPF weights to reflect actual traffic loads is presented by Fortz and Thorup (2000).

where

$$R = (N-2)! \sum_{j=N-1}^{\infty} \frac{1}{j!} = \sum_{j=0}^{\infty} \frac{(N-2)!}{(N-1+j)!}$$

$$= \frac{1}{N-1} + \frac{1}{(N-1)N} + \frac{1}{(N-1)N(N+1)} + \cdots$$

$$< \sum_{j=1}^{\infty} \left(\frac{1}{N-1} \right)^j = \frac{1}{N-2}$$

implying that for $N \geq 3$, $R < 1$. But M_N is an integer. Hence, the total number of paths in K_N is exactly equal to

$$M_N = [e(N-2)!] \qquad (15.3)$$

where $e = 2.718\,281...$ and $[x]$ denotes the largest integer smaller than or equal to x. Since any graph is a subgraph of the complete graph, the maximum total number of paths between two nodes in any graph is upper bounded by $[e(N-2)!]$.

15.3 The degree of a node in a graph

The degree d_j of a node j in a graph $G(N, L)$ equals the number of its neighboring nodes and $0 \leq d_j \leq N-1$. Clearly, the node j is disconnected from the rest of the graph if $d_j = 0$. Hence, in connected graphs, $1 \leq d_j \leq N-1$. The basic law for the degree (see also Appendix (B.2)) is

$$\sum_{j=1}^{N} d_j = 2L,$$

since each link belongs to precisely two nodes and, hence, is counted twice. In directed graphs, the in(out)-degree is defined as the number of the in(out)-going links at a node, while the sum of in- and out-degree equals the degree. The minimum nodal degree in the graph G is denoted by $d_{\min} = \min_{j \in G} d_j$. The average degree of a graph is defined as $d_a = \frac{1}{N} \sum_{j=1}^{N} d_j = \frac{2L}{N}$ which is, for a connected graph, bounded by $2 - \frac{2}{N} \leq d_a \leq N-1$. The lower bound is obtained for any spanning tree, a graph that connects all nodes and that contains no cycles and where $L = L_{\min} = N-1$. The upper bound is reached in the complete graph K_N with $L_{\max} = \frac{N(N-1)}{2}$. Graphs where $d_{\min} = d_a$ such as K_N and the ring topology in Fig. 15.1 are called regular graphs since any node has precisely d_a links.

Sometimes networks are classified either as dense if d_a is high or as sparse

Fig. 15.2. Degree graph with $\tau = 2.4$ and $N = 300$. All nodes are drawn on a circle.

if d_a is small. For instance, the Internet is sparse with average degree $d_a \approx$ 3, although some backbone routers may have a much higher degree, even exceeding 100. The distribution of the degree D_{Internet} of an arbitrary node in the Internet is shown to be approximately polynomial (Siganos *et al.*, 2003),

$$\Pr\left[D_{\text{Internet}} = k\right] \approx \frac{k^{-\tau}}{\zeta\left(\tau\right)} \qquad (15.4)$$

with[3] $\tau \in (2.2, 2.5)$ and $\zeta\left(s\right) = \sum_{k=1}^{\infty} k^{-s}$ for $\text{Re}(s) > 1$ is the Riemann Zeta function (Titchmarsh and Heath-Brown, 1986). A graph of this class is called a degree graph. Figures 15.2 and 15.3 show two instances of a degree graph.

Also the web graph consisting of websites and hyperlinks features a power law for the in-degree. David Aldous has given the following argument why a power law of the in-degree of the web graph is natural. To a good approximation, the number of websites is growing exponentially at rate $\rho > 0$. This means that the lifetime T of a random website satisfies $\Pr\left[T > t\right] \approx e^{-\rho t}$.

---

[3] A more general expression than (15.4) is $\Pr\left[d_j = k\right] = ck^{\alpha}g(k)$, where c is a normalization constant and where $g(k)$ is a slowly varying function (Feller, 1971, pp. 275-284) with basic property that $\lim_{t \to \infty} \frac{g(tx)}{g(t)} = 1$, for every $x > 0$.

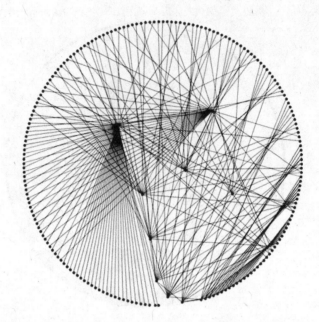

Fig. 15.3. Degree graph with $\tau = 2.4$ and $N = 200$. The higher degree nodes are put inside the circle.

Let $l(u)$ denote the number of links into a site at time u after its creation. At observation time t, the distribution of the number of links X into a random website is, by the law of total probability,

$$\Pr[X > k] = \int_0^t \Pr[X > k | T = u] \frac{d\Pr[T \leq u]}{du} du$$

$$\approx \rho \int_0^t e^{-\rho u} \Pr[X > k | T = u] du$$

$$\approx \rho \int_0^t e^{-\rho u} 1_{\{l(u) > k\}} du = \rho \int_{l^{-1}(k)}^t e^{-\rho u} du = -e^{-\rho t} + e^{-\rho l^{-1}(k)}$$

Only if l increases exponentially fast as $l(u) \sim e^{\beta u}$ for some $\beta < \rho$, a power law behavior of the in-degree

$$\Pr[X > k] \approx k^{-\frac{\rho}{\beta}}$$

arises for sufficiently large t. For a polynomial growth $l(u) \sim u^\beta$ and large t,

$$\Pr[X > k] \approx e^{-\rho k^{\frac{1}{\beta}}}$$

The large difference in the decrease of $\Pr[X > k]$ with k between both examples illustrates the importance of the growth law of $l(u)$. The argument shows that a polynomial scaling law, commonly referred to as a power law, is a natural consequence of exponential growth. An exponential growth possesses the property that $\frac{dl(u)}{du} = \beta l(u)$ which is established by preferential attachment. Preferential attachment means that new links are on average added to sites proportional to their size. The more links a site has, the larger the probability that a new link attaches to this site. For example, already popular websites are increasingly more often linked to than small or less popular websites. Since many aspects of the Internet, such as the number of IP packets, number of users, number of websites, number of routers, etc., are currently growing approximately exponentially fast, the often observed power laws are more or less expected.

15.4 Connectivity and robustness

A graph G is connected if there is a path between each pair of nodes and disconnected otherwise. A telecommunication network should be connected. Moreover, it is essential that the network should be robust: it should still operate if some of the links between routers or switches are broken or temporarily blocked by other calls. Hence, the network graph should possess a redundancy of links. The minimum number of links to connect all nodes in the network equals $N-1$. This minimum configuration is called redundancy level 1. In general, a redundancy level of D is defined by Baran (2002) as the link-to-node ratio in an infinite D-lattice[4]. A redundancy level of at least 3 is regarded as a highly robust network. A consequence of this insight has been employed in the design of the early Internet (Arpanet): it would be theoretically possible to build extremely reliable communication networks out of unreliable links by the proper use of redundancy. Another more timely application of the same principle is the design of reliable ad-hoc and sensor networks.

---

[4] A D-lattice is a graph where each nodal position corresponds to a point with integer coordinates within a D dimensional hyper-cube with size Z. Apart from the border nodes, each node has a same degree equal to $2D$. The number of nodes equals $N = Z^D$. From (B.2), the link-to-node ratio follows as

$$\frac{L}{N} = \frac{1}{2N} \sum_{j=1}^{N} d_j = D - r$$

where the correction $r = O\left(N^{\frac{1}{D}-1}\right)$ is due to the border nodes. For an infinite D-lattice, where the limit $Z \to \infty$ (which implies $N \to \infty$), we obtain

$$\lim_{Z \to \infty} \frac{L}{N} = D$$

There exist interesting results from graph theory that help to dimension a reliable telecommunication network. Instead of the redundancy level, the edge and vertex connectivity seem more natural quantifiers from which robustness can be derived. The edge connectivity $\lambda(G)$ of a connected graph G is the smallest number of edges (links) whose removal disconnects G. The vertex connectivity $\kappa(G)$ of a connected graph different[5] from the complete graph K_N is the smallest number of vertices (nodes) whose removal disconnects G.

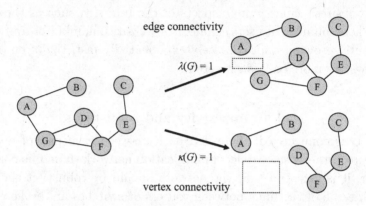

Fig. 15.4. An example of the edge and the vertex connectivity of a graph.

These definitions are illustrated in Fig. 15.4. For any connected graph G holds that

$$\kappa(G) \leq \lambda(G) \leq d_{\min}(G) \tag{15.5}$$

In particular, if G is the complete graph K_N, then $\kappa(K_N) = \lambda(K_N) = d_{\min}(K_N) = N - 1$. Due to the importance of the inequality[6] (15.5), it deserves some more discussion. Let us concentrate on a connected graph G that is not a complete graph. Since $d_{\min}(G)$ is the minimum degree of a node, say n, in G, by removing all links of node n, G is disconnected. By definition, since $\lambda(G)$ is the minimum number of links that leads to disconnectivity, it follows that $\lambda(G) \leq \delta(G)$ and $\lambda(G) \leq N - 2$ because G is not a complete graph and consequently the minimum nodal degree is at most $N - 2$. Furthermore, the definition of $\lambda(G)$ implies that there exists a set S of $\lambda(G)$ links whose removal splits the graph G into two connected subgraphs G_1 and G_2, as illustrated in Fig. 15.5. Any link of that set S

---

[5] The complete graph K_N cannot be disconnected by removing nodes and we define $\kappa(K_N) = N - 1$ for $N \geq 3$.

[6] A second general inequality (B.23) relates the second smallest eigenvalue of the Laplacian to the edge and vertex connectivity (see Section B.4).

connects a node in G_1 to a node in G_2. Indeed, adding again an arbitrary link of that set makes G again connected. But G can be disconnected into the same two connected subgraphs by removing nodes in G_1 and/or G_2. Since possible disconnectivity inside either G_1 or G_2 can occur before $\lambda(G)$ nodes are removed, it follows that $\kappa(G)$ cannot exceed $\lambda(G)$, which establishes the inequality (15.5).

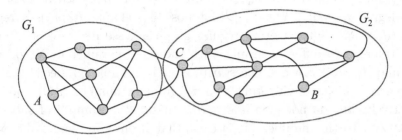

Fig. 15.5. A graph G with $N = 16$ nodes and $L = 32$ links. Two connected subgraphs G_1 and G_2 are shown. The graph's connectivity parameters are $\kappa(G) = 1$ (removal of node C), $\lambda(G) = 2$ (removal of links from C to G_1), $d_{\min}(G) = 3$ and $d_a = \frac{2L}{N} = 4$.

Let us proceed to find the number of link-disjoint paths between A and B in a connected graph G. Suppose that H is *a* set of links whose removal separates A from B. Thus, the removal of all links in the set H destroys all paths from A to B. The maximum number of link-disjoint paths between A and B cannot exceed the number of links in H. However, this property holds for any set H, and thus also for the set with the smallest possible number of links. A similar argument applies to node-disjoint paths. Hence, we end up with Theorem 15.4.1:

Theorem 15.4.1 (Menger's Theorem) *The maximum number of link-(node)-disjoint paths between A and B is equal to the minimum number of links (nodes) separating A and B.*

Recall that the edge connectivity $\lambda(G)$ (analogously vertex connectivity $\kappa(G)$) is the minimum number of links (nodes) whose removal disconnects G. By Menger's Theorem, it follows that there are at least $\lambda(G)$ link-disjoint paths and at least $\kappa(G)$ node-disjoint paths between any pair of nodes in G.

In order to dimension the graph G of a robust telecommunications network, the goal is to maximize both $\kappa(G)$ and $\lambda(G)$. Of course, the most reliable graph is the complete graph; however, it is also the most expensive. Usually, since the cost of digging and of installing/connecting the fibres is around 70% of the total network cost, the total number of links L

is minimized. Since the minimum cannot exceed the average, we have that $d_{\min} \leq d_a = \frac{2L}{N}$. From (15.5), it follows that the best possible reliability is achieved if the network graph is designed such that

$$\kappa(G) = \frac{2L}{N}$$

The optimum implies that $d_{\min}(G) = d_a = \frac{2L}{N}$ or that each node has the same degree $d_j = d_a$. Hence, a best possible reliable graph is a regular graph $(d_j = d_a)$, but not every regular graph necessarily obeys $\kappa(G) = d_a$. Furthermore, two different graphs with the same parameters N, L, $\kappa(G)$, $\lambda(G)$ and $d_{\min}(G)$ are not necessarily equally reliable. Indeed, the edge and vertex disconnectivity only give a minimum number $\lambda(G)$ and $\kappa(G)$ respectively, but do not give information about the number of subsets in G that lead to this number. It is clear that if only one vulnerable set of nodes is responsible for a low $\kappa(G)$, while in another graph there are more such sets, that the first graph is more reliable than the second one. In summary, the presented simplified analysis gives some insights, but more details (e.g. the number of vulnerable sets or subgraphs) must be considered in the dimensioning study.

15.5 Graph metrics

An important challenge in the modeling of a network is to determine the class of graphs that represents best the global and local structure of the network. Most of the valuable networks like the Internet, road infrastructures, neural networks in the human brain, social networks, etc. are large and changing over time. In order to classify graphs a set of distinguishing properties, called metrics, needs to be chosen. These metrics are in general function of the graph's structure $G(N, L)$. Natural metrics of a graph are the degree distribution and the hopcount distribution of an arbitrary path. Beside quantities such as the diameter and the complexity of a graph defined in algebraic graph theory (Appendix B), some other metrics are the clustering coefficient, the expansion, the resilience, the distortion and the betweenness.

The clustering coefficient $c_G(v)$ characterizes the density of connections in the environment of a node v and is defined as the ratio of the number of links y connecting the d_v neighbors of v over the total possible $\frac{d_v(d_v-1)}{2}$,

$$c_G(v) = \frac{2y}{d_v(d_v - 1)} \tag{15.6}$$

The expansion $e_G(h)$ of a graph reflects the number of nodes that can be reached in h hops from a node v,

$$e_G(h) = \frac{1}{N^2} \sum_{v \in \mathcal{N}} |C_v(h)| \tag{15.7}$$

where $C_v(h)$ is the set of nodes that can be reached in h hops from a node v and $|A|$ represents the number of elements in the set A. We can interpret $C_v(h)$ geometrically as a ball centered at node v with radius h.

The resilience $r_G(m)$ measures the connectivity or robustness of a graph. Let $m = |C_v(h)|$ denote the number of nodes in a ball centered at node v and with radius h, and define $l(v, m)$ as the number of links that needs to be removed to split $C_v(h)$ into two sets with roughly equal numbers of nodes (around $m/2$). The resilience $r_G(m)$ of a graph is

$$r_G(m) = \frac{1}{L} \sum_{v \in \mathcal{N}} l(v, m) \tag{15.8}$$

The distortion $t_G(m)$ measures how closely the graph resembles a tree and is defined as

$$t_G(m) = \frac{1}{N} \sum_{v \in \mathcal{N}} w\left(C_v(h)\right) \tag{15.9}$$

where $w(G)$ is the value of the minimum spanning tree in G with unit link weight $w(i \rightarrow j) = 1$ for each link of G.

Consider a flow with a unit amount of traffic between each pair of nodes in the graph G. Each flow between a node pair follows the shortest path between that node pair. The betweenness B of a link (node) is defined as the number of shortest paths between all possible pairs of nodes in G that traverse the link (node). If $H_{i \rightarrow j}$ denotes the number of hops in the shortest path from $i \rightarrow j$, then the total number of hops H_G in all shortest paths in G is $H_G = \sum_{i=1}^{N} \sum_{j=i+1}^{N} H_{i \rightarrow j}$. This number is also equal to $H_G = \sum_{l=1}^{L} B_l$, where B_l is the betweenness of a link l in G. Taking the expectation of both relations gives the average betweenness of a link in terms of the average hopcount

$$E[B] = \frac{\binom{N}{2}}{L} E[H_N] \geq E[H_N]$$

with equality only for the complete graph.

15.6 Random graphs

Besides the regular topologies in Fig. 15.1, the class of random graphs constitutes an attractive set of topologies to analyze network performance. The theory of random graphs originated from a series of papers by Erdös and Rényi in the late 1940s. There exists an astonishingly large amount of literature on random graphs. The standard work on random graphs is the book by Bollobas (2001). We also mention the work of Janson *et al.* (1993) on evolutionary processes in random graphs.

The two most frequently occurring models for random graphs are the Erdös-Rényi random graphs $G_p(N)$ and $G_r(N, L)$. The class of random graphs denoted by $G_p(N)$ consists of all graphs with N nodes in which the links are chosen independently and with probability p. In the class $G_p(N)$ the total number of links is not deterministic, but on average equal to $E[L] = p L_{\max}$ where $L_{\max} = \binom{N}{2}$. Since $p = \frac{E[L]}{L_{\max}}$ we also call p the link density of $G_p(N)$. An instance of the class $G_{0.013}(300)$ is drawn in Fig. 15.6. Related to $G_p(N)$ are the *geometric* random graphs $G_{\{p_{ij}\}}(N)$ where the links are still chosen independently but where the probability of $i \rightarrow j$ being an edge is p_{ij}. An example of $G_{\{p_{ij}\}}(N)$ is the Waxman graph (Waxman, 1998; Van Mieghem, 2001) with $p_{ij} = \exp(-a|\vec{r}_i - \vec{r}_j|)$ and where the vector $\vec{r}_i$ represents the position of node i and a a real, non-negative number. Geometric random graphs are good models for ad-hoc wireless networks where the probability $p_{ij} = f(r_{ij})$ that there is a wireless link between node i and j is specified by the radio propagation that is briefly explained at the end of Section 3.5.

The class $G_r(N, L)$ is the set of random graphs with N nodes and L links. In total, we can construct $\binom{L_{\max}}{L}$ different graphs, which corresponds to the number of ways we can distribute a set of L ones in the $L_{\max}$ possible places in the upper triangular part above the diagonal of the adjacency matrix A. Each of the possible $L_{\max}$ links has equal probability to belong to a random graph of the class $G_r(N, L)$. The probability that an element in the adjacency matrix A is $a_{ij} = 1$ equals $p = \frac{L}{L_{\max}}$. As opposed to the class $G_p(N)$, the number of non-zero elements in A in each random graph of $G_r(N, L)$ is precisely $2L$ (see Appendix B.1, art. **2**), which induces weak dependence between links in $G_r(N, L)$. The latter also explains why more computations are easier in $G_p(N)$ than in $G_r(N, L)$.

The average number of paths with j hops between two arbitrary nodes in $G_p(N)$ follows from (15.1) and (2.13) as

$$E[X_j] = \frac{(N-2)!}{(N-j-1)!} p^j \tag{15.10}$$

for $1 \leq j \leq N-1$. The average total number of paths between two arbitrary

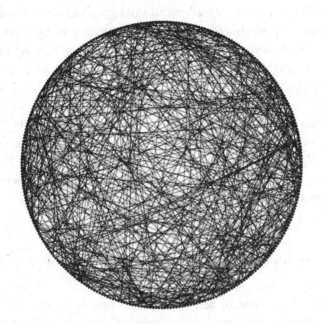

Fig. 15.6. A connected random graph $G_p(N)$ with $N = 300$ and $p = 0.013$ drawn on a circle.

nodes A and B equals

$$E\left[\sum_{j=1}^{N-1} X_j\right] = (N-2)!\,p^{N-1} \sum_{l=0}^{N-2} \frac{p^{-l}}{l!} \le (N-2)!\,p^{N-1} e^{\frac{1}{p}}$$

where the latter bound is closely approached for large N. Moreover, when the random graph reduces to the complete graph ($p = 1$), we again obtain (15.3). Since the degree d_j of a node j is the number of links incident with that node, it follows directly from the definition of $G_p(N)$ that the probability density function of the degree D_{rg} of an arbitrary node in $G_p(N)$ equals

$$\Pr\left[D_{\mathrm{rg}} = k\right] = \binom{N-1}{k} p^k (1-p)^{N-1-k} \qquad (15.11)$$

The interest in random graphs is fueled by the fact that the topology of the Internet is inaccurately known and also that good models[7] are lacking. In some sense, the Internet can be regarded as a growing and changing organism. Such complex networks also arise in other fields. Increased interest

---

[7] A detailed discussion on difficulties in modeling or simulating the Internet is presented by Floyd and Paxson (2001).

from different disciplines to understand network behavior resulted in a new
wave of science, which may be termed "the physics of networks" and which
was recently reviewed by Strogatz (2001). Random graphs are an elegant
vehicle to thoroughly analyze the performance of, for example, routing algorithms. Some constructed overlay networks such as Gnutella and mobile
wireless ad-hoc networks seem reasonably well modeled by $G_p(N)$. However, the class $G_p(N)$ does not describe the Internet topology well, and the
degree distribution especially deviates significantly. The degree distribution
(15.11) in $G_p(N)$ is binomially distributed, while that of the Internet is
close to a power law (15.4). Hence, there is a discrepancy between Internet
measurements and properties of the random graph $G_p(N)$.

15.6.1 The number $C(N, L)$ of connected random graphs in the class $G_r(N, L)$

From the point of view of telecommunications networks, by far the most
interesting graphs are those with connected topology. This limitation restricts the value of the link density p from below by a critical threshold p_c.
For large N, the critical threshold is $p_c \sim \frac{\log N}{N}$, as shown in Section 15.6.3.
In the theory of random graphs, the problem to determine the number of
connected random graphs $C(N, L)$ in the class $G_r(N, L)$ has been inten-
sively studied. Gilbert (1956) has presented an exact recursion formula for
$C(N, L)$ via the technique of enumeration. Erdös and Rényi (1959, 1960)
have determined the asymptotic behavior of random graphs via the probabilistic method, largely introduced by Erdös himself. Since the analysis[8] of
Gilbert is both exact and simple, we will review his results here and those
of Erdös and Rényi in the next section.

Consider a particular random graph J of the class of random graphs
$G_r(N + 1, L)$, which is constructed from the class $G_r(N, L)$ by adding one
node labelled $N + 1$. Suppose that the node labelled $N + 1$ in the random
graph J belongs to a connected component K that possesses v other nodes
and some number μ of links. The remaining part of J has $N - v$ nodes and
$L - \mu$ links. There are $\binom{N}{v}$ ways in which the v nodes of K can be chosen
out of the N nodes in $G_r(N, L)$. On the other hand, there are $C(v + 1, \mu)$
ways of picking a connected random graph K while there are $\binom{\binom{N-v}{2}}{L-\mu}$ ways
of constructing the remaining part of J. Hence, since the number of ways
we can construct a graph J equals $\binom{\binom{N+1}{2}}{L}$, we obtain Gilbert's recursion

---

[8] A different, less admissible approach is found in Goulden and Jackson (1983).

formula,

$$\binom{\binom{N+1}{2}}{L} = \sum_{v=0}^{N} \binom{N}{v} \sum_{\mu=v}^{\binom{v+1}{2}} C(v+1,\mu) \binom{\binom{N-v}{2}}{L-\mu} \qquad (15.12)$$

Gilbert (1956) further derives the generating function for $C(N,L)$ as

$$\sum_{N=1}^{\infty} \sum_{L=1}^{\infty} \frac{C(N,L)}{N!} x^N y^L = \log\left(1 + \sum_{k=1}^{\infty} \frac{(1+y)^{\binom{k}{2}} x^k}{k!}\right) \qquad (15.13)$$

which converges for $-2 \le y \le 0$ and all x. So far, no other explicit formulae for $C(N,L)$ exist.

In 1889, Caley (see Appendix B.1, art. **3**) proved that, in the special case where $L = N - 1$, there holds $C(N, N-1) = N^{N-2}$. In the other extreme where $L = L_{\max}$ corresponding with a full mesh, we have $C(N, L_{\max}) = 1$. Actually, when $\binom{N}{2} - N - 1 < L \le L_{\max}$, the graph is always connected because the adjacency matrix A has necessarily at least one non-zero element per row. This means that $C(N,L) = \binom{\binom{N}{2}}{L}$ for $\binom{N}{2} - N - 1 < L \le L_{\max}$. In all cases where $L < N - 1$, the random graphs are necessarily Disconnected, leading to $C(N,L) = 0$.

For computational purposes, we rewrite (15.12) as

$$\binom{\binom{N+1}{2}}{L} = \sum_{\mu=N}^{\binom{N+1}{2}} C(N+1,\mu) \binom{\binom{0}{2}}{L-\mu} + \sum_{v=0}^{N-1} \binom{N}{v} \sum_{\mu=v}^{\binom{v+1}{2}} C(v+1,\mu) \binom{\binom{N-v}{2}}{L-\mu}$$

Since $\binom{0}{L-\mu} = \delta_{\mu,L}$, we arrive after a substitution of $N \to N-1$ at the recursion formula,

$$C(N,L) = \binom{\binom{N}{2}}{L} - \sum_{v=0}^{N-2} \binom{N-1}{v} \sum_{\mu=v}^{\binom{v+1}{2}} C(v+1,\mu) \binom{\binom{N-1-v}{2}}{L-\mu} \qquad (15.14)$$

Below we list a few values:

$C(2,1) = 1$			
$C(3,2) = 3$	$C(3,3) = 1$		
$C(4,3) = 16$	$C(4,4) = 15$	$C(4,5) = 6$	$C(4,6) = 1$
$C(5,4) = 125$	$C(5,5) = 222$	$C(5,6) = 205$	$C(5,7) = 120$
$C(5,8) = 45$	$C(5,9) = 10$	$C(5,10) = 1$	
$C(6,5) = 1296$	$C(6,6) = 3660$	$C(6,7) = 5700$	$C(6,8) = 6165$
$C(6,9) = 4945$	$C(6,10) = 2997$	$C(6,11) = 1365$	$C(6,12) = 455$
$C(6,13) = 105$	$C(6,14) = 15$	$C(6,15) = 1$	
$C(7,6) = 16807$	$C(7,7) = 68295$	$C(7,8) = 156555$	$C(7,9) = 258125$
$C(7,10) = 331506$	$C(7,11) = 343140$	$C(7,12) = 290745$	$C(7,13) = 202755$
$C(7,14) = 116175$	$C(7,15) = 54257$	$C(7,16) = 20349$	$C(7,17) = 5985$
$C(7,18) = 1330$	$C(7,19) = 210$	$C(7,20) = 21$	$C(7,21) = 1$

15.6.2 The Erdös and Rényi asymptotic analysis

In a classical paper, Erdös and Rényi (1959) proved that

$$\lim_{N \to \infty} \Pr\left[G_r\left(N, \left[\frac{1}{2} N \log N + xN \right] \right) \text{ is connected} \right] = e^{-e^{-2x}} \quad (15.15)$$

Ignoring the integral part [.] operator and eliminating x using the number of links $L = \frac{1}{2} N \log N + xN$ gives, for large N,

$$\Pr[G_r(N, L) = \text{connected}] \sim e^{-N e^{-\frac{2L}{N}}} \quad (15.16)$$

which should be compared with the exact result,

$$\Pr[G_r(N, L) = \text{connected}] = \frac{C(N, L)}{\binom{N}{2}} \quad (15.17)$$

In contrast to the unattractive computation of the exact $C(N, L)$ via recursion (15.14), the Erdös and Rényi asymptotic expression (15.16) is simple. The accuracy for relatively small N is shown in Fig. 15.7.

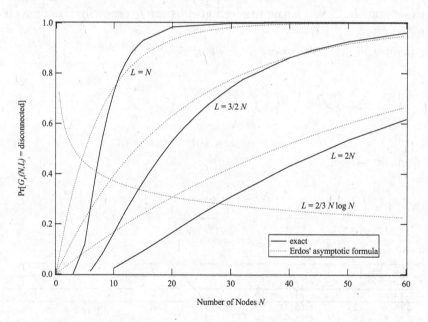

Fig. 15.7. The probability that a random graph $G_r(N, L)$ is *disconnected*: a comparison between the exact result (15.17) and Erdos' asymptotic formula (15.16) for $L = N$, $L = \frac{3}{2}N$, $L = 2N$ and $L = \frac{2}{3}N \log N$.

The key observation of Erdös and Rényi (1959) is that a phase transition in random graphs with N nodes occurs when the number of links L is around

$L_c = \frac{1}{2}N \log N$. Phase transitions are well-known phenomena in physics. For example, at a certain temperature, most materials possess a solid-liquid transition and at a higher temperature a second liquid-gas transition. Below that critical temperature, most properties of the material are completely different than above that temperature. Some materials are superconductive below a certain critical temperature T_c, but normally conductive (or even isolating) above T_c. Erdös and Rényi concentrated on the property A_k that a random graph $G_r(N, L)$ with $L_x = \left[\frac{1}{2}N \log N + xN\right]$ consists of $N - k$ connected nodes and k isolated nodes for fixed k. If A_k^c means the absence of property A_k, they proved that, for all fixed k, $\Pr[A_k^c] \to 0$ if $N \to \infty$ which means that for a large number of nodes N, almost all random graphs $G_r(N, L_x)$ possess property A_k. This result is equivalent to a result proved in Section 15.6.3 that the class of random graphs $G_p(N)$ is almost surely disconnected if the link density p is below $p_c \sim \frac{\log N}{N}$ and connected for $p > p_c$. In view of the analogy with physics, it is not surprising that corresponding sharp transitions also are observed for other properties than just A_k.

In the sequel, we will show that, for the random graph $G_r(N, L_x)$, the probability that the largest connected component, called the giant component $GC(N, L_x)$, has $N - k$ nodes is, for large N, Poisson distributed with · mean e^{-2x},

$$\lim_{N \to \infty} \Pr[\text{number of nodes in } GC(N, L_x) = N - k] = \frac{(e^{-2x})^k e^{-e^{-2x}}}{k!}$$
$$(15.18)$$

If $k = 0$, then all nodes belong to the giant component and the graph is completely connected in which case (15.18) leads to (15.16).

The total number of graphs $G_r(N, L_x)$ with $k \geq 1$ isolated nodes equals $\binom{N}{k}\binom{\binom{N-k}{2}}{L_x}$, the number of ways in which k isolated nodes can be chosen out of the total of N nodes multiplied by the number of graphs that can be constructed with $N - k$ nodes and L_x links. Observe that this total number also includes those graphs where not all the $N - k$ nodes are necessarily connected. In other words, this total number includes the graphs that do not possess property A_k. The total number of graphs T_0 without isolated node follows from the inclusion-exclusion formula (2.10) as

$$T_0(N, L_x) = \sum_{k=0}^{N} (-1)^k \binom{N}{k}\binom{\binom{N-k}{2}}{L_x}$$

where the index $k = 0$ equals the total number of graphs with N nodes and L_x links, i.e. the total number of elements in the sample space. Evidently, the total number $C(N, L_x)$ of connected random graphs of the class $G_r(N, L_x)$ is smaller than $T_0(N, L_x)$ because all of them must obey property A_0 as well.

Since[9]

$$\lim_{N \to \infty} \frac{\binom{N}{k}\binom{\binom{N-k}{2}}{L_x}}{\binom{\binom{N}{2}}{L_x}} = \frac{(e^{-2x})^k}{k!}$$

we obtain

$$\lim_{N \to \infty} \frac{T_0(N, L_x)}{\binom{\binom{N}{2}}{L_x}} = \sum_{k=0}^{\infty} (-1)^k \frac{(e^{-2x})^k}{k!} = e^{-e^{-2x}}$$

But, if $N \to \infty$, the difference

$$\frac{T_0(N, L_x) - C(N, L_x)}{\binom{\binom{N}{2}}{L_x}} \le \Pr[A_0^c] \to 0$$

which demonstrates (15.18) for $k = 0$. The remaining case for $k \ge 0$ in (15.18) follows from the observation that the number of graphs in $G_r(N, L_x)$

---

[9] It is convenient to take the logarithm of

$$t_k = \frac{\binom{N}{k}\binom{\binom{N-k}{2}}{L_x}}{\binom{\binom{N}{2}}{L_x}} = \frac{1}{k!}\prod_{j=0}^{k-1}(N-j)\prod_{j=0}^{L_x-1}\frac{\binom{N-k}{2}-j}{\binom{N}{2}-j}$$

$$= \frac{N^k}{k!}\prod_{j=0}^{k-1}\left(1-\frac{j}{N}\right)\left(1-\frac{k}{N}\right)^{L_x-1}\left(1-\frac{k}{N-1}\right)^{L_x-1}\prod_{j=0}^{L_x-1}\frac{1-\frac{2j}{(N-k)(N-1-k)}}{1-\frac{2j}{N(N-1)}}$$

which is

$$\log(k!t_k) = k\log N + \sum_{j=0}^{k-1}\log\left(1-\frac{j}{N}\right) + (E_x-1)\left(\log\left(1-\frac{k}{N}\right)+\log\left(1-\frac{k}{N-1}\right)\right)$$

$$+ \sum_{j=0}^{E_x-1}\log\left(1-\frac{2j}{(N-k)(N-1-k)}\right) - \log\left(1-\frac{2j}{N(N-1)}\right)$$

For large N and using the expansion $\log(1-z) = -z + O(z^2)$, we have for fixed k with

$$\log\left(1-\frac{2j}{(N-k)(N-1-k)}\right) = \log\left(1-\frac{2j}{N(N-1)}+O(N^{-3})\right)$$

that

$$\log(k!t_k) = k\log N + O(N^{-1}) - L_x\frac{2k}{N} + O(L_x^2 N^{-3})$$

In order to have a finite limit $\lim_{N\to\infty}\log(k!t_k) = c \in \mathbb{R}$, we must require that $k\log N - L_x\frac{2k}{N} = c$ which implies that $L_x = \frac{N}{2}\log N - \frac{Nc}{2k}$. For this scaling the order term $O(L_x^2 N^{-3})$ indeed vanishes if $N \to \infty$. By choosing $x = -\frac{c}{2k}$, we arrive at the correct scaling of $L_x = \frac{1}{2}N\log N + xN$ postulated above and $c = -2kx$.

with property A_k is equal to $\binom{N}{k}$ multiplied by the number of connected graphs with $N - k$ nodes and L_x links, which is approximately

$$\frac{\binom{N}{k}\binom{\binom{N-k}{2}}{L_x}}{\binom{\binom{N}{2}}{L_x}}\frac{T_0\,(N-k,L_x)}{\binom{\binom{N-k}{2}}{L_x}} \rightarrow \frac{(e^{-2x})^k}{k!}e^{-e^{-2x}}$$

where the limit gives the correct result because the small difference between the total number and that without property A_k tends to zero.

15.6.3 Connectivity and degree

There is an interesting relation between the connectivity of a graph, a global property, and the degree D of an arbitrary node, a local property. The implication $\{G \text{ is connected}\} \implies \{D_{\min} \geq 1\}$ where $D_{\min} = \min_{\text{all nodes } \in G} D$ is always true. The opposite implication is not always true, however, because a network can consists of separate, disconnected clusters containing nodes each with minimum degree larger than 1. A random graph can be generated from a set of labelled N nodes by randomly assigning a link with probability p to each pair of nodes. During this construction process, initially separate clusters originate, but at a certain moment, one of those clusters starts dominating (and swallowing) the other clusters. This largest cluster becomes the *giant* component. For large N and a certain p_N which depends on N, the implication $\{D_{\min} \geq 1\} \implies \{G_p\,(N) \text{ is connected}\}$ is almost surely (a.s.) correct. A rigorous mathematical proof is fairly complex and omitted. Thus, for large random graphs $G_p\,(N)$ holds the equivalence $\{G_p\,(N) \text{ is connected}\} \iff \{D_{\min} \geq 1\}$ almost surely such that

$$\Pr\left[G_p\,(N) \text{ is connected}\right] = \Pr\left[D_{\min} \geq 1\right] \qquad \text{a.s.}$$

From (3.32) and (15.11), we have that

$$\Pr[D_{\min} \geq 1] = (\Pr[D_{rg} \geq 1])^N = (1 - \Pr[D_{rg} = 0])^N = \left(1 - (1-p)^{N-1}\right)^N$$

which shows that $\Pr\left[D_{\min} \geq 1\right]$ rapidly tends to one for fixed $0 < p < 1$ and large N. Therefore, the asymptotic behavior of $\Pr\left[G_p(n) \text{ is connected}\right]$

requires the investigation of the influence of p as a function of N,

$$\Pr\left[G_p\left(N\right) \text{ is connected}\right] = \exp\left(N\log\left(1 - \left(1 - p_N\right)^{N-1}\right)\right)$$

$$= \exp\left(-N\sum_{j=1}^{\infty}\frac{\left(1 - p_N\right)^{j(N-1)}}{j}\right)$$

$$= e^{-N(1-p_N)^{N-1}}\exp\left(-N\sum_{j=2}^{\infty}\frac{\left(1 - p_N\right)^{(N-1)j}}{j}\right)$$

If we denote $c_N \triangleq N \cdot \left(1 - p_N\right)^{N-1}$, then

$$N\sum_{j=2}^{\infty}\frac{\left(1 - p_N\right)^{(N-1)j}}{j} = \sum_{j=2}^{\infty}\frac{c_N^j}{jN^{j-1}}$$

can be made arbitrarily small for large N provided we choose $c_N = O\left(N^{\beta}\right)$ with $\beta < \frac{1}{2}$. Thus, for large N, we have that

$$\Pr\left[G_p\left(N\right) \text{ is connected}\right] = e^{-cN^{\beta}}\left(1 + O\left(N^{2\beta-1}\right)\right)$$

which tends to 0 for $0 < \beta < \frac{1}{2}$ and to 1 for $\beta < 0$. Hence, the critical exponent where a sharp transition occurs is $\beta = 0$. In that case, $c_N = c$ (a real positive constant) and

$$p_N = 1 - \exp\left(\frac{\log\frac{c}{N}}{N-1}\right) = \frac{\log N}{N} + O\left(\frac{\log c}{N}\right).$$

In summary, for large N,

$$\Pr\left[G_p\left(N\right) \text{ is connected}\right] \longrightarrow \begin{array}{ll} 0 & \text{if } p < \frac{\log N}{N} \\ 1 & \text{if } p > \frac{\log N}{N} \end{array} \qquad (15.19)$$

with a transition region around $p_c \sim \frac{\log N}{N}$ with a width of $O(\frac{1}{N})$. Notice the agreement with (15.15) where $p_x = \frac{L_x}{L_{\max}} \simeq \frac{\log N}{N} + \frac{x}{N}$: for large $x < 0$, $e^{-e^{-2x}} \to 0$, while for large $x > 0$, $e^{-e^{-2x}} \to 1$ and the width of the transition region for the link density p is $O(\frac{1}{N})$.

15.6.4 Size of the giant component

Let $S = \Pr\left[n \in C\right]$ denote the probability that a node n in $G_p\left(N\right)$ belongs to the giant component C. If $n \notin C$, then none of the neighbors of node n

belongs to the giant component. The number of neighbors of a node n is the degree d_n of a node such that

$$\Pr[n \notin C] = \Pr[\text{all neighbor of } n \notin C]$$
$$= \sum_{k \geq 0} \Pr[\text{all } k \text{ neighbors of } n \notin C | d_n = k] \Pr[d_n = k]$$

Since in $G_p(N)$ all neighbors of n are independent[10], the conditional probability becomes, with $1 - S = \Pr[n \notin C]$,

$$\Pr[\text{all } k \text{ neighbors of } n \notin C | d_n = k] = (\Pr[n \notin C])^k = (1 - S)^k$$

Moreover, this probability holds for any node in $n \in G_p(N)$ such that, writing the random variable D_{rg} instead of an instance d_n,

$$1 - S = \sum_{k=0}^{\infty} (1 - S)^k \Pr[D_{\text{rg}} = k] = \varphi_{D_{\text{rg}}}(1 - S)$$

where $\varphi_{D_{\text{rg}}}(u) = E[u^{D_{\text{rg}}}]$ is the generating function of the degree D_{rg} in $G_p(N)$. For large N, the degree distribution in $G_p(N)$ is Poisson distributed with mean degree $\mu_{\text{rg}} = p(N-1)$ and $\varphi_{D_{\text{rg}}}(u) \simeq e^{\mu_{\text{rg}}(u-1)}$. For large N, the fraction S of nodes in the giant component in the random graph satisfies an equation similar to that in (12.13) of the extinction probability in a branching process,

$$S = 1 - e^{-\mu_{\text{rg}} S} \tag{15.20}$$

and the average size of the giant component is NS. For $\mu_{\text{rg}} < 1$ the only solution is $S = 0$ whereas for $\mu_{\text{rg}} > 1$ there is a non-zero solution for the size of the giant component. The solution can be expressed as a Lagrange series using (5.34),

$$S(\mu_{\text{rg}}) = 1 - e^{-\mu_{\text{rg}}} \sum_{n=0}^{\infty} \frac{(n+1)^n}{(n+1)!} \left(\mu_{\text{rg}} e^{-\mu_{\text{rg}}}\right)^n \tag{15.21}$$

By reversing (15.20), the average degree in the random graph can be expressed in terms of the fraction S of nodes in the giant component,

$$\mu_{\text{rg}}(S) = -\frac{\log(1 - S)}{S} \tag{15.22}$$

---

[10] This argument is not valid, for example, for a two-dimensional lattice $\mathbb{Z}_p^2$ in which each link between adjacent nodes at integer value coordinates in the plane exists with probability p. The critical link density for connectivity in $\mathbb{Z}_p^2$ is $p_c = \frac{1}{2}$, a famous result proved in the theory of percolation (see, for example, Grimmett (1989)).

15.7 The hopcount in a large, sparse graph with unit link weights

Routers in the Internet forward IP packets to the next hop router, which is found by routing protocols (such as OSPF and BGP). Intra-domain routing as OSPF is based on the Dijkstra *shortest path* algorithm, while inter-domain routing with BGP is policy-based, which implies that BGP does not minimize a length criterion. Nevertheless, end-to-end paths in the Internet are shortest paths in roughly 70% of the cases. Therefore, we consider the shortest path between two *arbitrary* nodes because (a) the IP address does not reflect a precise geographical location and (b) uniformly distributed world wide communication, especially, on the web seems natural since the information stored in servers can be located in places unexpected and unknown to browsing users. The Internet type of communication is different from classical telephony because (a) telephone numbers have a direct binding with a physical location and (b) the intensity of average human interaction rapidly decreases with distance. We prefer to study the *hopcount* H_N because it is simple to measure via the trace-route utility, it is an integer, dimensionless, and the quality of service (QoS) measures (such as packet delay, jitter and packet loss) depend on the hopcount, the number of traversed routers. In this section, we first investigate the hopcount in a sparse, but connected graph where all links have unit weight. Chapter 16 treats graphs with other link weight structures.

15.7.1 Bi-directional search

The basic idea of a bi-directional search to find the shortest path is by starting the discovery process (e.g. using Dijkstra's algorithm) from A and B simultaneously. When both subsections from A and from B meet, the concatenation forms the shortest path from A to B. In case all link weights are equal, $w(i \to j) = 1$ for any link $i \to j$ in a graph G, the shortest path from A and B is found when the discovery process from A and that from B have precisely one node of the graph in common.

Denote by $C_A(l)$, respectively $C_B(l)$, the set of nodes that can be reached from A, respectively B, in l or less hops. We define $C_A(0) = \{A\}$ and $C_B(0) = \{B\}$. The hopcount is larger than $2l$ if and only if $C_A(l) \cap C_B(l)$ is empty. Conditionally on $|C_A(l)| = n_A$, respectively $|C_B(l)| = n_B$, the sets $C_A(l)$ and $C_B(l)$ do not possess a common node with probability

$$\Pr\left[C_A(l) \cap C_B(l) = \varnothing \,\middle|\, |C_A(l)| = n_A, |C_B(l)| = n_B\right] = \frac{\binom{N-1-n_A}{n_B}}{\binom{N-1}{n_B}}$$

which consists of the ratio of all combinations in which the n_B nodes around B can be chosen out of the remaining nodes that do not belong to the set C_A over all combinations in which n_B nodes can be chosen in the graph with N nodes except for node A. Furthermore,

$$\frac{\binom{N-1-n_A}{n_B}}{\binom{N-1}{n_B}} = \frac{(N-n_A-1)(N-n_A-2)\cdots(N-n_A-n_B)}{(N-1)(N-2)\cdots(N-n_B)}$$

$$= \frac{(1-\frac{n_A+1}{N})(1-\frac{n_A+2}{N})\cdots(1-\frac{n_A+n_B}{N})}{(1-\frac{1}{N})(1-\frac{2}{N})\cdots(1-\frac{n_B}{N})}$$

For large N, we apply the Taylor series around $x = 0$ of $\log(1-x) = -x - \sum_{j=2}^{\infty} \frac{x^j}{j}$,

$$\log \frac{\binom{N-1-n_A}{n_B}}{\binom{N-1}{n_B}} = \sum_{k=1}^{n_B} \log\left(1 - \frac{n_A+k}{N}\right) - \log\left(1 - \frac{k}{N}\right)$$

$$= -\sum_{k=1}^{n_B}\left(\frac{n_A+k}{N} - \frac{k}{N}\right) - \sum_{j=2}^{\infty}\frac{1}{jN^j}\left(\sum_{k=1}^{n_B}(n_A+k)^j - k^j\right)$$

$$= -\frac{n_A n_B}{N} - \left(\frac{n_A n_B}{N}\right)^2\left(\frac{1}{2n_A} + \frac{1}{2n_B} + \frac{1}{2n_A n_B}\right) - R$$

where the remainder is

$$R = \sum_{j=3}^{\infty}\frac{1}{jN^j}\sum_{m=0}^{j-1}\binom{j}{m}n_A^{j-m}\sum_{k=1}^{n_B}k^m = O\left(\left(\frac{n_A n_B}{N}\right)^3\right)$$

After exponentiation

$$\Pr\left[H_N > 2l \big| |C_A(l)| = n_A, |C_B(l)| = n_B\right] = e^{-\frac{n_A n_B}{N}}\left(1 + O\left(\left(\frac{n_A n_B}{N}\right)^2\right)\right)$$

By the law of total probability (2.47) and up to $O\left(\frac{E[|C_A(l)|^2|C_B(l)|^2]}{N^2}\right)$ for large N, we obtain

$$\Pr\left[H_N > 2l\right] \approx E\left[\exp\left(-\frac{|C_A(l)|\,|C_B(l)|}{N}\right)\right] \tag{15.23}$$

This probability (15.23) holds for any large graph with a unit link weight structure provided $E\left[|C_A(l)|^2 |C_B(l)|^2\right] = o(N^2)$. Formula (15.23) becomes increasingly accurate for decreasing $|C_A(l)|$ and $|C_B(l)|$, and so for sparser large graphs.

15.7.2 Sparse large graphs and a branching process

In order to proceed, the number of nodes in the sets $C_A(l)$ and $C_B(l)$ needs to be determined, which is difficult in general. Therefore, we concentrate here on a special class of graphs in which the discovery process from A and B is reasonably well modeled by a branching process (Chapter 12). A branching process evolves from a given set $C_A(l-1)$ in the next l-th discovery cycle (or generation) to the set $C_A(l)$ by including only new nodes, not those previously discovered. The application of a branching process implies that the newly discovered nodes do not possess links to any previously discovered node of $C_A(l-1)$ except for its parent node in $C_A(l-1)$. Hence, only for large and sparse graphs or tree-like graphs, this assumption can be justified, provided that the number of links that point backwards to early discovered nodes in $C_A(l-1)$ is negligibly small.

Assuming that a branching process models the discovery process well, we will compute the number of nodes that can be reached from A and similarly from B in l hops from a branching process with production Y specified by the degree distribution of the nodes in the graph. The additional number of nodes X_l discovered during the l-th cycle of a branching process that are included in the set $C_A(l)$ is described by the basic law (12.1). Thus, $|C_A(l)| = \sum_{k=0}^{l} X_k$ with $X_0 = 1$ (namely node A). In terms of the scaled random variable $W_k = \frac{X_k}{\mu^k}$ with unit mean $E[W_k] = 1$,

$$|C_A(l)| = \sum_{k=0}^{l} W_k \mu^k$$

and where $\mu = E[Y] - 1 > 1$ denotes the average degree minus 1, i.e. the outdegree, in the graph. Only the root has $E[Y]$ equal to the mean degree. Immediately, the average size of the set of nodes reached from A in l hops is with $E[W_k] = 1$,

$$E[|C_A(l)|] = \sum_{k=0}^{l} \mu^k = \frac{\mu^{l+1} - 1}{\mu - 1}$$

which equally holds for $E[|C_B(l)|]$.

Applying Jensen's inequality (5.5) to (15.23) yields

$$\exp\left(-\frac{E[C_A(l)]E[C_B(l)]}{N}\right) \leq E\left[\exp\left(-\frac{C_A(l)C_B(l)}{N}\right)\right]$$

such that

$$\Pr[H_N > k] \geq \exp\left(-\frac{\mu^2}{N(\mu-1)^2}\mu^k\right)$$

With the tail probability expression (2.36) for the average, we arrive at the lower bound for the expected hopcount in large graphs,

$$E[H_N] = \sum_{k=0}^{\infty} \Pr[H_N > k] \geq \sum_{k=0}^{\infty} \exp\left(-\frac{\mu^2}{N(\mu-1)^2} \mu^k\right)$$

The sum $S_1(t) = \sum_{k=0}^{\infty} \exp(-t\mu^k)$ can be evaluated exactly[11] as

$$S_1(t) = \frac{1}{2} - \frac{\log t + \gamma}{\log \mu} + \frac{2}{\sqrt{\log \mu}} \sum_{k=1}^{\infty} \frac{\cos\left[\frac{2k\pi}{\log \mu} \log\left(\frac{\mu}{t}\right) + \arg \Gamma\left(\frac{2k\pi i}{\log \mu}\right)\right]}{\sqrt{2k \sinh \frac{2k\pi^2}{\log \mu}}}$$

$$+ \sum_{k=1}^{\infty} \left(1 - e^{-t\mu^{-k}}\right)$$

Furthermore,

$$\left|\sum_{k=1}^{\infty} \frac{\cos\left[\frac{2k\pi}{\log \mu} \log\left(\frac{\mu}{t}\right) + \arg \Gamma\left(\frac{2k\pi i}{\log \mu}\right)\right]}{\sqrt{2k \sinh \frac{2k\pi^2}{\log \mu}}}\right| \leq \sum_{k=1}^{\infty} \frac{1}{\sqrt{2k \sinh \frac{2k\pi^2}{\log \mu}}} = b(\mu)$$

and the function $T(\mu) = \frac{2b(\mu)}{\sqrt{\log \mu}}$ is increasing, but for $1 < \mu \leq 5$ its maximum value $T(5)$ is smaller than 0.0035. Since $t = \frac{\mu^2}{N(\mu-1)^2}$ is small and $\mu > 1$, we approximate

$$S_1(t) \approx \frac{1}{2} - \frac{\log t + \gamma}{\log \mu} \qquad (15.24)$$

---

[11] For $\sigma = Re(s) > 0$ and $Re(p) \geq 0$, we have

$$\frac{\Gamma(s)}{\mu^{ps}} = \int_0^{\infty} t^{s-1} e^{-\mu^p t} dt$$

and

$$\Gamma(s) \sum_{k=0}^{\infty} \frac{1}{\mu^{ks}} = \int_0^{\infty} t^{s-1} \sum_{k=0}^{\infty} e^{-\mu^k t} dt$$

or

$$\int_0^{\infty} t^{s-1} S_1(t) dt = \Gamma(s) \frac{\mu^s}{\mu^s - 1}$$

By Mellin inversion, for $c > 0$,

$$S_1(t) = \frac{1}{2\pi i} \int_{c-i\infty}^{c+i\infty} \frac{\Gamma(s)}{\mu^s - 1} \left(\frac{\mu}{t}\right)^s ds$$

By moving the line of integration to the left, we encounter a double pole at $s = 0$ from $\Gamma(s)$ and $\frac{1}{\mu^s-1}$ and simple poles at $s = \frac{2k\pi i}{\log \mu}$ from $\frac{1}{\mu^s-1}$. Invoking Cauchy's residue theorem leads to the result.

and arrive, for large N, at

$$E\left[H_N\right] \geq \frac{\log N}{\log \mu} + \frac{1}{2} - \frac{\log \frac{\mu^2}{(\mu-1)^2} + \gamma}{\log \mu} \approx \frac{\log N}{\log \mu}$$

This shows that in large, sparse graphs for which the discovery process is well modeled by a branching process, it holds that $E\left[H_N\right]$ scales as $\frac{\log N}{\log \mu}$ where $\mu = E\left[Y\right] - 1 > 1$ is the average degree minus 1 in the graph.

We can refine the above analysis. Let us now assume that the convergence of $W_k \to W$ is sufficiently fast for large N and that $W > 0$ such that,

$$\left|C_A(l)\right| \sim W_A \sum_{k=0}^{l} \mu^k = W_A \frac{\mu^{l+1} - 1}{\mu - 1} \approx W_A \frac{\mu^{l+1}}{\mu - 1}$$

is a good approximation (and similarly for $\left|C_B(l)\right|$). The verification of this approximation is difficult in general. Theorem 12.3.2 states that $\Pr\left[W = 0\right] = \pi_0$ and equivalently $\Pr\left[W > 0\right] = 1 - \pi_0$ where the extinction probability π_0 obeys the equation (12.13). Using this approximation, we find from (15.23)

$$\Pr\left[H_N > 2l\right] \approx E\left[\exp\left(-\frac{W_A W_B}{N} \frac{\mu^{2l+2}}{(\mu - 1)^2}\right) \middle| W_A, W_B > 0\right]$$

where the condition on $W > 0$ is required else there are no clusters $C_A(l)$ and $C_B(l)$ nor a path. Since the same asymptotics also holds for odd values of the hopcount, we finally arrive, for $k \geq 1$ and large N, at

$$\Pr\left[H_N > k\right] \approx E\left[\exp\left(-Z\mu^k\right) \middle| W_A, W_B > 0\right]$$

where the random variable

$$Z = \frac{W_A W_B}{N} \frac{\mu^2}{(\mu - 1)^2}$$

and $W_A \stackrel{d}{=} W_B \stackrel{d}{=} W$. A more explicit computation of $\Pr\left[H_N > k\right]$ requires the knowledge of the limit random variable W, which strongly depends on the nodal degree Y.

The average hopcount $E\left[H_N\right]$ is found similarly as in the analysis above by using (15.24) with $t = Z$,

$$E\left[H_N\right] \approx E\left[S_1(Z) \middle| W_A, W_B > 0\right]$$

$$= E\left[\frac{1}{2} - \frac{2\log W - \log N + 2\log \frac{\mu}{(\mu-1)} + \gamma}{\log \mu} \middle| W > 0\right]$$

$$= \frac{1}{2} + \frac{\log N - 2\log \frac{\mu}{(\mu-1)} - \gamma}{\log \mu} - 2\frac{E\left[\log W \middle| W > 0\right]}{\log \mu}$$

In sparse graphs with average degree $E[Y]$ equal to μ and for a large number of nodes N, the average hopcount is well approximated[12] by

$$E[H_N] = \frac{\log N}{\log \mu} + \frac{1}{2} - \frac{\gamma - 2\log(\mu - 1)}{\log \mu} - 2\frac{E[\log W \,|\, W > 0]}{\log \mu} \quad (15.25)$$

This expression (15.25) for the average hopcount – which is more refined than the commonly used estimate $E[H_N] \approx \frac{\log N}{\log \mu}$ – contains the curious average $E[\log W \,|\, W > 0]$ where W is the limit random variable of the branching process produced by the graph's degree distribution Y.

Application to $\mathbf{G_p(N)}$ The above analysis holds for fixed $E[Y] = p(N-1)$ such that, for large N, we require that $p = \frac{\mu}{N}$ where μ is approximately equal to the average degree. Since the binomial distribution (15.11) for the degree in $G_p(N)$ is very well approximated by the Poisson distribution $\Pr[D_{\mathrm{rg}} = k] \approx \frac{\mu^k}{k!}e^{-\mu}$ for large N and constant μ, formula (15.25) requires the computation of $E[\log W \,|\, W > 0]$ in a Poisson branching process, which is presented in Hooghiemstra and Van Mieghem (2005) but here summarized in Fig. 15.8. The numerical evaluation of average hopcount (15.25) in a

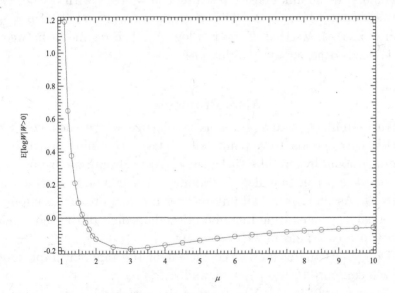

Fig. 15.8. The quantity $E[\log W \,|\, W > 0]$ of a Poisson branching process versus the average degree μ.

---

[12] A more rigorous derivation that stochastically couples the graph's growth specified by a certain degree distribution to a corresponding branching process is found in van der Hofstad *et al.* (2005). In particular, the analysis is shown to be valid for any randomly constructed graph with a finite variance of the degree. More details on the result for the average hopcount are presented in Hooghiemstra and Van Mieghem (2005).

random graph of the class $G_p(N)$ for small average degree μ and large N shows that (15.25) is much more accurate than only its first term $\log_\mu N$.

At the other end of the scale for a constant link density $p = c < 1$, which corresponds to an average degree $E[Y] = c(N-1)$, the above analysis no longer applies for such large values of the average degree $E[Y]$. Fortunately, in that case, an exact asymptotic analysis is possible (see Problem (iii)):

$$\Pr[H_N = 1] = p$$
$$\Pr[H_N = 2] = (1-p)\left(1 - (1-p^2)^{N-2}\right) \qquad (15.26)$$

Values of H_N higher than 2 are extremely unlikely since $\Pr[H_N > 2] = (1-p)\left[1 - p^2\right]^{N-2}$ tends to zero rapidly for sufficiently large N. Hence, $E[H_N] \simeq \Pr[H_N = 1] + 2\Pr[H_N = 2] \simeq 2 - p$ and, similarly, we find $\mathrm{Var}[H_N] \simeq p(1-p)$. This asymptotic analysis even holds for a larger link density regime $p = cN^{-\frac{1}{2}+\epsilon}$ with $\epsilon > 0$ because

$$\lim_{N\to\infty} \Pr[H_N > 2] = \lim_{N\to\infty} (1 - cN^{-\frac{1}{2}+\epsilon})\left[1 - cN^{-1+2\epsilon}\right]^{N-2} = 0$$

but for $\epsilon = 0$, it holds that $\lim_{N\to\infty} \Pr[H_N > 2] = e^{-c} > 0$.

In summary, if the link density p scales as $p = cN^{-\alpha}$ with $\alpha \in [0, \frac{1}{2})$, the average hopcount $E[H_N] \simeq 2 - p$ is constant and very small. If $p = \frac{\lambda}{N^{1-\epsilon}}$, equation (15.25) shows that $E[H_N] \approx \log_\lambda N$. The regime in between for $\alpha \in [\frac{1}{2}, 1)$ needs other analysis techniques.

15.8 Problems

(i) An extremely regular graph is a d-lattice where each nodal position corresponds to a point with integer coordinates within a d-dimensional hyper-cube with size Z. Apart from border nodes, each node has a constant degree (number of neighbors), precisely equal to $2d$. Assuming that all link metrics are equal to one, compute the probability generating function of the hopcount of the shortest path between two uniformly chosen points.

(ii) If $c_{G_p(N)}$ is the clustering coefficient of the random graph $G_p(N)$, then compute $\Pr\left[c_{G_p(N)} \le x\right]$ and $E\left[c_{G_p(N)}\right]$.

(iii) Derive (15.26) in $G_p(N)$ with unit link weights.

16

The Shortest Path Problem

The shortest path problem asks for the computation of the path from a source to a destination node that minimizes the sum of the positive weights[1] of its constituent links. The related shortest path tree (SPT) is the union of the shortest paths from a source node to a set of m other nodes in the graph with N nodes. If $m = N - 1$, the SPT connects all nodes and is termed a spanning tree. The SPT belongs to the fundamentals of graph theory and has many applications. Moreover, powerful shortest path algorithms like that of Dijkstra exist. Section 15.7 studied the hopcount, the number of hops (links) in the shortest path, in sparse graphs with unit link weights. In this chapter, the influence of the link weight structure on the properties of the SPT will be analyzed. Starting from one of the simplest possible graph models, the complete graph with i.i.d. exponential link weight, the characteristics of the shortest path will be derived and compared to Internet measurements.

The link weights seriously impact the path properties in QoS routing (Kuipers and Van Mieghem, 2003). In addition, from a traffic engineering perspective, an ISP may want to tune the weight of each link such that the resulting shortest paths between a particular set of in- and egresses follow the desirable routes in its network. Thus, apart from the topology of the graph, the link weight structure clearly plays an important role. Often, as in the Internet or other large infrastructures, both the topology and the link weight structure are not accurately known. This uncertainty about the precise structure leads us to consider both the underlying graph and each of the link weights as random variables.

---

[1] A zero link weight is regarded as the coincidence of two nodes (which we exclude), while an infinite link weight means the absence of a link.

16.1 The shortest path and the link weight structure

Since the shortest path is mainly sensitive to the smaller, positive link weights, the probability distribution of the link weights around zero will dominantly influence the properties of the resulting shortest path. A *regular* link weight distribution $F_w(x) = \Pr[w \le x]$ has a Taylor series expansion around $x = 0$,

$$F_w(x) = f_w(0)\, x + O\left(x^2\right)$$

since $F_w(0) = 0$ and $F'_w(0) = f_w(0)$ exists. A regular link weight distribution is thus linear around zero. The factor $f_w(0)$ only scales all link weights, but does not influence the shortest path. The simplest distribution of the link weight w with a distinct different behavior for small values is the polynomial distribution

$$F_w(x) = x^\alpha 1_{x\in[0,1]} + 1_{x\in[1,\infty)}, \quad \alpha > 0, \tag{16.1}$$

The corresponding density is $f_w(x) = \alpha x^{\alpha-1} 1_{x\in[0,1]}$. The exponent

$$\alpha = \lim_{x\downarrow 0} \frac{\log F_w(x)}{\log x}$$

is called the *extreme value index* of the probability distribution of w and $\alpha = 1$ for regular distributions. By varying the exponent α over all non-negative real values, any extreme value index can be attained and a large class of corresponding SPTs, in short α-trees, can be generated.

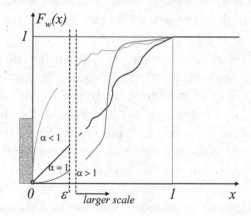

Fig. 16.1. A schematic drawing of the distribution of the link weights for the three different α-regimes. The shortest path problem is mainly sensitive to the small region around zero. The scaling invariant property of the shortest path allows us to divide all link weights by the largest possible such that $F_w(1) = 1$ for all link weight distributions.

Figure 16.1 illustrates schematically the probability distribution of the link weights around zero $(0, \epsilon]$, where $\epsilon > 0$ is an arbitrarily small, positive real number. The larger link weights in the network will hardly appear in a shortest path provided the network possesses enough links. These larger link weights are drawn in Fig. 16.1 from the double dotted line to the right. The nice advantage that only small link weights dominantly influence the property of the resulting shortest path tree implies that the remainder of the link weight distribution (denoted by the arrow with larger scale in Fig. 16.1) only plays a second order role. To some extent, it also explains the success of the simple SPT model based on the complete graph K_N with i.i.d. exponential link weights, which we derive in Section 16.2. A link weight structure effectively thins the complete graph K_N – any other graph is a subgraph of K_N – to the extent that a specific shortest path tree can be constructed.

Finally, we assume the independence of link weights, which we deem a reasonable assumption in large networks, such as the Internet with its many independent autonomous systems (ASs). Apart from the Section 16.7, we will mainly consider the case for $\alpha = 1$, which allows an exact analysis.

16.2 The shortest path tree in K_N with exponential link weights

16.2.1 The Markov discovery process

Let us consider the shortest path problem in the complete graph K_N, where each node in the graph is connected to each other node. The problem of finding the shortest path between two nodes A and B in K_N with exponentially distributed link weights with mean 1 can be rephrased in terms of a Markov discovery process. The discovery process evolves as a function of time and stops at a random time T when node B is found. The process is shown in Fig. 16.2.

The evolution of the discovery process can be described by a continuous-time Markov chain $X(t)$, where $X(t)$ denotes the number of discovered nodes at time t, because the characteristics of a Markov chain (Theorem 10.2.3) are based on the exponential distribution and the memoryless property. Of particular interest here is the property (see Section 3.4.1) that the minimum of n independent exponential variables each with parameter α_i is again an exponential variable with parameter $\sum_{i=1}^{n} \alpha_i$.

The discovery process starts at time $t = T_0$ with the source node A and for the initial distribution of the Markov chain, we have $\Pr[X(T_0) = 1] = 1$. The state space of the continuous Markov chain is the set S_N consisting of all positive integers (nodes) n with $n \leq N$. For the complete graph K_N, the transition rates are given by

$$\lambda_n = n(N - n), \qquad\qquad n \in S_N \qquad\qquad (16.2)$$

Indeed, initially there is only the source node A with label[2] 0, hence $n = 1$. From this first node A precisely $N - 1$ new nodes can be reached in the complete graph K_N. Alternatively one can say that $N - 1$ nodes are competing with each other each with exponentially distributed strength to be discovered and the winner amongst them, say C with label 1, is the one reached in shortest time which corresponds to an exponential variable with rate $N - 1$.

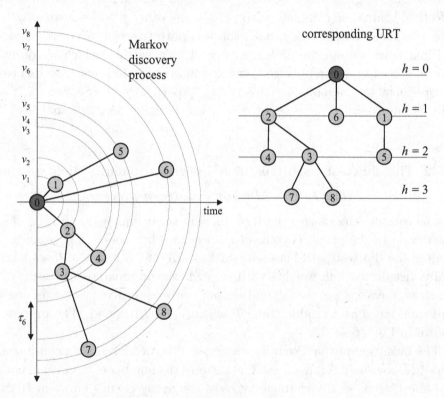

Fig. 16.2. On the left, the Markov discovery process as function of time in a graph with $N = 9$ nodes. The circles centered at the discovering node A with label 0 present equi-time lines and v_k is the discovering time of the k-th node, while $\tau_k = v_k - v_{k-1}$ is the k-th interattachment time. The set of discovered nodes redrawn per level are shown on the right, where a level gives the number of hops h from the source node A. The tree is a uniform recursive tree (URT).

---

[2] When continuous measures such as time and weight of a path are computed, the source node is most conveniently labeled by zero, whereas in counting processes, such as the number of hops of a path, the source node is labeled by one.

After having reached C from A at hitting time v_1, two nodes $n = 2$ are found and the discovery process restarts from both A and C. Although at time v_1 we were already progressed a certain distance towards each of the $N - 2$ other, not yet discovered, nodes, the memoryless property of the exponential distribution tells us that the remaining distance to these $N - 2$ nodes is again exponentially distributed with the same parameter 1. Hence, this allows us to restart the process from A and C by erasing the previously partial distance to any other not yet discovered node as if we ignore that it were ever travelled. From the discovery time v_1 of the first node on, the discovery process has double strength to reach precisely $N - 2$ new nodes. Hence, the next winner, say D labeled by 2, is reached at v_2 in the minimum time out of $2(N - 2)$ traveling times. This node D has equal probability to be attached to A or C because of symmetry. When D is attached to A (the argument below holds similarly for attachment to C), symmetry appears to be broken, because D and C have only one link used, whereas A has already two links used. However, since we are interested in the shortest path problem and since the direct link from A to D is shorter than the path $A \rightarrow C \rightarrow D$, we exclude the latter in the discovery process, hereby establishing again the full symmetry in the Markov chain. This exclusion also means that the Markov chain maintains single paths from A to each newly discovered node and this path is also the shortest path. Hence, there are no cycles possible. Furthermore, similar to Dijkstra's shortest path algorithm, each newly reached node is withdrawn from the next competition round, which guarantees that the Markov chain eventually terminates. Besides terminating by extinction of all available nodes, after each transition when a new node is discovered, the Markov chain stops with probability equal to $\frac{1}{N-n}$, since each of the n already discovered nodes has precisely 1 possibility out of the remaining $N - n$ to reach B and only one of them is the discoverer. The stopping time T is defined as the infimum for $t \geq 0$ at which the destination node B is discovered. In summary, the described Markov discovery process, a pure birth process with birth rate $\lambda_n = n(N - n)$, models exactly the shortest path for all values of N.

16.2.2 The uniform recursive tree

A uniform recursive tree (URT) of size N is a random tree rooted at A. At each stage a new node is attached uniformly to one of the existing nodes until the total number of nodes is equal to N. The hopcount h_N (equivalent to the depth or distance) is the smallest number of links between the root A and a destination chosen uniformly from all nodes $\{1, 2, \ldots, N\}$.

Denote by $\left\{X_N^{(k)}\right\}$ the k-th level set of a tree T, which is the set of nodes in the tree T at hopcount k from the root A in a graph with N nodes, and by $X_N^{(k)}$ the number of elements in the set $\left\{X_N^{(k)}\right\}$. Then, we have $X_N^{(0)} = 1$ because the zeroth level can only contain the root node A itself. For all $k > 0$, it holds that $0 \leq X_N^{(k)} \leq N - 1$ and that

$$\sum_{k=0}^{N-1} X_N^{(k)} = N \tag{16.3}$$

Another consequence of the definition is that, if $X_N^{(n)} = 0$ for some level $n < N - 1$, then all $X_N^{(j)} = 0$ for levels $j > n$. In such a case, the longest possible shortest path in the tree has a hopcount of n. The level set

$$L_N = \left\{1, X_N^{(1)}, X_N^{(2)}, \ldots, X_N^{(N-1)}\right\}$$

of a tree T is defined as the set containing the number of nodes $X_N^{(k)}$ at each level k. An example of a URT organized per level k is drawn on the right in Fig. 16.2 and in Fig. 16.3. A basic theorem for URTs proved in van der Hofstad *et al.* (2002b), is the following:

Theorem 16.2.1 *Let* $\{Y_N^{(k)}\}_{k,N\geq 0}$ *and* $\{Z_N^{(k)}\}_{k,N\geq 0}$ *be two independent copies of the vector of level sets of two sequences of independent URTs. Then*

$$\{X_N^{(k)}\}_{k\geq 0} \stackrel{d}{=} \{Y_{N_1}^{(k-1)} + Z_{N-N_1}^{(k)}\}_{k\geq 0}, \tag{16.4}$$

where on the right-hand side the random variable N_1 *is uniformly distributed over the set* $\{1, 2, \ldots, N - 1\}$.

Theorem 16.2.1 also implies that a subtree rooted at a direct child of the root is a URT. For example, in Fig. 16.3, the tree rooted at node 5 is a URT of size 13 as well as the original tree without the tree rooted at node 5. By applying Theorem 16.2.1 to the URT subtree, any subtree rooted at a member of a URT is also a URT.

An arbitrary URT U consisting of N nodes and with the root labeled by 1 can be represented as

$$U = (n_2 \longleftarrow 2)(n_3 \longleftarrow 3)\ldots(n_N \longleftarrow N) \tag{16.5}$$

where $(n_j \longleftarrow j)$ means that the j-th node is attached to node $n_j \in [1, j-1]$ and $n_2 = 1$. Hence, n_j is the predecessor of j and the predecessor relation is indicated by the arrow "$\longleftarrow$". Moreover, n_j is a discrete uniform random variable on $[1, j-1]$ and all $n_2, n_3, \ldots, n_N$ are independent.

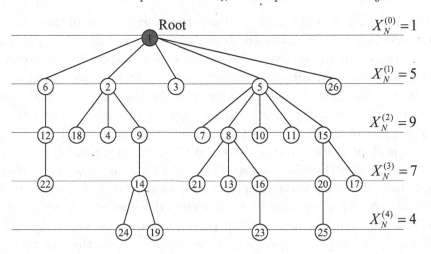

Fig. 16.3. An instance of a uniform recursive tree with $N = 26$ nodes organized per level $0 \le k \le 4$. The node number (inside the circle) indicates the order in which the nodes were attached to the tree.

Theorem 16.2.2 *The total number of URTs with N nodes is $(N-1)!$*

Proof: **(a)** Let the nodes be labeled in the order of attachment to the URT and assign label 1 for the root. The URT growth law indicates that node 2 can only be attached in one way, node 3 in two ways, namely to node 1 and node 2 with equal probability. The k-th node can be attached in $k-1$ possible nodes. Each of these possible constructions leads to a URT.

(b) By summing over all allowable configurations in (16.5), we obtain

$$\sum_{n_2=1}^{1} \sum_{n_3=1}^{2} \cdots \sum_{n_N=1}^{N-1} 1 = (N-1)!$$

and this proves the theorem.   □

In general, Cayley's Theorem (Appendix B.1 art. **3**) states that there are N^{N-2} labeled trees possible. The URT is a subset of the set of all possible labeled trees. Not all labeled trees are URTs, because the nodes that are further away from the root must have larger labels.

The shortest path tree from the source or root A to other nodes in the complete graph is the tree associated with the Markov discovery process, where the number of nodes $X(t)$ at time t is constructed as follows. Just as the discovery process, the associated tree starts at the root A. We now investigate the embedded Markov chain (Section 10.4) of the continuous-time discovery process. After each transition in the continuous-time Markov chain, $X(t) \rightarrow$

$X(t)+1$, an edge of *unit* length is attached *randomly* to one of the n already discovered nodes in the associated tree because a new edge is equally likely to be attached to any of the n discovering nodes. Hence, the construction of the tree associated with the Markov discovery process and illustrated in Fig. 16.2 on the right demonstrates that the shortest path tree in the complete graph K_N with exponential link weights is an *uniform recursive tree*. This property of the shortest path tree in K_N with exponential link weights is an important motivation to study the URT. More generally, in van der Hofstad *et al.* (2001) we have proved that, for a fixed link density p and sufficiently large N, the shortest path tree in the *class RGU*, the class of random graphs $G_p(N)$ with exponential or uniformly distributed link weights, is a URT. Smythe and Mahmoud (1995) have reviewed a number of results on recursive trees that have appeared in the literature from the late 1960s up to 1995.

16.3 The hopcount h_N in the URT

16.3.1 Theory

The hopcount h_N from the root to an arbitrary chosen node in the URT equals the number of links or hops from the root to that node. We allow the arbitrary node to coincide with the root in which case $h_N = 0$.

Theorem 16.3.1 *The probability generating function of the hopcount in the URT with N nodes is*

$$\varphi_{h_N}(z) = E\left[z^{h_N}\right] = \frac{\Gamma(N+z)}{\Gamma(N+1)\Gamma(z+1)} \tag{16.6}$$

Proof: Since the number of nodes at hopcount k from the root (or at level k) is $X_N^{(k)}$, a node uniformly chosen out of N nodes in the URT has probability $\dfrac{E\left[X_N^{(k)}\right]}{N}$ of having hopcount k,

$$\Pr[h_N = k] = \frac{E\left[X_N^{(k)}\right]}{N} \tag{16.7}$$

If the size of the URT grows from n to $n+1$ nodes, each node at hopcount $k-1$ from the root can generate a node at hopcount k with probability $1/n$. Hence, for $k \geq 1$,

$$E\left[X_N^{(k)}\right] = \sum_{n=k}^{N-1} \frac{E\left[X_n^{(k-1)}\right]}{n}$$

With (16.7), a recursion for $\Pr[h_N = k]$ follows for $k \geq 1$ as

$$\Pr[h_N = k] = \frac{1}{N} \sum_{n=k}^{N-1} \Pr[h_n = k-1]$$

The generating function of h_N equals

$$\varphi_{h_N}(z) = E\left[z^{h_N}\right] = \Pr[h_N = 0] + \sum_{k=1}^{N-1} \Pr[h_N = k] z^k$$

$$= \frac{1}{N} + \frac{1}{N} \sum_{k=1}^{N-1} \sum_{n=k}^{N-1} \Pr[h_n = k-1] z^k$$

$$= \frac{1}{N} + \frac{1}{N} \sum_{n=1}^{N-1} \sum_{k=1}^{n} \Pr[h_n = k-1] z^k = \frac{1}{N} + \frac{z}{N} \sum_{n=1}^{N-1} \varphi_{h_n}(z)$$

Taking the difference between $(N+1)\varphi_{h_{N+1}}(z)$ and $N\varphi_{h_N}(z)$ results in the recursion

$$(N+1)\varphi_{h_{N+1}}(z) = (N+z)\varphi_{h_N}(z)$$

Iterating this recursion starting from $\varphi_{h_1}(z) = E\left[z^{h_1}\right] = E\left[z^0\right] = 1$ leads to (16.6).   □

Corollary 16.3.2 *The probability density function of the hopcount in the URT with N nodes is*

$$\Pr[h_N = k] = \frac{(-1)^{N-(k+1)} S_N^{(k+1)}}{N!} \tag{16.8}$$

Proof: The probability generating function $\varphi_{h_N}(z)$ in (16.6) is also the generating function of the Stirling numbers $S_N^{(k)}$ of the first kind (Abramowitz and Stegun, 1968, 24.1.3) such that the probability that a uniformly chosen node in the URT has hopcount k equals (16.8).   □

The explicit form of the generating function shows that the average hopcount h_N in a URT of size N equals

$$E[h_N] = \varphi_{h_N}'(1) = \frac{d}{dz} \log \varphi_{h_N}(z)\Big|_{z=1} = \sum_{l=2}^{N} \frac{1}{l} \tag{16.9}$$

$$= \psi(N+1) + \gamma - 1$$

where $\psi(z) = \frac{\Gamma'(z)}{\Gamma(z)}$ is the digamma function (Abramowitz and Stegun, 1968, Section 6.3) and the Euler constant is $\gamma = 0.57721\ldots$. Similarly, the variance (2.27) follows from the logarithm of the generating function $L_{h_N}(z) = \log \Gamma(N+z) - \log \Gamma(N+1) - \log \Gamma(z+1)$ as

$$\text{Var}[h_N] = \psi'(N+1) - \psi'(2) + \psi(N+1) + \gamma - 1$$

$$= \psi(N+1) + \gamma - \frac{\pi^2}{6} + \psi'(N+1)$$

Using the asymptotic formulae for the digamma function leads to

$$E[h_N] = \log N + \gamma - 1 + O\left(\frac{1}{N}\right) \tag{16.10}$$

$$\text{Var}[h_N] = \log N + \gamma - \frac{\pi^2}{6} + O\left(\frac{1}{N}\right) \tag{16.11}$$

For large N, we apply an asymptotic formula of the Gamma function (Abramowitz and Stegun, 1968, Section 6.1.47) to the generating function of the hopcount (16.6),

$$\varphi_{h_N}(z) = \frac{N^{z-1}}{\Gamma(z+1)}\left(1 + O\left(\frac{1}{N}\right)\right)$$

Introducing the Taylor series of $\frac{1}{\Gamma(z)} = \sum_{k=1}^{\infty} c_k z^k$ where the coefficients c_k are listed in Abramowitz and Stegun (1968, Section 6.1.34), we obtain with $N^z = e^{z \log N}$,

$$\varphi_{h_N}(z) = \frac{1}{N} \sum_{k=1}^{\infty} c_k z^{k-1} \sum_{k=0}^{\infty} \frac{\log^k N}{k!} z^k \left(1 + O\left(\frac{1}{N}\right)\right)$$

$$= \frac{1 + O\left(\frac{1}{N}\right)}{N} \sum_{k=0}^{\infty} \sum_{m=0}^{k} c_{m+1} \frac{\log^{k-m} N}{(k-m)!} z^k$$

With the definition (2.18) of the probability generating function, we conclude that the asymptotic form of the probability density function (16.8) of the hopcount in the URT is

$$\Pr[h_N = k] = \frac{1 + O\left(\frac{1}{N}\right)}{N} \sum_{m=0}^{k} c_{m+1} \frac{\log^{k-m} N}{(k-m)!} \tag{16.12}$$

Since the coefficients c_k are rapidly decreasing, approximating the sum in (16.12) by its first term ($m = 0$) yields to first order in N,

$$\Pr[h_N = k] \sim \frac{(\log N)^k}{N k!} \tag{16.13}$$

which is recognized as a Poisson distribution (3.9) with mean $\log N$. Hence, for large N and to first order, the average and variance of the hopcount in the URT are approximately $E[h_N] \sim \text{Var}[h_N] \sim \log N$. The accuracy of the Poisson approximation can be estimated by comparison with the average (16.10) and the variance (16.11) found above up to second order in N. For example, if the URT has $N = 10^4$ nodes, the Poisson approximation yields $E[h_N] = \text{Var}[h_N] = 9.21034$, while the average (16.10) is $E[h_N] = 8.78756$ accurate up to 10^{-4} and the variance (16.11) is $\text{Var}[h_N] = 8.14262$. The exact results are $E[h_N] = 8.78761$ and $\text{Var}[h_N] = 8.14277$.

16.3.2 Application of the URT to the hopcount in the Internet

In trace-route measurements explained in Van Mieghem (2004a), we are interested in the hopcount H_N denoted with capital H, which equals h_N in the URT excluding the event $h_N = 0$. In other words, the source and the destination are different nodes in the graph. Since from (16.8) $\Pr[h_N = 0] = \frac{(-1)^{N-1}S_N^{(1)}}{N!} = \frac{1}{N}$ we obtain, for $1 \leq k \leq N-1$,

$$
\Pr[H_N = k] = \Pr[h_N = k|h_N \neq 0] = \frac{\Pr[h_N = k, h_N \neq 0]}{\Pr[h_N \neq 0]}
$$

$$
= \frac{N}{N-1} \Pr[h_N = k]
$$

Using (16.8), we find

$$
\Pr[H_N = k] = \frac{N}{N-1} \frac{(-1)^{N-(k+1)} S_N^{(k+1)}}{N!} \tag{16.14}
$$

with corresponding generating function,

$$
\varphi_{H_N}(z) = \sum_{k=1}^{N-1} \Pr[H_N = k] z^k
$$

$$
= \frac{N}{N-1} \sum_{k=0}^{N-1} \Pr[h_N = k] z^k - \frac{N}{N-1} \Pr[h_N = 0]
$$

$$
= \frac{N}{N-1} \left(\varphi_{h_N}(z) - \frac{1}{N} \right)
$$

The average hopcount $E[H_N] = E[h_N|h_N \neq 0]$ is

$$
E[H_N] = \frac{N}{N-1} \sum_{l=2}^{N-1} \frac{1}{l} \tag{16.15}
$$

Hence, for large N and in practice, we find that

$$\Pr[H_N = k] = \Pr[h_N = k] + O\left(\frac{1}{N}\right)$$

which allows us to use the previously derived expressions (16.12), (16.10) and (16.11).

The histogram of the number of traversed routers in the Internet measured between two arbitrary communicating parties seems reasonably well modeled by the pdf (16.12). Figure 16.4 shows both the histogram of the hopcount deduced from paths in the Internet measured via the trace-route utility and the fit with (16.12). From the fit, we find a rather high number of nodes

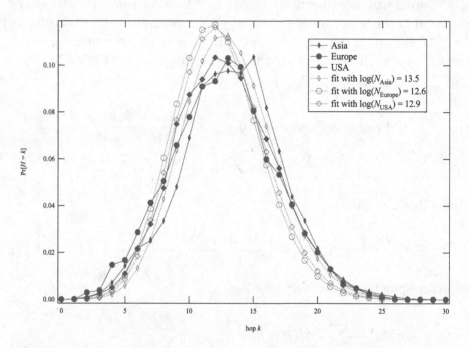

Fig. 16.4. The histograms of the hopcount derived from the trace-route measurement in three continents from CAIDA in 2004 are fitted by the pdf (16.12) of the hopcount in the URT.

$e^{12.6} \approx 3\ 10^5 \le N \le e^{13.5} \approx 7\ 10^5$, which points to the approximate nature of modeling the Internet hopcount by that deduced from a URT. The relation between Internet measurements and the properties of the URT is further analyzed in a series of articles (Van Mieghem *et al.*, 2000; van der Hofstad *et al.*, 2001; Van Mieghem *et al.*, 2001b; Janic *et al.*, 2002; van der Hofstad *et al.*, 2002b). At the time of writing, an accurate model of the hopcount in the Internet is not available.

16.4 The weight of the shortest path

The weight – sometimes also called the length – of the shortest path is defined as the sum of the link weights that constitute the shortest path. In Section 16.2.1, the shortest path tree in the complete graph with exponential link weights was shown to be a URT. In this section, we confine ourselves to the same type of graph and require that the source node A (or root) is different from the destination node B.

By Theorem 10.2.3 of a continuous-time Markov chain, the discovery time of the k-th node from node A equals $v_k = \sum_{n=1}^{k} \tau_n$, where $\tau_1, \tau_2, \ldots, \tau_k$ are independent, exponentially distributed random variables with parameter $\lambda_n = n(N-n)$ with $1 \le n \le k$. We call τ_j the interattachement time between the discovery or the attachment to the URT of the $j - 1$-th and j-th node in the graph. The Laplace transform of v_k is

$$E\left[e^{-zv_k}\right] = \int_0^\infty e^{-zt} \frac{d}{dt} \Pr\left[v_k \le t\right]$$

For a sum of independent exponential random variables, using the probability generating function (3.16), we have

$$E\left[e^{-zv_k}\right] = E\left[\exp\left(-z\sum_{n=1}^{k} \tau_n\right)\right] = \prod_{n=1}^{k} E\left[e^{-z\tau_n}\right] = \prod_{n=1}^{k} \frac{n(N-n)}{z + n(N-n)}$$

(16.16)

The probability generating function[3] $\varphi_{W_N}(z) = E\left[e^{-zW_N}\right]$ of the weight W_N of the shortest path equals

$$\varphi_{W_N}(z) = \sum_{k=1}^{N-1} E\left[e^{-zv_k}\right] \Pr\left[B \text{ is } k\text{-th attached node in URT}\right]$$

$$= \frac{1}{N-1} \sum_{k=1}^{N-1} \prod_{n=1}^{k} \frac{n(N-n)}{z + n(N-n)}$$

(16.17)

because any node apart from the root A but including the destination node B has equal probability to be the k-th attached node.

The average weight is

$$E\left[W_N\right] = -\left.\frac{d\varphi_{W_N}(z)}{dz}\right|_{z=0} = -\frac{1}{N-1} \sum_{k=1}^{N-1} \frac{d}{dz} \prod_{n=1}^{k} \frac{n(N-n)}{z + n(N-n)}\bigg|_{z=0}$$

---

[3] If the link weights have mean $\frac{1}{a}$ (instead of 1), then W_N is multiplied by a as explained in Sections 16.2.1 and 3.4.1. The weight of the scaled shortest path $W_{N,a}$ has pgf

$$\varphi_{W_{N,a}}(z) = E\left[e^{-zaW_N}\right] = \varphi_{W_N}(az)$$

Using the logarithmic derivative of the product,

$$\frac{d}{dz}\prod_{n=1}^{k}\frac{n(N-n)}{z+n(N-n)}\Bigg|_{z=0} = \prod_{n=1}^{k}\frac{n(N-n)}{z+n(N-n)}\frac{d}{dz}\left(\sum_{n=1}^{k}\log\frac{n(N-n)}{z+n(N-n)}\right)\Bigg|_{z=0}$$

$$= -\sum_{n=1}^{k}\frac{1}{n(N-n)}$$

gives

$$E\left[W_N\right] = \frac{1}{N-1}\sum_{k=1}^{N-1}\sum_{n=1}^{k}\frac{1}{n(N-n)} = \frac{1}{N-1}\sum_{n=1}^{N-1}\frac{1}{n(N-n)}\sum_{k=n}^{N-1}1$$

$$= \frac{1}{N-1}\sum_{n=1}^{N-1}\frac{N-n}{n(N-n)}$$

The average weight is

$$E\left[W_N\right] = \frac{1}{N-1}\sum_{n=1}^{N-1}\frac{1}{n} = \frac{\psi(N)+\gamma}{N-1} \qquad (16.18)$$

For large N,

$$E\left[W_N\right] = \frac{\log N + \gamma}{N} + O\left(\frac{1}{N^2}\right)$$

Similarly, the variance is computed (see problem (ii) in Section 16.9) as,

$$\text{Var}\left[W_N\right] = \frac{3}{N(N-1)}\sum_{n=1}^{N-1}\frac{1}{n^2} - \frac{\left(\sum_{n=1}^{N-1}\frac{1}{n}\right)^2}{(N-1)^2 N} \qquad (16.19)$$

and for large N,

$$\text{Var}\left[W_N\right] = \frac{\pi^2}{2N^2} + O\left(\frac{\log^2 N}{N^3}\right)$$

By inverse Laplace transform of (16.17), the distribution $\Pr\left[W_N \leq t\right]$ can be computed. The asymptotic distribution for the weight of the shortest path is (see problem (iii) in Section 16.9)

$$\lim_{N\to\infty}\Pr\left[NW_N - \log N \leq x\right] = e^{-e^{-x}} \qquad (16.20)$$

A related but slightly more complex analysis is presented in Section 16.5.1 where we study the flooding time. The interest of such an asymptotic analysis is that it often leads to tractable solutions that are physically more appealing to interpret. Moreover, it turns out that results for finite, not too small N are reasonably approximated by the asymptotic law.

Since W_N equals the sum of the link weights of the shortest path from the root to an arbitrary node and since $H_N = h_N | h_N > 0$ is the number of links in that shortest path (where the arbitrary destination node is different from the root), one may wonder whether there is a relation between them. Although the shortest path has precisely H_N hops, the destination node of that path is not necessarily the H_N-th attached node to the URT grown at the root. The destination node cannot be discovered sooner than the H_N-th attached node, otherwise the hopcount of the shortest path would be shorter than H_N. Hence, the destination node is the k-th discovered node and attached to the URT somewhere in between the $H_N - 1$-th and the last attached node. Thus, $k \in [H_N, N-1]$. If $k = H_N$, then all previously discovered nodes belong to the shortest path and the j-th attached node in the URT is linked to the $j - 1$-th, for all $j \le k$. If $k > H_N$, precisely $k - H_N$ of the attached nodes do not belong to the shortest path. Hence $W_N = W_k$ provided $k - H_N$ nodes in the URT discovered so far do not belong to the path and precisely H_N do. The latter condition requires the determination of all structurally favorable possibilities which is rather complex.

Curiously, the probability that the shortest path consists of the direct link between source and destination is, with (16.14), (16.18) and $S_N^{(2)} = (-1)^N (N-1)! \sum_{k=1}^{N-1} \frac{1}{k}$,

$$\Pr[H_N = 1] = \frac{1}{N-1} \sum_{k=1}^{N-1} \frac{1}{k} = E[W_N]$$

16.5 The flooding time T_N

The most commonly used process that informs each node (router) about changes in the network topology is called *flooding*: the source node initiates the flooding process by sending the packet with topology information to all adjacent neighbors and every router forwards the packet on all interfaces except for the incoming one and duplicate packets are discarded. Flooding is particularly simple and robust since it progresses, in fact, along all possible paths from the emitting node to the receiving node. Hence, a flooded packet reaches a node in the network in the shortest possible time (if overheads in routers are ignored). Therefore, an interesting problem lies in the determination of the flooding time T_N, which is the minimum time needed to inform all nodes in a network with N nodes. Only after a time T_N, all topology databases at each router in the network are again synchronized, i.e. all routers possess the same topology information. The flooding time T_N

is defined as the minimum time needed to reach all $N - 1$ remaining nodes from a source node over their respective shortest paths.

We will here consider the flooding time T_N in the complete graph containing N nodes and with independent, exponentially distributed link weights with mean 1. The generalization to the random graph $G_p(N)$ with i.i.d. exponential (or uniform[4]) distributed link weight is treated in van der Hofstad *et al.* (2002a).

The flooding time T_N equals the absorption time, starting from state $n = 1$ of the birth-process with rates (16.2). The probability generating function follows directly from (16.16) with $k = N - 1$,

$$\varphi_{T_N}(x) = E[e^{-xT_N}] = \int_0^\infty e^{-xt} f_{T_N}(t)\, dt = \prod_{n=1}^{N-1} \frac{n(N-n)}{n(N-n)+x} \quad (16.21)$$

The average flooding time equals

$$E[T_N] = \sum_{n=1}^{N-1} E[\tau_n] = \sum_{n=1}^{N-1} \frac{1}{n(N-n)} = \frac{2}{N} \sum_{n=1}^{N-1} \frac{1}{n} = \frac{2}{N}(\psi(N) + \gamma) \quad (16.22)$$

Using the asymptotic expansion (Abramowitz and Stegun, 1968, Section 6.3.18) of the diagamma function, we conclude that

$$E[T_N] \sim \frac{2 \log N}{N}$$

which demonstrates that the average flooding time in the complete graph with exponential link weights with mean 1 decreases to zero when $N \to \infty$. Also, the average flooding time is about twice as long as the average weight of an arbitrary shortest path (16.18). The variance of T_N equals

$$\mathrm{Var}[T_N] = \sum_{n=1}^{N-1} \mathrm{Var}[\tau_n] = \sum_{n=1}^{N-1} \frac{1}{n^2(N-n)^2} = \frac{2}{N^2} \sum_{n=1}^{N-1} \frac{1}{n^2} + \frac{4}{N^3} \sum_{n=1}^{N-1} \frac{1}{n}$$
$$(16.23)$$

For large N, we have that $\mathrm{Var}[T_N] = \frac{\pi^2}{3N^2} + O\left(\frac{\log N}{N^3}\right)$.

16.5.1 *The asymptotic law for flooding time* T_N

The exact expression $f_{T_N}(t)$ for probability density function of the flooding time T_N derived in van der Hofstad *et al.* (2002a), does not provide much

---

[4] Both the exponential and uniform distribution are regular distributions with extreme value index $\alpha = 1$. This means that the small link weights that are most likely included in the shortest path are almost identically distributed for all regular distributions with same $f_w(0)$.

insight. Because we are interested in the flooding time in *large* networks, we investigate the asymptotic distribution of T_N, for N large. We rewrite (16.21) as

$$\varphi_{T_N}(x) = \frac{[(N-1)!]^2}{\prod_{n=1}^{N-1}\left[x + \frac{N^2}{4} - \left(n - \frac{N}{2}\right)^2\right]} \qquad (16.24)$$

For $N = 2M$, using $\frac{\Gamma(z+m)}{\Gamma(z+1)} = \prod_{n=1}^{m-1}(n+z)$, we deduce that

$$\varphi_{T_{2M}}(x) = \left(\frac{\Gamma(2M)\Gamma(1 + \sqrt{x + M^2} - M)}{\Gamma(M + \sqrt{x + M^2})}\right)^2 \qquad (16.25)$$

For large M, there holds $\sqrt{x + M^2} \sim M + \frac{x}{2M}$, provided $|x| < 2M$. After substitution of $x = 2My$ in (16.25), with $|y| < 1$, we obtain

$$\varphi_{T_{2M}}(2My) \sim \Gamma^2(1 + y)\frac{\Gamma^2(2M)}{\Gamma^2(2M+y)} \sim \Gamma^2(1 + y)(2M)^{-2y}$$

from which follows the asymptotic relation

$$\lim_{N\to\infty} N^{2y}\varphi_{T_N}(Ny) = \Gamma^2(1 + y), \qquad |y| < 1 \qquad (16.26)$$

Equivalently, we have for $|y| < 1$,

$$\lim_{N\to\infty} E[e^{-y(NT_N - 2\log N)}] = \lim_{N\to\infty} \frac{1}{N}\int_{-\infty}^{\infty} e^{-yt} f_{T_N}\left(\frac{t + 2\log N}{N}\right) dt$$
$$= \Gamma^2(1 + y)$$

This limit demonstrates that the probability distribution function of the random variable $NT_N - 2\log N$ converges to a probability distribution with Laplace transform $\Gamma^2(1 + y)$. Let us define the normalized density function

$$g_N(t) = \frac{1}{N} f_{T_N}\left(\frac{t + 2\log N}{N}\right) \qquad (16.27)$$

We can prove convergence in *density*, i.e. $\lim_{N\to\infty} g_N(t) = g(t)$ and that the latter exists. By the inversion theorem for Laplace transforms we obtain for $t \in \mathbb{R}$,

$$\lim_{N\to\infty} g_N(t) = \lim_{N\to\infty} \frac{1}{2\pi i}\int_{c-i\infty}^{c+i\infty} e^{yt} N^{2y}\varphi_{T_N}(Ny) dy$$

where $0 < c < 1$. Since $\Gamma(z)$ is analytic over the entire complex plane except for simple poles at the points $z = -n$ for $n = 0, 1, 2, ...$, we find that $N^{2y}\varphi_{T_N}(Ny)$ is analytic whenever the real part of y is non-negative. Evaluation along the line $\text{Re}(y) = c = 0$ then gives

$$\lim_{N\to\infty} g_N(t) = \lim_{N\to\infty} \frac{1}{2\pi}\int_{-\infty}^{\infty} e^{itu} N^{2iu}\varphi_{T_N}(iNu) du$$

As dominating function we take

$$\left| e^{itu} N^{2iu} \varphi_{T_N}(iNu) \right| = \left| \varphi_{T_N}(iNu) \right| \le \frac{1+u^2}{u^4}$$

when $|u| > 1$, and $|\varphi_{T_N}(iNu)| \le 1$, for $|u| \le 1$. This follows from the first equality in (16.24), using only the factors in the product with $n = 1$ and $n = N - 1$, and bounding the other factors using

$$\frac{n(N-n)}{|n(N-n) + iNu|} \le 1$$

The Dominated Convergence Theorem 6.1.4 allows us to interchange the limit and integration operator such that

$$\lim_{N \to \infty} g_N(t) = \frac{1}{2\pi} \int_{-\infty}^{\infty} e^{itu} \lim_{N \to \infty} N^{2iu} \varphi_{T_N}(iNu) du = \frac{1}{2\pi i} \int_{-i\infty}^{i\infty} e^{ty} \lim_{N \to \infty} N^{2y} \varphi_{T_N}(Ny) dy$$

$$= \frac{1}{2\pi i} \int_{-i\infty}^{i\infty} e^{ty} \Gamma^2(1+y) dy \qquad (16.28)$$

The right-hand side of (16.26) is a perfect square, which indicates that the limit distribution is a two-fold convolution. Now, the Mellin transform (Titchmarsh, 1948) of the exponential function is

$$e^{-t} = \frac{1}{2\pi i} \int_{c-i\infty}^{c+i\infty} t^{-y} \Gamma(y) \, dy, \qquad c > 0$$

and thus with $t = e^{-u}$,

$$\frac{d}{du} \left(e^{-e^{-u}} \right) = \frac{1}{2\pi i} \int_{c-i\infty}^{c+i\infty} e^{yu} \Gamma(y+1) \, dy$$

which shows that (16.28) is the two-fold convolution of the probability density function $\frac{d}{dt} \Lambda(t)$, where $\Lambda(t) = e^{-e^{-t}}$ is the Gumbel distribution (3.37). Furthermore, the two-fold convolution is given by

$$\frac{d}{dt} \left(\Lambda^{(2*)}(t) \right) = e^{-t} \int_{-\infty}^{\infty} e^{-e^{-u}} e^{-e^{-(t-u)}} du$$

$$= e^{-t} \int_{-\infty}^{\infty} \exp \left[-2e^{-t/2} \cosh \left(\frac{t}{2} - u \right) \right] du$$

$$= 2e^{-t} \int_{0}^{\infty} \exp \left[-2e^{-t/2} \cosh(u) \right] du = 2e^{-t} K_0 \left(2e^{-t/2} \right)$$

where $K_\nu(x)$ denotes the modified Bessel function (Abramowitz and Stegun, 1968, Section 9.6) of order ν.

In summary,

$$\lim_{N \to \infty} g_N(t) = g(t) = \frac{d}{dt} \left(\Lambda^{(2*)}(t) \right) = 2e^{-t} K_0 \left(2e^{-t/2} \right) \qquad (16.29)$$

and the corresponding distribution function is

$$\lim_{N \to \infty} \Pr[NT_N - 2\log N \le z] = 2 \int_{-\infty}^{z} e^{-t} K_0(2e^{-t/2}) dt = 2e^{-z/2} K_1\left(2e^{-z/2}\right)$$

(16.30)

The right-hand side of (16.29) is maximal for $t = 0.506357$, which is slightly smaller than $\gamma = 0.577261$, but still in accordance with $E[T_N]$ given by (16.22). The asymmetry shows that $\{NT_N \ge 2\log N + z\}$ is much more

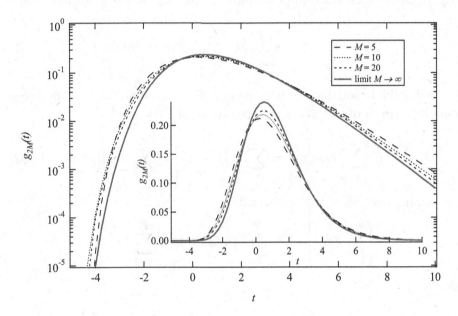

Fig. 16.5. The scaled density $g_N(t)$ for three values of $N = 2M$ (dotted lines) and the asymptotic result (full line) on a log-lin scale. The insert is drawn on a lin-lin scale.

likely than the event $\{NT_N \le 2\log N - z\}$, which confirms the intuition that the flooding time can be much longer than the average $E[T_N]$, but not so much shorter than $E[T_N]$. Figure 16.5 illustrates the convergence of $g_N(t)$ to the limit in (16.29). When comparing (16.26) with the corresponding result (C.6) for the weight of the shortest path, we observe that, for large N, the random variable $NT_N - 2\log N$ consists of the sum of $NW_{N;1} - \log N + NW_{N;2} - \log N$, where both $NW_{N;j} - \log N$ are i.i.d. random variables. Intuitively, we can say that the flooding time consists of the time to travel from a left-hand corner of the graph to the center and from the center to a right-hand corner of the graph.

The asymptotic distribution (16.30) is a beautiful example of a sum of N

independent random variables that clearly does not converge to a Gaussian and, hence, does not obey the (extended) Central Limit Theorem 6.3.1.

16.6 The degree of a node in the URT

Let us denote by $\left\{D_N^{(k)}\right\}$ the set of nodes with degree k in a graph with N nodes and by $D_N^{(k)}$ the cardinality (the number of elements) of this set $\left\{D_N^{(k)}\right\}$. Since each node appears only in one set, it holds for any graph that

$$\sum_{k=1}^{N-1} D_N^{(k)} = N \tag{16.31}$$

In a probabilistic setting, we may investigate the event that the degree k occurs in a graph of size N. The expectation of that event is

$$E\left[D_N^{(k)}\right] = E\left[\sum_{j=1}^{N} 1_{\{d_j = k\}}\right] = \sum_{j=1}^{N} E\left[1_{\{d_j = k\}}\right] = \sum_{j=1}^{N} \Pr\left[d_j = k\right] \tag{16.32}$$

By summing over all k, we verify that

$$E\left[\sum_{k=1}^{N-1} D_N^{(k)}\right] = \sum_{j=1}^{N} \sum_{k=1}^{N-1} \Pr\left[d_j = k\right] = \sum_{j=1}^{N} 1 = N$$

which is again (16.31).

16.6.1 Recursion for $\Pr\left[D_N^{(k)} = j\right]$ in the URT

The growth law of URTs dictates the way a specific tree of size N transforms to the tree of size $N + 1$ by adding the node with label $N + 1$ at random. Based on this growth law, the set of nodes with degree k in a specific tree of size $N + 1$ consists of:

(i)   the same set of nodes with degree k in the ancestor tree of size N provided the new node n_{N+1} is *not* attached to any of the nodes of this set nor to any of the nodes with degree $k - 1$;

(ii) the same set of nodes with degree k except for one, say node n_l, provided the new node n_{N+1} is attached to that node n_l;

(iii) the same set of nodes with degree k and one additional node of the set of $k - 1$ degree nodes provided the new node n_{N+1} is attached to a node of the set of degree $k - 1$.

The evolution scenario in three parts is generally applicable for any class of trees that possess a growth law. It does not hold for graphs in general because only in a tree, a node has one well-defined parent node and the in-degree is one. Using the law of total probability (2.46) yields,

$$\Pr\left[D_{N+1}^{(k)} = j\right] = \Pr\left[D_N^{(k)} = j \mid n_{N+1} \notin \left\{D_N^{(k)}, D_N^{(k-1)}\right\}\right] \Pr\left[n_{N+1} \notin \left\{D_N^{(k)}, D_N^{(k-1)}\right\}\right]$$
$$+ \Pr\left[D_N^{(k)} = j+1 \mid n_{N+1} \in \left\{D_N^{(k)}\right\}\right] \Pr\left[n_{N+1} \in \left\{D_N^{(k)}\right\}\right]$$
$$+ \Pr\left[D_N^{(k)} = j-1 \mid n_{N+1} \in \left\{D_N^{(k-1)}\right\}\right] \Pr\left[n_{N+1} \in \left\{D_N^{(k-1)}\right\}\right]$$

If the process of attaching a new node $N+1$ does not depend on the way the N previous nodes are attached but rather on their number, there holds $\Pr\left[D_N^{(k)} = j \mid n_{N+1}\right] = \Pr\left[D_N^{(k)} = j\right]$. This property holds for the URT. We obtain a three point recursion for $k > 1$,

$$\Pr\left[D_{N+1}^{(k)} = j\right] = \Pr\left[D_N^{(k)} = j\right] \Pr\left[n_{N+1} \notin \left\{D_N^{(k)}, D_N^{(k-1)}\right\}\right]$$
$$+ \Pr\left[D_N^{(k)} = j+1\right] \Pr\left[n_{N+1} \in \left\{D_N^{(k)}\right\}\right]$$
$$+ \Pr\left[D_N^{(k)} = j-1\right] \Pr\left[n_{N+1} \in \left\{D_N^{(k-1)}\right\}\right]$$

The probability generating function

$$\varphi_D(z, N; k) = E\left[z^{D_N^{(k)}}\right] = \sum_{j=0}^{\infty} \Pr\left[D_N^{(k)} = j\right] z^j$$

is obtained after multiplication by z^j and summing over all j,

$$\varphi_D(z, N+1; k) = \Pr\left[n_{N+1} \notin \left\{D_N^{(k)}, D_N^{(k-1)}\right\}\right] \varphi_D(z, N; k)$$
$$+ \Pr\left[n_{N+1} \in \left\{D_N^{(k)}\right\}\right] \frac{\varphi_D(z, N; k) - \varphi_D(0, N; k)}{z}$$
$$+ \Pr\left[n_{N+1} \in \left\{D_N^{(k-1)}\right\}\right] z\varphi_D(z, N; k)$$

Now $\varphi_D(0, N; k) = \Pr\left[D_N^{(k)} = 0\right]$ is the probability of the event that there are no nodes with degree k for $1 \le k \le k_{\max} \le N - 1$. Since the normalization of the generating function requires that $\varphi_D(1, N; k) = 1$ and since

$$\Pr\left[n_{N+1} \notin \left\{D_N^{(k)}, D_N^{(k-1)}\right\}\right] + \Pr\left[n_{N+1} \in \left\{D_N^{(k)}\right\}\right] + \Pr\left[n_{N+1} \in \left\{D_N^{(k-1)}\right\}\right] = 1$$

it follows that $\Pr\left[D_N^{(k)} = 0\right] \Pr\left[n_{N+1} \in \left\{D_N^{(k)}\right\}\right] = 0$. Further,

$$\Pr\left[n_{N+1} \in \left\{D_N^{(k)}\right\}\right] \ne 0$$

for any $k \in [1, k_{\max}]$ because the attachment of the node n_{N+1} is possible to any non-empty set, this means that the absence of nodes with degree $k \in [1, k_{\max}]$ cannot occur in URTs, thus $\Pr\left[D_N^{(k)} = 0\right] = 0$. A consequence is that the probability generating function $\varphi_D(z, N; k)$ is at least $O(z)$ as $z \to 0$ (for $N > 1$). After using $\varphi_D(0, N; k) = 0$ and eliminating $\Pr\left[n_{N+1} \notin \left\{D_N^{(k)}, D_N^{(k-1)}\right\}\right]$, the recursion relation for the probability generating function becomes[5]

$$\frac{\varphi_D(z, N+1; k)}{\varphi_D(z, N; k)} = 1 + \left(\frac{\Pr\left[n_{N+1} \in \left\{D_N^{(k)}\right\}\right]}{z} - \Pr\left[n_{N+1} \in \left\{D_N^{(k-1)}\right\}\right]\right)(1-z)$$

(16.33)

The special case for $k = 1$ and $N > 1$ is

$$\varphi_D(z, N+1; 1) = \left(1 + \frac{\Pr\left[n_{N+1} \in \left\{D_N^{(1)}\right\}\right](1-z)}{z}\right)\varphi_D(z, N; 1)$$

16.6.2 The Average Number of Degree k Nodes in the URT

In the URT, a new node n_{N+1} is attached uniformly to any of N previously attached nodes such that

$$\Pr\left[n_{N+1} \in \left\{D_N^{(k)}\right\}\right] = \frac{E\left[D_N^{(k)}\right]}{N}$$

Also, the probability that an arbitrary node in a URT of size N has degree k equals $\frac{E\left[D_N^{(k)}\right]}{N}$. We obtain from (16.33)

$$\varphi_D(z, N+1; k) = \left(1 + \left(\frac{E\left[D_N^{(k)}\right]}{Nz} - \frac{E\left[D_N^{(k-1)}\right]}{N}\right)(1-z)\right)\varphi_D(z, N; k)$$

(16.34)

---

[5] With the initialization $\varphi_D(z, m; k) = E\left[z^{D_m^{(k)}}\right] = 1$ for all $m \leq k$ because $D_m^{(k)} = 0$ if $1 < m \leq k$, after iterating (16.33) we arrive at

$$\varphi_D(z, N; k) = \prod_{m=k}^{N-1} \left(1 - \Pr\left[n_{m+1} \in \left\{D_m^{(k)}\right\}\right]\left(1 - \frac{1}{z}\right) - \Pr\left[n_{m+1} \in \left\{D_m^{(k-1)}\right\}\right](1-z)\right)$$

By taking the derivative of both sides in (16.34) with respect to z and evaluating at $z = 1$, a recursion for the average is found,

$$E\left[D_{N+1}^{(k)}\right] = \frac{N-1}{N} E\left[D_N^{(k)}\right] + \frac{E\left[D_N^{(k-1)}\right]}{N} \tag{16.35}$$

Let $r_N^{(k)} = (N-1)E\left[D_N^{(k)}\right]$, then the recursion valid for $1 < k \le N - 2$ becomes

$$r_{N+1}^{(k)} = r_N^{(k)} + \frac{r_N^{k-1}}{N-1} \tag{16.36}$$

Theorem 16.6.1 *In the URT, the average number of degree k nodes is given by*

$$E\left[D_N^{(k)}\right] = \frac{N}{2^k} + \frac{(-1)^{N+k-1} S_{N-1}^{(k)}}{(N-1)!} + \frac{(-1)^N}{2^k (N-1)!} \sum_{j=1}^{k-1} S_{N-1}^{(j)} (-2)^j \tag{16.37}$$

Proof: See Section 16.8. □

For large N and using the asymptotics of the Stirling numbers $S_N^{(k)}$ of the first kind (Abramowitz and Stegun, 1968, Section 24.1.3.III), the asymptotic law is

$$\Pr\left[D_{\mathrm{URT}} = k\right] = \frac{E\left[D_N^{(k)}\right]}{N} = \frac{1}{2^k} + O\left(\frac{\log^{k-1} N}{N^2}\right) \tag{16.38}$$

The ratio of the average number of nodes with degree k over the total number of nodes, which equals the probability that an arbitrary node in a URT of size N has degree k, decreases exponentially fast with rate $\ln 2$.

The variance $\mathrm{Var}\left[D_N^{(k)}\right]$ is most conveniently computed from the logarithm of the probability generating function with (2.27). By taking the logarithm of both sides in (16.34) and differentiating twice and adding (16.35), we obtain

$$\mathrm{Var}\left[D_{N+1}^{(k)}\right] = f(N;k) + \mathrm{Var}\left[D_N^{(k)}\right]$$

where

$$f(N;k) = -\left(-\frac{E\left[D_N^{(k)}\right]}{N} + \frac{E\left[D_N^{(k-1)}\right]}{N}\right)^2 + \left(\frac{E\left[D_N^{(k)}\right]}{N} + \frac{E\left[D_N^{(k-1)}\right]}{N}\right)$$

Since $\mathrm{Var}\left[D_m^{(k)}\right] = 0$ for $m \le k$, the general solution is

$$\mathrm{Var}\left[D_N^{(k)}\right] = \sum_{j=k}^{N} f\left(j; k\right)$$

For large N, using (16.38), we observe that

$$\frac{\mathrm{Var}\left[D_N^{(k)}\right]}{N} = \left(\frac{3}{2^k} - \frac{1}{2^{2k}}\right) + O\left(\frac{\log^{2k-2} N}{N^2}\right) \tag{16.39}$$

In practice, if we use the estimator $\hat{t}_N^{(k)} = \frac{D_N^{(k)}}{N}$ for the probability that the degree of a node equals k, then (a) the estimator is unbiased because the mean of the estimator $E\left[\hat{t}_N^{(k)}\right]$ equals the correct mean $\frac{E\left[D_N^{(k)}\right]}{N}$ and (b) the variance $\mathrm{Var}\left[\hat{t}_N^{(k)}\right] = \mathrm{Var}\left[\frac{D_N^{(k)}}{N}\right] = \frac{\mathrm{Var}\left[D_N^{(k)}\right]}{N^2} \to 0$ as $O\left(\frac{1}{N}\right)$ for large N.

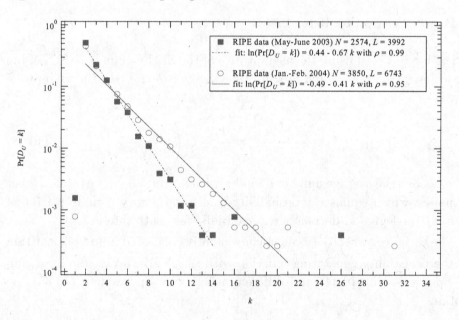

Fig. 16.6. The histogram of the degree D_U derived from the graph G_U formed by the union of paths measured via trace-route in the Internet. Both measurements in 2003 and 2004 are fitted on a log-lin plot and the correlation coefficient ρ quantifies the quality of the fit.

The law (16.38) is observed in Fig. 16.6, which plots the histogram of the degree D_U in the graph G_U. The graph G_U is obtained from the union of trace-routes from each RIPE measurement box to any other box positioned mainly in the European part of the Internet. For about 50 measurement boxes in 2003, the correspondence is striking because the slope of the fit on a log-lin scale equals -0.668 while the law (16.38) gives $-\ln 2 = -0.693$. Ignoring in Fig. 16.6 the leave nodes with $k = 1$ suggests that the graph G_U is URT-like. For 72 measurement boxes in 2004 which obviously results in a larger graph G_U, deviations from the URT law (16.38) are observed. If measurements between a larger full mesh of boxes were possible and if the measurement boxes were more homogeneously spread over the Internet, a power law behavior is likely to be expected as mentioned in Section 15.3. However, these earlier reported trace-route measurements that lead to power law degrees have been performed from a relatively small number of sources to a large number of destinations. These results question the observability of the Internet: how accurate are Internet properties such as hopcount and degree that are derived from incomplete measurements, i.e. from a selected small subset of nodes at which measurement boxes are placed?

16.6.3 The degree of the shortest path tree in the complete graph with i.i.d. exponential link weights

In the complete graph K_N with i.i.d. exponential link weights, any node n possesses equal properties in probability because of symmetry. If we denote by d_n the degree of node n in the shortest path tree rooted at that node n, the symmetry implies that $\Pr[d_n = k] = \Pr[d_i = k]$ for any node n and i. In fact, we consider here the degree of a URT as an overlay tree in a complete graph. Concentrating on a node with label 1, we obtain from (16.32)

$$E\left[D_N^{(k)}\right] = N\Pr[d_1 = k] = N\Pr\left[X_N^{(1)} = k\right]$$

The latter follows from the fact that the degree of a node is equal to the number of its direct neighbors, the nodes at level 1, $X_N^{(1)}$. By definition of the URT, the second node surely belongs to the level set 1, while node 3 has equal probability to be attached to the root or to node 2. In general, when attaching a node j to a URT of size $j - 1$, the probability that node j is attached to the root equals $\frac{1}{j-1}$. Thus, the number of nodes at level 1 in the URT (constructed upon the complete graph) is in distribution equal to the sum of $N - 1$ independent Bernoulli random variables each with different mean $\frac{1}{j-1}$,

$$X_N^{(1)} \overset{d}{=} \sum_{j=2}^{N} \text{Bernoulli}\left(\frac{1}{j-1}\right) = \sum_{j=1}^{N-1} \text{Bernoulli}\left(\frac{1}{j}\right)$$

because each node in the complete graph is connected to $N - 1$ neighbors. The generating function is

$$E\left[z^{X_N^{(1)}}\right] = E\left[z^{\sum_{j=1}^{N-1} \text{Bernoulli}\left(\frac{1}{j}\right)}\right] = \prod_{j=1}^{N-1} E\left[z^{\text{Bernoulli}\left(\frac{1}{j}\right)}\right]$$

Using the probability generating function (3.1) of a Bernoulli random variable, $E\left[z^{\text{Bernoulli}\left(\frac{1}{j}\right)}\right] = 1 - \frac{1}{j} + \frac{z}{j}$, yields

$$E\left[z^{X_N^{(1)}}\right] = \prod_{j=1}^{N-1} \left(\frac{z+j-1}{j}\right) = \frac{\Gamma(z+N-1)}{\Gamma(z)\Gamma(N)}$$

Compared to the generating function (16.6) of the hopcount h_N, we recognize that

$$E\left[z^{X_N^{(1)}}\right] = z\varphi_{N-1}(z) = \sum_{k=0}^{N-2} \Pr[h_{N-1} = k]z^{k+1} = \sum_{k=1}^{N-1} \Pr[h_{N-1} = k-1]z^k$$

from which we deduce, for $1 \leq k \leq N - 1$,

$$\Pr\left[X_N^{(1)} = k\right] = \Pr[h_{N-1} = k-1]$$

Using (16.7), we arrive at the curious result

$$\Pr\left[X_N^{(1)} = k\right] = \frac{E\left[X_{N-1}^{(k-1)}\right]}{N-1} \qquad \text{for } 1 \leq k \leq N - 1.$$

The probability that the number of level 1 nodes in the shortest path tree in the complete graph with i.i.d. exponential link weights is k equals the average number of nodes on level $k - 1$ in a URT of size $N - 1$ divided by that size $N - 1$. In other words, the "horizontal" distribution at level 1 is related to the "vertical" distribution of the size of the level sets. In summary[6], in the complete graph with i.i.d. exponential link weights, the probability that an arbitrary node n as root of a shortest path tree has degree k is

$$\Pr\left[d_n = k\right] = \Pr[h_{N-1} = k-1] = \frac{(-1)^{N-1-k}S_{N-1}^{(k)}}{(N-1)!} \qquad (16.40)$$

The degree of an arbitrary node in the union of all shortest paths trees in the complete graph K_N with i.i.d. exponential link weights is also given by (16.40) because in that union each node n is once a root and further plays,

---

[6] This result is due to Remco van der Hofstad (private communication).

by symmetry, the role of the j-th attached node in the URT rooted at any other node in K_N.

16.7 The minimum spanning tree

From an algorithmic point of view, the shortest path problem is closely related to the computation of the minimum spanning tree (MST). The Dijkstra shortest path algorithm is similar to Prim's minimum spanning tree algorithm (Cormen *et al.*, 1991). In this section, we compute the average weight of the MST in a graph with a general link weight structure.

16.7.1 The Kruskal growth process of the MST

Since the link weights in the underlying complete graph are chosen independently and assigned randomly to links in the complete graph, the resulting graph is probabilistically the same if we first order the set of link weights and assign them in increasing order randomly to links in the complete graph. In the latter construction process, only the order statistics or the ranking of the link weights suffice to construct the graph because the precise link weight can be unambiguously associated to the rank of a link. This observation immediately favors the Kruskal algorithm for the MST over Prim's algorithm. Although the Prim algorithm leads to the same MST, it gives a more complicated, long-memory growth process, where the attachment of each new node depends stochastically on the whole growth history so far. Pietronero and Schneider (1990) illustrate that in our approach Prim, in contrast with Kruskal, leads to a very complicated stochastic process for the construction of the MST.

The Kruskal growth process described here is closely related to a growth process of the random graph $G_r(N, L)$ with N nodes and L links. The construction or growth of $G_r(N, L)$ starts from N individual nodes and in each step an arbitrary, not yet connected random pairs is connected. The only difference with Kruskal's algorithm for the MST is that, in Kruskal, links generating loops are forbidden. Those forbidden links are the links that connect nodes within the same connected component or "cluster". As a result, the internal wiring of the clusters differs, but the cluster size statistics (counted in nodes, not links) is exactly the same as in the corresponding random graph. The metacode of the Kruskal growth process for the construction of the MST is shown in Fig. 16.7.

The growth process of the random graph $G_p(N)$, which is asymptotically equal to that of $G_r(N, L)$, is quantified in Section 15.6.4 for large N. The

---

KRUSKALGROWTHMST

1. start with N disconnected nodes
2. **repeat until** all nodes are connected
3.   randomly select a node pair (i, j)
4.   **if** a path $\mathcal{P}_{i \to j}$ does not exist
5.   **then** connect i to j

Fig. 16.7. Kruskal growth process

fraction of nodes S in the giant component of $G_p(N)$ is related to the average degree or to the link density p because $\mu_{\mathrm{rg}} = p(N-1)$ in $G_p(N)$ by (15.20). For large N, the size of the giant cluster in the forest is thus determined as a function of the number of added links that increase μ_{rg}.

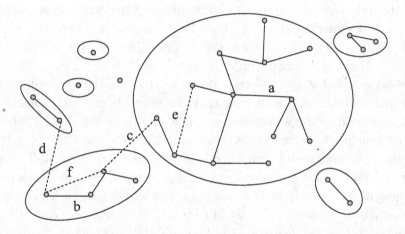

Fig. 16.8. Component structure during the Kruskal growth process.

We will now transform the mean degree μ_{rg} in the random graph $G_p(N)$ to the mean degree μ_{MST} in the corresponding stage in Kruskal growth process of the MST. In early stages of the growth each selected link will be added with high probability such that $\mu_{\mathrm{MST}} = \mu_{\mathrm{rg}}$ almost surely. After some time the probability that a selected link is forbidden increases, and thus μ_{rg} exceeds μ_{MST}. In the end, when connectivity of all N nodes is reached, $\mu_{\mathrm{MST}} = 2$ (since it is a tree) while $\mu_{\mathrm{rg}} = O(\log N)$, as follows from (15.19) and the critical threshold $p_c \sim \frac{\log N}{N}$.

Consider now an intermediate stage of the growth as illustrated in Fig. 16.8. Assume there is a giant component of average size NS and $n_l = N(1-S)/s_l$ small components of average size s_l each. Then we can distinguish six types of links labelled a-f in Fig. 16.8. Types a and b are links that have been chosen earlier in the giant component (a) and in the small components (b) respectively. Types c and d are eligible links between the giant component and a small component (c) and between small components (d) respectively. Types e and f are forbidden links connecting nodes within the giant component (e), respectively within a small component (f). For large N, we can enumerate the average number of links L_x of each type x:

$$
\begin{array}{l|l}
L_a + L_b = \frac{1}{2}\mu_{MST}N & L_e = \frac{1}{2}(SN)^2 - SN \\
L_c = SN \cdot (1-S)N & L_f = \frac{1}{2}n_l s_l(s_l - 1) - n_l(s_l - 1) \\
L_d = \frac{1}{2}n_l^2 \cdot s_l^2 &
\end{array}
$$

To highest order in $O(N^2)$, we have

$$
L_c = N^2 S(1-S), \quad L_d = \frac{1}{2}N^2(1-S)^2, \quad L_e = \frac{1}{2}N^2 S^2
$$

The probability that a randomly selected link is eligible is $q = \frac{L_c + L_d}{L_c + L_d + L_e + L_f}$ or, to order $O(N^2)$,

$$
q = 1 - S^2 \tag{16.41}
$$

In contrast with the growth of the random graph $G_p(N)$ where at each stage a link is added with probability p, in the Kruskal growth of the MST we are only successful to add one link (with probability 1) per $\frac{1}{q}$ stages on average. Thus the average number of links added in the random graph corresponding to one link in the MST is $\frac{1}{q} = \frac{1}{1-S^2}$. This provides an asymptotic mapping between μ_{rg} and μ_{MST} in the form of a differential equation,

$$
\frac{d\mu_{\mathrm{rg}}}{d\mu_{\mathrm{MST}}} = \frac{1}{1 - S^2}
$$

By using (15.22), we find

$$
\frac{d\mu_{\mathrm{MST}}}{dS} = \frac{d\mu_{\mathrm{MST}}}{d\mu_{\mathrm{rg}}} \frac{d\mu_{\mathrm{rg}}}{dS} = \frac{(1+S)(S + (1-S)\log(1-S))}{S^2}
$$

Integration with the initial condition $\mu_{\mathrm{MST}} = 2$ at $S = 1$, finally gives the average degree μ_{MST} in the MST as function of the fraction S of nodes in the giant component

$$
\mu_{\mathrm{MST}}(S) = 2S - \frac{(1-S)^2}{S}\log(1-S) \tag{16.42}
$$

As shown in Fig. 16.9, the asymptotic result (16.42) agrees well with the simulation (even for a single sample), except in a small region around the transition $\mu_{\mathrm{MST}} = 1$ and for relatively small N.

The key observation is that all transition probabilities in the Kruskal

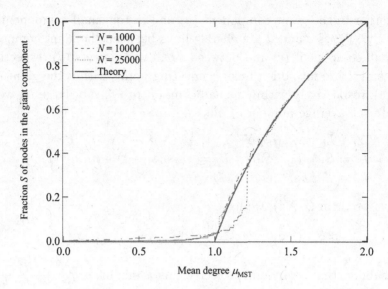

Fig. 16.9. Size of the giant component (divided by N) as a function of the mean degree μ_{MST}. Each simulation for a different number of nodes N consists of one MST sample.

growth process asymptotically depend on merely one parameter S, the fraction of nodes in the giant component, and S is called an *order parameter* in statistical physics. In general, the expectation of an order parameter distinguishes the qualitatively different regimes (states) below and above the phase transition. In higher dimensions, fluctuations of the order parameter around the mean can be neglected and the mean value can be computed from a self-consistent mean-field theory. In our problem, the underlying complete (or random) graph topology makes the problem effectively infinite-dimensional. The argument leading to (15.20) is essentially a mean-field argument.

16.7.2 The average weight of the minimum spanning tree

By definition, the weight of the MST is

$$W_{\mathrm{MST}} = \sum_{j=1}^{L} w_{(j)} 1_{j \in \mathrm{MST}} \qquad (16.43)$$

where $w_{(j)}$ is the j-th smallest link weight. The average MST weight is

$$E\left[W_{\mathrm{MST}}\right] = \sum_{j=1}^{L} E\left[w_{(j)} 1_{j \in \mathrm{MST}}\right]$$

The random variables $w_{(j)}$ and $1_{j \in \text{MST}}$ are independent because the j-th smallest link weight $w_{(j)}$ only depends on the link weight distribution and the number of links L, while the appearance of the j-th link in the MST only depends on the graph's topology, as shown in Section 16.7.1. Hence,

$$E\left[w_{(j)} 1_{j \in \text{MST}}\right] = E\left[w_{(j)}\right] E\left[1_{j \in \text{MST}}\right] = E\left[w_{(j)}\right] \Pr\left[j \in \text{MST}\right]$$

such that the average weight of the MST is

$$E\left[W_{\text{MST}}\right] = \sum_{j=1}^{L} E\left[w_{(j)}\right] \Pr\left[j \in \text{MST}\right] \tag{16.44}$$

In general for independent link weights with probability density function $f_w(x)$ and distribution function $F_w(x) = \Pr\left[w \le x\right]$, the probability density function of the j-th order statistic follows from (3.36) as

$$f_{w_{(j)}}(x) = \frac{j f_w(x)}{F_w(x)} \binom{L}{j} (F_w(x))^j (1 - F_w(x))^{L-j} \tag{16.45}$$

The factor $\binom{L}{j} (F_w(x))^j (1 - F_w(x))^{L-j}$ is a binomial distribution with mean $\mu = F_w(x) L$ and variance $\sigma^2 = L F_w(x)(1 - F_w(x))$ that, by the Central Limit Theory 6.3.1, tends for large L to a Gaussian $\frac{1}{\sigma\sqrt{2\pi}} e^{-\frac{(j-\mu)^2}{2\sigma^2}}$, which peaks at $j = \mu$. For large N and fixed $\frac{j}{L}$, we have[7] $x_j = E\left[w_{(j)}\right] \simeq F_w^{-1}(\frac{j}{L})$.

We found before in (16.41) that the link ranked j appears in the MST with probability

$$\Pr\left[j \in \text{MST}\right] = 1 - S_j^2$$

where S_j is the fraction of nodes in the giant component during the construction process of the random graph at the stage where the number of links precisely equals j. Since links are added independently, that stage in fact establishes the random graph $G_r(N, L = j)$. Our graph under consideration is the complete graph K_N such that we add in total $L = \binom{N}{2}$ links.

---

[7] In general, it holds that $w_{(k)} = F_w^{-1}(U_{(k)})$ and

$$E\left[w_{(k)}\right] = E\left[F_w^{-1}(U_{(k)})\right] \ne F_w^{-1}(E\left[U_{(k)}\right])$$

but, for a large number of order statistics L, the Central Limit Theorem 6.3.1 leads to

$$E\left[w_{(k)}\right] \simeq F_w^{-1}\left(\frac{j}{L}\right) \simeq F_w^{-1}(E\left[U_{(k)}\right])$$

because for a uniform random variable U on [0,1] the average weight of the j-th smallest link is exactly

$$E\left[w_{(i)}\right] = \frac{j}{L+1} \simeq \frac{j}{L}$$

With (15.22) and $\mu_{\rm rg} = \frac{2L}{N}$, it follows that

$$\frac{2j}{N} = -\frac{\log(1 - S_j)}{S_j} \tag{16.46}$$

Hence,

$$E\left[W_{\rm MST}\right] \simeq \sum_{j=1}^{L} F_w^{-1}\left(\frac{j}{L}\right)\left(1 - S_j^2\right)$$

We now approximate the sum by an integral,

$$E\left[W_{\rm MST}\right] \simeq \int_1^L F_w^{-1}\left(\frac{u}{L}\right)\left(1 - S_u^2\right) du$$

Substituting $x = \frac{2u}{N}$ (which is the average degree in any graph $G\left(N, u\right)$) yields for large N where $L \simeq \frac{N^2}{2}$,

$$E\left[W_{\rm MST}\right] \simeq \frac{N}{2} \int_{\frac{2}{N}}^{N-1} F_w^{-1}\left(\frac{x}{N}\right)\left(1 - S_{\frac{N}{2}x}^2\right) dx \simeq \frac{N}{2} \int_0^N F_w^{-1}\left(\frac{x}{N}\right)\left(1 - S^2(x)\right) dx$$

It is known (Janson *et al.*, 1993) that, if the number of links in the growth process of the random graph is below $\frac{N}{2}$, with high probability (and ignoring a small onset region just below $\frac{N}{2}$), there is no giant component such that $S\left(x\right) = 0$ for $x \in [0, 1]$. Thus, we arrive at the general formula valid for large N,

$$E\left[W_{\rm MST}\right] \simeq \frac{N}{2} \int_0^1 F_w^{-1}\left(\frac{x}{N}\right) dx + \frac{N}{2} \int_1^N F_w^{-1}\left(\frac{x}{N}\right)\left(1 - S^2(x)\right) dx \tag{16.47}$$

The first term is the contribution from the smallest $N/2$ links in the graph, which are included in the MST almost surely. The remaining part comes from the more expensive links in the graph, which are included with diminishing probability since $1 - S^2(x)$ decreases exponentially for large x as can be deduced from (15.21). The rapid decrease of $1 - S^2(x)$ makes only relatively small values of the argument $F_w^{-1}\left(\frac{x}{N}\right)$ contribute to the second integral.

At this point, the specifics of the link weight distribution needs to be introduced. The Taylor expansion of $\frac{N}{2}F_w^{-1}\left(\frac{x}{N}\right)$ for large N to first order is

$$\frac{N}{2}F_w^{-1}\left(\frac{x}{N}\right) = \frac{N}{2}F_w^{-1}(0) + \frac{x}{2f_w(0)} + O\left(\frac{1}{N}\right) = \frac{x}{2f_w(0)} + O\left(\frac{1}{N}\right)$$

since we require that link weights are positive such that $F_w^{-1}(0) = 0$. This expansion is only useful provided f_w is regular, i.e. $f_w(0)$ is neither zero nor infinity. These cases occur, for example, for polynomial link weights with $f_w(x) = \alpha x^{\alpha-1}$ and $\alpha \neq 1$. For polynomial link weights, however, holds that $\frac{N}{2} F_w^{-1}\left(\frac{x}{N}\right) = \frac{N^{1-\frac{1}{\alpha}}}{2} x^{\frac{1}{\alpha}}$. Formally, this latter expression reduces to the first order Taylor approach for $\alpha = 1$, apart from the constant factor $\frac{1}{f_w(0)}$. Therefore, we will first compute $E[W_{\mathrm{MST}}]$ for polynomial link weights and then return to the case in which the Taylor expansion is useful.

16.7.2.1 Polynomial link weights

The average weight of the MST for polynomial link weights follows[8] from (16.47) as

$$E[W_{\mathrm{MST}}(\alpha)] \simeq \frac{N^{1-\frac{1}{\alpha}}}{2}\left(\frac{1}{\frac{1}{\alpha}+1} + \int_1^N x^{\frac{1}{\alpha}}\left(1 - S^2(x)\right)dx\right)$$

Let $y = S(x)$ and use (15.22), then $x = S^{-1}(y) = -\frac{\log(1-y)}{y}$ and $dx = -\frac{d}{dy}\left(\frac{\log(1-y)}{y}\right)dy$ while $y = S(1) = 0$ and $y = S(N) = 1$, such that

$$I = \int_1^N x^{\frac{1}{\alpha}}\left(1 - S^2(x)\right)dx$$

$$= \int_0^1 \left(-\frac{\log(1-y)}{y}\right)^{\frac{1}{\alpha}}(1-y^2)\frac{d}{dy}\left(-\frac{\log(1-y)}{y}\right)dy$$

After partial integration, we have

$$I = -\frac{1}{\frac{1}{\alpha}+1} + \frac{2}{\frac{1}{\alpha}+1}\int_0^\infty x^{\frac{1}{\alpha}+1}\frac{e^{-x}}{(1-e^{-x})^{\frac{1}{\alpha}}}dx$$

Finally, we end up with

$$E[W_{\mathrm{MST}}(\alpha)] \simeq N^{1-\frac{1}{\alpha}}\left(\frac{1}{\frac{1}{\alpha}+1}\int_0^\infty x^{\frac{1}{\alpha}+1}\frac{e^{-x}}{(1-e^{-x})^{\frac{1}{\alpha}}}dx\right) \qquad (16.48)$$

---

[8] Since the average of the k-th smallest link weight can be computed from (3.36) as

$$E[w_{(k)}] = \frac{E!}{\Gamma(E+1+\frac{1}{\alpha})}\frac{\Gamma(k+\frac{1}{\alpha})}{\Gamma(k)}$$

the exact formula (16.44) reduces to

$$E[W_{\mathrm{MST}}(\alpha)] = \frac{E!}{\Gamma(E+1+\frac{1}{\alpha})}\sum_{j=1}^E \frac{\Gamma(j+\frac{1}{\alpha})}{\Gamma(j)}\left(1 - S_j^2\right)$$

Analogously to the above manipulations, after converstion to an integral, substituting $x = \frac{2u}{N}$ and using (Abramowitz and Stegun, 1968, Section 6.1.47), for large z, that $\frac{\Gamma(z+\frac{1}{\alpha})}{\Gamma(z)} = (z)^{\frac{1}{\alpha}}\left(1 + O\left(\frac{1}{z}\right)\right)$, we arrive at the same formula.

If $\alpha < 1$, then $E[W_{\mathrm{MST}}(\alpha)] \to 0$ for $N \to \infty$, while for $\alpha > 1$, $E[W_{\mathrm{MST}}(\alpha)] \to \infty$. In particular, $\lim_{\alpha \to \infty} E[W_{\mathrm{MST}}(\alpha)] = N - 1$. Only for $\alpha = 1$, $E[W_{MST}(1)]$ is finite for large N. More precisely,

$$E[W_{\mathrm{MST}}(1)] = \zeta(3) = 1.202\ldots \tag{16.49}$$

where we have used (Abramowitz and Stegun, 1968, Section 23.2.7) the integral of the Riemann Zeta function $\Gamma(s)\zeta(s) = \int_0^\infty \frac{u^{s-1}}{e^u - 1} du$, which is convergent for $\mathrm{Re}(s) > 1$. This particular case for $\alpha = 1$ has been proved earlier by Frieze (1985) based on a different method.

16.7.2.2 Generalizations

We now return to the Taylor series valid for link weights where $0 < f_w(0) < \infty$. The above result for $\alpha = 1$ immediately yields

$$E[W_{\mathrm{MST}}] = \frac{\zeta(3)}{f_w(0)} \tag{16.50}$$

This result is for the complete graph K_N. A random graph $G_p(N)$ with $p < 1$ and weight density $f_w(x)$ is equivalent to K_N with a fraction $1 - p$ of infinite link weights. Thus the effective link weight distribution is $pf_w(x) + (1 - p)\delta_{w,\infty}$, and we can simply replace $f_w(0)$ by $pf_w(0)$ in the expression (16.50) to obtain the average weight of the MST in the random graph $G_p(N)$.

16.8 The proof of the degree Theorem 16.6.1 of the URT

16.8.1 The case $k = N$: $E\left[D_N^{(N-1)}\right]$

If $k = N$, the recursion (16.35) becomes with $D_N^{(N)} = 0$,

$$E\left[D_{N+1}^{(N)}\right] = \frac{E\left[D_N^{(N-1)}\right]}{N}$$

With initial value $D_2^{(1)} = 2$, the solution

$$E\left[D_N^{(N-1)}\right] = \frac{2}{(N-1)!} \tag{16.51}$$

is readily verified. Since for any URT, it holds that $\mathrm{Pr}\left[D_N^{(k)} = j\right] = 0$ for $j > N - k$, we have that $E\left[D_N^{(N-1)}\right] = \mathrm{Pr}\left[D_N^{(N-1)} = 1\right]$. Since there exists in total $(N-1)!$ different URTs of size N, this result (16.51) means that there are precisely two possible URTs with a node of degree $N - 1$. Indeed, one is the root with $N - 1$ children and the other is the root with one child of degree $N - 1$ that in turn possesses $N - 2$ children. Also,

$$r_N^{(N-1)} = (N-1)E\left[D_N^{(N-1)}\right] = \frac{2}{(N-2)!} \tag{16.52}$$

16.8.2 The case $k = 1$: $E\left[D_N^{(1)}\right]$

If $k = 1$ and $N \geq 3$, the recursion (16.35) is slightly different because the newly attached node n_{N+1} necessarily belongs to the set of degree 1 nodes in the URT of size $N + 1$ such that

$$E\left[D_{N+1}^{(1)}\right] = \frac{N-1}{N} E\left[D_N^{(1)}\right] + 1$$

with $D_1^{(0)} = 1$ and $D_2^{(1)} = 2$. Hence, $E\left[D_3^{(1)}\right] = 2$. With $r_N^{(1)} = (N-1)E\left[D_N^{(1)}\right]$, the recursion for $k = 1$ becomes

$$r_{N+1}^{(1)} = r_N^{(1)} + N$$

The particular solution is $r_{N;p}^{(1)} = aN^2 + bN + c$. Substitution of $r_N^{(1)} = aN^2 + bN + c$ into the difference equation yields

$$aN^2 + (b + 2a)N + a + b + c = aN^2 + (b+1)N + c$$

or, by equating corresponding power in N, we find the conditions $b + 2a = b + 1$ and $a + b = 0$ from which $a = \frac{1}{2}$, $b = -\frac{1}{2}$. Thus,

$$r_N^{(1)} = (N-1)E\left[D_N^{(1)}\right] = \frac{N}{2}(N-1) + c$$

and

$$E\left[D_N^{(1)}\right] = \frac{N}{2} + \frac{c}{N-1}$$

Using $E\left[D_3^{(1)}\right] = 2$ shows that $c = 1$ such that, for $N > 2$,

$$E\left[D_N^{(1)}\right] = \frac{N}{2} + \frac{1}{N-1} \tag{16.53}$$

16.8.3 The general case: $E\left[D_N^{(k)}\right]$

Let us denote

$$R(x, y) = \sum_{N=3}^{\infty} \sum_{k=2}^{N-1} r_N^{(k)} x^k y^N \tag{16.54}$$

then the recursion (16.36) is transformed into

$$\sum_{N=3}^{\infty} \sum_{k=2}^{N-1} (N-1) r_{N+1}^{(k)} x^k y^N = \sum_{N=3}^{\infty} \sum_{k=2}^{N-1} (N-1) r_N^{(k)} x^k y^N + \sum_{N=3}^{\infty} \sum_{k=2}^{N-1} r_N^{(k-1)} x^k y^N$$

Now, the left-hand side is

$$\sum_{N=3}^{\infty} \sum_{k=2}^{N-1} (N-1) r_{N+1}^{(k)} x^k y^N = \frac{1}{y} \sum_{N=4}^{\infty} \sum_{k=2}^{N-2} (N-2) r_N^{(k)} x^k y^N = \frac{1}{y} \sum_{N=3}^{\infty} \sum_{k=2}^{N-2} (N-2) r_N^{(k)} x^k y^N$$

$$= \frac{1}{y} \sum_{N=3}^{\infty} \sum_{k=2}^{N-1} N r_N^{(k)} x^k y^N - \frac{1}{y} \sum_{N=3}^{\infty} N r_N^{(N-1)} x^{N-1} y^N$$

$$- \frac{2}{y} \sum_{N=3}^{\infty} \sum_{k=2}^{N-1} r_N^{(k)} x^k y^N + \frac{2}{y} \sum_{N=3}^{\infty} r_N^{(N-1)} x^{N-1} y^N$$

Using (16.54) yields

$$\sum_{N=3}^{\infty} \sum_{k=2}^{N-1} (N-1) r_{N+1}^{(k)} x^k y^N = \frac{\partial R(x,y)}{\partial y} - \frac{2}{y} R(x,y)$$

$$- \frac{1}{y} \sum_{N=3}^{\infty} N r_N^{(N-1)} x^{N-1} y^N + \frac{2}{y} \sum_{N=3}^{\infty} r_N^{(N-1)} x^{N-1} y^N$$

Invoking (16.52) yields

$$\frac{2}{y} \sum_{N=3}^{\infty} r_N^{(N-1)} x^{N-1} y^N = \frac{4}{xy} \sum_{N=3}^{\infty} \frac{(xy)^N}{(N-2)!} = 4xy \sum_{N=1}^{\infty} \frac{(xy)^N}{N!} = 4xy \left(e^{xy} - 1 \right)$$

$$\frac{1}{y} \sum_{N=3}^{\infty} N r_N^{(N-1)} x^{N-1} y^N = \frac{2}{xy} \sum_{N=3}^{\infty} \frac{N (xy)^N}{(N-2)!} = 2xy \sum_{N=1}^{\infty} \frac{(N+2)(xy)^N}{N!}$$

$$= 2 (xy)^2 \sum_{N=0}^{\infty} \frac{(xy)^N}{N!} + 4xy \sum_{N=1}^{\infty} \frac{(xy)^N}{N!} = 2 (xy)^2 e^{xy} + 4xy \left(e^{xy} - 1 \right)$$

such that

$$\sum_{N=3}^{\infty} \sum_{k=2}^{N-1} (N-1) r_{N+1}^{(k)} x^k y^N = \frac{\partial R(x,y)}{\partial y} - \frac{2}{y} R(x,y) - 2 (xy)^2 e^{xy}$$

Similarly,

$$\sum_{N=3}^{\infty} \sum_{k=2}^{N-1} r_N^{(k-1)} x^k y^N = x \sum_{N=3}^{\infty} \sum_{k=1}^{N-2} r_N^{(k)} x^k y^N$$

$$= x \sum_{N=3}^{\infty} \sum_{k=2}^{N-1} r_N^{(k)} x^k y^N + x^2 \sum_{N=3}^{\infty} r_N^{(1)} y^N - x \sum_{N=3}^{\infty} r_N^{(N-1)} x^{N-1} y^N$$

$$= x R(x,y) + x^2 \sum_{N=3}^{\infty} r_N^{(1)} y^N - x \sum_{N=3}^{\infty} r_N^{(N-1)} x^{N-1} y^N$$

Using both (16.52) and $r_N^{(1)} = \frac{N}{2}(N-1) + 1$ leads to

$$x^2 \sum_{N=3}^{\infty} r_N^{(1)} y^N = x^2 \sum_{N=3}^{\infty} \frac{N}{2}(N-1) y^N + x^2 \sum_{N=3}^{\infty} y^N = \frac{(xy)^2}{2} \frac{d^2}{dy^2} \left(\frac{y^3}{1-y} \right) + \frac{x^2 y^3}{1-y}$$

$$x \sum_{N=3}^{\infty} r_N^{(N-1)} x^{N-1} y^N = 2 \sum_{N=3}^{\infty} \frac{(xy)^N}{(N-2)!} = 2 (xy)^2 \sum_{N=1}^{\infty} \frac{(xy)^N}{N!} = 2 (xy)^2 \left(e^{xy} - 1 \right)$$

such that

$$\sum_{N=3}^{\infty} \sum_{k=2}^{N-1} r_N^{(k-1)} x^k y^N = x R(x,y) + \frac{(xy)^2}{2} \frac{d^2}{dy^2} \left(\frac{y^3}{1-y} \right) + \frac{x^2 y^3}{1-y} - 2 (xy)^2 \left(e^{xy} - 1 \right)$$

Combining all transforms the recursion (16.36) to a first order linear partial differential equation

$$(1-y) \frac{\partial R(x,y)}{\partial y} + \left(1 - x - \frac{2}{y} \right) R(x,y) = x^2 y^3 \frac{3 - 3y + y^2}{(1-y)^3} + \frac{x^2 y^3}{1-y} + 2 (xy)^2$$

$$= x^2 y^2 \left[\frac{1}{1-y} + \frac{1}{(1-y)^3} \right]$$

with boundary equations $R(x,0) = R(0,y) = 0$. Further,

$$\int \frac{R(x,y)}{y^2} dy = \sum_{N=3}^{\infty} \sum_{k=2}^{N-1} \frac{r_N^{(k)}}{N-1} x^k y^{N-1} = \sum_{N=3}^{\infty} \sum_{k=2}^{N-1} E\left[D_N^{(k)}\right] x^k y^{N-1}$$

Hence, if $x = 1$,

$$\int \frac{R(1,y)}{y^2} dy = \sum_{N=3}^{\infty} \sum_{k=2}^{N-1} E\left[D_N^{(k)}\right] y^{N-1} = \sum_{N=3}^{\infty} \left(N - E\left[D_N^{(1)}\right]\right) y^{N-1}$$

$$= \sum_{N=3}^{\infty} N y^{N-1} - \sum_{N=3}^{\infty} E\left[D_N^{(1)}\right] y^{N-1} = \sum_{N=3}^{\infty} N y^{N-1} - \sum_{N=3}^{\infty} \frac{N}{2} y^{N-1} - \sum_{N=3}^{\infty} \frac{y^{N-1}}{N-1}$$

$$= \frac{1}{2} \frac{1}{(1-y)^2} - \frac{1}{2}(1+2y) - \int \frac{y}{1-y} dy$$

or

$$R(1,y) = \frac{y^2}{(1-y)^3} - \frac{y^2}{(1-y)} \tag{16.55}$$

It is more convenient to consider the differential equation as an ordinary differential equation in y and to regard the variable x as a parameter. The homogeneous differential equation,

$$(1-y) \frac{\partial R_h(x,y)}{\partial y} = \left(x + \frac{2}{y} - 1\right) R_h(x,y)$$

is solved after integration with respect to y,

$$\ln R_h(x,y) = \int \frac{\left(x + \frac{2}{y} - 1\right)}{1-y} dy = (x-1) \int \frac{dy}{1-y} + 2 \int \frac{dy}{y(1-y)}$$

$$= (1-x)\ln(1-y) + 2(\ln y - \ln(1-y)) = \ln\left[(1-y)^{-x-1} y^2\right]$$

or $R_h(x,y) = (1-y)^{-x-1} y^2$. The particular solution is of the form $R(x,y) = C(y) R_h(x,y)$ where $C(y)$ obeys

$$\frac{\partial C(y)}{\partial y} = x^2 (1-y)^x \left(\frac{1}{1-y} + \frac{1}{(1-y)^3}\right)$$

or

$$C(y) = x^2 \int \left((1-y)^{x-3} + (1-y)^{x-1}\right) dy + c(x)$$

$$= -\frac{x^2(1-y)^{x-2}}{x-2} - \frac{x^2(1-y)^x}{x} + c(x)$$

where $c(x)$ is a function of x, independent of y, to be determined later. The solution is

$$R(x,y) = -\frac{(xy)^2}{(x-2)(1-y)^3} - \frac{xy^2}{(1-y)} + c(x)(1-y)^{-x-1} y^2$$

The initial condition $R(0,y) = 0$ shows that $c(0) = 0$, while the boundary condition (16.55) implies that $c(1) = 0$. Expanding this solution in a power series around $x = 0$ and $y = 0$ yields

$$R_h(x,y) = (1-y)^{-x-1} y^2 = \sum_{N=0}^{\infty} \binom{-x-1}{N} (-1)^N y^{N+2}$$

From the generating function of the Stirling numbers of the first kind (Abramowitz and Stegun, 1968, Section 24.1.3),

$$\frac{\Gamma(x+1)}{\Gamma(x+1-n)} = \sum_{j=0}^{n} S_n^{(j)} x^j \tag{16.56}$$

we observe that

$$\binom{-x-1}{N} = \frac{\Gamma(-x)}{N!\,\Gamma(-x-N)} = \sum_{k=0}^{N} \frac{S_{N+1}^{(k+1)} (-1)^k}{N!} x^k$$

such that

$$R_h(x,y) = \sum_{N=0}^{\infty} \sum_{k=0}^{N} \frac{S_{N+1}^{(k+1)}}{N!} (-1)^{N+k} x^k y^{N+2} = \sum_{N=2}^{\infty} \sum_{k=0}^{N-2} \frac{S_{N-1}^{(k+1)}}{(N-2)!} (-1)^{N+k} x^k y^N$$

Hence,

$$R(x,y) = \sum_{N=2}^{\infty} \sum_{k=2}^{\infty} \frac{1}{2^{k-1}} \binom{N}{2} x^k y^N - x \sum_{N=2}^{\infty} y^N + c(x) \sum_{N=2}^{\infty} \sum_{k=0}^{N-2} \frac{S_{N-1}^{(k+1)}}{(N-2)!} (-1)^{N+k} x^k y^N$$

It remains to determine $c(x)$ by equating the corresponding powers in x and y at both sides. With the definition (16.54), equating the second power $(N=2)$ in y yields

$$0 = \sum_{k=2}^{\infty} \frac{1}{2^{k-1}} x^k - x + c(x)$$

which indicates that

$$c(x) = x - \frac{x^2}{2-x}$$

agreeing with $c(0) = c(1) = 0$. The Taylor series around $x = 0$ is $c(x) = \sum_{k=0}^{\infty} c_k x^k$ with $c_0 = 0$, $c_1 = 1$ and $c_k = -\frac{1}{2^{k-1}}$ for $k > 1$. Equating the power $N > 2$ in y,

$$\sum_{k=2}^{N-1} r_N^{(k)} x^k = \sum_{k=2}^{\infty} \frac{1}{2^{k-1}} \binom{N}{2} x^k - x + c(x) \sum_{k=0}^{N-2} \frac{S_{N-1}^{(k+1)}}{(N-2)!} (-1)^{N+k} x^k$$

$$= \sum_{k=2}^{\infty} \frac{1}{2^{k-1}} \binom{N}{2} x^k - x + \sum_{k=0}^{\infty} c_k x^k \sum_{k=0}^{\infty} \frac{S_{N-1}^{(k+1)}}{(N-2)!} (-1)^{N+k} x^k$$

$$= \sum_{k=2}^{\infty} \frac{1}{2^{k-1}} \binom{N}{2} x^k - x + \sum_{k=1}^{\infty} \left(\sum_{j=0}^{k-1} c_{k-j} \frac{S_{N-1}^{(j+1)}}{(N-2)!} (-1)^{N+j} \right) x^k$$

$$= \sum_{k=2}^{\infty} \frac{1}{2^{k-1}} \binom{N}{2} x^k - x + \left(c_1 \frac{(-1)^N S_{N-1}^{(1)}}{(N-2)!} \right) x + \sum_{k=2}^{\infty} \left(\sum_{j=0}^{k-1} c_{k-j} \frac{(-1)^{N+j} S_{N-1}^{(j+1)}}{(N-2)!} \right) x^k$$

which, by using $S_{N-1}^{(1)} = (-1)^N (N-2)!$ and $c_1 = 1$, equals

$$\sum_{k=2}^{N-1} r_N^{(k)} x^k = \sum_{k=2}^{\infty} \frac{1}{2^{k-1}} \binom{N}{2} x^k + \sum_{k=2}^{\infty} \left(\sum_{j=0}^{k-1} c_{k-j} \frac{S_{N-1}^{(j+1)}}{(N-2)!} (-1)^{N+j} \right) x^k$$

Finally, by equating the corresponding powers in x, leads to

$$r_N^{(k)} = \frac{1}{2^{k-1}} \binom{N}{2} + \sum_{j=0}^{k-1} c_{k-j} \frac{S_{N-1}^{(j+1)}}{(N-2)!} (-1)^{N+j}$$

$$= \frac{1}{2^{k-1}} \binom{N}{2} + \frac{(-1)^{N+k-1} S_{N-1}^{(k)}}{(N-2)!} + \frac{(-1)^N}{2^k (N-2)!} \sum_{j=1}^{k-1} S_{N-1}^{(j)} (-2)^j$$

or to (16.37). As a check, using (16.56) the generating function $\frac{(-1)^n \Gamma(n-x)}{\Gamma(-x)} = \sum_{j=0}^n S_n^{(j)} x^j$, reveals that

$$E\left[D_N^{(N-1)}\right] = \frac{N}{2^{N-1}} + \frac{1}{(N-1)!} + \frac{(-1)^N}{2^{N-1}(N-1)!} \sum_{j=1}^{N-2} S_{N-1}^{(j)} (-2)^j$$

$$= \frac{N}{2^{N-1}} + \frac{1}{(N-1)!} + \frac{(-1)^N}{2^{N-1}(N-1)!} \left[\frac{(-1)^{N-1} \Gamma(N-1+2)}{\Gamma(2)} - S_{N-1}^{(N-1)} (-2)^{N-1} \right]$$

$$= \frac{N}{2^{N-1}} + \frac{1}{(N-1)!} - \frac{N!}{2^{N-1}(N-1)!} + \frac{1}{(N-1)!} = \frac{2}{(N-1)!}$$

Also $E\left[D_N^1\right] = \frac{N}{2} + \frac{1}{N-1}$ is readily verified.

16.9 Problems

(i) *Comparison of simulations with exact results.* Many of the theoretical results are easily verified by simulations. Consider the following standard simulation: (a) Construct a graph of a certain class, e.g. an instance of the random graphs $G_p(N)$ with exponentially distributed link weights (b) Determine in that graph a desired property, e.g. the hopcount of the shortest path between two different arbitrary nodes, (c) Store the hopcount in a histogram and (d) repeat the sequence (a)-(c) n times with each time a different graph instance in (a). Estimate the relative error of the simulated hopcount in $G_p(N)$ with $p = 1$ for $n = 10^4, 10^5$ and 10^6.

(ii) Given the probability generating function (16.17) of the weight of the shortest path in a complete graph with independent exponential link weights, compute the variance of W_N.

(iii) Prove the asymptotic law (16.20) of the weight of the shortest path in a complete graph with i.i.d. exponential link weights.

(iv) In a communication network often two paths are computed for each important flow to guarantee sufficient reliability. Apart from the shortest path between a source A and a destination B, a second path between A and B is chosen that does not travel over any intermediate router of the shortest path. We call such a path node-disjoint to the shortest path. Derive a good approximation for the distribution of the hopcount of the shortest node-disjoint path to the shortest path in the complete graph with exponential link weights with mean 1.

17

The efficiency of multicast

The efficiency or gain of multicast in terms of network resources is compared to unicast. Specifically, we concentrate on a one-to-many communication, where a source sends a same message to m different, uniformly distributed destinations along the shortest path. In unicast, this message is sent m times from the source to each destination. Hence, unicast uses on average $f_N(m) = mE[H_N]$ link-traversals or hops, where $E[H_N]$ is the average number of hops to a uniform location in the graph with N nodes. One of the main properties of multicast is that it economizes on the number of link-traversals: the message is only copied at each branch point of the multicast tree to the m destinations. Let us denote by $H_N(m)$ the number of links in the shortest path tree (SPT) to m uniformly chosen nodes. If we define the multicast gain $g_N(m) = E[H_N(m)]$ as the average number of hops in the SPT rooted at a source to m randomly chosen distinct destinations, then $g_N(m) \leq f_N(m)$. The purpose here is to quantify the multicast gain $g_N(m)$. We present general results valid for *all* graphs and more explicit results valid for the random graph $G_p(N)$ and for the k-ary tree. The analysis presented here may be valuable to derive a business model for multicast: "How many customers m are needed to make the use of multicast for a service provider profitable?"

Two modeling assumptions are made. First, the multicast process is assumed[1] to deliver packets along the shortest path from a source to each of the m destinations. As most of the current Internet protocols forward packets based on the (reverse) shortest path, the assumption of SPT delivery is quite realistic. The second assumption is that the m multicast group member nodes are uniformly chosen out of the total number of nodes N. This assumption has been discussed by Phillips *et al.* (1999). They concluded

---

[1] The assumption ignores shared tree multicast forwarding such as core-based tree (CBT, see RFC2201).

that, if m and N are large, deviations from the uniformity assumption are negligibly small. Also the Internet measurements of Chalmers and Almeroth (2001) seem to confirm the validity of the uniformity assumption.

17.1 General results for $g_N(m)$

Theorem 17.1.1 *For any connected graph with N nodes,*

$$m \leq g_N(m) \leq \frac{Nm}{m+1} \tag{17.1}$$

Proof: We need at least one edge for each different user; therefore $g_N(m) \geq m$ and the lower bound is attained in a star topology with the source at the center.

We will next show that an upper bound is obtained in a line topology. It is sufficient to consider trees, because multicast only uses shortest paths without cycles. If the tree has not a line topology, then at least one node has degree 3 or the root has degree 2. Take the node closest to the root with this property and cut one of the branches at this node; we paste that branch to a node at the deepest level. Through this procedure the multicast function $g_N(m)$ stays unaltered or increases. Continuing in this fashion until we reach a line topology demonstrates the claim.

For the line topology we place the source at the origin and the other nodes at the integers $1, 2, \ldots, N-1$. The links of the graph are given by $(i, i+1)$, $i = 0, 1, \ldots, N-2$. The multicast gain $g_N(m)$ equals $E[M]$, where M is the maximum of a sample of size m, without replacement, from the integers $1, 2, \ldots, N-1$. Thus,

$$\Pr[M \leq k] = \frac{\binom{k}{m}}{\binom{N-1}{m}}, \quad m \leq k \leq N-1$$

from which $g_N(m) = E[M]$ is

$$g_N(m) = \sum_{k=m}^{N-1} k \frac{\binom{k}{m} - \binom{k-1}{m}}{\binom{N-1}{m}} = \sum_{k=m}^{N-1} k \frac{\binom{k-1}{m-1}}{\binom{N-1}{m}}$$

$$= m \sum_{k=m}^{N-1} \frac{\binom{k}{m}}{\binom{N-1}{m}} = \frac{mN}{m+1} \sum_{k=m}^{N-1} \frac{\binom{k}{m}}{\binom{N}{m+1}} = \frac{mN}{m+1}$$

where we have used that $\sum_{k=m}^{N-1} \binom{k}{m} / \binom{N}{m+1} = 1$, because it is a sum of probabilities over *all* possible disjoint outcomes. $\square$

Figure 17.1 shows the allowable space for $g_N(m)$.

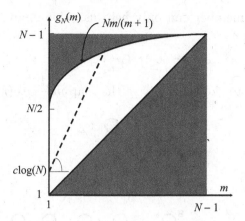

Fig. 17.1. The allowable region (in white) of $g_N(m)$. For exponentially growing graphs, $E[H_N] = c \log N$, implying that the allowable region for these graphs is smaller and bounded at the left (in dotted line) by the straight line $m(c \log N)$.

Theorem 17.1.2 *For any connected graph with N nodes, the map $m \mapsto g_N(m)$ is concave and the map $m \mapsto \frac{g_N(m)}{f_N(m)}$ is decreasing.*

Proof: Define Y_m to be the random variable giving the additional number of hops necessary to reach the m-th user when the first $m-1$ users are already connected. Then we have that

$$E[Y_m] = g_N(m) - g_N(m-1)$$

Moreover, let Y'_m be the random number of additional hops necessary to reach the m-th multicast group member, when we discard all extra hops of the $(m-1)$-st group member. An example is illustrated in Fig. 17.2. The random variable Y'_m has the same distribution as Y_{m-1}, because both the $(m-1)$-st and the m-th group member are chosen uniformly from the remaining $N - m - 1$ nodes. In general, $Y'_m \neq Y_{m-1}$, but, for each k, $\Pr[Y'_m = k] = \Pr[Y_{m-1} = k]$ and, hence,

$$E[Y'_m] = E[Y_{m-1}] \tag{17.2}$$

Furthermore, we have by construction that $Y_m \leq Y'_m$ with probability 1, implying that

$$E[Y_m] \leq E[Y'_m] \tag{17.3}$$

Indeed, attaching the m-th group member to the reduced tree takes at least as many hops as attaching that same group member to the non-reduced tree because the former is contained in the latter and the extra hops added by the $m - 1$ group member can only help us. Combining (17.2) and (17.3) immediately gives

$$g_N(m) - g_N(m-1) = E[Y_m] \le E[Y'_m] = g_N(m-1) - g_N(m-2) \quad (17.4)$$

This is equivalent to the concavity of the map $m \mapsto g_N(m)$.

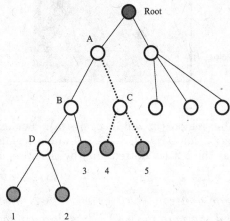

Fig. 17.2. A multicast session with $m = 5$ group members where $Y_5 = 1$ (namely link C-5). To construct Y'_5 the three dotted lines must be removed and we observe that $Y'_5 = 2$ (A-C-5), which is referred to as the reduced tree. In this example, $Y'_5 = Y_4 = 2$ because A-C-4 and A-C-5 both consist of 2 hops. In general, they are equal in distribution because the role of group member 4 and 5 are identical in the reduced tree.

In order to show that $\frac{g_N(m)}{f_N(m)}$ is decreasing it suffices to show that $m \mapsto \frac{g_N(m)}{m}$ is decreasing, since $f_N(m)$ is proportional to m. Defining $g_N(0) = 0$, we can write $g_N(m)$ as a telescoping sum

$$g_N(m) = \sum_{k=1}^{m} \{g_N(k) - g_N(k-1)\} = \sum_{k=1}^{m} x_k$$

where $x_k = g_N(k) - g_N(k-1)$, $k = 1, \ldots, m$. Then,

$$\frac{g_N(m)}{m} = \frac{1}{m} \sum_{k=1}^{m} x_k$$

is the mean of a sequence of m positive numbers x_k. By (17.4) the sequence $x_k \le x_{k-1}$ is decreasing and, hence,

$$\frac{g_N(m)}{m} = \frac{1}{m} \sum_{k=1}^{m} x_k \le \frac{1}{m-1} \sum_{k=1}^{m-1} x_k = \frac{g_N(m-1)}{m-1}$$

This proves that $m \mapsto g_N(m)/m$ is decreasing. □

Next, we will give a representation for $g_N(m)$ that is valid for all graphs. Let X_i be the number of joint hops that *all i* uniformly chosen and different group members have in common, then the following general theorem holds,

Theorem 17.1.3 *For any connected graph with N nodes,*

$$g_N(m) = \sum_{i=1}^{m} \binom{m}{i} (-1)^{i-1} E[X_i] \tag{17.5}$$

Note that

$$g_N(1) = f_N(1) = E[X_1] = E[H_N]$$

so that the decrease in average hops or the "gain" by using multicast over unicast is precisely

$$g_N(m) - f_N(m) = \sum_{i=2}^{m} \binom{m}{i} (-1)^{i-1} E[X_i]$$

However, computing $E[X_i]$ for general graphs is difficult.

Proof of Theorem 17.1.3: Let $A_1, A_2, \ldots, A_m$ be sets where A_i consists of all links that constitute the shortest path from the source to multicast group member i. Denote by $|A_i|$ the number of elements in the set A_i. The multicast group members are chosen uniformly from the set of all nodes except for the root. Hence,

$$E[X_1] = E[|A_i|], \qquad \text{for } 1 \le i \le N$$

and

$$E[X_2] = E[|A_i \cap A_j|], \qquad \text{for } 1 \le i < j \le N$$

etc.. Now, $g_N(m) = E[|A_1 \cup A_2 \cup \cdots \cup A_m|]$. Since $Q(A) = E[|A|]/\binom{N}{2}$ is a probability measure on the set of all links, we obtain from the inclusion-exclusion formula (2.3) applied to Q and multiplied with $\binom{N}{2}$ afterwards,

$$E[|A_1 \cup A_2 \cup \cdots \cup A_m|] = \sum_{i=1}^{m} E[|A_i|] - \sum_{i<j} E[|A_i \cap A_j|] + \cdots$$

$$+ (-1)^{m-1} E[|A_1 \cap A_2 \cap \cdots \cap A_m|]$$

$$= mE[X_1] - \binom{m}{2} E[X_2] + \cdots + (-1)^{m-1} E[X_m]$$

This proves Theorem 17.1.3. □

Corollary 17.1.4 *For any connected graph with N nodes,*

$$E[X_m] = \sum_{i=1}^{m} \binom{m}{i} (-1)^{i-1} g_N(i) \qquad (17.6)$$

The corollary is a direct consequence of the inversion formula for the binomial (Riordan, 1968, Chapter 2). Alternatively, in view of the Gregory-Newton interpolation formula (Lanczos, 1988, Chapter 4, Section 2) for $g_N(m) = \sum_{i=1}^{\infty} \binom{m}{i} \Delta^i g_N(0)$, we can write $E[X_i] = (-1)^{i-1} \Delta^i g_N(0)$ where Δ is the difference operator, $\Delta f(0) = f(1) - f(0)$.

Corollary 17.1.5 *For any connected graph, the multicast efficiency $g_N(m)$ is bounded by*

$$\frac{f_N(m)}{g_N(m)} \leq E[H_N] \qquad (17.7)$$

where $E[H_N]$ is the average number of hops in unicast.

Proof: We give two demonstrations. (a) From $g_N(N-1) = N-1$ (all nodes, source plus $N-1$ destinations, of the graph are spanned by a tree consisting of $N-1$ links) and the monotonicity of $m \mapsto \frac{g_N(m)}{f_N(m)}$ (see Theorem 17.1.2) we obtain:

$$\frac{g_N(m)}{f_N(m)} \geq \frac{g_N(N-1)}{f_N(N-1)} = \frac{N-1}{(N-1)E[H_N]} = \frac{1}{E[H_N]}$$

(b) Alternatively, Theorem 17.1.1 indicates that $g_N(m) \geq m$, which, with the identity $f_N(m) = mE[H_N]$, immediately leads to (17.7). $\qquad \square$

Corollary 17.1.5 means that for any *connected* graph, including the graph describing the Internet, the ratio of the unicast over multicast efficiency is bounded by the expected hopcount in unicast. In order words, the maximum savings in resources an operator can gain by using multicast (over unicast) never exceeds $E[H_N]$, which is roughly about 15 in the current Internet.

17.2 The random graph $G_p(N)$

In this section, we confine to the class RGU, the random graphs of the class $G_p(N)$ with independent identically and exponentially distributed link weights w with mean $E[w] = 1$ and where $\Pr[w \leq x] = 1 - e^{-x}$, $x > 0$. In Section 16.2, we have shown that the corresponding SPT is, asymptotically, a URT. The analysis below is exact for the complete graph K_N while asymptotically correct for connected random graphs $G_p(N)$.

17.2.1 The hopcount of the shortest path tree

Based on properties of the URT, the complete probability density function of the number of links $H_N(m)$ in the SPT to m uniformly chosen nodes can be determined. We first derive a recursion for the probability generating function $\varphi_{H_N(m)}(z) = E\left[z^{H_N(m)}\right]$ of the number of links $H_N(m)$ in the SPT to m uniformly chosen nodes in the complete graph K_N.

Lemma 17.2.1 *For $N > 1$ and all $1 \leq m \leq N - 1$,*

$$\varphi_{H_N(m)}(z) = \frac{(N - m - 1)(N - 1 + mz)}{(N - 1)^2}\varphi_{H_{N-1}(m)}(z) + \frac{m^2 z}{(N - 1)^2}\varphi_{H_{N-1}(m-1)}(z) \tag{17.8}$$

Proof: To prove (17.8), we use the recursive growth of URTs: a URT of size N is a URT of size $N - 1$, where we add an additional link to a uniformly chosen node.

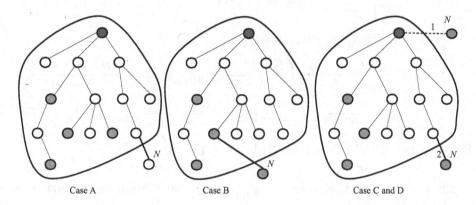

Fig. 17.3. The several possible cases in which the N-th node can be attached uniformly to the URT of size $N - 1$. The root is dark shaded while the m multicast member nodes are lightly shaded.

In order to obtain a recursion for $H_N(m)$ we distinguish between the m uniformly chosen nodes all being in the URT of size $N - 1$ or not. The probability that they all belong to the tree of size $N - 1$ is equal to $1 - \frac{m}{N-1}$ (case A in Fig. 17.3). If they all belong to the URT of size $N - 1$, then we have that $H_N(m) = H_{N-1}(m)$. Thus, we obtain

$$\varphi_{H_N(m)}(z) = \left(1 - \frac{m}{N - 1}\right)\varphi_{H_{N-1}(m)}(z) + \frac{m}{N - 1}E\left[z^{1+L_{N-1}(m)}\right] \tag{17.9}$$

where $L_{N-1}(m)$ is the number of links in the subtree of the URT of size $N - 1$ spanned by $m - 1$ uniform nodes and the "one" refers to the link from the added N-th node to its ancestor in the URT of size $N-1$. We complete the proof by investigating the generating function of $L_{N-1}(m)$. Again, there are two cases. In the first case (B in Fig. 17.3), the ancestor of the added N-th node is one of the $m-1$ previous nodes (which can only happen if it is unequal to the root), else we get one of the cases C and D in Fig. 17.3. The probability of the first event equals $\frac{m-1}{N-1}$, the probability of the latter equals $1-\frac{m-1}{N-1}$. If the ancestor of the added N-th node is one of the $m-1$ previous nodes, then the number of links $L_{N-1}(m)$ equals $H_{N-1}(m-1)$, otherwise the generating function of the number of additional links equals

$$\left(1-\frac{1}{N-m}\right)\varphi_{H_{N-1}(m)}(z) + \frac{1}{N-m}\varphi_{H_{N-1}(m-1)}(z)$$

The first contribution comes from the case where the ancestor of the added N-th node is not the root, and the second from where it is equal to the root, which has probability $\frac{1}{N-1-(m-1)} = \frac{1}{N-m}$. Therefore,

$$\begin{aligned}
E\left[z^{L_{N-1}(m)}\right] &= \frac{m-1}{N-1}\varphi_{H_{N-1}(m-1)}(z) \\
&\quad + \frac{N-m}{N-1}\left(\frac{N-m-1}{N-m}\varphi_{H_{N-1}(m)}(z) + \frac{\varphi_{H_{N-1}(m-1)}(z)}{N-m}\right) \\
&= \frac{m}{N-1}\varphi_{H_{N-1}(m-1)}(z) + \frac{N-m-1}{N-1}\varphi_{H_{N-1}(m)}(z) \quad (17.10)
\end{aligned}$$

Substitution of (17.10) into (17.9) leads to (17.8). $\qquad\square$

Since $g_N(m) = E[H_N(m)] = \varphi'_{H_N(m)}(1)$, we obtain the recursion for $g_N(m)$,

$$g_N(m) = \left(1 - \frac{m^2}{(N-1)^2}\right)g_{N-1}(m) + \frac{m^2}{(N-1)^2}g_{N-1}(m-1) + \frac{m}{N-1} \tag{17.11}$$

Theorem 17.2.2 *For all $N \geq 1$ and $1 \leq m \leq N-1$,*

$$\varphi_{H_N(m)}(z) = E\left[z^{H_N(m)}\right] = \frac{m!(N-1-m)!}{((N-1)!)^2}\sum_{k=0}^{m}\binom{m}{k}(-1)^{m-k}\frac{\Gamma(N+kz)}{\Gamma(1+kz)} \tag{17.12}$$

Consequently,

$$\Pr[H_N(m) = j] = \frac{m!(-1)^{N-(j+1)}S_N^{(j+1)}S_j^{(m)}}{(N-1)!\binom{N-1}{m}} \tag{17.13}$$

where $S_N^{(j+1)}$ and $\mathcal{S}_j^{(m)}$ denote the Stirling numbers of first and second kind (Abramowitz and Stegun, 1968, Section 24.1).

Proof: By iterating the recursion (17.8) for small values of m, the computations given in van der Hofstad *et al.* (2006a, Appendix) suggest the solution (17.12) for (17.8). One can verify that (17.12) satisfies (17.8). This proves (17.12) of Theorem 17.2.2. Using (Abramowitz and Stegun, 1968, Section 24.1.3.B), the Taylor expansion around $z = 0$ equals

$$\varphi_{H_N(m)}(z) = \frac{m!N(N-1-m)!}{(N-1)!} \sum_{k=0}^m \binom{m}{k}(-1)^{m-k} \left(\frac{\Gamma(N+kz)}{N!\Gamma(1+kz)} - \frac{1}{N} \right)$$

$$= \frac{m!N(N-1-m)!}{(N-1)!} \sum_{k=0}^m \binom{m}{k}(-1)^{m-k} \sum_{j=1}^{N-1} \frac{(-1)^{N-(j+1)}S_N^{(j+1)}}{N!} k^j z^j$$

$$= \frac{m!N(N-1-m)!}{(N-1)!} \sum_{j=1}^{N-1} \frac{(-1)^{N-(j+1)}S_N^{(j+1)}}{N!} \left(\sum_{k=0}^m \binom{m}{k}(-1)^{m-k}k^j \right) z^j$$

Using the definition of Stirling numbers of the second kind (Abramowitz and Stegun, 1968, 24.1.4.C),

$$m!\mathcal{S}_j^{(m)} = \sum_{k=0}^m \binom{m}{k}(-1)^{m-k}k^j$$

for which $\mathcal{S}_j^{(m)} = 0$ if $j < m$, gives

$$\varphi_{H_N(m)}(z) = \frac{(m!)^2 (N-1-m)!}{((N-1)!)^2} \sum_{j=1}^{N-1}(-1)^{N-(j+1)} S_N^{(j+1)} \mathcal{S}_j^{(m)} z^j$$

This proves (17.13) and completes the proof of Theorem 17.2.2. $\quad\square$

Figure 17.4 plots the probability density function of $H_{50}(m)$ for different values of m.

Corollary 17.2.3 *For all $N \geq 1$ and $1 \leq m \leq N-1$,*

$$g_N(m) = E[H_N(m)] = \frac{mN}{N-m} \sum_{k=m+1}^N \frac{1}{k} \tag{17.14}$$

and

$$\text{Var}[H_N(m)] = \frac{N-1+m}{N+1-m}g_N(m) - \frac{g_N^2(m)}{(N+1-m)} - \frac{m^2 N^2 \sum_{k=m+1}^N \frac{1}{k^2}}{(N-m)(N+1-m)} \tag{17.15}$$

The formula (17.14) is proved in two different ways. The earlier proof presented in Section 17.6 below does not rely on the recursion in Lemma 17.2.1 nor on Theorem 17.2.2. The shorter proof is presented here. Formula (17.14) can be expressed in terms of the digamma function $\psi(x)$ as

$$g_N(m) = mN \left(\frac{\psi(N) - \psi(m)}{N-m} \right) - 1 \tag{17.16}$$

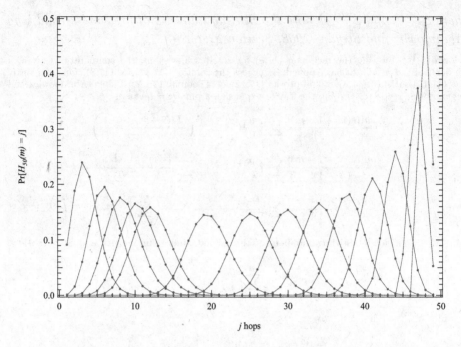

Fig. 17.4. The pdf of $H_{50}(m)$ for $m = 1, 2, 3, 4, 5, 10, 15, 20, 25, 30, 35, 40, 45, 47$.

Proof of Corollary 17.2.3: The expectation and variance of $H_N(m)$ will not be obtained using the explicit probabilities (17.13), but by rewriting (17.12) as

$$\varphi_{H_N(m)}(z) = \frac{\Gamma(m+1)\Gamma(N-m)}{\Gamma^2(N)} \sum_{k=0}^{m} \binom{m}{k} (-1)^{m-k} \partial_t^{N-1} \left[t^{N-1+kz} \right]_{t=1}$$

$$= \frac{\Gamma(m+1)\Gamma(N-m)}{\Gamma^2(N)} (-1)^m \partial_t^{N-1} \left[t^{N-1}(1-t^z)^m \right]_{t=1} \qquad (17.17)$$

Indeed,

$$E[H_N(m)] = \frac{\Gamma(m+1)\Gamma(N-m)}{\Gamma^2(N)} (-1)^m \partial_z \partial_t^{N-1} \left[t^{N-1}(1-t^z)^m \right]_{t=z=1}$$

$$= \frac{\Gamma(m+1)\Gamma(N-m)}{\Gamma^2(N)} m(-1)^{m-1} \partial_t^{N-1} \left[t^N \log t (1-t)^{m-1} \right]_{t=1},$$

$$E[H_N(m)(H_N(m)-1)] = \frac{\Gamma(m+1)\Gamma(N-m)}{\Gamma^2(N)} (-1)^m \partial_z^2 \partial_t^{N-1} \left[t^{N-1}(1-t^z)^m \right]_{t=z=1}$$

$$= \frac{\Gamma(m+1)\Gamma(N-m)}{\Gamma^2(N)} m(-1)^{m-1}$$

$$\times \partial_t^{N-1} \left[t^N \log^2 t (1-t)^{m-2} \left[-(m-1)t + (1-t) \right] \right]_{t=1}$$

We will start with the former. Using $\partial_t^i (1-t)^j|_{t=1} = j!(-1)^j \delta_{i,j}$ and Leibniz' rule, we find

$$E[H_N(m)] = \frac{\Gamma(m+1)\Gamma(N-m)}{\Gamma^2(N)} m! \binom{N-1}{m-1} \partial_t^{N-m} \left[t^N \log t \right]_{t=1}$$

Since

$$\partial_t^k [t^n \log t]_{t=1} = \frac{n!}{(n-k)!} \sum_{j=n-k+1}^{n} \frac{1}{j}$$

we obtain expression (17.14) for $E[H_N(m)]$.

We now extend the above computation to $E[H_N(m)(H_N(m)-1)]$ that we write as

$$E[H_N(m)\,(H_N(m)-1)] = \frac{\Gamma(m+1)\Gamma(N-m)}{\Gamma^2(N)}\,(R_1 + R_2) \tag{17.18}$$

where

$$R_1 = m(m-1)(-1)^{m-2}\partial_t^{N-1}\left[t^{N+1}\log^2 t(1-t)^{m-2}\right]_{t=1}$$

$$R_2 = m(-1)^{m-1}\partial_t^{N-1}\left[t^N \log^2 t(1-t)^{m-1}\right]_{t=1}$$

Using

$$\partial_t^k [t^n \log^2 t]_{t=1} = 2\frac{n!}{(n-k)!}\sum_{i=n-k+1}^{n}\sum_{j=i+1}^{n}\frac{1}{ij} = \frac{n!}{(n-k)!}\left[\left(\sum_{i=n-k+1}^{n}\frac{1}{i}\right)^2 - \sum_{i=n-k+1}^{n}\frac{1}{i^2}\right]$$

we obtain,

$$R_1 = \binom{N-1}{m-2}m(m-1)(m-2)!\partial_t^{N-m+1}\left[t^{N+1}\log^2 t\right]_{t=1}$$

$$= (N+1)!\binom{N-1}{m-2}\left[\left(\sum_{k=m+1}^{N+1}\frac{1}{k}\right)^2 - \sum_{k=m+1}^{N+1}\frac{1}{k^2}\right]$$

Similarly,

$$R_2 = \binom{N-1}{m-1}m(m-1)!\partial_t^{N-m}\left[t^N \log^2 t\right]_{t=1}$$

$$= N!\binom{N-1}{m-1}\left[\left(\sum_{k=m+1}^{N}\frac{1}{k}\right)^2 - \sum_{k=m+1}^{N}\frac{1}{k^2}\right]$$

Substitution into (17.18) leads to

$$E[H_N(m)(H_N(m)-1)] = \frac{m^2 N^2}{(N+1-m)(N-m)}\left[\left(\sum_{k=m+1}^{N}\frac{1}{k}\right)^2 - \sum_{k=m+1}^{N}\frac{1}{k^2}\right]$$

$$+ \frac{2m(m-1)N}{(N+1-m)(N-m)}\sum_{k=m+1}^{N}\frac{1}{k}$$

From $g_N(m) = E[H_N(m)]$ and $\mathrm{Var}\,[H_N(m)] = E[H_N(m)(H_N(m)-1)] + g_N(m) - g_N^2(m)$, we obtain (17.15). This completes the proof of Corollary 17.2.3. $\qquad\square$

For $N = 1000$, Fig. 17.5 illustrates the typical behavior for large N of the expectation $g_N(m)$ and the standard deviation $\sigma_N(m) = \sqrt{\mathrm{Var}\,[H_N(m)]}$ for all values of m. For any spanning tree, the number of links $H_N(N-1)$ is precisely $N-1$, so that $\mathrm{Var}[H_N(N-1)] = 0$.

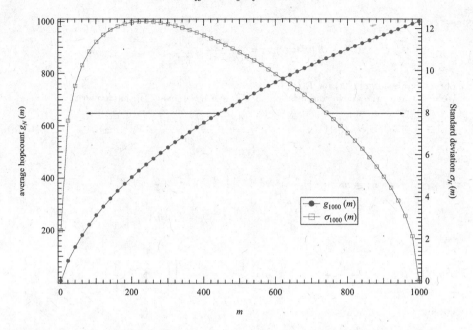

Fig. 17.5. The average number of hops $g_N(m)$ (left axis) in the SPT and the corresponding standard deviation $\sigma_N(m)$ (right axis) as a function of the number m of multicast group members in the complete graph with $N = 1000$.

Figure 17.5 also indicates that the standard deviation $\sigma_N(m)$ of $H_N(m)$ is much smaller than the average, even for $N = 1000$. In fact, we obtain from (17.15) that

$$\operatorname{Var}[H_N(m)] \leq \frac{N-1+m}{N+1-m}g_N(m) \leq \frac{2Ng_N(m)}{N-m} = \frac{2g_N^2(m)}{m\sum_{k=m+1}^{N}\frac{1}{k}} = o(g_N^2(m))$$

This bound implies that with probability converging to 1 for every $m = 1, \ldots, N-1$,

$$\left|\frac{H_N(m)}{g_N(m)} - 1\right| \leq \varepsilon$$

In van der Hofstad *et al.* (2006a), the scaled random variable $\frac{H_N(m)-g_N(m)}{\sqrt{g_N(m)}}$ is proved to tend to a Gaussian random variable, i.e. $\frac{H_N(m)-g_N(m)}{\sqrt{g_N(m)}} \xrightarrow{d}$ $N(0,1)$, for all $m = o(\sqrt{N})$. For large graphs of the size of the Internet and larger, this observation implies that the mean $g_N(m) = E[H_N(m)]$ is a good approximation for the random variable $H_N(m)$ itself because the variations of $H_N(m)$ around the mean are small. Consequently, it underlines the importance of $g_N(m)$ as a significant measure for multicast.

17.2.2 The weight of the shortest path tree

In this section, we summarize results on the weight $W_N(m)$ of the SPT and omit derivations, but refer to van der Hofstad *et al.* (2006a). For all $1 \le m \le N - 1$, the average weight of the SPT is

$$E\left[W_N(m)\right] = \sum_{j=1}^{m} \frac{1}{N-j} \sum_{k=j}^{N-1} \frac{1}{k} \tag{17.19}$$

In particular, if the shortest path tree spans the whole graph, then for all $N \ge 2$,

$$E\left[W_N(N-1)\right] = \sum_{n=1}^{N-1} \frac{1}{n^2} \tag{17.20}$$

from which $E\left[W_N(N-1)\right] < \zeta(2) = \frac{\pi^2}{6}$ for any finite N. The variance is

$$\mathrm{Var}\left[W_N(N-1)\right] = \frac{4}{N} \sum_{k=1}^{N-1} \frac{1}{k^3} + 4 \sum_{j=1}^{N-1} \frac{1}{j^3} \sum_{k=1}^{j} \frac{1}{k} - 5 \sum_{j=1}^{N-1} \frac{1}{j^4} \tag{17.21}$$

or asymptotically, for large N,

$$\mathrm{Var}\left[W_N(N-1)\right] = \frac{4\zeta(3)}{N} + O\left(\frac{\log N}{N^2}\right) \tag{17.22}$$

Asymptotically for large N, the average weight of a shortest path tree is $\zeta(2) = 1.645\ldots$, while the average weight of the minimum spanning tree, given by (16.49), is $\zeta(3) < \zeta(2)$. This result has an interesting implication. The Steiner tree is the minimum weight tree that connects a set of m members out of N nodes in the graph. The Steiner tree problem is NP-complete, which means that it is unfeasible to compute for large N. If $m = 2$, the weight of the Steiner tree equals that of the shortest path, $W_{\mathrm{Steiner},N}(2) = W_N$, while for $m = N$, we have $W_{\mathrm{Steiner},N}(N) = W_{\mathrm{MST}}$. Hence, for any $m < N$ and N, $E[W_{\mathrm{Steiner},N}(m)] \le \zeta(3)$ because the weight of the Steiner tree does not decrease if the number of members m increases. The ratio $\frac{\zeta(2)}{\zeta(3)} = 1.368$ indicates that the use of the SPT (computationally easy) never performs on average more than 37% worse than the optimal Steiner tree (computationally unfeasible). In a broader context and referring to the concept of the "Prize of Anarchy", which is broadly explained in Robinson (2004), the SPT used in a communications network is related to the Nash equilibrium, while the Steiner tree gives the hardly achievable global optimum.

Simulations – even for small N, which allow us to cover the entire m-range

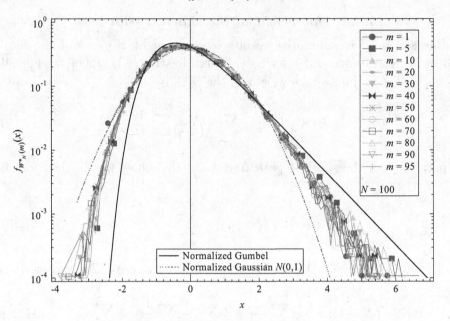

Fig. 17.6. The pdf of the normalized random variable $W_N^*(m)$ for $N = 100$.

as illustrated in Fig. 17.6 – indicate that the normalized random variable $W_N^*(m) = \frac{W_N(m) - E[W_N(m)]}{\sqrt{var[W_N(m)]}}$ lies between a normalized Gaussian $N(0,1)$ and a normalized Gumbel (see Theorem 6.4.1). Fig. 17.6 may suggest that, for all $m < N$,

$$\lim_{N \to \infty} \Pr[W_N^*(m) \le x] \approx e^{-e^{-\frac{\pi}{\sqrt{6}}x - \gamma}} \qquad (17.23)$$

where $\gamma = 0.57721...$ is Euler's constant. For the particular case of $m = 1$, the relation to the Gumbel distribution has been shown in Section 16.4 where the correct limit law is given in (16.20). However, van der Hofstad *et al.* (2006b) show that the weight of the shortest path tree for $m = N - 1$ tends to a Gaussian,

$$\sqrt{N}\left(W_N(N-1) - \zeta(2)\right) \xrightarrow{d} N\left(0, \sigma_{\text{SPT}}^2\right)$$

with $\sigma_{\text{SPT}}^2 = 4\zeta(3) \approx 4.80823$ as follows from (17.22). This shows that simulations alone may be inadequate to deduce asymptotic behavior. Finally, Janson (1995) gave the related result for the minimum spanning tree. He extended Frieze's result (16.49) by proving that the scaled weight of the minimum spanning tree also tends to a Gaussian for large N,

$$\sqrt{N}\left(W_{\text{MST}} - \zeta(3)\right) \xrightarrow{d} N\left(0, \sigma_{\text{MST}}^2\right)$$

where[2] $\sigma_{\mathrm{MST}}^2 = 6\zeta(4) - 4\zeta(3) \approx 1.6857$.

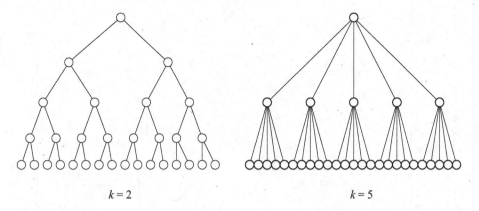

$k = 2$ $k = 5$

Fig. 17.7. The left hand side tree ($k = 2$) has $N = 31$ and $D = 4$, while the right hand side ($k = 5$) has $N = 31$ and $D = 2$.

17.3 The k-ary tree

In this section, we consider the k-ary tree of depth[3] D with the source at the root of the tree and m receivers at randomly chosen nodes (see Fig. 17.7). In a k-ary tree the total number of nodes satisfies

$$N = 1 + k + k^2 + \cdots + k^D = \frac{k^{D+1} - 1}{k - 1} \tag{17.24}$$

Theorem 17.3.1 *For the k-ary tree,*

$$g_{N,k}(m) = N - 1 - \sum_{j=0}^{D-1} k^{D-j} \frac{\binom{N-1-\frac{k^{j+1}-1}{k-1}}{m}}{\binom{N-1}{m}} \tag{17.25}$$

Proof: See Section 17.7.   □

Unfortunately, the j-summation seems difficult to express in closed form. Observe that $g_N(N-1) = N - 1$, because all binomials vanish. The sum extends over all levels $j \le D-1$, for which the remaining number of nodes in

---

[2] Wästlund (2005) succeeded in computing the triple sum in Janson's original result

$$\sigma_{\mathrm{MST}}^2 = \frac{\pi^4}{45} - 2 \sum_{i=0}^{\infty} \sum_{j=1}^{\infty} \sum_{k=1}^{\infty} \frac{(i+k-1)!\,k^k\,(i+j)^{i-2}j}{i!\,k!\,(i+j+k)^{i+k+2}}$$

[3] The depth D is equal to the number of hops from the root to a node at the leaves.

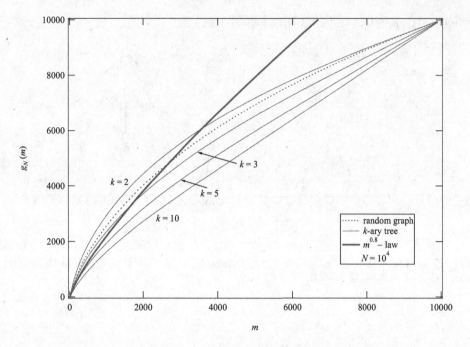

Fig. 17.8. The multicast gain $g_N(m)$ computed for the k-ary tree with four values of k, the random graph (with "effective" $k_{\text{rg}} = e = 2.718...$), and the Chuang-Sirbu power law for $N = 10^4$ on a linear scale where the prefactor $E[H_N]$ is given by (16.10).

the lower levels l (i.e. $D \geq l > j$) is larger than m nodes. In some sense, we may regard (17.25) as an (exact) expansion around $m = N - 1$. Explicitly,

$$g_{N,k}(m) = N - 1 - k^D \left(1 - \frac{m}{N-1}\right) - k^{D-1} \prod_{q=0}^{k} \left(1 - \frac{m}{N-1-q}\right)$$

$$- \sum_{j=2}^{D-1} k^{D-j} \prod_{q=0}^{\frac{k^{j+1}-1}{k-1}-1} \left(1 - \frac{m}{N-1-q}\right) \tag{17.26}$$

which shows that $g_{N,k}(m)$ is a polynomial in m of degree $\leq \frac{N-1}{k}$. Moreover, the terms in the j-sum rapidly decrease; their ratio equals

$$\frac{\prod_{q=\frac{k^j-1}{k-1}}^{\frac{k^{j+1}-1}{k-1}-1} \left(1 - \frac{m}{N-1-q}\right)}{k} < \frac{1}{k} \left(1 - \frac{m}{N-1-\frac{k^j-1}{k-1}}\right)^{k^j} << 1$$

Figure 17.8 indicates that formula (17.25), although derived subject to (17.24), also seems valid when $D = \left\lfloor \frac{\log[1+N(k-1)]}{\log k} - 1 \right\rfloor$, where $\lfloor x \rfloor$ is the largest integer smaller than or equal to x. This suggests that the deepest level D need not be filled completely to count k^D nodes and that (17.25) may extend to "incomplete" k-ary trees. As further observed from Fig. 17.8, $g_{N,k}(m)$ is monotonously decreasing in k. Hence, it is quite likely that he map $k \mapsto g_{N,k}(m)$ is decreasing in $k \in [1, N-1]$. Intuitively, this conjecture can be understood from Fig. 17.7. Both the $k = 2$ and $k = 5$ trees have an equal number of nodes. We observe that the deeper D (or the smaller k), the more overlap is possible, hence, the larger $g_{N,k}(m)$.

Theorem 17.1.1 can also be deduced from (17.25). The lower bound is attained in a star topology where $k = N - 1$, $D = 1$ and $E[H_N] = 1$. The upper bound is attained in a line topology where $k = 1$, $D = N - 1$ and $E[H_N] = \frac{N}{2}$. Furthermore, for real values of $k \in [1, N-1]$, the set of curves specified by (17.25) covers the total allowable space of $g_{N,k}(m)$, as shown in Fig. 17.1. This suggests to consider (17.25) for estimating k in real topologies.

Since $g_N(1) = E[H_N]$, the average hopcount in a k-ary tree follows from (17.25) as

$$E[H_N] = N - 1 - \sum_{j=0}^{D-1} k^{D-j} \frac{N - 1 - \frac{k^{j+1}-1}{k-1}}{N - 1} = \frac{1}{N-1} \sum_{j=0}^{D-1} k^{D-j} \frac{k^{j+1}-1}{k-1}$$

$$= \frac{ND}{N-1} + \frac{D}{(N-1)(k-1)} - \frac{1}{k-1} \qquad (17.27)$$

For large N, we find with

$$D = \left\lfloor \frac{\log[1 + N(k-1)]}{\log k} - 1 \right\rfloor \sim \log_k N + \log_k(1 - 1/k) + O(1/N)$$

that

$$E[H_N] = \log_k N + \log_k(1 - 1/k) - \frac{1}{k-1} + O\left(\frac{\log_k N}{N}\right) \qquad (17.28)$$

Comparing (17.28) with the average hopcount in the random graph (16.10) shows equality to first order if $k_{\rm rg} = e$. Moreover, both the second order terms $\gamma - 1 = -0.42$ and $\log(1 - 1/e) - \frac{1}{e-1} = -1.04$ are $O(1)$ and independent of N. This shows that the multicast gain in the random graph is well approximated by $g_{N,e}(m)$.

17.4 The Chuang–Sirbu law

We discuss the empirical Chuang–Sirbu scaling law, which states that $g_N(m) \approx E[H_N] m^{0.8}$ for the Internet. Based on Internet measurements, Chuang and Sirbu (1998) observed that $g_N(m) \approx E[H_N] m^{0.8}$. Subsequently, Phillips *et al.* (1999) dubbed this observation the Chuang–Sirbu law.

Corollary 17.1.5 implies that the empirical law of Chuang–Sirbu cannot hold true for all $m \leq N$. Indeed, if $g_N(m) = E[H_N] m^{0.8}$, we obtain from the inequality (17.7) and the identity $f_N(m) = mE[H_N]$, that $m^{0.2} \leq E[H_N]$. Write $m = xN$ for a fixed $0 < x < 1$ and x independent of N. Hence, we have shown that

Corollary 17.4.1 *For all graphs satisfying the condition that $\frac{E[H_N]}{N^{0.2}} \to 0$, for large N, the empirical Chuang–Sirbu law does not hold in the region $m = xN$ with $0 < x \leq 1$ and sufficiently large N.*

The most realistic graph models for the Internet assume that $E[H_N] \approx c \log N$, since this implies that the number of routers that can be reached from any starting destination grows exponentially with the number of hops. For these realistic graphs, Corollary 17.4.1 states that empirical Chuang–Sirbu law does not hold for all m. On the other hand, there are more regular graphs (such as a d-lattice, where $E[H_N] \simeq \frac{d}{3} N^{1/d}$) with $E[H_N] \sim N^{0.2+\epsilon}$ (and $\epsilon > 0$) for which the mathematical condition $m^{0.2} \leq E[H_N]$ is satisfied for all m and N. As shown in Van Mieghem *et al.* (2000), however, these classes of graphs, in contrast to random graphs, are not leading to good models for SPTs in the Internet.

17.4.1 Validity range of the Chuang–Sirbu law

For the random graph $G_p(N)$, the SPT is very close to a URT for large N and with (16.10), we obtain

$$f_N(m) \sim m(\log N + \gamma - 1)$$

From the exact $g_N(m)$ formula (17.16) for the random graph $G_p(N)$, the asymptotic for large N and m follows as

$$g_N(m) \sim \frac{mN}{N - m} \log\left(\frac{N}{m}\right) - \frac{1}{2} \tag{17.29}$$

The above scaling explains the empirical Chuang–Sirbu law for $G_p(N)$: for m small with respect to N, the graphs of $(\log N + \gamma - 1)m^{0.8}$ and $\frac{mN}{N-m} \log\left(\frac{N}{m}\right) - \frac{1}{2}$ look very alike in a log-log plot, as illustrated in Fig. 17.9.

Using the asymptotic properties of the digamma function ψ, we obtain (17.29) as an excellent approximation for large N (and all m) or, in normalized form with $m = xN$ and $0 < x < 1$,

$$\frac{g_N(xN) + 0.5}{N} \sim \frac{x \log x}{x - 1} \qquad (17.30)$$

The normalized Chuang–Sirbu law is $\frac{g_N(xN)}{N} = \frac{E[H_N]}{N^{0.2}} x^{0.8}$. It is interesting to note that the Chuang–Sirbu law is "best" if $\frac{E[H_N]}{N^{0.2}} = 1$, since then both endpoints $x = 0$ and $x = 1$ coincide with (17.30). This optimum is achieved when $N \approx 250\,000$, which is of the order of magnitude of the estimated number of routers in the current Internet. This observation may explain the fairly good correspondence on a less sensitive log-log scale with Internet measurements. At the same time, it shows that for a growing Internet, the fit of the Chuang–Sirbu law will deteriorate. For $N \geq 10^6$, the Chuang–Sirbu law underestimates $g_N(m)$ for all m.

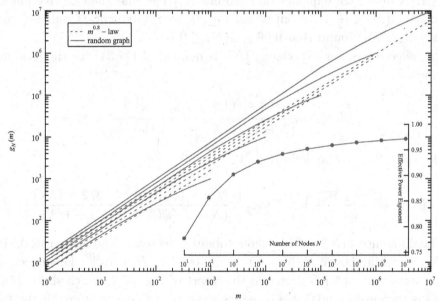

Fig. 17.9. The multicast efficiency for $N = 10^j$ with $j = 3, 4, ..., 7$. The endpoint of each curve $g_N(N - 1) = N - 1$ determines N. The insert shows the effective power exponent versus N.

17.4.2 The effective power exponent $\beta(N)$

For small to moderate values of m, $g_N(m)$ is very close to a straight line in a log-log plot. This "power law behavior" implies that $\log g_N(m) \approx$

$\log E(H_N) + \beta(N) \log m$, which is a first order Taylor expansion of $\log g_N(m)$ in $\log m$. This observation suggests the computation[4] of the effective power exponent $\beta(N)$ as

$$\beta(N) = \left. \frac{d \log g_N(m)}{d \log m} \right|_{m=1} \tag{17.31}$$

Only for a straight line, the differential operator can be replaced by the difference operator such that $\beta(N) \equiv \beta^*(N)$, where

$$\beta^*(N) = \frac{\log \frac{g_N(2)}{E[H_N]}}{\log 2} \tag{17.32}$$

In general, for small m, the effective power exponent (17.31) is not a constant 0.8 as in the Chuang–Sirbu law, but dependent on N. Since $g_N(m)$ is concave by Theorem 17.1.2, $\beta(N)$ is the maximum possible value for $\frac{d \log g_N(m)}{d \log m}$ at any $m \geq 1$. A direct consequence of Theorem 17.1.1 is that the effective power exponent $\beta(N) \in [\frac{1}{2}, 1]$. From recent Internet measurements, Chalmers and Almeroth (2001) found that $0.66 \leq \beta(N) \leq 0.7$.

The effective power exponent $\beta(N)$ as defined in (17.31) for the random graph is

$$\beta(N) = \frac{N \left(\psi(N) + \gamma - \frac{\pi^2}{6} + \frac{\pi^2}{6N} \right)}{(N-1) \left(\psi(N) + (\gamma - 1) + \frac{1}{N} \right)}$$

while, according to the definition (17.32),

$$\beta^*(N) = \frac{\log \frac{g_N(2)}{E[H_N]}}{\log 2} = 1 + \log_2 \left[\frac{(N-1)(\psi(N) + \gamma - 3/2 + 1/N)}{(N-2)(\psi(N) + \gamma - 1 + 1/N)} \right]$$

The difference $\beta(N) - \beta^*(N)$ monotonously decreases and is largest, 0.048 at $N = 3$ while 0.0083 at $N = 10^5$ and 0.0037 at $N = 10^{10}$. This effective power exponent $\beta(N)$ is drawn in the insert of Fig. 17.9, which shows that $\beta(N)$ is increasing and not a constant close to 0.8. More interestingly, for large N, we find with (16.10) and (16.11) that $\beta(N) \sim \frac{\mathrm{Var}[H_N]}{E[H_N]}$ and that $\lim_{N \to \infty} \beta(N) = 1$. In Van Mieghem *et al.* (2000), the ratio $\alpha = \frac{\mathrm{Var}[H_N]}{E[H_N]}$ pops up naturally as the extreme value index of the distribution of the link weights in a topology. Since measurements of the hopcount in Internet indicate that $\frac{\mathrm{Var}[H_N]}{E[H_N]} \approx 1$, which corresponds to a regular distribution, this extreme value index strongly favors the model of the hopcount based on

---

[4] Although (17.5) only has meaning for integer m, analytic continuation to a complex variable is possible and, hence, differentiation can be defined.

shortest paths in $G_p(N)$, although random graphs do *not* model the Internet topology well.

Thus, if the number of nodes in the Internet is still growing, we suggest, *only for small to moderate values of* m, the consideration of a power law approximation for the multicast gain

$$g_N(m) \approx E[H_N] \, m^{\frac{\mathrm{Var}[H_N]}{E[H_N]}}$$

instead of the Chuang–Sirbu law.

In summary, many properties in nature seem linear on an insensitive log-log scale. However, deriving from these plots simple and attractive power laws for complicated matter, seems a little oversimplified[5].

17.5 Stability of a multicast shortest path tree

We now turn to the problem of quantifying the stability in a multicast tree. Inspired by Poisson arrival processes, at a single instant of time, we assume that either one or zero group members can leave. In the sequel, we do not make any further assumption about the time-dependent process of leaving/joining a multicast group and refrain from dependencies on time. The number of links in the tree that change after one multicast group member leaves the group has been chosen as measure for the stability of the multicast tree. If we denote this quantity by $\Delta_N(m)$, then, by definition of $g_N(m)$, the average number of changes equals

$$E[\Delta_N(m)] = g_N(m) - g_N(m-1) \tag{17.33}$$

Since $g_N(m)$ is concave (Theorem 17.1.2), $E[\Delta_N(m)]$ is always positive and decreasing in m. If the scope of m is extended to real numbers, $E[\Delta_N(m)] \approx g'_N(m)$ which simplifies further estimates.

The situation where on average less than one link changes if one multicast group member leaves may be regarded as a stable regime. Since $E[\Delta_N(m)]$ is always positive and decreasing in m, this stable regime is reached when the group size m exceeds m_1, which satisfies $E[\Delta_N(m_1)] = 1$. For example, for the URT that is asymptotically the SPT for the class RGU defined in Section 16.2.2, this condition approximately follows from (17.29) as

$$E[\Delta_N(m)] \sim \frac{mN}{N-m} \log\left(\frac{N}{m}\right) - \frac{(m-1)N}{N-m+1} \log\left(\frac{N}{m-1}\right) \tag{17.34}$$

---

[5] Many recent articles devote attention to power law behavior but most of them seem prudent: just recall the immense interest (hype?) a few years ago in the long range and self-similar nature of Internet traffic and the relation to the "simple" power law with only the Hurst parameter (comparable to $\beta(N)$ here) in the exponent.

Let $x = \frac{m}{N}$, then $0 < x < 1$ and

$$\frac{E\left[\Delta_N(m)\right]}{N} \sim \frac{-x}{1-x} \log x + \frac{(x-1/N)}{1-(x-1/N)} \log\left(x - \frac{1}{N}\right)$$

After expanding the second term in a Taylor series around x to first order in $\frac{1}{N}$,

$$E\left[\Delta_N(xN)\right] \sim \frac{x-1-\log x}{(1-x)^2} + O\left(\frac{1}{N}\right)$$

For large N, $E\left[\Delta_N(x_1 N)\right] \sim 1$ occurs when $x_1 = 0.3161$, which is the solution in x of $\frac{x-1-\log x}{(1-x)^2} = 1$. For the class RGU, a stable tree as defined above is obtained when the multicast group size m is larger than $m_1 = 0.3161N \approx \frac{N}{3}$. In the sequel, since m_1 is high and of less practical interest, we will focus on multicast group sizes smaller than m_1. The computation of m_1 for other graph types turns out to be difficult. Since, as mentioned above, the comparison with Internet measurement (Van Mieghem *et al.*, 2001a) shows that formula (17.29) provides a fairly good estimate, we expect that $m_1 \approx \frac{N}{3}$ also approximates well the stable regime in the Internet.

The following theorem quantifies the stability in the class RGU.

Theorem 17.5.1 *For sufficiently large N and fixed m, the number of changed edges $\Delta_N(m)$ in a random graph $G_p(N)$ with uniformly distributed link weights tends to a Poisson distribution,*

$$\Pr\left[\Delta_N(m) = k\right] \sim e^{-E[\Delta_N(m)]} \frac{\left(E\left[\Delta_N(m)\right]\right)^k}{k!} \qquad (17.35)$$

where $E\left[\Delta_N(m)\right] = g_N(m) - g_N(m-1)$ and $g_N(m)$ is given by (17.16) or approximately by (17.29).

Proof: In Section 16.2.1 we have mentioned that the SPT in the class RGU is an URT for large N. In addition, the random variable for the number of hops H_N from the root to an arbitrary node tends, for large N, to a Poisson random variable with mean $E\left[H_N\right] \sim \log N + \gamma - 1$ as shown in Section 16.3.1. Now, $\Delta_N(m) = H_N(m) - H_N(m-1)$ is the positive discrete random variable that counts the absolute value of the difference between the hopcount $h_{R \to m}$ from the root (source) to user m and the hopcount $h_{R \to m-1}$ from the root to the user closest in the tree to m, which we here relabel by $m - 1$. Both users m and $m - 1$ are not independent nor the two random variables $h_{R \to m}$ and $h_{R \to m-1}$ are independent in general due to possible overlap in their paths.

If the shortest paths from the root to each of the two users m and $m - 1$

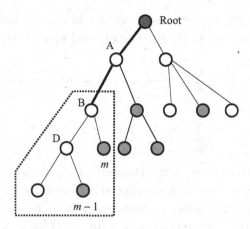

Fig. 17.10. A sketch of a uniform recursive tree, where $h_{R \to m} = 3$ and $h_{R \to m-1} = 4$ and the number of links in common is two (shown in bold Root-A-B).

overlap, there always exists a node in the SPT, say node B as illustrated in Fig. 17.10, that sees the partial shortest paths from itself to m and $m - 1$ as non-overlapping and independent. Since the SPT is a URT, the subtree rooted at that node B (enclosed in dotted line in Fig. 17.10) is again a URT as follows from Theorem 16.2.1. With respect to B, the nodes m and $m - 1$ are uniformly chosen and the number of links $\Delta_N(m)$ that change if the m-th node leaves is just its hopcount with respect to B (instead of the original root). We denote the unknown number of nodes in that subtree rooted at B by $\nu(m) \leq N$. We have that $\nu(m) \leq \nu(m - 1)$ because by adding a group member, the size of the subtree can only decrease. For large N and small m, $\nu(m)$ is large such that the above mentioned asymptotic law of the hopcount applies. If both m and N are large, $\nu(m)$ will become too small for the asymptotic law to apply. Thus, for fixed m and large N, this implies that $\Delta_N(m)$ tends to a Poisson random variable with mean $E[\Delta_N(m)]$. $\qquad \square$

Simulations in Van Mieghem and Janic (2002) indicate that the Poisson law seems more widely valid than just in the asymptotic regime ($N \to \infty$).

The proof can be extended to a general topology. Assume for a certain class of graphs that the pdf of the hopcount $\Pr[H_N = k]$ and the multicast efficiency $g_N(m)$ can be computed for all sizes N. The subtree rooted at B is again a SPT in a subcluster of size $\nu(m)$, which is an unknown random variable. The argument similar as the one in the proof above shows that

$$\Pr[\Delta_N(m) = k] = \Pr[H_{\nu(m)} = k]$$

This argument implicitly assumes that all multicast users are uniformly distributed over the graph. By the law of total probability,

$$\Pr\left[H_{\nu(m)} = k\right] = \sum_{n=1}^{N} \Pr\left[H_{\nu(m)} = k | \nu(m) = n\right] \Pr\left[\nu(m) = n\right]$$

$$= \sum_{n=1}^{N} \Pr\left[H_n = k\right] \Pr\left[\nu(m) = n\right]$$

which, unfortunately shows that the pdf of $\nu(m)$ is required to specify $\Pr\left[\Delta_N(m) = k\right]$. However, we can proceed further in an approximate way by replacing the unknown random variable $\nu(m)$ by its best estimate, $E\left[\nu(m)\right]$. In that approximation, the average size $E\left[\nu(m)\right]$ of the shortest path subtree rooted at B can be specified, at least in principle, with the use of (17.33). Indeed, since $E\left[H_{E[\nu(m)]}\right] = \sum_{k=1}^{E[\nu(m)]-1} k \Pr\left[H_{E[\nu(m)]} = k\right]$, by equating

$$E\left[H_{E[\nu(m)]}\right] = g_N(m) - g_N(m-1)$$

a relation in one unknown $E\left[\nu(m)\right]$ is found and can be solved for $E\left[\nu(m)\right]$. In conclusion, we end up with the approximation

$$\Pr\left[\Delta_N(m) = k\right] \approx \Pr\left[H_{E[\nu(m)]} = k\right]$$

which roughly demonstrates that, in general, $\Pr\left[\Delta_N(m) = k\right]$ is likely related to the hopcount distribution in that certain class of graphs.

Unfortunately, for very few types of graphs, both the pdf $\Pr\left[H_N = k\right]$ and the multicast gain $g_N(m)$ can be computed. This fact augments the value of Theorem 17.5.1, although the class RGU is *not* a good model for the *graph* of the Internet. Fortunately, the *shortest path tree* deduced from that class seems a reasonable approximation as shown in Fig. 16.4 and sufficient to provide first order estimates.

17.6 Proof of (17.16): $g_N(m)$ for random graphs

Before embarking with the proof of formula (17.16), we first prove the following lemma.

Lemma 17.6.1 *For $a > b$,*

$$S(a,b) = \sum_{k=1}^{b} \frac{(a-k)!}{(b-k)!} \frac{1}{k} = \frac{a!}{b!}\left[\psi(a+1) - \psi(a-b+1)\right]$$

and

$$S(b,b) = \sum_{k=1}^{b} \frac{1}{k} = \psi(b+1) + \gamma$$

Proof: We start by writing

$$S(a,b) = \sum_{k=1}^{b} \frac{(a-k)\cdots(b-k+1)}{k}$$

$$= a\sum_{k=1}^{b} \frac{(a-1-k)\cdots(b-k+1)}{k} - \sum_{k=1}^{b}(a-1-k)\cdots(b-k+1)$$

Since $(a-1-k)\cdots(b-k+1) = (a-b-1)!\binom{a-1-k}{b-k}$ and by the recurrence for the binomial $\sum_{k=1}^{b}\binom{a-1-k}{b-k} = \binom{a-1}{b-1}$, we have that

$$S(a,b) = aS(a-1,b) - \frac{1}{a-b}\frac{(a-1)!}{(b-1)!}$$

After p iterations, we have

$$S(a,b) = a(a-1)\cdots(a-p+1)S(a-p,b) - \frac{a!}{(b-1)!}\sum_{j=0}^{p-1}\frac{1}{(a-j)(a-j-b)}$$

and, if $p = a - b$, the recursions stops with result,

$$S(a,b) = \frac{a!}{b!}\sum_{k=1}^{b}\frac{1}{k} - \frac{a!}{(b-1)!}\sum_{j=0}^{a-b-1}\frac{1}{(a-j)(a-j-b)}$$

$$= \frac{a!}{b!}\sum_{k=1}^{b}\frac{1}{k} - \frac{a!}{b!}\left(\sum_{k=1}^{a-b}\frac{1}{k} - \sum_{k=b+1}^{a}\frac{1}{k}\right) = \frac{a!}{b!}\left(\sum_{k=1}^{a}\frac{1}{k} - \sum_{k=1}^{a-b}\frac{1}{k}\right)$$

from which the lemma follows. $\square$

Proof of equation (17.16): We will investigate $E[X_i] = E\left[X_i^{(N)}\right]$ in the URT with N nodes. Here $E[X_i]$ is the number of joint hops in a multicast SPT from the root to i uniformly chosen nodes in the URT and where all the group member nodes are different from the root. Let $E\left[\tilde{X}_i\right]$ be the same quantity where we allow the group member nodes to be the root. Then,

$$E\left[\tilde{X}_i\right] = \frac{N-i}{N}E[X_i]$$

since there are i possibilities each with probability $\frac{1}{N}$ that one of the nodes equals the root, in which case $X_i = 0$.

The average number of joint hops $E\left[\tilde{X}_i\right]$ is deduced from Fig. 17.11, where two clusters are shown each with respectively k and $N-k$ nodes. The first cluster with k nodes does not possess the root (dark shaded), but it contains the i multicast group members (light shaded). There is already at least 1 joint hop because the link between the root and node A, that can be viewed as the root of the first cluster, is used by all i group members lying in the first cluster. Given the size k of the first cluster, the probability that all i uniformly chosen group members belong to the first cluster equals $\frac{k(k-1)\cdots(k-i+1)}{N(N-1)\cdots(N-i+1)}$ because the probability that the first group member belongs to that cluster, which is $\frac{k}{N}$, the probability that the second group member also belongs to the first cluster, which is $\frac{k-1}{N-1}$ and so on. Since the size of the first cluster connected to the root is uniform in between 1 and $N-1$, the probability that the size is k equals $\frac{1}{N-1}$. When all i nodes are in that first cluster of size k, X_i is at least 1, and the problem restarts, but with N replaced by k and A being the root. Hence, if all i group members belong to the first cluster, the average number of joint hops is $\frac{1}{N-1}\sum_{k=1}^{N-1}\frac{k(k-1)\cdots(k-i+1)}{N(N-1)\cdots(N-i+1)}\left(1 + E\left[\tilde{X}_i^{(k)}\right]\right)$ because we must sum over all possible sizes for the first cluster. If *not* all i group member nodes are in the first cluster, the group member nodes are divided over the two clusters. But, in that case, we have no joint overlaps or $X_i = 0$.

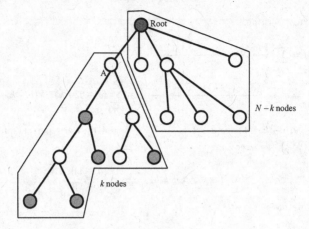

Fig. 17.11. The two contributing clusters leading to the $E\left[\tilde{X}_i^{(N)}\right]$-recursion.

Thus, if not all i group members nodes are in the first cluster, the only way that there are possible joint overlaps ($X_i > 0$), is that all i group member nodes are in the second cluster. However, by removing the first cluster, we are left again with a uniform recursive tree of size $N - k$. The average number of joint hops in this case is $\frac{1}{N-1}\sum_{k=1}^{N-1}\frac{(N-k)(N-k-1)\cdots(N-k-i+1)}{N(N-1)\cdots(N-i+1)}E\left[\tilde{X}_i^{(N-k)}\right]$. Adding both contributions results in the recursion formula

$$E\left[\tilde{X}_i^{(N)}\right] = \frac{1}{N-1}\sum_{k=1}^{N-1}\frac{k(k-1)\cdots(k-i+1)}{N(N-1)\cdots(N-i+1)}\left(1 + 2E\left[\tilde{X}_i^{(k)}\right]\right) \qquad (17.36)$$

We next write

$$\alpha_i^{(N)} = N(N-1)\cdots(N-i+1)E\left[\tilde{X}_i^{(N)}\right] = \frac{N!}{(N-i)!}E\left[\tilde{X}_i^{(N)}\right]$$

then the above recurrence equation (17.36) turns into

$$\alpha_i^{(N)} = \frac{1}{N-1}\sum_{k=1}^{N-1}[k(k-1)\cdots(k-i+1)+2\alpha_i^{(k)}]$$

$$= \frac{1}{N-1}\sum_{k=i}^{N-1}k(k-1)\cdots(k-i+1) + \frac{2}{N-1}\sum_{k=1}^{N-1}\alpha_i^{(k)}$$

Subtracting

$$(N-1)\alpha_i^{(N)} - (N-2)\alpha_i^{(N-1)} = \frac{(N-1)!}{(N-i-1)!} + 2\alpha_i^{(N-1)}$$

from which we obtain

$$\frac{\alpha_i^{(N)}}{N} = \frac{(N-2)!}{N(N-i-1)!} + \frac{\alpha_i^{(N-1)}}{N-1} \qquad (17.37)$$

Iterating (17.37) gives

$$\frac{\alpha_i^{(N)}}{N} = \sum_{j=0}^{k-1}\frac{(N-2-j)!}{(N-j)(N-i-1-j)!} + \frac{\alpha_i^{(N-k)}}{N-k}$$

Since $\alpha_i^{(i)} = E\left[\tilde{X}_i^{(i)}\right] = 0$, because the root is then always one of the group member nodes, we finally obtain,

$$\alpha_i^{(N)} = N \sum_{j=0}^{N-i-1} \frac{(N-2-j)!}{(N-j)(N-i-1-j)!} = N \sum_{k=i+1}^{N} \frac{(k-2)!}{k(k-i-1)!} \tag{17.38}$$

It can be shown that, for large N, $\alpha_i^{(N)} \sim \frac{N}{i-1} \frac{(N-2)!}{(N-i-1)!}$.

Because

$$E\left[X_i^{(N)}\right] = \frac{N}{N-i} E\left[\tilde{X}_i^{(N)}\right] = \frac{(N-i-1)!}{(N-1)!}\alpha_i^{(N)}$$

we have that

$$E\left[X_i^{(N)}\right] = \frac{(N-i-1)!N}{(N-1)!} \sum_{k=i+1}^{N} \frac{(k-2)!}{k(k-i-1)!} \tag{17.39}$$

and, for large N, $E\left[X_i^{(N)}\right] \sim \frac{1}{i-1}\frac{N}{(N-1)} \sim \frac{1}{i-1}$.

Invoking Theorem 17.1.3, the average number of multicast hops for m uniformly chosen, distinct group members is

$$g_N(m) = \sum_{i=1}^{m} \binom{m}{i}(-1)^{i-1} \frac{(N-i-1)!N}{(N-1)!} \sum_{k=2}^{N} \frac{(k-2)!}{k(k-i-1)!}$$

$$= \frac{-N}{(N-1)!} \sum_{s=0}^{N-2} \frac{(N-2-s)!}{N-s} \sum_{i=1}^{m} \binom{m}{i}(-1)^i \frac{(N-i-1)!}{(N-i-1-s)!}$$

The i-summation can be executed as follows. Consider $x^{N-1}(1-1/x)^m = \sum_{i=0}^{m} \binom{m}{i}(-1)^i x^{N-i-1}$. Differentiating s times yields

$$\sum_{i=0}^{m} \binom{m}{i}(-1)^i \frac{(N-i-1)!}{(N-i-1-s)!} x^{N-i-s-1} = \frac{d^s}{dx^s}\left[x^{N-1-m}(x-1)^m\right]$$

Expanding the right-hand side around $x = 1$ gives

$$\frac{d^s}{dx^s}\left[x^{N-1-m}(x-1)^m\right] = \sum_{k=0}^{\infty} \binom{N-1-m}{k} \frac{d^s}{dx^s}(x-1)^{k+m}$$

$$= \sum_{k=0}^{\infty} \binom{N-1-m}{k} \frac{(k+m)!}{(k+m-s)!}(x-1)^{k+m-s}$$

Evaluation at $x = 1$ only leads to a non-zero contribution if $k + m - s = 0$. Hence,

$$\sum_{i=1}^{m} \binom{m}{i}(-1)^i \frac{(N-i-1)!}{(N-i-1-s)!} = \binom{N-1-m}{s-m}s! - \frac{(N-1)!}{(N-1-s)!}$$

and

$$g_N(m) = \frac{-N(N-1-m)!}{(N-1)!} \sum_{s=0}^{N-2} \frac{s!}{(N-s)(s-m)!(N-1-s)} + N \sum_{s=0}^{N-2} \frac{1}{(N-s)(N-1-s)}$$

$$= \frac{-N(N-1-m)!}{(N-1)!} \left[\sum_{s=m}^{N-2} \frac{s!}{(s-m)!(N-1-s)} - \sum_{s=m}^{N-2} \frac{s!}{(s-m)!(N-s)} \right]$$

$$+ N \left[\sum_{k=1}^{N-1} \frac{1}{k} - \sum_{k=2}^{N} \frac{1}{k} \right]$$

$$= \frac{-N(N-1-m)!}{(N-1)!} \left[\sum_{k=1}^{N-m-1} \frac{(N-k-1)!}{(N-k-1-m)!k} - \sum_{k=2}^{N-m} \frac{(N-k)!}{(N-k-m)!k} \right] + N - 1$$

Rewrite the first summation as

$$\sum_{k=1}^{N-m-1} \frac{(N-k-1)!}{(N-k-1-m)!k} = \frac{(N-2)!}{(N-2-m)!} + \sum_{k=2}^{N-m} \frac{(N-k-1)!(N-k-m)}{(N-k-m)!k}$$

$$= \frac{(N-2)!}{(N-2-m)!} + \sum_{k=2}^{N-m} \frac{(N-k)!}{(N-k-m)!k} - m \sum_{k=2}^{N-m} \frac{(N-k-1)!}{(N-k-m)!k}$$

Then,

$$g_N(m) = \frac{-N(N-1-m)!}{(N-1)!} \left[\frac{(N-2)!}{(N-2-m)!} - m \sum_{k=2}^{N-m} \frac{(N-k-1)!}{k(N-k-m)!} \right] + N - 1$$

$$= \frac{N(m-1)+1}{(N-1)} + \frac{mN(N-1-m)!}{(N-1)!} \sum_{k=2}^{N-m} \frac{(N-k-1)!}{k(N-k-m)!}$$

$$= -1 + \frac{mN(N-1-m)!}{(N-1)!} \sum_{k=1}^{N-m} \frac{(N-k-1)!}{k(N-k-m)!}$$

Using Lemma 17.6.1

$$\sum_{k=1}^{N-m} \frac{(N-k-1)!}{(N-k-m)!} \frac{1}{k} = \frac{(N-1)!}{(N-m)!} \left[\psi(N) - \psi(m) \right] \qquad (17.40)$$

finally leads to (17.16).   □

17.7 Proof of Theorem 17.3.1: $g_N(m)$ for k-ary trees

Let $\tilde{X}_i$ be the number of joint hops for i *different* multicast group members (we allow the root to be a user in which case $\tilde{X}_i = 0$). Then,

$$\Pr\left[\tilde{X}_i \geq 1\right] = \Pr\left[\text{All group members belong to the same cluster connected to the root}\right]$$

$$= k \cdot \Pr\left[\text{All group members belong to the first cluster connected to the root}\right]$$

$$= k \frac{\binom{(N-1)/k}{i}}{\binom{N}{i}} = k \frac{\binom{1+k+\cdots+k^{D-1}}{i}}{\binom{1+k+\cdots+k^{D}}{i}} \qquad (17.41)$$

By self-similarity of k-ary trees we obtain

$$\Pr\left[\tilde{X}_i \geq 2 | \tilde{X}_i \geq 1\right] = p_i^{(D-1)} = k\frac{\binom{1+k+\cdots+k^{D-2}}{i}}{\binom{1+k+\cdots+k^{D-1}}{i}}$$

because each cluster extending from the root is itself a k-ary tree of depth $D-1$. In general, we have $\Pr\left[\tilde{X}_i \geq j\right] = \Pr\left[\tilde{X}_i \geq j | \tilde{X}_i \geq j-1\right] \Pr\left[\tilde{X}_i \geq j-1\right]$. Hence, by iteration,

$$\Pr\left[\tilde{X}_i \geq j\right] = \prod_{n=D-j+1}^{D} p_i^{(n)}, \quad j = 1, 2, \ldots, D-1 \tag{17.42}$$

Note that for $i \geq 2$ the probability $\Pr\left[\tilde{X}_i \geq D\right] = 0$, because if $\tilde{X}_i = D$ some destinations must be identical. From (2.36) we obtain for $i \geq 2$,

$$E\left[\tilde{X}_i\right] = \sum_{j=1}^{D-1}\prod_{n=D-j+1}^{D} p_i^{(n)} = \sum_{j=1}^{D-1}\frac{k^j\binom{1+\cdots+k^{D-j}}{i}}{\binom{1+\cdots+k^D}{i}} = \sum_{j=1}^{D-1}\frac{k^{D-j}\binom{1+\cdots+k^j}{i}}{\binom{1+\cdots+k^D}{i}} \tag{17.43}$$

Since $E[X_i] = \frac{N}{N-i}E\left[\tilde{X}_i\right]$, we find

$$E[X_i] = \frac{N}{N-i}\sum_{j=1}^{D-1}\frac{k^{D-j}\binom{1+\cdots+k^j}{i}}{\binom{1+\cdots+k^D}{i}}, \quad i \geq 2 \tag{17.44}$$

For the value of $E\left[\tilde{X}_1\right]$ and $E[X_1]$ we find

$$E\left[\tilde{X}_1\right] = \frac{1}{N}\sum_{j=1}^{D} k^j(1 + \cdots + k^{D-j}) = \frac{1}{N(k-1)}\left\{Dk^{D+1} - (N-1)\right\}$$

and

$$E[X_1] = \frac{1}{N-1}\sum_{j=1}^{D} k^j(1 + \cdots + k^{D-j}) = \frac{N}{N-1}\sum_{j=1}^{D-1}\frac{k^{D-j}\binom{1+\cdots+k^j}{1}}{\binom{1+\cdots+k^D}{1}} + \frac{k^D}{N-1}$$

Invoking Theorem 17.1.3 yields

$$g_{N,k}(m) = \frac{mk^D}{N-1} - \sum_{i=1}^{m}\binom{m}{i}(-1)^i\frac{N}{N-i}\sum_{j=1}^{D-1}\frac{k^{D-j}\binom{1+\cdots+k^j}{i}}{\binom{1+\cdots+k^D}{i}}$$

Writing $A_j = \frac{k^{j+1}-1}{k-1}$ and reversing the i- and j-summation yields, using (17.24),

$$g_{N,k}(m) = \frac{mk^D}{N-1} - N\sum_{j=1}^{D-1} k^{D-j}\frac{A_j!}{N!}\sum_{i=1}^{m}\binom{m}{i}(-1)^i\frac{(N-i-1)!}{(A_j-i)!}$$

Concentrating on the inner sum with lower sum bound $i = 0$, denoted as S_j, and substituting $k = m - i$, we have

$$S_j = \sum_{k=0}^{m}\binom{m}{k}(-1)^{m-k}\frac{\Gamma(N-m+k)}{\Gamma(A_j-m+k+1)}$$

Invoking the Taylor series of the hypergeometric function (Abramowitz and Stegun, 1968, Section 15.1.1),

$$F(a, b; c; z) = \frac{\Gamma(c)}{\Gamma(a)\Gamma(b)}\sum_{n=0}^{\infty}\frac{\Gamma(a+n)\Gamma(b+n)}{\Gamma(c+n)n!}z^n$$

$m!S_j$ is the coefficient in z^m of the Cauchy product of

$$(1-z)^m = \sum_{k=0}^{\infty} \binom{m}{k}(-1)^k z^k$$

and

$$\frac{\Gamma(N-m)}{\Gamma(A_j-m+1)}F(1,N-m;A_j-m+1;z) = \sum_{k=0}^{\infty} \frac{\Gamma(N-m+k)}{\Gamma(A_j-m+1+k)} z^k$$

Hence,

$$S_j = \frac{1}{m!}\frac{\Gamma(N-m)}{\Gamma(A_j-m+1)}\frac{d^m}{dz^m}\left[(1-z)^m F(1,N-m;A_j-m+1;z)\right]|_{z=0}$$

Invoking the differentiation formula (Abramowitz and Stegun, 1968, Section 15.2.7),

$$\frac{d^m}{dz^m}\left[(1-z)^{a+m-1}F(a,b;c;z)\right] = \frac{(-1)^m \Gamma(a+m)\Gamma(c-b+m)\Gamma(c)}{\Gamma(a)\Gamma(c-b)\Gamma(c+m)}(1-z)^{a-1}F(a+m,b;c+m;z)$$

we have, since $a=1$ and $F(a,b;c;0)=1$,

$$S_j = \frac{(-1)^m \Gamma(N-m)\Gamma(A_j+1-N+m)}{\Gamma(A_j+1-N)\Gamma(A_j+1)}$$

Thus,

$$g_{N,k}(m) = \frac{mk^D}{N-1} - N\sum_{j=1}^{D-1} k^{D-j}\frac{A_j!}{N!}\left(\frac{(-1)^m(N-m-1)!(A_j-N+m)!}{(A_j-N)!A_j!} - \frac{(N-1)!}{A_j!}\right)$$

$$= \frac{mk^D}{N-1} + \sum_{j=1}^{D-1} k^{D-j} + \frac{(-1)^{m-1}(N-m-1)!}{(N-1)!}\sum_{j=1}^{D-1} k^{D-j}\frac{(A_j-N+m)!}{(A_j-N)!}$$

from which (17.25) is immediate.   □

17.8 Problem

(i) Compute the effective power exponent $\beta^*(N)$ for the k-ary tree.

18

The hopcount to an anycast group

In this chapter, the probability density function of the number of hops to the most nearby member of the anycast group consisting of m members (e.g. servers) is analyzed. The results are applied to compute a performance measure η of the efficiency of anycast over unicast and to the server placement problem. The server placement problem asks for the number of (replicated) servers m needed such that any user in the network is not more than j hops away from a server of the anycast group with a certain prescribed probability. As in Chapter 17 on multicast, two types of shortest path trees are investigated: the regular k-ary tree and the irregular uniform recursive tree treated in Chapter 16. Since these two extreme cases of trees indicate that the performance measure $\eta \approx 1 - a \log m$ where the real number a depends on the details of the tree, it is believed that for trees in real networks (as the Internet) a same logarithmic law applies. An order calculus on exponentially growing trees further supplies evidence for the conjecture that $\eta \approx 1 - a \log m$ for small m.

18.1 Introduction

IPv6 possesses a new address type, anycast, that is not supported in IPv4. The anycast address is syntactically identical to a unicast address. However, when a set of interfaces is specified by the same unicast address, that unicast address is called an anycast address. The advantage of anycast is that a group of interfaces at different locations is treated as one single address. For example, the information on servers is often duplicated over several secondary servers at different locations for reasons of robustness and accessibility. Changes are only performed on the primary servers, which are then copied onto all secondary servers to maintain consistency. If both the primary and all secondary servers have a same anycast address, a query from some source towards that anycast address is routed towards the closest server of the group. Hence, instead of routing the packet to the root server (primary server) anycast is more efficient.

Suppose there are m (primary plus all secondary) servers and that these m servers are uniformly distributed over the Internet. The number of hops from the querying device A to the closest server is the minimum number of hops, denoted by $h_N(m)$, of the set of shortest paths from A to these m servers in a network with N nodes. In order to solve the problem, the shortest path tree rooted at node A, the querying device, needs to be investigated. We assume in the sequel that one of the m uniformly distributed servers can possibly coincide with the same router to which the querying machine A is attached. In that case, $h_N(m) = 0$. This assumption is also reflected in the notation, small h, according to the convention made in Section 16.3.2 that capital H for the hopcount excludes the event that the hopcount can be zero.

Clearly, if $m = 1$, the problem reduces to the hopcount of the shortest path from A to one uniformly chosen node in the network and we have that

$$h_N(1) = h_N,$$

where h_N is the hopcount of the shortest path in a graph with N nodes. The other extreme for $m = N$ leads to

$$h_N(N) = 0$$

because all nodes in the network are servers. In between these extremes, it holds that

$$h_N(m) \leq h_N(m - 1)$$

since one additional anycast group member (server) can never increase the minimum number of hops from an arbitrary node to that larger group.

The hopcount to an anycast group is a stochastic problem. Even if the network graph is exactly known, an arbitrary node A views the network along a tree. Most often it is a shortest path tree. Although the sequel emphasizes "shortest path trees", the presented theory is equally valid for any type of tree. The node A's perception of the network is very likely different from the view of another node A'. Nevertheless, shortest path trees in the same graph possess to some extent related structural properties that allow us to treat the problem by considering certain types or classes of shortest path trees. Hence, instead of varying the arbitrary node A over all possible nodes in the graph and computing the shortest path tree at each different node, we vary the structure of the shortest path tree rooted at A over all possible shortest path trees of a certain type. Of course, the confinement of the analysis then lies in the type of tree that is investigated. We will only consider the regular k-ary tree and the irregular URT . It seems reasonable to assume that "real" shortest path trees in the Internet possess a structure somewhere in between these extremes and that scaling laws observed in both the two extreme cases may also apply to the Internet.

The presented analysis allows us to address at least two different issues. First, for a same class of trees, the efficiency of anycast over unicast defined in terms of a performance measure η,

$$\eta = \frac{E\left[h_N(m)\right]}{E\left[h_N(1)\right]} \leq 1$$

is quantified. The performance measure η indicates how much hops (or link traversals or bandwidth consumption) can be saved, on average, by anycast. Alternatively, η also reflects the gain in end-to-end delay or how much faster than unicast, anycast finds the desired information. Second, the so-called *server placement problem* can be treated. More precisely, the question "How many servers m are needed to guarantee that any user request can access the information within j hops with probability $\Pr\left[h_N(m) > j\right] \leq \epsilon$, where ϵ is certain level of stringency," can be answered. The server placement problem is expected to gain increased interest especially for real-time services where end-to-end QoS (e.g. delay) requirements are desirable. In the most general setting of this server placement problem, all nodes are assumed to be equally important in the sense that users' requests are generated equally likely at any router in the network with N nodes. As mentioned in Chapter 17, the validity of this assumption has been justified by Phillips *et al.* (1999). In the case of uniform user requests, the best strategy is to place servers also uniformly over the network. Computations of $\Pr\left[h_N(m) > j\right] < \epsilon$ for given stringency ϵ and hop j, allow the determination of the minimum number m of servers. The solution of this server placement problem may be regarded as an instance of the general quality of service (QoS) portfolio of an network operator. When the number of servers for a major application offered by the service provider are properly computed, the service provider may announce levels ϵ of QoS (e.g. via $\Pr\left[h_N(m) > j\right] < \epsilon$) and accordingly price the use of the application.

18.2 General analysis

Let us consider a particular shortest path tree T rooted at node A with the level set $L_N = \left\{X_N^{(k)}\right\}_{1 \leq k \leq N-1}$ as defined in Section 16.2.2. Suppose that the result of uniformly distributing m anycast group members over the graph leads to a number $m^{(k)}$ of those anycast group member nodes that are k hops away from the root. These $m^{(k)}$ distinct nodes all belong to the k-th level set $\left\{X_N^{(k)}\right\}$. Similarly as for $X_N^{(k)}$, some relations are immediate. First, $m^{(0)} = 0$ means that none of the m anycast group members coincides with the root node A or $m^{(0)} = 1$ means that one of them (and at most one) is attached to the same router A as the querying device. Also, for all $k > 0$, it holds that $0 \le m^{(k)} \le X_N^{(k)}$ and that

$$\sum_{k=0}^{N-1} m^{(k)} = m \qquad (18.1)$$

Given the tree T specified by the level set L_N and the anycast group members specified by the set $\left\{m^{(0)}, m^{(1)}, \ldots, m^{(N-1)}\right\}$, we will derive the lowest non-empty level $m^{(j)}$, which is equivalent to $h_N(m)$.

Let us denote by e_j the event that all first $j + 1$ levels are not occupied by an anycast group member,

$$e_j = \left\{m^{(0)} = 0\right\} \cap \left\{m^{(1)} = 0\right\} \cap \cdots \cap \left\{m^{(j)} = 0\right\}$$

The probability distribution of the minimum hopcount, $\Pr\left[h_N(m) = j | L_N\right]$, is then equal to the probability of the event $e_{j-1} \cap \left\{m^{(j)} > 0\right\}$. Since the event $\left\{m^{(j)} > 0\right\} = \left\{m^{(j)} = 0\right\}^c$, using the conditional probability yields

$$\Pr\left[h_N(m) = j | L_N\right] = \Pr\left[\left.\left\{m^{(j)} > 0\right\}\right| e_{j-1}\right] \Pr\left[e_{j-1}\right]$$
$$= \left(1 - \Pr\left[\left.\left\{m^{(j)} = 0\right\}\right| e_{j-1}\right]\right) \Pr\left[e_{j-1}\right] \qquad (18.2)$$

Since $e_j = e_{j-1} \cap \left\{m^{(j)} = 0\right\}$, the probability of the event e_j can be decomposed as

$$\Pr\left[e_j\right] = \Pr\left[\left.\left\{m^{(j)} = 0\right\}\right| e_{j-1}\right] \Pr\left[e_{j-1}\right] \qquad (18.3)$$

The assumption that all m anycast group members are uniformly distributed enables to compute $\Pr\left[\left.\left\{m^{(j)} = 0\right\}\right| e_{j-1}\right]$ exactly. Indeed, by the uniform assumption, the probability equals the ratio of the favorable possibilities over the total possible. The total number of ways to distribute m items over $N - \sum_{k=0}^{j-1} X_N^{(k)}$ positions – the latter constraints follows from the condition e_{j-1} – equals $\binom{N - \sum_{k=0}^{j-1} X_N^{(k)}}{m}$. Likewise, the favorable number of ways to distribute m items over the remaining levels higher than j, leads to

$$\Pr\left[\left\{m^{(j)}=0\right\}\Big|e_{j-1}\right]=\frac{\binom{N-\sum_{k=0}^{j}X_N^{(k)}}{m}}{\binom{N-\sum_{k=0}^{j-1}X_N^{(k)}}{m}} \tag{18.4}$$

The recursion (18.3) needs an initialization, given by

$$\Pr\left[e_0\right]=\Pr\left[m^{(0)}=0\right]=1-\frac{m}{N}$$

which follows from $\Pr\left[m^{(0)}=0\right]=\frac{\binom{N-1}{m}}{\binom{N}{m}}$ and equals $\Pr\left[\left\{m^{(0)}=0\right\}|e_{-1}\right]$ (although the event e_{-1} is meaningless). Observe that $\Pr\left[m^{(0)}=1\right]=\frac{m}{N}$ holds for any tree such that

$$\Pr\left[h_N(m)=0\right]=\frac{m}{N}$$

By iteration of (18.3), we obtain

$$\Pr\left[e_j\right]=\prod_{s=0}^{j}\frac{\binom{N-\sum_{k=0}^{s}X_N^{(k)}}{m}}{\binom{N-\sum_{k=0}^{s-1}X_N^{(k)}}{m}}=\frac{\binom{N-\sum_{k=0}^{j}X_N^{(k)}}{m}}{\binom{N}{m}} \tag{18.5}$$

where the convention in summation is that $\sum_{k=a}^{b}f_k=0$ if $a>b$. Finally, combining (18.2) with (18.4) and (18.5), we arrive at the general conditional expression for the minimum hopcount to the anycast group,

$$\Pr\left[h_N(m)=j|L_N\right]=\frac{\binom{N-\sum_{k=0}^{j-1}X_N^{(k)}}{m}-\binom{N-\sum_{k=0}^{j}X_N^{(k)}}{m}}{\binom{N}{m}} \tag{18.6}$$

Clearly, while $\Pr\left[h_N(0)=j|L_N\right]=0$ since there is no path, we have for $m=1$,

$$\Pr\left[h_N(1)=j|L_N\right]=\frac{X_N^{(j)}}{N}$$

It directly follows from (18.6) that

$$\Pr\left[h_N(m)\le n|L_N\right]=1-\frac{\binom{N-\sum_{k=0}^{n}X_N^{(k)}}{m}}{\binom{N}{m}} \tag{18.7}$$

If $N-\sum_{k=0}^{n}X_N^{(k)}<m$ or, equivalently, $\sum_{k=n+1}^{N-1}X_N^{(k)}<m$, then equation (18.7) shows that $\Pr\left[h_N(m)>n|L_N\right]=0$. The maximum possible hopcount of a shortest path to an anycast group strongly depends on the specifics of the shortest path tree or the level set L_N. A general result is worth mentioning:

Theorem 18.2.1 *For any graph, it holds that*

$$\Pr[h_N(m) > N - m] = 0$$

In words, the longest shortest path to an anycast group with m members can never possess more than $N - m$ hops.

Proof: This general theorem follows from the fact that the line topology is the tree with the longest hopcount $N - 1$ and only in the case that all m last positions (with respect to the source or root) are occupied by the m anycast group members, is the maximum hopcount $N - m$. $\quad\square$

For the URT , $\Pr[h_N(m) = N - m]$ is computed exactly in (18.12).

Corollary 18.2.2 *For any graph, it holds that*

$$\Pr[h_N(N - 1) = 1] = \frac{1}{N}$$

Proof: This corollary follows from Theorem 18.2.1 and the law of total probability. Alternatively, if there are $N - 1$ anycast members in a network with N nodes, the shortest path can only consist of one hop if none of the anycast members coincides with the root node. This probability is precisely $\frac{1}{N}$. $\quad\square$

Using the tail probability formula (2.36) for the average, it follows from (18.7) that

$$E\left[h_N(m)|L_N\right] = \frac{1}{\binom{N}{m}} \sum_{n=0}^{N-2} \binom{N - \sum_{k=0}^{n} X_N^{(k)}}{m} \qquad (18.8)$$

from which we find,

$$E\left[h_N(1)|L_N\right] = \frac{1}{N} \sum_{k=1}^{N-1} k X_N^{(k)}$$

Thus, given L_N, a performance measure η for anycast over unicast can be quantified as

$$\eta = \frac{E\left[h_N(m)|L_N\right]}{E\left[h_N(1)|L_N\right]} \leq 1$$

Using the law of total probability, the distribution of the minimum hopcount to the anycast group is

$$\Pr\left[h_N(m) = j\right] = \sum_{\text{all } L_N} \Pr\left[h_N(m) = j|L_N\right] \Pr\left[L_N\right] \qquad (18.9)$$

or explicitly,

$$\Pr[h_N(m)=j] = \sum_{\sum_{k=1}^{N-1} x_k = N-1} \frac{\binom{\sum_{k=j}^{N-1} x_k}{m} - \binom{\sum_{k=j+1}^{N-1} x_k}{m}}{\binom{N}{m}} \Pr\left[X_N^{(1)} = x_1,\ldots,X_N^{(N-1)} = x_{N-1}\right]$$

where the integers $x_k \geq 0$ for all k. This expression explicitly shows the importance of the level structure L_N of the shortest path tree T. The level set L_N entirely determines the *shape* of the tree T. Unfortunately, a general form for $\Pr[L_N]$ or $\Pr[h_N(m) = j]$ is difficult to obtain. In principle, via extensive trace-route measurements from several roots, the shortest path tree and $\Pr[L_N]$ can be constructed such that a (rough) estimate of the level set L_N in the Internet can be obtained.

18.3 The k-ary tree

For regular trees, explicit expressions are possible because the summation in (18.9) simplifies considerably. For example, for the k-ary tree defined in Section 17.3,

$$X_N^{(j)} = k^j$$

Provided the set L_N only contains these values of $X_N^{(j)}$ for each j, we have that $\Pr[L_N] = 1$, else it is zero (because then L_N is not consistent with a k-ary tree). Summarizing, for the k-ary tree with $N = \frac{k^{D+1}-1}{k-1}$ and D levels, the distribution of the minimum hopcount to the anycast group is

$$\Pr[h_N(m) = j] = \frac{\binom{N-\frac{k^j-1}{k-1}}{m} - \binom{N-\frac{k^{j+1}-1}{k-1}}{m}}{\binom{N}{m}} \qquad (18.10)$$

Extension of the integer k to real numbers in the formula (18.10) is expected to be of value as suggested in Section 17.3. When a k-ary tree was used to fit corresponding Internet multicast measurements (Van Mieghem *et al.*, 2001a), a remarkably accurate agreement was found for the value $k \approx 3.2$, which is about the average degree of the Internet graph. Hence, if we were to use the k-ary tree as model for the hopcount to an anycast group, we expect that $k \approx 3.2$ is the best value for Internet shortest path trees. However, we feel we ought to mention that the hopcount distribution of the shortest path between two arbitrary nodes is definitely not a k-ary tree, because $\Pr[h_N(1) = j]$ increases with the hopcount j, which is in conflict with Internet trace-route measurements (see, for example, the bell-shape curve in Fig. 16.4).

Figure 18.1 displays $\Pr[h(m) \leq j]$ for a k-ary with outdegree $k = 3$ and

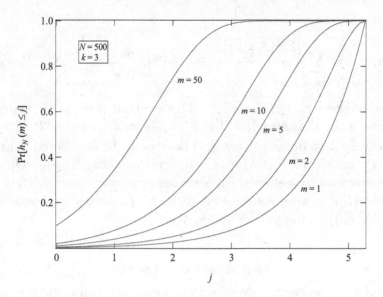

Fig. 18.1. The distribution function of $h_{500}(m)$ versus the hops j for various sizes of the anycast group in a k-ary tree with $k = 3$ and $N = 500$

$N = 500$. This type of plot allows us to solve the "server placement problem". For example, assuming that the k-ary tree is a good model and the network consists of $N = 500$ nodes, Fig. 18.1 shows that at least $m = 10$ servers are needed to assure that any user is not more than four hops separated from a server with a probability of 93%. More precisely, the equation $\Pr[h_{500}(m) > 4] < 0.07$ is obeyed if $m \geq 10$.

Figure 18.2 gives an idea how the performance measure η decreases with the size of the anycast group in k-ary trees (all with outdegree $k = 3$), but with different size N. For values of m up to around 20% of N, we observe that η decreases logarithmically in m.

18.4 The uniform recursive tree (URT)

Chapter 16 motivates the interest in the URT. The URT is believed to provide a reasonable, first order estimate for the hopcount problem to an anycast group in the Internet.

18.4.1 Recursion for $\Pr[h(m) = j]$

Usually, a combinatorial approach such as (18.9) is seldom successful for URTs while structural properties often lead to results. The basic Theo-

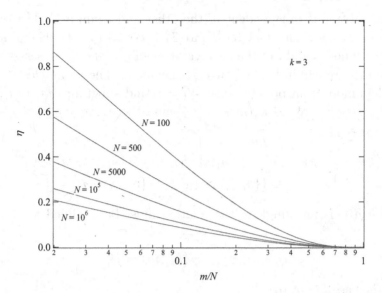

Fig. 18.2. The performance measure η for several sizes of k-ary trees (with $k = 3$) as a function of the ratio of anycast nodes over the total number of nodes.

rem 16.2.1 of the URT, applied to the anycast minimum hop problem, is illustrated in Fig. 18.3.

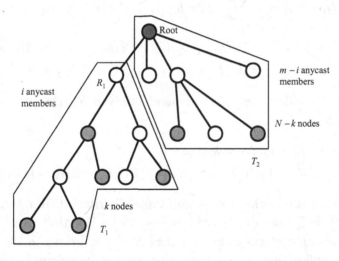

Fig. 18.3. A uniform recursive tree consisting of two subtrees T_1 and T_2 with k and $N - k$ nodes respectively. The first cluster contains i anycast members while the cluster with $N - k$ nodes contains $m - i$ anycast members.

Figure 18.3 shows that any URT can be separated into two subtrees T_1 and T_2 with k and $N - k$ nodes respectively. Moreover, Theorem 16.2.1 states that each subtree is independent of the other and again a URT. Consider now a specific separation of a URT T into $T_1 = t_1$ and $T_2 = t_2$, where the tree t_1 contains k nodes and i of the m anycast members and t_2 possesses $N - k$ nodes and the remaining $m - i$ anycast members. The event $\{h_T(m) = j\}$ equals the union of all possible sizes $N_1 = k$ and subgroups $m_1 = i$ of the event $\{h_{t_1}(i) = j - 1\} \cap \{h_{t_2}(m - i) \geq j\}$ and the event $\{h_{t_1}(i) > j - 1\} \cap \{h_{t_2}(m - i) = j\}$,

$$\{h_T(m) = j\} = \cup_k \cup_i \{\{h_{t_1}(i) = j - 1\} \cap \{h_{t_2}(m - i) \geq j\}\}$$
$$\cup \{\{h_{t_1}(i) > j - 1\} \cap \{h_{t_2}(m - i) = j\}\}$$

Because $h_N(0)$ is meaningless, the relation must be modified for the case $i = 0$ to

$$\{h_T(m) = j\} = \{h_{t_2}(m) = j\}$$

and for the case $i = m$ to

$$\{h_T(m) = j\} = \{h_{t_1}(m) = j - 1\}$$

This decomposition holds for any URT T_1 and T_2, not only for the specific ones t_1 and t_2. The transition towards probabilities becomes

$$\Pr[h_T(m) = j] = \sum_{\text{all } t_1, t_2, k, i} (\Pr[h_{t_1}(i) = j - 1] \Pr[h_{t_2}(m - i) \geq j]$$
$$+ \Pr[h_{t_1}(i) \geq j - 1] \Pr[h_{t_2}(m - i) = j])$$
$$\times \Pr[T_1 = t_1, T_2 = t_2, N_1 = k, m_1 = i]$$

Since T_1 and T_2 and also m_1 are independent given N_1, the last probability l simplifies to

$$l = \Pr[T_1 = t_1, T_2 = t_2, N_1 = k, m_1 = i]$$
$$= \Pr[T_1 = t_1 | N_1 = k] \Pr[T_2 = t_2 | N_1 = k] \Pr[m_1 = i | N_1 = k] \Pr[N_1 = k]$$

Theorem 16.2.1 states that N_1 is uniformly distributed over the set with $N - 1$ nodes such that $\Pr[N_1 = k] = \frac{1}{N-1}$. The fact that i out of the m anycast members, uniformly chosen out of N nodes, belong to the recursive subtree T_1 implies that $m - i$ remaining anycast members belong to T_2. Hence, analogous to a combinatorial problem outlined by Feller (1970, p. 43) that leads to the hypergeometric distribution, we have

$$\Pr[m_1 = i | N_1 = k] = \frac{\binom{k}{i}\binom{N-k}{m-i}}{\binom{N}{m}}$$

because all favorable combinations are those $\binom{k}{i}$ to distribute i anycast members in T_1 with k nodes multiplied by all favorable $\binom{N-k}{m-i}$ to distribute the remaining $m - i$ in T_2 containing $N - k$ nodes. The total way to distribute m anycast members over N nodes is $\binom{N}{m}$. Finally, we remark that the hopcount of the shortest path to m anycast members in a URT only depends on its size. This means that the sum over all t_1 of $\Pr[T_1 = t_1 | N_1 = k]$, which equals 1, disappears and likewise also the sum over all t_2. Combining the above leads to

$$\Pr[h_N(m) = j] = \sum_{k=1}^{N-1}\sum_{i=1}^{m-1}(\Pr[h_k(i) = j - 1]\Pr[h_{N-k}(m - i) \geq j]$$

$$+ \Pr[h_k(i) > j - 1]\Pr[h_{N-k}(m - i) = j])\frac{\binom{k}{i}\binom{N-k}{m-i}}{(N-1)\binom{N}{m}}$$

$$+ \sum_{k=1}^{N-1}\frac{\binom{N-k}{m}\Pr[h_{N-k}(m) = j] + \binom{k}{m}\Pr[h_k(m) = j - 1]}{(N-1)\binom{N}{m}}$$

By substitution of $k' = N - k$ and $m' = m - i$, we obtain the recursion,

$$\Pr[h_N(m) = j] = \sum_{k=1}^{N-1}\sum_{i=1}^{m-1}\frac{\binom{k}{i}\binom{N-k}{m-i}(\Pr[h_k(i) = j - 1] + \Pr[h_k(i) = j])}{(N-1)\binom{N}{m}}$$

$$\times \sum_{q=j}^{N-k-1}\Pr[h_{N-k}(m - i) = q]$$

$$+ \sum_{k=1}^{N-1}\frac{\binom{k}{m}(\Pr[h_k(m) = j] + \Pr[h_k(m) = j - 1])}{(N-1)\binom{N}{m}}$$

$$\tag{18.11}$$

This recursion (18.11) is solved numerically for $N = 20$. The result is shown in Fig. 18.4, which demonstrates that $\Pr[h(m) > N - m] = 0$ or that the path with the longest hopcount to an anycast group of m members consists of $N - m$ links.

Since there are $(N - 1)!$ possible recursive trees (Theorem 16.2.2) and there is only one line tree with $N - 1$ hops where each node has precisely one child node, the probability to have precisely $N - 1$ hops from the root is $\frac{1}{(N-1)!}$ (which also is $\Pr[h_N = N - 1]$ given in (16.8)). The longest possible hopcount from a root to m anycast members occurs in the line tree where all m anycast members occupy the last m positions. Hence, the probability for the longest possible hopcount equals

$$\Pr\left[h_N(m) = N - m\right] = \frac{m!}{(N-1)!\binom{N}{m}} \qquad (18.12)$$

because there are $m!$ possible ways to distribute the m anycast members at the m last positions in the line tree while there are $\binom{N}{m}$ possibilities to distribute m anycast members at arbitrary places in the line tree.

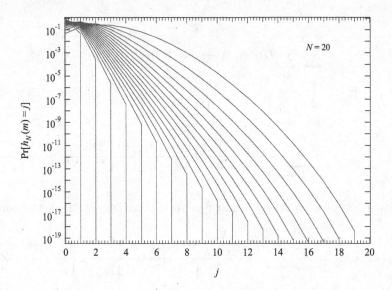

Fig. 18.4. The pdf of $h_N(m)$ in a URT with $N = 20$ nodes for all possible m. Observe that $\Pr[h_N(m) > N - m] = 0$. This relation connects the various curves to the value for m.

Figure 18.4 allows us to solve the "server placement problem". For example, consider the scenario in which a network operator announces that any user request will reach a server of the anycast group in no more than $j = 4$ hops in 99.9% of the cases. Assuming his network has $N = 20$ routers and the shortest path tree is a URT, the network operator has to compute the number of anycast servers m he has to place uniformly spread over the $N = 20$ routers by solving $\Pr\left[h_{20}(m) > 4\right] < 10^{-3}$. Figure 18.4 shows that the intersection of the line $j = 4$ and the line $\Pr\left[h_{20}(m) = 4\right] = 10^{-3}$ is the curve for $m = 7$. Since the curves for $m \geq 7$ are exponentially decreasing, $\Pr\left[h_{20}(m) > 4\right]$ is safely[1] approximated by $\Pr\left[h_{20}(m) = 4\right]$, which leads to the placing of $m = 7$ servers. When following the line $j = 4$, we also observe that the curves for $m = 5, 6, 7, 8$ lie near to that of $m = 7$. This means that

---

[1] More precisely, since $\Pr\left[h_{20}(4) > 4\right] = 0.001\,06$ and $\Pr\left[h_{20}(5) > 4\right] = 0.000\,32$, only $m = 5$ servers are sufficient.

placing a server more does not considerably change the situation. It is a manifestation of the law $\eta \approx 1 - a \log m$, which tells us that by placing m servers the gain measured in hops with respect to the single server case is slowly, more precisely logarithmically, increasing. The performance measure η for the URT is drawn for several sizes N in Fig. 18.5.

18.4.2 Analysis of the recursion relation

The product of two probabilities in the double sum in (18.11) seriously complicates a possible analytic treatment. A relation for a generating function of $\Pr[h_N(m) = j]$ and other mathematical results are derived in Van Mieghem (2004b). Here, we summarize the main results.

(a) Let us check $\Pr[h_N(m) = 0] = \frac{m}{N}$. Using $\Pr[h_k(i) \geq -1] = 1$, the convention that $\Pr[h_k(i) = -1] = 0$ and $\Pr[h_{N-k}(m-i) = 0] = \frac{m-i}{N-k}$, the right hand side of (18.11), denoted by r, simplifies to

$$
\begin{aligned}
r &= \frac{1}{(N-1)\binom{N}{m}} \sum_{k=1}^{N-1} \sum_{i=0}^{m} \frac{m-i}{N-k} \binom{k}{i} \binom{N-k}{m-i} \\
&= \frac{1}{(N-1)\binom{N}{m}} \sum_{k=1}^{N-1} \sum_{i=0}^{m-1} \binom{k}{i} \binom{N-1-k}{m-1-i} \\
&= \frac{1}{(N-1)\binom{N}{m}} \sum_{k=1}^{N-1} \binom{N-1}{m-1} = \frac{m}{N}
\end{aligned}
$$

(b) Observe that $\Pr[h_N(N) = j] = 0$ for $j > 0$.

(c) For $m = 1$,

$$
\Pr[h_N = j] = \frac{1}{N-1} \sum_{k=1}^{N-1} (\Pr[h_k = j] + \Pr[h_k = j-1]) \frac{k}{N}
$$

Multiplying both sides by z^j, summing over all j leads to the recursion for the generating function (16.6)

$$
(N+1)\varphi_{N+1}(z) = (z+N)\varphi_N(z)
$$

(d) The case $m = 2$ is solved in Van Mieghem (2004b, Appendix) as

$$\Pr[h_N(2) = j] = \frac{2(-1)^{N-1-j}}{N!} S_N^{(j+1)} + \frac{2(-1)^{N-j}}{N!(N-1)} \sum_{k=1}^{j} (-1)^k \binom{2k-1}{k} S_N^{(k+j+1)}$$

$$+ \frac{2(-1)^{N-j}}{N!(N-1)} \sum_{k=0}^{j-1} \left[\binom{j+k+1}{j} - \binom{2k+1}{k} \right] (-1)^k S_N^{k+j+1}$$

$$(18.13)$$

In van der Hofstad *et al.* (2002b) we have demonstrated that the covariance between the number of nodes at level r and j for $r \leq j$ in the URT is

$$E\left[X_N^{(r)} X_N^{(j)}\right] = \frac{(-1)^{N-1}}{(N-1)!} \sum_{k=0}^{r} (-1)^{k+j} \binom{2k+j-r}{k} S_N^{(k+j+1)}$$

For $j - r = 1$, the last term in (18.13) is recognized as $\dfrac{E\left[X_N^{(j-1)} X_N^{(j)}\right]}{\binom{N}{2}}$. Since $\binom{2k-1}{k} = \frac{1}{2}\binom{2k}{k}$, the first sum in (18.13) is

$$\frac{2(-1)^{N-j}}{N!(N-1)} \sum_{k=1}^{j} (-1)^k \binom{2k-1}{k} S_N^{(k+j+1)} = \frac{2(-1)^{N-j-1}}{(N-1)} \frac{S_N^{(j+1)}}{N!} - \frac{E\left[\left(X_N^{(j)}\right)^2\right]}{2\binom{N}{2}}$$

With $\dfrac{2(-1)^{N-1-j}}{N!} S_N^{(j+1)} = 2\Pr[h_N = j]$, we obtain

$$\Pr[h_N(2) = j] = \frac{2N}{N-1} \Pr[h_N = j] + \frac{E\left[X_N^{(j-1)} X_N^{(j)}\right]}{\binom{N}{2}} - \frac{E\left[\left(X_N^{(j)}\right)^2\right]}{2\binom{N}{2}}$$

$$+ \frac{2(-1)^{N-1}}{N!(N-1)} \sum_{k=1}^{j} \binom{j+k}{k} (-1)^{k+j} S_N^{k+j}$$

It would be of interest to find an interpretation for the last sum.

Without proof[2], we mention the following exact results:

$$\sum_{m=1}^{N} \binom{N}{m} \Pr[h_N(m) = N-2] = \sum_{m=1}^{N-1} \binom{N-1}{m} \frac{\Pr[h_{N-1}(m) = N-3]}{N-1} + \frac{1}{(N-2)!}$$

For $m \leq N - 3$, it holds that

$$\Pr[h_N(m) = N-m-1] = \frac{m!}{(N-1)!\binom{N}{m}} \left[\binom{N}{2} + (m-1)(m/2+1) + \sum_{k=2}^{m} \frac{m+1}{k} \right]$$

---

[2] By substitution into the recursion (18.11), one may verify these relations.

18.5 Approximate analysis

Since the general solution (18.9) is in many cases difficult to compute as shown for the URT in Section 18.4, we consider a simplified version of the above problem where each node in the tree has equal probability $p = \frac{m}{N}$ to be a server. Instead of having precisely m servers, the simplified version considers on average m servers and the probability that there are precisely m servers is $\binom{N}{m}p^m(1-p)^{N-m}$. In the simplified version, the associated equations to (18.4) and (18.3) are

$$\Pr\left[\left\{m^{(j)} = 0\right\}\Big| e_{j-1}\right] = \Pr\left[\left\{m^{(j)} = 0\right\}\right] = (1-p)^{X_N^{(j)}}$$

$$\Pr\left[e_j\right] = \prod_{l=0}^{j}\Pr\left[\left\{m^{(j)} = 0\right\}\right] = (1-p)^{\sum_{l=0}^{j-1} X_N^{(l)}}$$

which implies that the probability that there are no servers in the tree is $(1-p)^N$. Since in that case, the hopcount is meaningless, we consider the conditional probability (18.2) of the hopcount given that the level set contains at least one server (which is denoted by $\tilde{h}_N(m)$) is

$$\Pr\left[\tilde{h}_N(m) = j|L_N\right] = \frac{\left(1 - (1-p)^{X_N^{(j)}}\right)(1-p)^{\sum_{l=0}^{j-1} X_N^{(l)}}}{1 - (1-p)^N}$$

Thus,

$$\Pr\left[\tilde{h}_N(m) \leq n|L_N\right] = \frac{1 - (1-p)^{\sum_{l=0}^{n} X_N^{(l)}}}{1 - (1-p)^N}$$

Finally, to avoid the knowledge of the entire level set L_N, we use $E\left[X_N^{(l)}\right] = N\Pr\left[h_N(1) = l\right]$ from (16.7) as the best estimate for each $X_N^{(l)}$ and obtain the approximate formula

$$\Pr\left[\tilde{h}_N(m) = j\right] = \frac{\left(1 - (1-p)^{E\left[X_N^{(j)}\right]}\right)(1-p)^{\sum_{l=0}^{j-1} E\left[X_N^{(l)}\right]}}{1 - (1-p)^N} \tag{18.14}$$

In the dotted lines in Fig. 18.5, we have added the approximate result for the URT where $E\left[h_N(m)\right]$ is computed based on (18.14), but where $E[h_N(1)]$ is computed exactly. For $m = 1$, the approximate analysis (18.14) is not well suited: Fig. 18.5 illustrates this deviation in the fact that $\eta_{\mathrm{appr}}(1) = E\left[\tilde{h}_N(1)\right]/E\left[h_N(1)\right] < 1$. For higher values of m we observe a fairly good correspondence. We found that the probability (18.14) reasonably approximates the exact result plotted on a linear scale. Only the tail behavior (on log-scale) and the case for $m = 1$ deviate significantly. In summary for the URT, the approximation (18.14) for $\Pr[h_N(m) = j]$ is much faster to compute than the exact recursion and it seems appropriate for the computation of η for $m > 1$. However, it is less adequate to solve the server placement problem that requires the tail values $\Pr[h_N(m) > j]$.

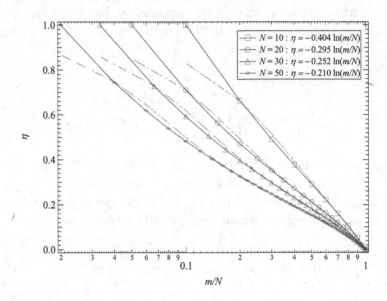

Fig. 18.5. The performance measure η for several sizes N of URTs as a function of the ratio m/N

18.6 The performance measure η in exponentially growing trees

In this section, we investigate the observed law $\eta \approx 1 - a \log m$ for a much larger class of trees, namely the class of exponentially growing trees to which both the k-ary tree and the URT belong. Also most trees in the Internet are exponentially growing trees. A tree is said to grow exponentially in the number of nodes N with degree κ if $\lim_{j \to \infty} \left(X_N^{(j)} \right)^{1/j} = \kappa$ or, equivalently, $X_N^{(j)} \sim \kappa^j$, for large j. The fundamental problem with this definition is that it only holds for infinite graphs $N = \infty$. For real (finite) graphs, there must exist some level $j = l$ for which the sequence $X_N^{(l+1)}, X_N^{(l+2)}, \ldots, X_N^{(N-1)}$ ceases to grow because $\sum_{j=0}^{N-1} X_N^{(j)} = N < \infty$. This boundary effect complicates the definition of exponential growth in finite graphs. The second complication is that even in the finite set $X_N^{(0)}, X_N^{(1)}, \ldots, X_N^{(l)}$ not necessary all $X_N^{(j)}$ with $0 \leq j \leq l$ need to obey $X_N^{(j)} \sim \kappa^j$, but "enough" should.

Without the limit concept, we cannot specify the precise conditions of exponential growth in a finite shortest path tree. If we assume in finite graphs that $X_N^{(j)} \sim \kappa^j$ for $j \le l$, then $\sum_{j=0}^{l} X_N^{(j)} = \alpha N$ with $0 < \alpha < 1$. Indeed, for $\kappa > 1$, the highest hopcount level l possesses by far the most nodes since $\frac{\kappa^{l+1}-1}{\kappa-1} \approx \kappa^l$, which cannot be larger than a fraction αN of the total number of nodes.

We now present an order calculus to estimate η for exponentially growing trees based on relation (18.8). Let us denote

$$y = \frac{\binom{N-x}{m}}{\binom{N}{m}} = \prod_{j=0}^{m-1} \left(1 - \frac{x}{N-j}\right)$$

For large N and fixed m,

$$y = \exp\left(-\frac{xm}{N}\right)(1 + o(1))$$

In the case where the tree is exponentially growing for $j \le l$ as $X_N^{(j)} = \beta_j \kappa^j$ with β_j some slowly varying sequence, only very few levels Δl (bounded by a fixed number) around l obey $\sum_{k=0}^{n} X_N^{(k)} = O(N)$ where $n \in [l - \Delta l, l]$, while for all $j > l$, we have $\sum_{k=0}^{n} X_N^{(k)} = \mu_n N$ with some sequence $\mu_n < \mu_{n+1} < \mu_{\max n} = 1$. Applied to (18.8) where $x = \sum_{k=0}^{n} X_N^{(k)} < N$,

$$E\left[h_N(m)|L_N\right] \approx (1 + o(1)) \sum_{n=0}^{l} \exp\left(-\frac{m}{N}\beta_n \kappa^n\right) + \sum_{n=l+1}^{N-2} \frac{\binom{(1-\mu_n)N}{m}}{\binom{N}{m}}$$

If there are only a few levels more than l, the last series is much smaller than 1 and can be omitted. Since the slowly varying sequence β_n is unknown, we approximate $\beta_n = \beta$ and

$$\sum_{n=0}^{l} \exp\left(-\frac{m}{N}\beta_n \kappa^n\right) \approx \int_0^l \exp\left(-\frac{m}{N}\beta \kappa^n\right) dn = \frac{1}{\log \kappa} \int_{\frac{m}{N}\beta}^{\frac{m}{N}\beta \kappa^l} \frac{e^{-u}}{u} du$$

$$\approx \frac{1}{\log \kappa} \int_{\frac{m}{N}\beta}^{\infty} \frac{e^{-u}}{u} du - \frac{e^{-m}}{m \log \kappa}$$

$$= \frac{1}{\log \kappa}\left(-\gamma - \frac{e^{-m}}{m} - \log \frac{m}{N}\beta + O\left(\frac{m}{N}\right)\right)$$

where in the last step a series (Abramowitz and Stegun, 1968, Section 5.1.11) for the exponential integral is used. Thus,

$$\eta \approx (1 + o(1)) \frac{\left(1 + \frac{-\gamma - \frac{e^{-m}}{m} - \log m - \log \beta}{\log N} + O\left(\frac{m}{N}\right)\right)}{\left(1 - \frac{\gamma + e^{-1} + \log \beta}{\log N} + O\left(\frac{1}{N}\right)\right)}$$

$$= (1 + o(1))\left(1 - \frac{\log m}{\log N} - \frac{e^{-1} - \frac{e^{-m}}{m}}{\log N} + O\left(\frac{1}{\log^2 N}\right)\right)$$

Since by definition $\eta = 1$ for $m = 1$, we finally arrive at

$$\eta \approx 1 - \frac{\log m}{\log N} - \frac{e^{-1} - \frac{e^{-m}}{m}}{\log N} + O\left(\frac{1}{\log^2 N}\right)$$

which supplies evidence for the conjecture $\eta \approx 1 - a \log m$ that exponentially growing graphs possess a performance measure η that logarithmically decreases in m, which is rather slow.

Measurement data in the Internet seem to support this $\log m$-scaling law. Apart from the correspondence with figures in the work of Jamin *et al.* (2001), Fig. 6 in Krishnan *et al.* (2000) shows that the relative measured traffic flow reduction decreases logarithmically in the number of caches m.

Appendix A

Stochastic matrices

This appendix reviews the matrix theory for Markov chains. In-depth analyses are found in classical books by Gantmacher (1959a,b), Wilkinson (1965) and Meyer (2000).

A.1 Eigenvalues and eigenvectors

1. The algebraic eigenproblem consists in the determination of the eigenvalues λ and the corresponding eigenvectors x of a matrix A for which the set of n homogeneous linear equations in n unknowns

$$Ax = \lambda x \tag{A.1}$$

has a non-zero solution. Clearly, the zero vector $x = 0$ is always a solution of (A.1). A non-zero solution of (A.1) is only possible if and only if the matrix $A - \lambda I$ is singular, that is

$$\det (A - \lambda I) = 0 \tag{A.2}$$

This determinant can be expanded in a polynomial in λ of degree n,

$$c(\lambda) = (-1)^n \lambda^n + c_{n-1}\lambda^{n-1} + \cdots + c_1\lambda + c_0 = 0 \tag{A.3}$$

which is called the characteristic (eigenvalue) polynomial of the matrix A and where the coefficients

$$c_k = (-1)^k \sum_{all} M_{n-k} \tag{A.4}$$

and M_k is a principal minor[1]. Since a polynomial of degree n has n complex zeros, the matrix A possesses n eigenvalues λ_k, not all necessarily distinct.

---

[1] A principal minor M_k is the determinant of a principal $k \times k$ submatrix $M_{k \times k}$ obtained by deleting the same $n - k$ rows and columns in A. Hence, the main diagonal elements $(M_{k \times k})_{ii}$ are k elements of main diagonal elements $\{a_{ii}\}_{1 \leq i \leq n}$ of A.

In general, the characteristic polynomials can be written as

$$c(\lambda) = \prod_{k=1}^{n} (\lambda_k - \lambda) \tag{A.5}$$

Since $c(\lambda) = \det(A - \lambda I)$, it follows from (A.3) and (A.5) that for $\lambda = 0$

$$\det A = c_0 = \prod_{k=1}^{n} \lambda_k \tag{A.6}$$

Hence, if $\det A = 0$, there is at least one zero eigenvalue. Also,

$$(-1)^{n-1} c_{n-1} = \sum_{k=1}^{n} \lambda_k = \text{trace}(A) \tag{A.7}$$

If all $\lambda_k \geq 0$, we can apply the general theorem of the arithmetic and geometric mean (5.2) to (A.6) and (A.7) with $q_k = \frac{\alpha_k}{\sum_{j=1}^{n} \alpha_j}$,

$$\prod_{k=1}^{n} \lambda_k^{\alpha_k} \leq \left(\frac{\sum_{k=1}^{n} \alpha_k \lambda_k}{\sum_{j=1}^{n} \alpha_j} \right)^{\sum_{j=1}^{n} \alpha_j}$$

and by choosing $\alpha_j = 1$, we find the inequality

$$\det A \leq \left(\frac{\text{trace}(A)}{n} \right)^n$$

To any eigenvalue λ, the set (A.1) has at least one non-zero eigenvector x. Furthermore, if x is a non-zero eigenvector, also kx is a non-zero eigenvalue. Therefore, eigenvectors are often normalized, for instance, a probabilistic eigenvector has the sum of its components equal to 1 or a norm $\|x\|_1 = 1$ as defined in (A.23). If the rank of $A - \lambda I$ is less than $n - 1$, there will be more than one independent vector. Just these cases seriously complicate the eigenvalue problem. In the sequel, we omit the discussion on multiple eigenvalues and refer to Wilkinson (1965).

2. The eigenproblem of the transpose A^T,

$$A^T y = \lambda y \tag{A.8}$$

is of singular importance. Since the determinant of a matrix is equal to the determinant of its transpose, $\det(A^T - \lambda I) = \det(A - \lambda I)$ which shows that the eigenvalues of A and A^T are the same. However, the eigenvectors are, in general, different. Alternatively, we can write (A.8) as

$$y^T A = \lambda y^T \tag{A.9}$$

The vector y_j^T is therefore called the left-eigenvector of A belonging to the eigenvalue λ_j, whereas x_j is called the right-eigenvector belonging to the same eigenvalue λ_j. An important relation between left- and right-eigenvectors of a matrix A is, for $\lambda_j \neq \lambda_k$,

$$y_j^T x_k = 0 \tag{A.10}$$

Indeed, left-multiplying (A.1) with $\lambda = \lambda_k$ by y_j^T,

$$y_j^T A x_k = \lambda_k y_j^T x_k$$

and similarly right-multiplying (A.9) with $\lambda = \lambda_j$ by x_k

$$y_j^T A x_k = \lambda_j y_j^T x_k$$

leads, after subtraction to $0 = (\lambda_k - \lambda_j) y_j^T x_k$ and (A.10) follows. Since eigenvectors may be complex in general and since $y_j^T x_k = x_k^T y_j$, the expression $y_j^T x_k$ is not an inner-product that is always real and for which $y_j^T x_k = \left(x_k^T y_j \right)^*$ holds. However, (A.10) expresses that the sets of left- and right-eigenvectors are orthogonal if $\lambda_j \neq \lambda_k$.

3. If A has n distinct eigenvalues, then the n eigenvectors are linearly independent and span the whole n dimensional space. The proof is by *reductio ad absurdum*. Assume that s is the smallest number of linearly dependent eigenvectors labelled by the first s smallest indices. Linear dependence then means that,

$$\sum_{k=1}^{s} \alpha_k x_k = 0 \tag{A.11}$$

where $\alpha_k \neq 0$ for $1 \leq k \leq s$. Left-multiplying by A and using (A.1) yields

$$\sum_{k=1}^{s} \alpha_k \lambda_k x_k = 0 \tag{A.12}$$

On the other hand, multiplying (A.11) by λ_s and subtracting from (A.12) leads to

$$\sum_{k=1}^{s-1} \alpha_k (\lambda_k - \lambda_s) x_k = 0,$$

which, because all eigenvalues are distinct, implies that there is a smaller set of $s - 1$ linearly depending eigenvectors. This contradicts the initial hypothesis.

This important property has a number of consequences. First, it applies to left- as well as to right-eigenvectors. Relation (A.10) then shows that the sets of left- and right-eigenvectors form a bi-orthogonal system with $y_k^T x_k \neq 0$. For, if x_k were orthogonal to y_k (or $y_k^T x_k = 0$), (A.10) demonstrates that x_k would be orthogonal to all left-eigenvectors y_j. Since the set of left-eigenvectors span the n dimensional vector space, it would mean that the n dimensional vector x_k would be orthogonal to the whole n-space, which is impossible because x_k is not the null vector. Second, any n dimensional vector can be written in terms of either the left- or right-eigenvectors.

4. Let us denote by X the matrix with in column j the right-eigenvector x_j and by Y^T the matrix with in row k the left-eigenvector y^T. If the right- and left-eigenvectors are scaled such that, for all $1 \leq k \leq n$, $y_k^T x_k = 1$, then

$$Y^T X = I \qquad (A.13)$$

or, the matrix Y^T is the inverse of the matrix X. Furthermore, for any right-eigenvector, (A.1) holds, rewritten in matrix form, that

$$AX = X \, \mathrm{diag}(\lambda_k) \qquad (A.14)$$

Left-multiplying by $X^{-1} = Y^T$ yields the similarity transform of matrix A,

$$X^{-1}AX = Y^T AX = \mathrm{diag}(\lambda_k) \qquad (A.15)$$

Thus, when the eigenvalues of A are distinct, there exists a similarity transform $H^{-1}AH$ that reduces A to diagonal form. In many applications, similarity transforms are applied to simplify matrix problems. Observe that a similarity transform preserves the eigenvalues, because, if $Ax = \lambda x$, then $\lambda H^{-1}x = H^{-1}Ax = (H^{-1}AH)H^{-1}x$. The eigenvectors are transformed to $H^{-1}x$.

When A has multiple eigenvalues, it may be impossible to reduce A to a diagonal form by similarity transforms. Instead of a diagonal form, the most compact form when A has r distinct eigenvalues each with multiplicity m_j such that $\sum_{j=1}^{r} m_j = n$ is the Jordan canonical form C,

$$C = \begin{bmatrix} C_{m_1-a}(\lambda_1) & & & & \\ & C_a(\lambda_1) & & & \\ & & \vdots & & \\ & & & C_{m_{r-1}}(\lambda_{r-1}) & \\ & & & & C_{m_r}(\lambda_r) \end{bmatrix}$$

where $C_m(\lambda)$ is a $m \times m$ submatrix of the form,

$$C_m(\lambda) = \begin{bmatrix} \lambda & 1 & 0 & \cdots & 0 \\ 0 & \lambda & 1 & 0 & \cdots \\ \vdots & \vdots & \vdots & \vdots & \vdots \\ 0 & \cdots & 0 & \lambda & 1 \\ 0 & \cdots & 0 & 0 & \lambda \end{bmatrix}$$

The number of independent eigenvectors is equal to the number of sub-matrices. If an eigenvalue λ has multiplicity m, there can be one large submatrix $C_m(\lambda)$, but also a number k of smaller submatrices $C_{b_j}(\lambda)$ such that $\sum_{j=1}^{k} b_j = m$. This illustrates, as mentioned in art. **1**, the much higher complexity of the eigenproblem in case of multiple eigenvalues. For more details we refer to Wilkinson (1965).

5. The companion matrix of the characteristic polynomial (A.3) of A is defined as

$$C = \begin{bmatrix} (-1)^{n-1}c_{n-1} & (-1)^{n-1}c_{n-2} & \cdots & (-1)^{n-1}c_1 & (-1)^{n-1}c_0 \\ 1 & 0 & \cdots & 0 & 0 \\ 0 & 1 & \cdots & 0 & 0 \\ \vdots & \vdots & \vdots & \vdots & \vdots \\ 0 & 0 & \cdots & 1 & 0 \end{bmatrix}$$

Expanding $\det(C - \lambda I)$ in cofactors of the first row yields $\det(C - \lambda I) = c(\lambda)$. If A has distinct eigenvalues, A as well as C are similar to $\text{diag}(\lambda_i)$. It has been shown that the similarity transform H for A equals $H = X$. The similarity transform for C is the Vandermonde matrix $V(\lambda)$, where

$$V(x) = \begin{bmatrix} x_1^{n-1} & x_2^{n-1} & \cdots & x_{n-1}^{n-1} & x_n^{n-1} \\ x_1^{n-2} & x_2^{n-2} & \cdots & x_{n-1}^{n-2} & x_n^{n-2} \\ \vdots & \vdots & \cdots & \vdots & \vdots \\ x_1 & x_2 & \vdots & x_{n-1} & x_n \\ 1 & 1 & \cdots & 1 & 1 \end{bmatrix}$$

The Vandermonde matrix $V(\lambda)$ is clearly non-singular if all eigenvalues are distinct. Furthermore,

$$V(\lambda)\mathrm{diag}\,(\lambda_i) = \begin{bmatrix} \lambda_1^n & \lambda_2^n & \cdots & \lambda_{n-1}^n & \lambda_n^n \\ \lambda_1^{n-1} & \lambda_2^{n-1} & \cdots & \lambda_{n-1}^{n-1} & \lambda_n^{n-1} \\ \vdots & \vdots & \cdots & \vdots & \vdots \\ \lambda_1^2 & \lambda_2^2 & \vdots & \lambda_{n-1}^2 & \lambda_n^2 \\ \lambda_1 & \lambda_2 & \cdots & \lambda_{n-1} & \lambda_n \end{bmatrix}$$

while

$$CV(\lambda) = \begin{bmatrix} (-1)^{n-1}c(\lambda_1)+\lambda_1^n & (-1)^{n-1}c(\lambda_2)+\lambda_2^n & \cdots & (-1)^{n-1}c(\lambda_n)+\lambda_n^n \\ \lambda_1^{n-1} & \lambda_2^{n-1} & \cdots & \lambda_n^{n-1} \\ \vdots & \vdots & \cdots & \vdots \\ \lambda_1^2 & \lambda_2^2 & \vdots & \lambda_n^2 \\ \lambda_1 & \lambda_2 & \cdots & \lambda_n \end{bmatrix}$$

Since $c(\lambda_j) = 0$, it follows that $CV(\lambda) = V(\lambda)\mathrm{diag}(\lambda_i)$, which demonstrates the claim. Hence, the eigenvector x_k of C belonging to eigenvalue λ_k is

$$x_k^T = \begin{bmatrix} \lambda_k^{n-1} & \lambda_k^{n-2} & \cdots & \lambda_k & 1 \end{bmatrix}$$

6. When left-multiplying (A.1), we obtain

$$A^2 x = \lambda A x = \lambda^2 x$$

or, in general for any integer $k \geq 0$,

$$A^k x = \lambda^k x \tag{A.16}$$

Since any eigenvalue λ satisfies its characteristic polynomial $c(\lambda) = 0$, we directly find from (A.16) that the matrix A satisfies its own characteristic equation,

$$c(A) = 0 \tag{A.17}$$

This result is the Caley–Hamilton theorem. There exist several other proofs of the Caley–Hamilton theorem.

7. Consider an arbitrary matrix polynomial in λ,

$$F(\lambda) = \sum_{k=0}^{m} F_k \lambda^k$$

where all F_k are $n \times n$ matrices and $F_m \neq O$. Any matrix polynomial $F(\lambda)$ can be right and left divided by another (non-zero) matrix polynomial $B(\lambda)$ in a unique way as proved in Gantmacher (1959a, Chapter IV). Hence the left-quotient and left-remainder $F(\lambda) = B(\lambda)Q_L(\lambda) + L(\lambda)$ and the right-quotient and right-remainder $F(\lambda) = Q_R(\lambda)B(\lambda) + R(\lambda)$ are unique. Let us concentrate on the right-remainder in the case where $B(\lambda) = \lambda I - A$ is a linear polynomial in λ. Using Euclid's division scheme for polynomials,

$$F(\lambda) = F_m \lambda^{m-1} (\lambda I - A) + (F_m A + F_{m-1}) \lambda^{m-1} + \sum_{k=0}^{m-2} F_k \lambda^k$$

$$= \left[F_m \lambda^{m-1} + (F_m A + F_{m-1}) \lambda^{m-2} \right] (\lambda I - A)$$

$$+ \left(F_m A^2 + F_{m-1} A + F_{m-2} \right) \lambda^{m-2} + \sum_{k=0}^{m-3} F_k \lambda^k$$

and continued, we arrive at

$$F(\lambda) = \left[F_m \lambda^{m-1} + \cdots + \lambda^{k-1} \sum_{j=k}^{m} F_j A^{j-k} + \cdots + \sum_{j=1}^{m} F_j A^{j-1} \right] (\lambda I - A)$$

$$+ \sum_{j=0}^{m} F_j A^j$$

In summary, $F(\lambda) = Q_R(\lambda)(\lambda I - A) + R(\lambda)$ (and similarly for the left-quotient and left-remainder) with

$$Q_R(\lambda) = \sum_{k=1}^{m} \lambda^{k-1} \left(\sum_{j=k}^{m} F_j A^{j-k} \right) \quad Q_L(\lambda) = \sum_{k=1}^{m} \lambda^{k-1} \left(\sum_{j=k}^{m} A^{j-k} F_j \right)$$
$$R(\lambda) = \sum_{j=0}^{m} F_j A^j = F(A) \qquad L(\lambda) = \sum_{j=0}^{m} A^j F_j$$
$$\text{(A.18)}$$

and where the right-remainder is independent of λ. The Generalized Bézout Theorem states that the polynomial $F(\lambda)$ is divisible by $(\lambda I - A)$ on the right (left) if and only if $F(A) = O$ $(L(\lambda) = O)$.

By the Generalized Bézout Theorem, the polynomial $F(\lambda) = g(\lambda)I - g(A)$ is divisible by $(\lambda I - A)$ because $F(A) = g(A)I - g(A) = O$. If $F(\lambda)$ is an ordinary polynomial, the right- and left-quotient and remainder are equal. The Caley–Hamilton Theorem (A.17) states that $c(A) = 0$, which indicates that $c(\lambda)I = Q(\lambda)(\lambda I - A)$ and also $c(\lambda)I = (\lambda I - A)Q(\lambda)$. The matrix $Q(\lambda) = (\lambda I - A)^{-1} c(\lambda)$ is called the adjoint matrix of A. Explicitly, from (A.18),

$$Q(\lambda) = \sum_{k=1}^{n} \lambda^{k-1} \left(\sum_{j=k}^{n} c_j A^{j-k} \right)$$

and, with (A.6), $Q(0) = -(A)^{-1} \det A = \sum_{j=1}^{n} c_j A^{j-k}$. The main theoretical interest of the adjoint matrix stems from its definition $c(\lambda)I =$

$Q(\lambda)(\lambda I - A) = (\lambda I - A)Q(\lambda)$ in case $\lambda = \lambda_k$ is an eigenvalue of A. Then, $(\lambda_k I - A)Q(\lambda_k) = 0$, which indicates by (A.1) that every non-zero column of the adjoint matrix $Q(\lambda_k)$ is an eigenvector belonging to the eigenvalue λ_k. In addition, by differentiation with respect to λ, we obtain

$$c'(\lambda)I = (\lambda I - A)Q'(\lambda) + Q(\lambda)$$

This demonstrates that, if $Q(\lambda_k) \neq O$, the eigenvalue λ_k is a simple root of $c(\lambda)$ and, conversely, if $Q(\lambda_k) = O$, the eigenvalue λ_k has higher multiplicity.

The adjoint matrix $Q(\lambda) = (\lambda I - A)^{-1}c(\lambda)$ is computed by observing that, on the Generalized Bézout Theorem, $\frac{c(\lambda)-c(\mu)}{\lambda-\mu}$ is divisible without remainder. By replacing in this polynomial λ and μ by λI and A respectively, $Q(\lambda)$ readily follows as illustrated in Section A.4.2.

8. Consider the arbitrary polynomial of degree l,

$$g(x) = g_0 \prod_{j=1}^{l}(x - \mu_j)$$

Substitute x by A, then

$$g(A) = g_0 \prod_{j=1}^{l}(A - \mu_j I)$$

Since $\det(AB) = \det A \det B$ and $\det(kA) = k^n \det A$, we have

$$\det(g(A)) = g_0^n \prod_{j=1}^{l} \det(A - \mu_j I) = g_0^n \prod_{j=1}^{l} c(\mu_j)$$

With (A.5),

$$\det(g(A)) = g_0^n \prod_{j=1}^{l} \prod_{k=1}^{n}(\lambda_k - \mu_j) = \prod_{k=1}^{n} g_0 \prod_{j=1}^{l}(\lambda_k - \mu_j)$$

$$= \prod_{k=1}^{n} g(\lambda_k)$$

If $h(x) = g(x) - \lambda$, we arrive at the general result: For any polynomial $g(x)$, the eigenvalues values of $g(A)$ are $g(\lambda_1), \ldots, g(\lambda_n)$ and the characteristic polynomial is

$$\det(g(A) - \lambda I) = \prod_{k=1}^{n}(g(\lambda_k) - \lambda) \tag{A.19}$$

which is a polynomial in λ of degree at most n. Since the result holds for an arbitrary polynomial, it should not surprise that, under appropriate conditions of convergence, it can be extended to infinite polynomials, in particular to the Taylor series of a complex function. As proved in Gantmacher (1959a, Chapter V), if the power series of a function $f(z)$ around $z = z_0$

$$f(z) = \sum_{j=1}^{\infty} f_j(z_0)(z - z_0)^j \qquad \text{(A.20)}$$

converges for all z in the disc $|z - z_0| < R$, then $f(A) = \sum_{j=1}^{\infty} f_j(z_0)(A - z_0 I)^j$ provided all eigenvalues of A lie with the region of convergence of (A.20), i.e. $|\lambda - z_0| < R$. For example,

$$e^{Az} = \sum_{k=0}^{\infty} \frac{z^k A^k}{k!} \qquad \text{for all } A$$

$$\log A = \sum_{k=1}^{\infty} \frac{(-1)^{k-1}}{k}(A - I)^k \text{ for } |\lambda_k - 1| < 1, \text{ all } 1 \leq k \leq n$$

and, from (A.19), the eigenvalues of e^{Az} are $e^{z\lambda_1}, \ldots, e^{z\lambda_1}$. Hence, the knowledge of the eigenstructure of a matrix A allows us to compute any function of A (under the same convergence restrictions as complex numbers z).

A.2 Hermitian and real symmetric matrices

A Hermitian matrix A is a complex matrix that obeys $A^H = (A^T)^* = A$, where $a^H = (a_{ij})^*$ is the complex conjugate of a_{ij}. Hermitian matrices possess a number of attractive properties. A particularly interesting subclass of Hermitian matrices are real, symmetric matrices that obey $A^T = A$. The inner-product of vector y and x is defined as $y^H x$ and obeys $(y^H x)^* = (y^H x)^H = x^H y$. The inner-product $x^H x = \sum_{j=1}^{n} |x_j|^2$ is real and positive for all vectors except for the null vector.

9. The eigenvalues of a Hermitian matrix are all real. Indeed, left-multiplying (A.1) by x^H yields

$$x^H A x = \lambda x^H x$$

and, since $(x^H A x)^H = x^H A^H x = x^H A x$, it follows that $\lambda x^H x = \lambda^H x^H x$ or $\lambda = \lambda^H$ because $x^H x$ is a positive real number. Furthermore, since $A = A^H$, we have

$$A^H x = \lambda x$$

Taking the complex conjugate, yields

$$A^T x^* = \lambda x^*$$

In general, the eigenvectors of a Hermitian matrix are complex, but real for a real symmetric matrix since $A^H = A^T$. Moreover, the left-eigenvector y^T is the complex conjugate of the right-eigenvector x. Hence, the orthogonality relation (A.10) reduces, after normalization, to an inner-product

$$x_k^H x_j = \delta_{kj} \qquad (A.21)$$

where δ_{kj} is the Kronecker delta, which is zero if $k \neq j$ and else $\delta_{kk} = 1$. Consequently, (A.13) reduces to

$$X^H X = I$$

which implies that the matrix X formed by the eigenvectors is an unitary matrix ($X^{-1} = X^H$). For a real symmetric matrix A, the corresponding relation $X^T X = I$ implies that X is an orthogonal matrix ($X^{-1} = X^T$). Although the arguments so far (see Section A.1) have assumed that the eigenvalues of A are distinct, the theorem applies in general (as proved in Wilkinson (1965, Section 47)): For any Hermitian matrix A, there exists a unitary matrix U such that

$$U^H A U = \text{diag}\,(\lambda_j) \qquad \text{real } \lambda_j$$

and for any real symmetric matrix A, there exists an orthogonal matrix U such that

$$U^T A U = \text{diag}\,(\lambda_j) \qquad \text{real } \lambda_j$$

10. To a real symmetric matrix A, a bilinear form $x^T A y$ is associated, which is a scalar defined as

$$x^T A y = x A y^T = \sum_{i=1}^{n} \sum_{j=1}^{n} a_{ij} x_i y_j$$

We call a bilinear form a quadratic form if $y = x$. A necessary and sufficient condition for a quadratic form to be positive definite, i.e. $x^T A x > 0$ for all $x \neq 0$, is that all eigenvalues of A should be positive. Indeed, art. **9** shows the existence of an orthogonal matrix U that transforms A to a diagonal form. Let $x = Uz$, then

$$x^T A x = z^T U^T A U z = \sum_{k=1}^{n} \lambda_k z_k^2 \qquad (A.22)$$

which is only positive for all z_k provided $\lambda_k > 0$ for all k. From (A.6), a positive definite quadratic form $x^T A x$ possesses a positive determinant, $\det A > 0$. This analysis shows that the problem of determining an orthogonal matrix U (or the eigenvectors of A) is equivalent to the geometrical problem of determining the principal axes of the hyper-ellipsoid

$$\sum_{i=1}^{n} \sum_{j=1}^{n} a_{ij} x_i y_j = 1$$

Relation (A.22) illustrates that the eigenvalues λ_k are the squares of the principal axis. A multiple eigenvalue refers to an indeterminacy of the principal axes. For example if $n = 3$, an ellipsoid with two equal principal axis means that any section along the third axis is a circle. Any two perpendicular diameters of the largest circle orthogonal to the third axis are principal axis of that ellipsoid.

A.3 Vector and matrix norms

Vector and matrix norms, denoted by $\|x\|$ and $\|A\|$ respectively, provide a single number reflecting a "size" of the vector or matrix and may be regarded as an extension of the concept of the modulus of a complex number. A norm is a certain function of the vector components or matrix elements. All norms, vector as well as matrix norms, satisfy the three distance relations

(i) $\|x\| > 0$ unless $x = 0$
(ii) $\|\alpha x\| = |\alpha| \, \|x\|$ for any complex number α
(iii) $\|x + y\| \leq \|x\| + \|y\|$

In general, the Hölder q-norm of a vector x is defined as

$$\|x\|_q = \left(\sum_{j=1}^{n} |x_j|^q \right)^{1/q} \tag{A.23}$$

For example, the well-known Euclidean norm or length of the vector x is found for $q = 2$ and $\|x\|_2^2 = x^H x$. In probability theory where x denotes a discrete pdf, the law of total probability states that $\|x\|_1 = \sum_{j=1}^{n} x_j = 1$ and we will write $\|x\|_1 = \|x\|$. Finally, $\max |x_j| = \lim_{q \to \infty} \|x\|_q = \|x\|_\infty$. The unit-spheres $S_q = \{x| \; \|x\|_q = 1\}$ are, in three dimensions $n = 3$, for $q = 1$, an octahedron, for $q = 2$, a ball and for $q = \infty$, a cube. Furthermore, S_1 fits into S_2, which in turn fits into S_∞, implies that $\|x\|_1 \geq \|x\|_2 \geq \|x\|_\infty$ for any x.

The Hölder inequality proved in Section 5.5 states that, for $\frac{1}{p} + \frac{1}{q} = 1$ and real $p, q > 1$,

$$|x^H y| \le \|x\|_p \|y\|_q \tag{A.24}$$

A special case of the Hölder inequality where $p = q = 2$ is the Cauchy-Schwarz inequality

$$|x^H y| \le \|x\|_2 \|y\|_2 \tag{A.25}$$

The $q = 2$ norm is invariant under an unitary (hence also orthogonal) transformation U, where $U^H U = I$, because $\|Ux\|_2^2 = x^H U^H U x = x^H x = \|x\|_2$.

An other example of a non-homogeneous vector norm is the quadratic form $\sqrt{\|x\|_A} = \sqrt{x^T A x}$ provided A is positive definite. Relation (A.22) shows that, if not all eigenvalues λ_j of A are the same, not all components of the vector x are weighted similarly and, thus, in general, $\sqrt{\|x\|_A}$ is a non-homogeneous norm. The quadratic form $\|x\|_I$ equals the homogeneous Euclidean norm $\|x\|_2^2$.

A.3.1 Properties of norms

All norms are equivalent in the sense that there exist positive real numbers c_1 and c_2 such that, for all x,

$$c_1 \|x\|_p \le \|x\|_q \le c_2 \|x\|_p$$

For example,

$$\|x\|_2 \le \|x\|_1 \le \sqrt{n} \|x\|_2$$
$$\|x\|_\infty \le \|x\|_1 \le n \|x\|_\infty$$
$$\|x\|_\infty \le \|x\|_2 \le \sqrt{n} \|x\|_\infty$$

By choosing in the Hölder inequality (5.15) $p = q = 1$, $x_j \to \alpha_j x_j^s$ for real $s > 0$ and $y_j \to \alpha_j > 0$, we obtain with $0 < \theta < 1$ an inequality for the *weighted q-norm*

$$\left(\frac{\sum_{j=1}^n \alpha_j |x_j|^{s\theta}}{\sum_{j=1}^n \alpha_j} \right)^{\frac{1}{s\theta}} \le \left(\frac{\sum_{j=1}^n \alpha_j |x_j^s|}{\sum_{j=1}^n \alpha_j} \right)^{\frac{1}{s}}$$

For $\alpha_j = 1$, the weights α_j disappear such that the inequality for the Hölder q-norm becomes

$$\|x\|_{s\theta} \le \|x\|_s \, n^{\frac{1}{s}(\frac{1}{\theta} - 1)}$$

where $n^{\frac{1}{s}(\frac{1}{\theta}-1)} \geq 1$. On the other hand, with $0 < \theta < 1$ and for real $s > 0$,

$$\frac{\|x\|_s}{\|x\|_{s\theta}} = \frac{\left(\sum_{j=1}^n |x_j|^s\right)^{\frac{1}{s}}}{\left(\sum_{k=1}^n |x_k|^{s\theta}\right)^{\frac{1}{s\theta}}} = \left(\sum_{j=1}^n \frac{|x_j|^s}{\left(\sum_{k=1}^n |x_k|^{s\theta}\right)^{\frac{1}{\theta}}}\right)^{\frac{1}{s}} = \left(\sum_{j=1}^n \left(\frac{|x_j|^{s\theta}}{\sum_{k=1}^n |x_k|^{s\theta}}\right)^{\frac{1}{\theta}}\right)^{\frac{1}{s}}$$

Since $y = \frac{|x_j|^{s\theta}}{\sum_{k=1}^n |x_k|^{s\theta}} \leq 1$ and $\frac{1}{\theta} > 1$, it holds that $y^{\frac{1}{\theta}} \leq y$ and

$$\left(\sum_{j=1}^n \left(\frac{|x_j|^{s\theta}}{\sum_{k=1}^n |x_k|^{s\theta}}\right)^{\frac{1}{\theta}}\right)^{\frac{1}{s}} \leq \left(\sum_{j=1}^n \frac{|x_j|^{s\theta}}{\sum_{k=1}^n |x_k|^{s\theta}}\right)^{\frac{1}{s}} = \left(\frac{\sum_{j=1}^n |x_j|^{s\theta}}{\sum_{k=1}^n |x_k|^{s\theta}}\right)^{\frac{1}{s}} = 1$$

which leads to the opposite inequality (without normalization as $n^{\frac{1}{s}(\frac{1}{\theta}-1)}$),

$$\|x\|_s \leq \|x\|_{s\theta}$$

In summary, if $p > q > 0$, then the general inequality for Hölder q-norm is

$$\|x\|_p \leq \|x\|_q \leq \|x\|_p \, n^{\frac{1}{q}-\frac{1}{p}} \tag{A.26}$$

For $m \times n$ matrices A, the most frequently used norms are the Euclidean or Frobenius norm

$$\|A\|_F = \left(\sum_{i=1}^m \sum_{j=1}^n |a_{ij}|^2\right)^{1/2} \tag{A.27}$$

and the q-norm

$$\|A\|_q = \sup_{x \neq 0} \frac{\|Ax\|_q}{\|x\|_q} \tag{A.28}$$

On the second distance relation, $\frac{\|Ax\|_q}{\|x\|_q} = \left\|A\frac{x}{\|x\|_q}\right\|_q$, which shows that

$$\|A\|_q = \sup_{\|x\|_q=1} \|Ax\|_q \tag{A.29}$$

Furthermore, the matrix q-norm (A.28) implies that

$$\|Ax\|_q \leq \|A\|_q \|x\|_q \tag{A.30}$$

Since the vector norm is a continuous function of the vector components and since the domain $\|x\|_q = 1$ is closed, there must exist a vector x for which equality $\|Ax\|_q = \|A\|_q \|x\|_q$ holds. Since the k-th vector component of Ax is $(Ax)_i = \sum_{j=1}^n a_{ij}x_j$, it follows from (A.23) that

$$\|Ax\|_q = \left(\sum_{i=1}^m \left|\sum_{j=1}^n a_{ij}x_j\right|^q\right)^{1/q}$$

For example, for all x with $\|x\|_1 = 1$, we have that

$$\|Ax\|_1 = \sum_{i=1}^{m}\left|\sum_{j=1}^{n}a_{ij}x_j\right| \leq \sum_{i=1}^{m}\sum_{j=1}^{n}|a_{ij}||x_j| = \sum_{j=1}^{n}|x_j|\sum_{i=1}^{m}|a_{ij}|$$

$$\leq \sum_{j=1}^{n}|x_j|\left(\max_j\sum_{i=1}^{m}|a_{ij}|\right) = \max_j\sum_{i=1}^{m}|a_{ij}|$$

Clearly, there exists a vector x for which equality holds, namely, if k is the column in A with maximum absolute sum, then $x = e_k$, the k-th basis vector with all components zero, except for the k-th one, which is 1. Similarly, for all x with $\|x\|_\infty = 1$,

$$\|Ax\|_\infty = \max_i\left|\sum_{j=1}^{n}a_{ij}x_j\right| \leq \max_i\sum_{j=1}^{n}|a_{ij}||x_j| \leq \max_i\sum_{j=1}^{n}|a_{ij}|$$

Again, if r is the row with maximum absolute sum and $x_j = 1.\text{sign}(a_{rj})$ such that $\|x\|_\infty = 1$, then $(Ax)_r = \sum_{j=1}^{n}|a_{rj}| = \max_i\sum_{j=1}^{n}|a_{ij}| = \|Ax\|_\infty$. Hence, we have proved that

$$\|A\|_\infty = \max_i\sum_{j=1}^{n}|a_{ij}| \tag{A.31}$$

$$\|A\|_1 = \max_j\sum_{i=1}^{m}|a_{ij}| \tag{A.32}$$

from which

$$\|A^H\|_\infty = \|A\|_1$$

The $q = 2$ matrix norm, $\|Ax\|_2$, is obtained differently. Consider

$$\|Ax\|_2^2 = (Ax)^H Ax = x^H A^H A x$$

Since $A^H A$ is a Hermitian matrix, art. **9** shows that all eigenvalues are real and non-negative because a norm $\|Ax\|_2^2 \geq 0$. These ordered eigenvalues are denoted as $\sigma_1^2 \geq \sigma_2^2 \geq \cdots \geq \sigma_n^2 \geq 0$. Applying the theorem in art. **9**, there exists a unitary matrix U such that $x = Uz$ yields

$$x^H A^H A x = z^H U^H A^H A U z = z^H \text{diag}\left(\sigma_j^2\right) z \leq \sigma_1^2 z^H z = \sigma_1^2 \|z\|_2^2$$

Since the $q = 2$ norm is invariant under a unitary (orthogonal) transform $\|x\|_2 = \|z\|_2$, by the definition (A.28),

$$\|A\|_2 = \sup_{x\neq 0}\frac{\|Ax\|_2}{\|x\|_2} = \sigma_1 \tag{A.33}$$

where the supremum is achieved if x is the eigenvector of $A^H A$ belonging to σ_1^2. Meyer (2000, p. 279) proves the corresponding result for the minimum eigenvalue provided that A is non-singular,

$$\left\| A^{-1} \right\|_2 = \frac{1}{\min_{\|x\|_2 = 1} \|Ax\|_2} = \sigma_n^{-1}$$

The non-negative quantity σ_j is called the j-th singular value and σ_1 is the largest singular value of A. The importance of this result lies in an extension of the eigenvalue problem to non-square matrices which is called the singular value decomposition. A detailed discussion is found in Golub and Loan (1983). If A has real eigenvalues $\lambda_1 \geq \lambda_2 \geq \cdots \geq \lambda_n$, the above can be simplified and we obtain

$$\lambda_1 = \sup_{x \neq 0} \frac{x^T A x}{x^T x} \tag{A.34}$$

$$\lambda_n = \inf_{x \neq 0} \frac{x^T A x}{x^T x} \tag{A.35}$$

because, for any x, it holds that $\lambda_n x^T x \leq x^T A x \leq \lambda_1 x^T x$.

The Frobenius norm $\|A\|_F^2 = \operatorname{trace}(A^H A)$. With (A.7) and the analysis of $A^H A$ above,

$$\|A\|_F^2 = \sum_{k=1}^{n} \sigma_k^2 \tag{A.36}$$

In view of (A.33), the bounds $\|A\|_2 \leq \|A\|_F \leq \sqrt{n} \, \|A\|_2$ may be attained.

A.3.2 Applications of norms

(a) Since $\left\| A^k \right\| = \left\| A A^{k-1} \right\| \leq \|A\| \, \left\| A^{k-1} \right\|$, by induction, we have for any integer k, that

$$\left\| A^k \right\| \leq \|A\|^k$$

and

$$\lim_{k \to \infty} A^k = 0 \text{ if } \|A\| < 1$$

(b) By taking the norm of the eigenvalue equation (A.1), $\|Ax\| = |\lambda| \, \|x\|$ and with (A.30),

$$|\lambda| \leq \|A\|_q \tag{A.37}$$

Applied to $A^H A$, for any q-norm,

$$\sigma_1^2 \le \left\| A^H A \right\|_q \le \left\| A^H \right\|_q \left\| A \right\|_q$$

Choose $q = 1$ and with (A.33),

$$\left\| A \right\|_2^2 \le \left\| A^H \right\|_1 \left\| A \right\|_1 = \left\| A \right\|_\infty \left\| A \right\|_1$$

(c) Any matrix A can be transformed by a similarity transform H to a Jordan canonical form C (art. **4**) as $A = HCH^{-1}$, from which $A^k = HC^k H^{-1}$. A typical Jordan submatrix $(C_m(\lambda))^k = \lambda^{k-2} B$, where B is independent of k. Hence, for large k, $A^k \to 0$ if and only if $|\lambda| < 1$ for all eigenvalues.

A.4 Stochastic matrices

A probability matrix P is reducible if there is a relabeling of the states that leads to

$$\tilde{P} = \begin{bmatrix} P_1 & B \\ O & P_2 \end{bmatrix}$$

where P_1 and P_2 are square matrices. Relabeling amounts to permuting rows and columns in the same fashion. Thus, there exists a similarity transform H such that $P = H\tilde{P}H^{-1}$.

A.4.1 The eigenstructure

In this section, the basic theorem on the eigenstructure of a stochastic, irreducible matrix will be proved.

Lemma A.4.1 *If P is an irreducible non-negative matrix and if v is a vector with positive components, then the vector $z = (P+I)v$ has always fewer zero components than v.*

Proof: Denote

$$v = \begin{bmatrix} v_1 \\ 0 \end{bmatrix} \text{ and } z = \begin{bmatrix} z_1 \\ 0 \end{bmatrix} \text{ where } v_1 > 0, z_1 > 0$$

which is always possible by suitable renumbering of the states and

$$P = \begin{bmatrix} P_{11} & P_{12} \\ P_{21} & P_{22} \end{bmatrix}$$

The relation $z = (P + I)v$ is written as

$$\begin{bmatrix} z_1 \\ 0 \end{bmatrix} = \begin{bmatrix} P_{11}v_1 \\ P_{21}v_1 \end{bmatrix} + \begin{bmatrix} v_1 \\ 0 \end{bmatrix}$$

Since P is irreducible, $P_{21} \neq O$, such that $v_1 > 0$ implies that $P_{21}v_1 \neq 0$, which proves the lemma. □

Observe, in addition, that all components of z are never smaller than those of v. Also, transposing does not alter the result.

Theorem A.4.2 (Frobenius) *The modulus of all eigenvalues λ of an irreducible stochastic matrix P are less than or equal to 1. There is only one real eigenvalue $\lambda = 1$ and the corresponding eigenvector has positive components.*

Proof: The $q = \infty$ norm (A.31) of a probability matrix P with n states defined by (9.7) subject to (9.8) precisely equals $\|P\|_\infty = 1$. From (A.37), it follows that all eigenvalues are, in absolute value, smaller than or equal to 1. Since all elements $P_{ij} \in [0, 1]$ and because an irreducible matrix has no zero element rows, $v^T P$ has positive components if v^T has positive components. Thus, there always exists a scalar $0 < \mu_v = \min_{1 \leq k \leq n} \frac{(v^T P)_k}{(v^T)_k}$, such that $\mu_v v^T \leq v^T P$. By Lemma A.4.1, we can always transform the vector v to a vector z by right-multiplying both sides with $(I + P)$ such that

$$\mu_v v^T (I + P) \leq v^T P (I + P)$$
$$\mu_v z^T \leq z^T P$$

and, by definition of μ_v, $\mu_v \leq \mu_z$ since the components of z are never smaller than those of v. Hence, for any arbitrary vector v with positive components, the transform in Lemma A.4.1 leads to an increasing set $\mu_v \leq \mu_z \leq \cdots$, which is bounded by 1 because no eigenvalue can exceed 1. This shows that $\lambda = 1$ is the largest eigenvalue and the corresponding eigenvector y^T has positive components.

This eigenvector y^T is unique. For, if there were another linearly independent eigenvector w^T corresponding to the eigenvalue $\lambda = 1$, any linear combination $z^T = \alpha y^T + \beta w^T$ is also an eigenvector belonging to $\lambda = 1$. But α and β can always be chosen to produce a zero component which the transform method shows to be impossible. The fact that the eigenvector y^T is the only eigenvector belonging $\lambda = 1$, implies that the eigenvalue $\lambda = 1$ is a single zero of the characteristic polynomial of P. □

The theorem proved for stochastic matrices is a special case of the famous Frobenius theorem for non-negative matrices (see for a proof, e.g. Gantmacher (1959b, Chapter XIII)). We note that, in the theory of Markov chains, the interest lies in the determination of the left-eigenvector $y^T = \pi$ belonging to $\lambda = 1$, because the right-eigenvector x of P belonging to $\lambda = 1$ equals $u^T = \beta[1 \ 1 \ \cdots \ 1]$, where β is a scalar, because of the constraints (9.8). Recall (A.10) and (A.13), the proper normalization, $y^T u = 1$, precisely corresponds to the total law of probability. Using the interpretation of Markov chains, an alternative argument is possible. If all eigenvalues were $|\lambda| < 1$, application (c) in Section A.3.2 indicates that the steady-state would be non-existent because $P^k \to 0$ for $k \to \infty$. Since this is impossible, there must be at least one eigenvalue with $|\lambda| = 1$. Furthermore, (9.22) shows that at least one eigenvalue corresponding to the steady-state is real and precisely 1.

Corollary A.4.3 *An irreducible probability matrix P cannot have two linearly independent eigenvectors with positive components.*

Proof: Consider, apart from $y^T = \pi$ belonging to $\lambda = 1$, another eigenvector w^T belonging to the eigenvalue $\omega \neq 1$. On art. **3**, $w^T y = 0$, which is only possible if not all components of w^T are positive.   □

The corollary is important because no other eigenvector of P than $y^T = \pi$ can represent a (discrete) probability density. Since the null vector is never an eigenvector, the corollary implies that at least one component in the other eigenvectors must be negative.

Since the characteristic polynomial of P has real coefficients (because P_{ij} is real), the eigenvalues occur in complex conjugate pairs. Since $\lambda = 1$ is an eigenvalue, for an even number of state n, there must be at least another real eigenvalue obeying $-1 \leq \lambda < 1$. It has been proved that the boundary of the locations of the eigenvalues inside the unit disc consists of a finite number of points on the unit circle joined by certain curvilinear arcs.

There exist an interesting property of a rank-one update $\bar{P}$ of a stochastic matrix P. The lemma is of a general nature and also applies to reducible Markov chains with several eigenvalues $\lambda_j = 1$ for $1 < j \leq k$.

Lemma A.4.4 *If $\{1, \lambda_2, \lambda_3, \ldots, \lambda_n\}$ are the eigenvalues of the stochastic matrix P, then the eigenvalues of $\bar{P} = \alpha P + (1 - \alpha)uv^T$, where v^T is any probability vector, are $\{1, \alpha\lambda_2, \alpha\lambda_3, \ldots, \alpha\lambda_n\}$.*

Proof: We start from the eigenvalues equation (A.2)

$$\det\left(\bar{P} - \lambda I\right) = \det\left(\alpha P - \lambda I + (1-\alpha)uv^T\right)$$
$$= \det\left((\alpha P - \lambda I)\left(I + (\alpha P - \lambda I)^{-1}(1-\alpha)uv^T\right)\right)$$
$$= \det\left(\alpha P - \lambda I\right)\det\left(I + (1-\alpha)(\alpha P - \lambda I)^{-1}uv^T\right)$$

Applying the formula

$$\det\left(I + cd^T\right) = 1 + d^T c \tag{A.38}$$

which follows, after taking the determinant, from the matrix identity

$$\begin{pmatrix} I & 0 \\ d^T & 1 \end{pmatrix}\begin{pmatrix} I + cd^T & c \\ 0 & 1 \end{pmatrix}\begin{pmatrix} I & 0 \\ -d^T & 1 \end{pmatrix} = \begin{pmatrix} I & c \\ 0 & 1 + d^T c \end{pmatrix}$$

gives

$$\det\left(\bar{P} - \lambda I\right) = \det\left(\alpha P - \lambda I\right)\left(1 + v^T(1-\alpha)(\alpha P - \lambda I)^{-1}u\right)$$

Since the row sum of a stochastic matrix P is 1, we have that $Pu = u$ and, thus, $(\alpha P - \lambda I)u = (\alpha - \lambda)u$ from which $(\alpha P - \lambda I)^{-1}u = (\alpha - \lambda)^{-1}u$. Using this result leads to

$$1 + v^T(1-\alpha)(\alpha P - \lambda I)^{-1}u = 1 + \frac{1-\alpha}{\alpha - \lambda}v^T u = 1 + \frac{1-\alpha}{\alpha - \lambda} = \frac{1-\lambda}{\alpha - \lambda}$$

because a probability vector is normalized to 1, i.e. $v^T u = 1$. Hence, we end up with

$$\det\left(\bar{P} - \lambda I\right) = \det\left(\alpha P - \lambda I\right)\frac{1-\lambda}{\alpha - \lambda}$$

Invoking (A.19) yields

$$\det\left(\bar{P} - \lambda I\right) = \prod_{k=1}^{n}(\alpha\lambda_k - \lambda)\frac{1-\lambda}{\alpha - \lambda} = (1-\lambda)\prod_{k=2}^{n}(\alpha\lambda_k - \lambda)$$

which shows that the eigenvalues of $\bar{P}$ are $\{1, \alpha\lambda_2, \alpha\lambda_3, \ldots, \alpha\lambda_N\}$. $\qquad\square$

A similar property may occur in a special case where a Markov chain is supplemented by an additional state $n+1$ which connects to every other state and to which every other state is connected (such that $\bar{P}$ is irreducible). Then,

$$\bar{P} = \begin{pmatrix} \alpha P & (1-\alpha)u \\ v^T & 0 \end{pmatrix}$$

with corresponding eigenvalues $\{1, \alpha\lambda_2, \alpha\lambda_3, \ldots, \alpha\lambda_n, 0\}$. This result is similarly proved as Lemma A.4.4 using (Meyer, 2000, p. 475)

$$\det \begin{pmatrix} A & B \\ C & D \end{pmatrix} = \det A \det \left(D - CA^{-1}B \right) \qquad (A.39)$$

provided A^{-1} exists unless $C = 0$.

A.4.2 Example: the two-state Markov chain

The two-state Markov chain is defined by

$$P = \begin{bmatrix} 1 - p & p \\ q & 1 - q \end{bmatrix}$$

Observe that $\det P = 1 - p - q$. The eigenvalues of P satisfy the characteristic polynomial $c(\lambda) = \lambda^2 - (2 - p - q)\lambda + \det P = 0$, from which $\lambda_1 = 1$ and $\lambda_2 = 1 - p - q = \det P$. The adjoint matrix $Q(\lambda)$ is computed (art. **7**) via the polynomial $\frac{c(\lambda) - c(\mu)}{\lambda - \mu}$,

$$\frac{c(\lambda) - c(\mu)}{\lambda - \mu} = \lambda + \mu - (2 - p - q)$$

and after $\lambda \to \lambda I$ and $\mu \to P$

$$Q(\lambda) = \lambda I + P - (2 - p - q)I$$

$$= \begin{bmatrix} \lambda - 1 + q & p \\ q & \lambda - 1 + p \end{bmatrix}$$

The (unscaled) right- (left-) eigenvectors of P follow as the non-zero columns (rows) of $Q(\lambda)$. For $\lambda_1 = 1$, we find $x_1 = (1,1)$ and $y_1^T = (q,p)$. For $\lambda_2 = 1 - p - q$, the eigenvector $x_2 = (-p, q)$ and $y_2^T = (1, -1)$. Normalization (art. **4**) requires that $y_k^T x_k = 1$ or $x_1 = \frac{1}{p+q}(1,1)$ and $x_2 = \frac{1}{p+q}(p, -q)$. If the eigenvalues are distinct $(p + q \neq 0)$, the matrix P can be written as (art. **4**) $P = X \mathrm{diag}(\lambda_k) Y^T$,

$$P = \frac{1}{p+q} \begin{bmatrix} 1 & p \\ 1 & -q \end{bmatrix} \begin{bmatrix} 1 & 0 \\ 0 & 1 - p - q \end{bmatrix} \begin{bmatrix} q & p \\ 1 & -1 \end{bmatrix}$$

from which any power P^k is immediate as

$$P^k = \frac{1}{p+q} \begin{bmatrix} 1 & p \\ 1 & -q \end{bmatrix} \begin{bmatrix} 1 & 0 \\ 0 & (1 - p - q)^k \end{bmatrix} \begin{bmatrix} q & p \\ 1 & -1 \end{bmatrix}$$

$$= \frac{1}{p+q} \begin{bmatrix} q & p \\ q & p \end{bmatrix} + \frac{(1 - p - q)^k}{p+q} \begin{bmatrix} p & -p \\ -q & q \end{bmatrix} \qquad (A.40)$$

The steady-state matrix $P^\infty = \lim_{k \to \infty} P^k$ follows as

$$P^\infty = \frac{1}{p+q} \begin{bmatrix} q & p \\ q & p \end{bmatrix} \equiv \begin{bmatrix} \pi \\ \pi \end{bmatrix} \qquad (A.41)$$

because $|1 - p - q| < 1$.

Alternatively, the steady-state vector is a solution of (9.25),

$$\begin{bmatrix} -p & q \\ 1 & 1 \end{bmatrix} \begin{bmatrix} \pi_1 \\ \pi_2 \end{bmatrix} = \begin{bmatrix} 0 \\ 1 \end{bmatrix}$$

Applying Cramer's rule with $D = \det \begin{bmatrix} -p & q \\ 1 & 1 \end{bmatrix} = -(p + q)$, we obtain

$$\pi_1 = \tfrac{1}{D} \det \begin{bmatrix} 0 & q \\ 1 & 1 \end{bmatrix} \text{ and } \pi_2 = \tfrac{1}{D} \det \begin{bmatrix} -p & 0 \\ 1 & 1 \end{bmatrix} \text{ or}$$

$$\pi = \begin{bmatrix} \frac{q}{p+q} & \frac{p}{p+q} \end{bmatrix}$$

which indeed agrees with (A.41) and (9.37).

A.4.3 The tendency towards the steady-state

A stochastic matrix P and the corresponding Markov chain is regular if the only eigenvalue with $|\lambda| = 1$ is $\lambda = 1$. It is fully regular if, in addition, $\lambda = 1$ is a simple zero of the characteristic polynomial of P. The Frobenius Theorem A.4.2 indicates that a regular matrix is necessarily reducible. Application (c) in Section A.3.2 demonstrates that the steady-state only exists for regular Markov chains. Alternatively, a regular matrix P has the property that $P^k > O$ (for some k), i.e. all elements are strictly positive.

In the sequel, we concentrate on fully regular stochastic matrices P, where all eigenvalues lie within the unit circle, except for the largest one, $\lambda = 1$. If the N eigenvalues of the regular stochastic matrix P are ordered as $\lambda_1 = 1 > |\lambda_2| \geq \cdots \geq |\lambda_N| \geq 0$, the second largest eigenvalue λ_2 will determine the speed of convergence of the Markov chain towards the steady-state.

A.4.3.1 Example: the three-state Markov chain

The three-state Markov chain P is defined by (9.7) with $N = 3$. Assuming that P is irreducible, we determine the eigenvalues. Since the Frobenius Theorem A.4.2 already determines one eigenvalue $\lambda_1 = 1$, the remaining two λ_2 and λ_3 are found from (A.6) and (A.7). They obey the equations

$$\lambda_2 \lambda_3 = \det P$$
$$\lambda_2 + \lambda_3 = P_{11} + P_{22} + P_{33} - 1 = \text{trace}(P) - 1$$

or the quadratic equation $x^2 - (\lambda_2 + \lambda_3)\,x + \lambda_2\lambda_3 = 0$. The explicit solution is

$$\lambda_2 = \frac{1}{2}\left(\mathrm{trace}(P) - 1\right) + \frac{1}{2}\sqrt{\left(\mathrm{trace}(P) - 1\right)^2 - 4\det P}$$

$$\lambda_3 = \frac{1}{2}\left(\mathrm{trace}(P) - 1\right) - \frac{1}{2}\sqrt{\left(\mathrm{trace}(P) - 1\right)^2 - 4\det P}$$

All eigenvalues are real if the discriminant $\left(\mathrm{trace}(P) - 1\right)^2 - 4\det P$ is non-negative which leads to three cases:

(a) In case $\left(\mathrm{trace}(P) - 1\right)^2 > 4\det P$, the eigenvalues obey $1 > \lambda_2 > \lambda_3$, but not necessarily $1 > |\lambda_2| > |\lambda_3|$. The latter inequality is true if $\mathrm{trace}(P) > 1$, in which case the speed of convergence towards the steady-state is determined by the decay of $(\lambda_2)^k$ as $k \to \infty$. If $\mathrm{trace}(P) = 1$, then $\lambda_2 = -\lambda_3 = \sqrt{-\det P}$ and if $\mathrm{trace}(P) < 1$, $|\lambda_3|$ determines the speed of convergence. Notice that $\lambda_2 > \frac{1}{2}\left(\mathrm{trace}(P) - 1\right) \geq -\frac{1}{2}$.

(b) In case $\left(\mathrm{trace}(P) - 1\right)^2 < 4\det P$, there are two complex conjugate roots $\lambda_2 = \alpha + i\beta$ and $\lambda_3 = \alpha - i\beta$, both with the same modulus $|\lambda_2| = |\lambda_3|$ equal to $\alpha^2 + \beta^2 = \lambda_2\lambda_3 = \det P$ and with real part $\alpha = \frac{1}{2}\left(\mathrm{trace}(P) - 1\right)$. In this case, we have that $0 \leq \det P < 1$. Hence, the Markov chain converges towards the steady-state as $\left(\sqrt{\det P}\right)^k$ in the discrete-time k.

(c) In case $\left(\mathrm{trace}(P) - 1\right)^2 = 4\det P$, there is a double eigenvalue

$$\lambda = \lambda_2 = \lambda_3 = \frac{1}{2}\left(\mathrm{trace}(P) - 1\right) = \pm\sqrt{\det P}$$

and P cannot be reduced by a similarity transform H to a diagonal matrix (Section A.1, art. **4**) that but to the Jordan canonical form C such that $P^k = H^{-1}C^k H$. Since (Meyer, 2000, pp. 599–600)

$$C^k = \left(\begin{bmatrix} 1 & 0 & 0 \\ 0 & \lambda & 1 \\ 0 & 0 & \lambda \end{bmatrix}\right)^k = \begin{bmatrix} 1 & 0 & 0 \\ 0 & \lambda^k & k\lambda^{k-1} \\ 0 & 0 & \lambda^k \end{bmatrix}$$

the Markov chain converges towards the steady-state as $k\left(\sqrt{\det P}\right)^{k-1}$ in the discrete-time k. We observe that $-\frac{1}{2} \leq \lambda_2 = \lambda_3 < 1$ because $0 \leq \mathrm{trace}(P) < 3$. If $\mathrm{trace}(P) = 3$, then $P = I$, and P is not irreducible.

The fastest possible convergence occurs when $\lambda_2 = \lambda_3 = 0$ or when $\det P = 0$ and $\mathrm{trace}(P) = 1$ in which case P has rank 1. In any matrix of rank 1, all row vectors are linearly dependent. Since the column sum of a stochastic matrix P is 1 by (9.8), every row in P is precisely the same and (9.6) shows that after one discrete-time step, the steady-state

$\pi = \left[\begin{array}{cccc} \frac{1}{N} & \frac{1}{N} & \cdots & \frac{1}{N} \end{array} \right]$ is reached. As shown in Section 9.3.1, a transition probability matrix with constant rows can be regarded as a limit transition probability matrix $A = \lim_{k \to \infty} \check{P}^k$ of a Markov process with transition probability matrix $\check{P}$.

A.5 Special types of stochastic matrices

A.5.1 Doubly stochastic matrices

A doubly stochastic matrix P has both row and column sums equal to 1,

$$\sum_{k=1}^{N} P_{ik} = \sum_{k=1}^{N} P_{kj} = 1 \qquad \text{for all } i, j$$

If P is symmetric, $P = P^T$, then P is doubly stochastic, but the reverse implication is not true. As observed in Section A.4.1, the left-eigenvector $y^T = \pi$ and the right-eigenvector $x = u$ belonging to eigenvalue $\lambda = 1$ satisfy $y^T u = 1$. For doubly stochastic matrices, it holds that the role of left- and right-eigenvector can be reversed, which leads to $y = x$ or

$$\pi = \left[\begin{array}{cccc} \frac{1}{N} & \frac{1}{N} & \cdots & \frac{1}{N} \end{array} \right]$$

The example in Section A.4.3 illustrates that a steady-state vector equal to $\pi = \left[\begin{array}{cccc} \frac{1}{N} & \frac{1}{N} & \cdots & \frac{1}{N} \end{array} \right]$ does not necessarily imply that P is doubly stochastic.

A.5.2 Tri-diagonal bandmatrices

A.5.2.1 Tri-diagonal Toeplitz bandmatrix

A Toeplitz matrix has constant entries on each diagonal parallel to the main diagonal. Of particular interest is the $N \times N$ tri-diagonal Toeplitz matrix,

$$A = \begin{bmatrix} b & a & & & \\ c & b & a & & \\ & \ddots & \ddots & \ddots & \\ & & c & b & a \\ & & & c & b \end{bmatrix}$$

that arises in the Markov chain of the random walk and the birth and death process. Moreover, the eigenstructure of the tri-diagonal Toeplitz matrix A can be expressed in analytic form.

An eigenvector x corresponding to eigenvalue λ satisfies $(A - \lambda I)\, x = 0$ or, written per component,

$$(b - \lambda)x_1 + ax_2 = 0$$
$$cx_{k-1} + (b - \lambda)x_k + ax_{k+1} = 0 \qquad 2 \le k \le N - 1$$
$$cx_{N-1} + (b - \lambda)x_N = 0$$

We assume that $a \ne 0$ and $c \ne 0$ and rewrite the set with $x_0 = x_{N+1} = 0$ as

$$x_{k+2} + \left(\frac{b - \lambda}{a}\right) x_{k+1} + \left(\frac{c}{a}\right) x_k = 0 \qquad 0 \le k \le N - 1$$

which are second order difference equations with constant coefficients. The general solution of these equations is $x_k = \alpha r_1^k + \beta r_2^k$ where r_1 and r_2 are the roots of the corresponding polynomial $x^2 + \left(\frac{b-\lambda}{a}\right) x + \left(\frac{c}{a}\right) = 0$. If $r_1 = r_2$, the general solution is $x_k = \alpha r_1^k + \beta k r_1^k$, which is impossible since it implies that all $x_k = 0$ due to the fact that $x_0 = x_{N+1} = 0$, which forces r_1 to be zero. An eigenvector is never the zero vector. Thus, we have distinct roots $r_1 \ne r_2$ that satisfy

$$r_1 + r_2 = -\left(\frac{b - \lambda}{a}\right)$$
$$r_1 r_2 = \frac{c}{a}$$

The constants α and β follow from the boundary requirement $x_0 = x_{N+1} = 0$ as

$$\alpha + \beta = 0$$
$$\alpha r_1^{N+1} + \beta r_2^{N+1} = 0$$

Rewriting the last equation with $\alpha = -\beta$, yields $\left(\frac{r_1}{r_2}\right)^{N+1} = 1$ or $\frac{r_1}{r_2} = e^{\frac{2\pi i m}{N+1}}$ for some $1 \le m \le N$ (the root $m = 0$ must be rejected since $r_1 \ne r_2$). Substitution of $r_1 = r_2 e^{\frac{2\pi i m}{N+1}}$ into the last root equation yields

$$r_1 = \sqrt{\frac{c}{a}}\, e^{\frac{\pi i m}{N+1}} \quad \text{and} \quad r_2 = \sqrt{\frac{c}{a}}\, e^{-\frac{\pi i m}{N+1}}$$

The first root equation is only possible for special values of $\lambda = \lambda_m$ with $1 \le m \le N$, which are the eigenvalues,

$$\lambda_m = b + a\sqrt{\frac{c}{a}} \left(e^{-\frac{\pi i m}{N+1}} + e^{\frac{\pi i m}{N+1}}\right) = b + 2\sqrt{ac}\,\cos\left(\frac{\pi m}{N+1}\right)$$

Since there are precisely N different values of m, there are N distinct eigenvalues λ_m. The components x_k of the eigenvector belonging to λ_m are

$$x_k = \alpha \left(\frac{c}{a}\right)^{\frac{k}{2}} \left(e^{\frac{\pi i m k}{N+1}} - e^{-\frac{\pi i m k}{N+1}}\right) = 2i\alpha \left(\frac{c}{a}\right)^{\frac{k}{2}} \sin\left(\frac{\pi m k}{N+1}\right)$$

The scaling constant α follows from the normalization $\|x\|_1 = 1$ or

$$2i\alpha \sum_{k=1}^{N} \left(\frac{c}{a}\right)^{\frac{k}{2}} \sin\left(\frac{\pi m k}{N+1}\right) = 1$$

Since $\sin\left(\frac{\pi m k}{N+1}\right) = \text{Im}\left[e^{\frac{\pi i m k}{N+1}}\right]$ we have

$$\sum_{k=1}^{N} \left(\frac{c}{a}\right)^{\frac{k}{2}} \sin\left(\frac{\pi m k}{N+1}\right) = \text{Im}\left[\sum_{k=1}^{N}\left(\sqrt{\frac{c}{a}}\,e^{\frac{\pi i m}{N+1}}\right)^k\right]$$

$$= \text{Im}\left[\frac{1-\left(\sqrt{\frac{c}{a}}\,e^{\frac{\pi i m}{N+1}}\right)^{N+1}}{1-\sqrt{\frac{c}{a}}\,e^{\frac{\pi i m}{N+1}}} - 1\right]$$

$$= \frac{\left(1+(-1)^m \left(\sqrt{\frac{c}{a}}\right)^{N+1}\right)\sin\left(\frac{\pi m}{N+1}\right)}{1-2\sqrt{\frac{c}{a}}\cos\left(\frac{\pi m}{N+1}\right)+\frac{c}{a}} - 1$$

from which the scaling constant α is

$$2i\alpha = \left[\frac{\left(1+(-1)^m \left(\sqrt{\frac{c}{a}}\right)^{N+1}\right)\sin\left(\frac{\pi m}{N+1}\right)}{1-2\sqrt{\frac{c}{a}}\cos\left(\frac{\pi m}{N+1}\right)+\frac{c}{a}} - 1\right]^{-1}$$

Finally, the components x_k of the eigenvector x belonging to λ_m become, for $1 \le k \le N$,

$$x_k = \frac{\left(\frac{c}{a}\right)^{\frac{k}{2}} \sin\left(\frac{\pi m k}{N+1}\right)}{\frac{\left(1+(-1)^m\left(\sqrt{\frac{c}{a}}\right)^{N+1}\right)\sin\left(\frac{\pi m}{N+1}\right)}{1-2\sqrt{\frac{c}{a}}\cos\left(\frac{\pi m}{N+1}\right)+\frac{c}{a}} - 1}$$

Observe that for stochastic matrices $a + b + c = 1$ (see the general random walk in Section 11.2) and for the infinitesimal rate matrix $a + b + c = 0$ (see the birth and death process in Section 11.3), which only changes the eigenvalue through b.

A.5.2.2 Tri-diagonal AMS matrix

This section computes the exact spectrum of the tri-diagonal AMS matrix specified in (14.51). The analysis bears some resemblance to that of the birth and dead process with constant birth and death rates in Section 11.3.3.

The eigenvalue equation $D^{-1}Qx = \zeta x$ is rewritten for the j-th component of the right-eigenvector belonging to the eigenvalue ζ as

$$\lambda(N - j + 1)\, x_{j-1} - [(\zeta + 1 - \lambda)\, j + N\lambda - \zeta c]\, x_j + (j + 1)\, x_{j+1} = 0$$

for $0 \leq j \leq N$. This difference equation has linear coefficients whereas those in Section A.5.2.1 are constant. It is most conveniently solved using generating functions. Let $G(z) = \sum_{j=0}^{N} x_j z^j$, then the difference equation is transformed with $x_j = 0$ if $j \notin [0, N]$ to

$$\lambda N z \left(G(z) - x_N z^N \right) - \lambda z^2 \left(G'(z) - N x_N z^{N-1} \right) - (\zeta + 1 - \lambda)\, z G'(z)$$
$$- [N\lambda - \zeta c] G(z) + G'(z) = 0$$

from which the logarithmic derivative is

$$\frac{G'(z)}{G(z)} = \frac{\lambda N z + \zeta c - N\lambda}{\lambda z^2 + (\zeta + 1 - \lambda)\, z - 1}$$

The integration of the right-hand side requires a partial faction decomposition,

$$\frac{\lambda N z + \zeta c - N\lambda}{\lambda z^2 + (\zeta + 1 - \lambda)\, z - 1} = \frac{c_1}{z - r_1} + \frac{c_2}{z - r_2}$$

where r_1 and r_2 are the roots of the quadratic polynomial $\lambda z^2 + (\zeta + 1 - \lambda)\, z - 1$ and c_1 and c_2 are the residues computed for $k = 1, 2$ as

$$c_k = \lim_{z \to r_k} \frac{(z - r_k)\,(\lambda N z + \zeta c - N\lambda)}{\lambda\,(z - r_1)\,(z - r_2)}$$

and they obey $c_1 + c_2 = N$ and $c_1 r_2 + c_2 r_2 = \frac{\zeta c - N\lambda}{\lambda}$. Explicitly,

$$r_1 = \frac{-(\zeta + 1 - \lambda) + \sqrt{(\zeta + 1 - \lambda)^2 + 4\lambda}}{2\lambda} > 0$$

$$r_2 = \frac{-(\zeta + 1 - \lambda) - \sqrt{(\zeta + 1 - \lambda)^2 + 4\lambda}}{2\lambda} < 0 \qquad \text{(A.42)}$$

with $r_1 r_2 = -\frac{1}{\lambda}$ and $r_1 + r_2 = -\frac{\zeta + 1 - \lambda}{\lambda}$. Moreover, unless $\zeta = \lambda - 1 \pm 2i\sqrt{\lambda}$

in which case $r_1 = r_2 = \pm\frac{i}{\sqrt{\lambda}}$, the roots are distinct. The residues are

$$c_1 = \frac{\lambda N r_1 + \zeta c - N\lambda}{\lambda (r_1 - r_2)} \tag{A.43}$$

$$c_2 = \frac{\lambda N r_2 + \zeta c - N\lambda}{\lambda (r_2 - r_1)} = N - c_1$$

Integration now yields

$$\log G(z) = c_1 \log (z - r_1) + c_2 \log (z - r_2) + b$$

or

$$G(z) = e^b (z - r_1)^{c_1} (z - r_2)^{N - c_1}$$

The integration constant b is obtained from $\lim_{z \to \infty} \frac{G(z)}{z^N} = x_N$. Thus,

$$\lim_{z \to \infty} \frac{G(z)}{z^N} = e^b \lim_{z \to \infty} \left(\frac{z - r_1}{z - r_2} \right)^{c_1} \left(1 - \frac{r_2}{z} \right)^N = e^b$$

such that $e^b = x_N$. The obvious scaling for the eigenvector is to choose $x_N = 1$ and we arrive at

$$G(z) = \sum_{j=0}^{N} x_j z^j = (z - r_1)^{c_1} (z - r_2)^{N - c_1} \tag{A.44}$$

which shows that c_1 must be an integer $k \in [0, N]$ for $G(z)$ to be a polynomial of degree N. Expanding the binomials with $c_1 = k$ gives

$$G(z) = \sum_{j=0}^{k} \binom{k}{j} z^j (-r_1)^{k-j} \sum_{n=0}^{N-k} \binom{N-k}{n} z^n (-r_2)^{N-k-n}$$

$$= \sum_{j=0}^{\infty} \sum_{n=0}^{j} \binom{k}{n} \binom{N-k}{j-n} (-r_1)^{k-j} (-r_2)^{N-k-j+n} z^j$$

from which the eigenvector components belonging to $\zeta \to \zeta(k)$ are, for $0 \le j \le N$,

$$x_j(k) = (-1)^{N-j} \sum_{n=0}^{j} \binom{k}{n} \binom{N-k}{j-n} r_1^{k-j} r_2^{N-k-j+n} \tag{A.45}$$

The requirement on c_1 also leads to equations for the eigenvalues ζ. Indeed, equating $c_1 = k$ in (A.43) and substituting the explicit expressions for the roots r_1 and r_2, we obtain after squaring the quadratic equations for the eigenvalue $\zeta(k)$ for $0 \le k \le N$

$$A(k) \zeta^2(k) + B(k) \zeta(k) + C(k) = 0 \tag{A.46}$$

where

$$A(k) = (N/2 - k)^2 - (N/2 - c)^2$$
$$B(k) = 2(1 - \lambda)(N/2 - k)^2 - N(1 + \lambda)(N/2 - c)$$
$$C(k) = -(1 + \lambda)^2 [(N/2)^2 - (N/2 - k)^2]$$

Each of the $N+1$ quadratic equations (A.46) has two roots $\zeta_1(k)$ and $\zeta_2(k)$, thus in total $2(N+1)$, while there are only $N+1$ eigenvalues. The coefficients $A(k)$, $B(k)$ and $C(k)$ only depend on k via $(N/2 - k)^2$, which means that the quadratics (A.46) for which $k' = N - k$ are identical. This observation reduces the set $\{\zeta_1(k), \zeta_2(k)\}_{0 \le k \le N}$ of roots to precisely $N+1$ and confines the analysis to $0 \le k \le N/2$. We will show that all roots are real and distinct (except for $k = N/2$).

The discriminant $\Delta(k) = B^2(k) - 4A(k)C(k)$ is with $y = (N/2 - k)^2 \in [0, (N/2)^2]$,

$$\Delta(k) = -16\lambda y^2 + 4(1 + \lambda)(c^2(1 + \lambda) - 2\lambda cN + \lambda N^2)y$$

which shows that $\Delta(k)$ is concave in y because $\frac{d^2\Delta(k)}{dy^2} = -32\lambda < 0$, for $y = 0$, $\Delta(N/2) = 0$ and, for $y = (N/2)^2$, $\Delta(0) = N^2(c(1 + \lambda) - \lambda N)^2 > 0$ and, hence, $\Delta(k) \ge 0$ for $k \in [0, N/2]$. This means that, for $0 \le k < N/2$, the roots $\zeta_1(k)$ and $\zeta_2(k)$ are real and distinct and, for $k = N/2$ (only if N is even) where $\Delta(N/2) = 0$,

$$\zeta_1(N/2) = \zeta_2(N/2) = \frac{-B(N/2)}{2A(N/2)} = -\frac{1 + \lambda}{1 - 2\frac{c}{N}}$$

For $\kappa < k \le N/2$, the roots $\{\zeta_1(\kappa), \zeta_2(\kappa)\}$ are different from the roots $\{\zeta_1(k), \zeta_2(k)\}$ because $A(k)z^2 + B(k)z + C(k) < A(\kappa)z^2 + B(\kappa)z + C(\kappa)$ for all z. Indeed, $A(\kappa) - A(k) = (N/2 - \kappa)^2 - (N/2 - k)^2 > 0$ and the discriminant $(B(\kappa) - B(k))^2 - 4(A(\kappa) - A(k))(C(\kappa) - C(k)) < 0$ shows that there are no real solutions. Thus, an extreme eigenvalue occurs for $k = 0$ for which $C(0) = 0$ such that $\zeta_1(0) = 0$ and

$$\zeta_2(0) = -\frac{B(0)}{A(0)} = -\frac{1 + \lambda - \frac{\lambda N}{c}}{1 - \frac{c}{N}} \tag{A.47}$$

The stability requirement $\rho = \frac{N\lambda}{c(1 + \lambda)} < 1$ and $c < N$ shows that $\zeta_2(0) < 0$, and thus $\zeta_2(0)$ is the largest negative eigenvalue. The eigenvalues for other $0 < k \le N/2$ are either larger than 0 or smaller than $\zeta_2(0)$. We need to consider two different cases (a) $c < N/2$ and (b) $c > N/2$ while $C(k) < 0$ for all $k \in [0, N)$.

(a) If $c < N/2$ and if $0 \le k < c$ and , then $A(k) > 0$. Hence, the product

$\zeta_1(k)\zeta_2(k) = \frac{C(k)}{A(k)} < 0$ which means that $\zeta_1(k) > 0 > \zeta_2(k)$ and that there are precisely $[c]$ positive eigenvalues. Similarly, $A(k) < 0$ for $c < k < N/2$, such that $\zeta_1(k)\zeta_2(k) > 0$ while $\zeta_1(k) + \zeta_2(k) = -\frac{B(k)}{A(k)} < 0$ shows that both eigenvalues are negative because $B(k) < 0$. Indeed, if $\lambda \geq 1$ and $c < N/2$, the above expression immediately leads to $B(k) < 0$ while if $\lambda < 1$ and $c < N/2$, the expression

$$B(k) = 2(1-\lambda)\left[\left(\frac{N}{2}-k\right)^2 - \left(\frac{N}{2}-c\right)^2\right] - 2c\left(\frac{N}{2}-c\right)\left[\frac{\lambda N}{c}+1-\lambda\right]$$

shows that both terms are negative.

(b) If $c > N/2$, we see that $A(k) > 0$ for $0 < k < N - c$ leading to $\zeta_1(k) > 0 > \zeta_2(k)$. For $N - c < k < N/2$, we have $A(k) < 0$ and thus $\zeta_1(k)\zeta_2(k) = \frac{C(k)}{A(k)} > 0$ while their same sign follows from $\zeta_1(k) + \zeta_2(k) = -\frac{B(k)}{A(k)}$ requires us to consider the sign of $B(k)$. If $\lambda \leq 1$, then $B(k) > 0$. If $\lambda > 1$, then

$$B(k) = N(1+\lambda)\left(c - \frac{N}{2}\right) + 2(1-\lambda)(N/2 - k)^2$$

$$< N(1+\lambda)\left(c - \frac{N}{2}\right) + 2(1-\lambda)\left(c - \frac{N}{2}\right)^2$$

$$= 2c\left(c - \frac{N}{2}\right)\left(\frac{\lambda(N-c)}{c} + 1\right) > 0$$

which shows that $0 < \zeta_2(k) < \zeta_1(k)$. Hence, there are $N - [c] + 2(N/2 - N + [c]) = [c]$ positive eigenvalues.

In summary, there are $[c]$ positive eigenvalues, one $\zeta_1(0) = 0$ and $N - [c]$ negative eigenvalues. Relabel the eigenvalues as $(\zeta_k, \zeta_{N-k}) = (\zeta_1(k), \zeta_2(k))$ in increasing order $\zeta_{N-[c]-1} < \cdots < \zeta_1 < \zeta_0 < \zeta_N = 0 < \zeta_{N-1} < \cdots < \zeta_{N-[c]}$. This way of writing distinguishes between underload and overload eigenvalues. In terms of the discriminant by $\Delta(k) = B^2(k) - 4A(k)C(k)$, the non-positive eigenvalues are

(a) If $c < N/2$,

$$\zeta_1(k) = \frac{-B(k) - \sqrt{\Delta(k)}}{2A(k)} \qquad 0 \leq k \leq [c]$$

$$\zeta_{1,2}(k) = \frac{-B(k) \mp \sqrt{\Delta(k)}}{2A(k)} \qquad [c] + 1 \leq k \leq \frac{N}{2}$$

(b) If $c > N/2$,

$$\zeta_1(k) = \frac{-B(k) - \sqrt{\Delta(k)}}{2A(k)} \qquad 0 \leq k \leq N - [c] - 1$$

The eigenvector belonging to ζ_j follows from (A.45) where r_1 and r_2 are given in (A.42) and k is determined from (A.43) since $k = c_1$. The eigenvectors for $\zeta_1(k)$ and $\zeta_2(k)$ belonging to a same quadratic k must be different. Especially in this case, the corresponding $k = c_1$ values can be determined from (A.43). For example, for $\zeta_N = 0$, we find $r_1 = 1$, $r_2 = -\frac{1}{\lambda}$ and $k = 0$ and the eigenvector belonging to ζ_N is with (A.45),

$$x_j(0) = (-1)^{N-j} \binom{N}{j} r_1^{-j} r_2^{N-j} = \binom{N}{j} \frac{\lambda^j}{\lambda^N} \qquad (A.48)$$

After renormalization such that $\|x(0)\|_1 = 1$, i.e. by dividing each component by $\sum_{j=0}^{N} x_j(0) = \frac{1}{\lambda^N} \sum_{j=0}^{N} \binom{N}{j} \lambda^j = \frac{(1+\lambda)^N}{\lambda^N}$, the steady-state vector (14.52) is obtained. Similarly, for the largest negative eigenvalue ζ_0 in (A.47), we find with $r_1 = 1 - \frac{N}{c}$, $r_2 = \frac{1}{\lambda(\frac{N}{c}-1)}$ and $k = c_1 = N$ such that

$$x_j(N) = (-1)^{N-j} \binom{N}{j} r_1^{N-j} r_2^0 = \binom{N}{j} \left(\frac{N}{c} - 1\right)^{N-j} \qquad (A.49)$$

The left-eigenvectors y satisfy (A.9): $y^T D^{-1} Q = \zeta y^T$. The above approach is applicable. However, there is a more elegant method based on the observation that there exists a diagonal matrix $W = diag(W_0, \ldots, W_N)$ for which $W^{-1}QW = (W^{-1}QW)^T$, namely $W_j = \sqrt{\binom{N}{j}\lambda^j}$. Since $W^{-1}QW$ is symmetric, the left- and right-eigenvectors corresponding to the same eigenvalue are the same (Section A.2, art. **9**). Now $y^T D^{-1} Q = \zeta y^T$ is equivalent to

$$\zeta y^T W = y^T W W^{-1} D^{-1} W W^{-1} Q W = y^T W \left(W^{-1}DW\right)^{-1} \left(W^{-1}QW\right)$$

With $y_W^T = y^T W$, $D_W = W^{-1}DW$ and $Q_W = W^{-1}QW = Q_W^T$, we obtain $y_W^T D_W^{-1} Q_W = \zeta y_W^T$. The transpose $\zeta y_W = Q_W D_W^{-1} y_W$ is

$$\zeta W^2 y = Q D^{-1} W^2 y$$

which shows compared to $D^{-1}Qx = \zeta x$ that $x = W^2 y$ or, the vector components are, for $0 \le j \le N$,

$$x_j = \binom{N}{j} \lambda^j y_j \qquad (A.50)$$

A.5.2.3 General tri-diagonal matrices

Since tri-diagonal matrices of the form (11.1) frequently occur in Markov theory, we devote this section to illustrate how far the eigen-analysis can be pushed. For an eigenpair (the right-eigenvector x belonging to eigenvalue λ), the components in $(P - \lambda I)x = 0$ satisfy

$$(r_0 - \lambda)\, x_0 + p_0 x_1 = 0$$
$$q_j x_{j-1} + (r_j - \lambda)\, x_j + p_j x_{j+1} = 0 \qquad\qquad 1 \le j < N$$
$$q_N x_{N-1} + (r_N - \lambda)\, x_N = 0$$

If $p_j = p$ and $q_j = q$, the matrix P reduces to a Toeplitz form for which the eigenvalues and eigenvectors can be explicitly written, as shown in Appendix A.5.2.1. Here, we consider the general case and show how orthogonal polynomials enter the scene.

Using $r_j = 1 - q_j - p_j$, $r_0 = 1 - p_0$ and $r_N = 1 - q_N$, the set becomes, with $\xi = \lambda - 1$,

$$x_1(\xi) = \frac{p_0 + \xi}{p_0} x_0(\xi)$$
$$x_{j+1}(\xi) = \frac{p_j + q_j + \xi}{p_j} x_j(\xi) - \frac{q_j}{p_j} x_{j-1}(\xi) \qquad\qquad 1 \le j < N \quad \text{(A.51)}$$
$$x_N(\xi) = \frac{q_N}{q_N + \xi} x_{N-1}(\xi)$$

The dependence on the eigenvalue ξ is made explicit. Solving (A.51) iteratively for $j < N$,

$$x_2(\xi) = \frac{x_0(\xi)}{p_0 p_1}\left(\xi^2 + (q_1 + p_1 + p_0)\,\xi + p_1 p_0\right)$$
$$x_3(\xi) = \frac{x_0(\xi)}{p_2 p_1 p_0}\left(\xi^3 + (q_1 + q_2 + p_2 + p_1 + p_0)\,\xi^2\right)$$
$$+ (q_2 q_1 + q_2 p_0 + p_2 q_1 + p_2 p_1 + p_2 p_0 + p_1 p_0)\,\xi + p_2 p_1 p_0$$

reveals a polynomial of degree j in the eigenvalue $\xi = \lambda - 1$. By inspection, the general form of $x_j(\xi)$ for $j < N$ is

$$x_j(\xi) = \frac{x_0(\xi)}{\prod_{m=0}^{j-1} p_m} \sum_{k=0}^{j} c_k(j)\xi^k \qquad\qquad \text{(A.52)}$$

with

$$c_j(j) = 1$$

$$c_{j-1}(j) = \sum_{m=0}^{j-1} (p_m + q_m)$$

$$c_0(j) = \prod_{m=0}^{j-1} p_m$$

where $q_0 = p_N = 0$. By substituting (A.52) into (A.51),

$$\sum_{k=1}^{j-1} c_k(j+1)\xi^k = \sum_{k=1}^{j-1} [(q_j + p_j)\,c_k(j) - q_j p_{j-1} c_k(j-1) + c_{k-1}(j)]\,\xi^k$$

and equating the corresponding powers in ξ, a recursion relation for the coefficients $c_k(j)$ $(0 \le k < N)$ is obtained with $c_j(j) = 1$,

$$c_k(j+1) = (q_j + p_j)\,c_k(j) - q_j p_{j-1} c_k(j-1) + c_{k-1}(j)$$

from which all coefficients can be determined. Finally, for $j = N$, the explicit form of $x_N(\xi)$ follows from (A.51) as

$$x_N(\xi) = \frac{q_N}{q_N + \xi} x_{N-1}(\xi) = \frac{q_N}{q_N + \xi} \frac{x_0(\xi)}{\prod_{m=0}^{N-2} p_m} \sum_{k=0}^{N-1} c_k(N-1)\xi^k$$

We can always scale an eigenvector without effecting the corresponding eigenvalue. If we require a normalization of the eigenvector $\|x(\xi)\|_1 = 1$, then $x_0(\xi)$ is uniquely determined,

$$x_0(\xi) = \frac{1}{1 + \sum_{j=1}^{N-1} \left| \sum_{k=0}^{j} \frac{c_k(j)}{\prod_{m=0}^{j-1} p_m} \xi^k \right| + \frac{q_N}{|q_N + \xi|} \left| \sum_{k=0}^{N-1} \frac{c_k(N-1)}{\prod_{m=0}^{N-2} p_m} \xi^k \right|}$$

Another scaling consists of choosing $x_0(\xi) = 1$. Hence, apart from the eigenvalue ξ, all eigenvector components $x_j(\xi)$ are explicitly determined.

If $\lambda = 1$ or $\xi = 0$, the solution is $x_j(\xi) = x_0(0)$. If $\|x(\xi)\|_1 = 1$, then $x_j(\xi) = \frac{1}{N+1}$, which is, after proper scaling by $N + 1$ (art. **4** in Section A.1), the right-eigenvector $u = \begin{bmatrix} 1 & 1 & \cdots & 1 \end{bmatrix}$ belonging to the left-eigenvector π (see also Section A.4.1). If $x_0(\xi) = 1$, we immediate obtain u. Eigenvectors belonging to different eigenvalues $\xi' \ne \xi$ are linearly independent (art. **3** in Section A.1), but only orthogonal if $P = P^T$, i.e. if $p_j = q_{j+1}$. Only in the latter case (art. **9** in Section A.2), where also all eigenvalues are real, we have

$$\sum_{j=0}^{N} x_j\left(\xi\right) x_j\left(\xi'\right) = \|x(\xi)\|_2^2\, \delta_{\xi\xi'}$$

This orthogonality requirement determines the different eigenvalues ξ. Since $\xi' = 0$ is an eigenvalue, each other real eigenvalue $\xi \neq 0$ must obey

$$\sum_{j=0}^{N} x_j\left(\xi\right) = 0$$

while the normalization enforces $\|x(\xi)\|_1 = \sum_{j=0}^{N} |x_j\left(\xi\right)| = 1$. The scaling $x_0\left(\xi\right) = 1$ leads to the polynomial $\sum_{j=0}^{N} b_k \xi^k$ of degree N whose N zeros equal the eigenvalues $\xi \neq 0$ and whose coefficients are, with $p_j = q_{j+1}$ and for $2 \leq k \leq N - 2$,

$$b_0 = (N+1)\, q_N$$

$$b_1 = N + q_N \sum_{j=1}^{N-2} \frac{c_1(j)}{\prod_{m=0}^{j-1} p_m} + 2q_N \frac{c_1(N-1)}{\prod_{m=0}^{N-2} p_m}$$

$$b_k = \sum_{j=k}^{N-2} \frac{q_N c_k(j)}{\prod_{m=0}^{j-1} p_m} + \frac{2 q_N c_k(N-1)}{\prod_{m=0}^{N-2} p_m} + \sum_{j=k-1}^{N-1} \frac{c_{k-1}(j)}{\prod_{m=0}^{j-1} p_m}$$

$$b_{N-1} = \frac{p_{N-2} + 2q_N + c_{N-2}(N-1)}{\prod_{m=0}^{N-2} p_m}$$

$$b_N = \frac{1}{\prod_{m=0}^{N-1} p_m}$$

The Newton identities (B.9) relate these coefficients to the sum of integer powers of the real zeros $\xi \neq 0$.

Proceeding much further in the case that P is not symmetric is difficult. A similarity transform is needed to transform the linearly independent set of vectors $x\left(\xi\right)$ for different ξ to an orthogonal set from which the eigenvalues then follow, as in the symmetric case above. Karlin and McGregor (see Schoutens (2000, Chapter 3)) have shown the existence of a set of orthogonal polynomials (similar to our set $x_j(\xi)$) that obey an integral orthogonality condition (similar to Legendre or Chebyshev polynomials) instead of our summation orthogonality condition. Only in particular cases, however, were they able to specify this orthogonal set explicitly.

A.5.3 A triangular matrix complemented with one subdiagonal

The transition probability matrix P has the structure of a triangular matrix complemented with one subdiagonal,

$$
P = \begin{bmatrix}
P_{00} & P_{01} & P_{02} & \cdots & \cdots & P_{0N} \\
P_{10} & P_{11} & P_{12} & \cdots & \cdots & P_{1N} \\
0 & P_{21} & P_{22} & \cdots & \cdots & P_{2N} \\
\vdots & \vdots & \vdots & \cdots & \vdots & \vdots \\
0 & 0 & \cdots & P_{N-1,N-2} & P_{N-1;N-1} & P_{N-1;N} \\
0 & 0 & \cdots & 0 & P_{N;N-1} & P_{NN}
\end{bmatrix}
$$

Besides the normalization $\|\pi\|_1 = 1$, the steady-state vector π obeys the relation $\pi = \pi.P$, or per vector component (9.23),

$$
\pi_j = \sum_{k=0}^{j+1} P_{kj}\pi_k
$$

because $P_{kj} = 0$ if $k > j + 1$. Immediately we obtain an iterative equation that expresses π_{j+1} (for $j < N$) in terms of the π_k for $0 \le k \le j$ as

$$
\pi_{j+1} = \left(\frac{1 - P_{jj}}{P_{j+1;j}}\right)\pi_j - \sum_{k=0}^{j-1}\frac{P_{kj}}{P_{j+1;j}}\pi_k
$$

Let us consider the eigenvalue equation (A.1) that is written for stochastic matrices as $(P - \lambda I)^T x^T = 0$. The matrix $(P - \lambda I)^T$ is a $(N+1) \times (N+1)$ matrix of rank N because $\det(P - \lambda I)^T = 0$ (else all eigenvectors x are zero). When writing this set of equations in terms of x_0, we produce the following set of N equations,

$$
\begin{bmatrix}
P_{10} & 0 & 0 & \cdots & 0 & 0 \\
P_{11} - \lambda & P_{21} & 0 & \cdots & 0 & 0 \\
P_{12} & P_{22} - \lambda & P_{32} & \cdots & \vdots & \vdots \\
\cdots & \cdots & \cdots & \ddots & \cdots & \cdots \\
\vdots & \vdots & \vdots & \cdots & P_{N-1;N-2} & 0 \\
P_{1;N-1} & P_{2;N-1} & P_{3;N-1} & \cdots & P_{N-1;N-1} - \lambda & P_{N;N-1}
\end{bmatrix}
\cdot
\begin{bmatrix}
x_1 \\ x_2 \\ x_3 \\ \vdots \\ x_{N-1} \\ x_N
\end{bmatrix}
=
\begin{bmatrix}
\lambda - P_{00} \\ -P_{01} \\ -P_{02} \\ \vdots \\ -P_{0;N-2} \\ -P_{0;N-1}
\end{bmatrix}
x_0
$$

Since the right hand side matrix is a triangular matrix, the determinant equals the product of the diagonal elements or $\prod_{k=0}^{N-1} P_{k+1;k}$. By Cramer's rule, we find that

$$
\frac{x_j}{x_0} = \frac{\det \begin{bmatrix} P_{10} & 0 & \cdots & 0 & \lambda - P_{00} & 0 & \cdots & 0 \\ P_{11} - \lambda & P_{21} & \cdots & 0 & -P_{01} & 0 & \cdots & 0 \\ P_{12} & P_{22} - \lambda & \ddots & \vdots & \vdots & \vdots & \cdots & \vdots \\ \vdots & \vdots & \cdots & P_{j-1;j-2} & -P_{0,j-2} & \vdots & \cdots & \vdots \\ \vdots & \vdots & \cdots & P_{j-1;j-1} - \lambda & -P_{0;j-1} & 0 & \cdots & \vdots \\ P_{1j} & P_{2j} & \cdots & P_{j-1;j} & -P_{0j} & P_{j+1;j} & \cdots & \vdots \\ \vdots & \vdots & \vdots & \vdots & \vdots & \vdots & \ddots & 0 \\ P_{1;N-1} & P_{2;N-1} & \cdots & P_{j-1;N-1} & -P_{0;N-1} & P_{j+1;N-1} & \cdots & P_{N;N-1} \end{bmatrix}}{\prod_{k=0}^{N-1} P_{k+1;k}}
$$

The above determinant is of the form (Meyer, 2000, p. 467)

$$
\det \begin{bmatrix} A_{j \times j} & O_{j \times N-j} \\ B_{N-j \times j} & C_{N-j \times N-j} \end{bmatrix} = \det C \det A
$$

where $\det C = \prod_{k=j}^{N-1} P_{k+1;k}$. In the determinant $\det A$, we can change the j-th column with the $(j-1)$-th, and subsequently, the $(j-1)$-th with the $(j-2)$-th and so on until the last column is permuted to the first column, in total $j-1$ permutations. After changing the sign of that first column, the result is that $\det A = (-1)^j \det (P_{j \times j} - \lambda I_{j \times j})$ where $P_{j \times j}$ is the original transition probability matrix limited to j states (instead of $N+1$). Hence, for $1 \le j \le N$,

$$
x_j = \frac{x_0 (-1)^j \det (P_{j \times j} - \lambda I_{j \times j})}{\prod_{k=0}^{j-1} P_{k+1;k}}
$$

and the normalization of eigenvectors $\|x\|_1 = 1$ determines x_0 as

$$
x_0 = \frac{1}{1 + \sum_{j=1}^{N} \frac{(-1)^j \det(P_{j \times j} - \lambda I_{j \times j})}{\prod_{k=0}^{j-1} P_{k+1;k}}}
$$

If the $N+1$ eigenvalues λ are known, we observe that all eigenvectors can be expressed in terms of the original matrix P in a same way.

Appendix B
Algebraic graph theory

This appendix reviews the elementary basics of the matrix theory for graphs $G(N, L)$. The book by Cvetkovic *et al.* (1995) is the current standard work on algebraic graph theory.

B.1 The adjacency and incidence matrix

1. The *adjacency matrix* A of a graph G with N nodes is an $N \times N$ matrix with elements $a_{ij} = 1$ only if (i, j) is a link of G, otherwise $a_{ij} = 0$. Because the existence of a link implies that $a_{ij} = a_{ji}$, the adjacency matrix $A = A^T$ is a real symmetric matrix. It is assumed further that the graph G does not contain self-loops ($a_{ii} = 0$) nor multiple links between two nodes. The complement G^c of the graph G consists of the same set of nodes but with a link between (i, j) if there is *no* link (i, j) in G and vice versa. Thus, $(G^c)^c = G$ and the adjacency matrix A^c of the complement G^c is $A^c = J - I - A$ where J is the all-one matrix ($(J)_{ij} = 1$). Information about the direction

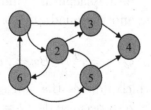

Fig. B.1. A graph with $N = 6$ and $L = 9$. The links are lexicographically ordered, $e_1 = 1 \rightarrow 2, e_2 = 1 \rightarrow 3, e_3 = 1 \longleftarrow 6$ etc.

of the links is specified by the *incidence matrix* B, an $N \times L$ matrix with elements

$$b_{ij} = \begin{cases} 1 \text{ if link } e_j = i \longrightarrow j \\ -1 \text{ if link } e_j = i \longleftarrow j \\ 0 \text{ otherwise} \end{cases}$$

Figure B.1 exemplifies the definition of A and B:

$$A = \begin{bmatrix} 0 & 1 & 1 & 0 & 0 & 1 \\ 1 & 0 & 1 & 0 & 1 & 1 \\ 1 & 1 & 0 & 1 & 0 & 0 \\ 0 & 0 & 1 & 0 & 1 & 0 \\ 0 & 1 & 0 & 1 & 0 & 1 \\ 1 & 1 & 0 & 0 & 1 & 0 \end{bmatrix} \quad B = \begin{bmatrix} 1 & 1 & -1 & 0 & 0 & 0 & 0 & 0 & 0 \\ -1 & 0 & 0 & 1 & -1 & -1 & 0 & 0 & 0 \\ 0 & -1 & 0 & -1 & 0 & 0 & 1 & 0 & 0 \\ 0 & 0 & 0 & 0 & 0 & 0 & -1 & -1 & 0 \\ 0 & 0 & 0 & 0 & 1 & 0 & 0 & 1 & -1 \\ 0 & 0 & 1 & 0 & 0 & 1 & 0 & 0 & 1 \end{bmatrix}$$

2. The relation between adjacency and incidence matrix is given by the *admittance* matrix or *Laplacian* Q,

$$Q = BB^T = \Delta - A$$

where $\Delta = \mathrm{diag}(d_1, d_2, \ldots, d_N)$ is the degree matrix. Indeed, if $i \neq j$ and noting that each column has only two non-zero elements at a different row,

$$q_{ij} = \left(BB^T\right)_{ij} = \sum_{k=1}^{L} b_{ik} b_{jk} = -1$$

If $i = j$, then $\sum_{k=1}^{L} b_{ik}^2 = d_i$, the number of links that have node i in common.

Also, by the definition of A, the row sum i of A equals the degree d_i of node i,

$$d_i = \sum_{k=1}^{N} a_{ik} \tag{B.1}$$

Consequently, each row sum $\sum_{k=1}^{N} q_{ik} = 0$ which shows that Q is singular implying that $\det Q = 0$. The Laplacian is symmetric $Q = Q^T$ because A and Δ are both symmetric and the quadratic form defined in Section A.2 art. **10**,

$$x^T Q x = x^T Q^T x = x^T B^T B x = \|Bx\|_2^2 \geq 0$$

is positive semidefinite, which implies that all eigenvalues of Q are non-negative and at least one is zero because $\det Q = 0$.

Since $\sum_{i=1}^{N} \sum_{k=1}^{N} a_{ik} = 2L$, the basic law for the degree follows as

$$\sum_{i=1}^{N} d_i = 2L \tag{B.2}$$

Notice that $P = \Delta^{-1}A$ is a stochastic matrix because all elements of P lie in the interval $[0, 1]$ and each row sum is 1.

3. Let J denote the all-one matrix with $(J)_{ij} = 1$ and $\xi(G)$ the total number of spanning trees in the graph G, also called the *complexity* of G, then

$$\text{adj}Q = \xi(G) J \tag{B.3}$$

where $Q^{-1} = \frac{\text{adj}Q}{\det Q}$. We omit the proof, but apply the relation (B.3) to the complete graph K_N where $Q = NI - J$. Equation (B.3) demonstrates that all elements of $\text{adj}Q$ are equal to $\xi(G)$. Hence, it suffices to compute one suitable element of $\text{adj}Q$, for example $(\text{adj}Q)_{11}$ that is equal to the determinant of the $(N-1) \times (N-1)$ principal submatrix of Q obtained by deleting the first row and column in Q,

$$(\text{adj}Q)_{11} = \det \begin{bmatrix} N-1 & -1 & \cdots & -1 \\ -1 & N-1 & \cdots & -1 \\ \vdots & & \ddots & \vdots \\ -1 & -1 & \cdots & N-1 \end{bmatrix}$$

Adding all rows to the first and subsequently adding this new first row to all other rows gives

$$(\text{adj}Q)_{11} = \det \begin{bmatrix} 1 & 1 & \cdots & 1 \\ -1 & N-1 & \cdots & -1 \\ \vdots & & \ddots & \vdots \\ -1 & -1 & \cdots & N-1 \end{bmatrix} = \det \begin{bmatrix} 1 & 1 & \cdots & 1 \\ 0 & N & \cdots & 0 \\ \vdots & & \ddots & \vdots \\ 0 & 0 & \cdots & N \end{bmatrix} = N^{N-2}$$

Hence, the total number of spanning trees in the complete graph K_N which is also the total number of possible spanning trees in any graph with N nodes equals N^{N-2}. This is a famous theorem of Cayley of which many proofs exist (van Lint and Wilson, 1996, Chapter 2).

4. The complexity of G is also given by

$$\xi(G) = \frac{\det(J+Q)}{N^2} \tag{B.4}$$

Indeed, observe that $JQ = (JB) B^T = 0$ since $JB = 0$. Hence,

$$(NI - J)(J + Q) = NJ + NQ - J^2 - JQ = NQ$$

and

$$\text{adj}((NI - J)(J + Q)) = \text{adj}(J + Q)\,\text{adj}(NI - J) = \text{adj}(NQ)$$

Since $Q_{K_N} = NI - J$ and as shown in art. **3**, $\mathrm{adj}(NI - J) = N^{N-2}J$ and since $\mathrm{adj}(NQ) = N^{N-1}\mathrm{adj}Q = N^{N-1}\xi(G)J$ where we have used (B.3),

$$\mathrm{adj}(J+Q)J = N\xi(G)J$$

Left-multiplication with $J+Q$ taking into account that $JQ = 0$ and $J^2 = NJ$ finally gives

$$(J+Q)\,\mathrm{adj}\,(J+Q)\,J = \det(J+Q)\,J = N^2\xi(G)\,J$$

which proves (B.4).

5. A walk of length k from node i to node j is a succession of k arcs of the form $(n_0 \to n_1)(n_1 \to n_2)\cdots(n_{k-1} \to n_k)$ where $n_0 = i$ and $n_k = j$. A path is a walk in which all vertices are different, i.e. $n_l \neq n_m$ for all $0 \leq l \neq m \leq k$.

Lemma B.1.1 *The number of walks of length k from node i to node j is equal to the element $\left(A^k\right)_{ij}$.*

Proof (by induction): For $k = 1$, the number of walks of length 1 between state i and j equals the number of direct links between i and j, which is by definition the element a_{ij} in the adjacency matrix A. Suppose the lemma holds for $k - 1$. A walk of length k consists of a walk of length $k - 1$ from i to some vertex r which is adjacent to j. By the induction hypothesis, the number of walks of length $k - 1$ from i to r is $\left(A^{k-1}\right)_{ir}$ and the number of walks with length 1 from r to j equals a_{rj}. The total number of walks from i to j with length k then equals $\sum_{r=1}^{N}\left(A^{k-1}\right)_{ir}a_{rj} = \left(A^k\right)_{ij}$ (by the rules of matrix multiplication). $\qquad\square$

Explicitly,

$$\left(A^k\right)_{ij} = \sum_{r_1=1}^{N}\sum_{r_2=1}^{N}\cdots\sum_{r_{k-1}=1}^{N} a_{ir_1}a_{r_1r_2}\cdots a_{r_{k-2}r_{k-1}}a_{r_{k-1}j}$$

As shown in Section 15.2, the number of paths with k hops between node i and node j is

$$X_k(i \to j; N) = \sum_{r_1\neq\{i,j\}}\sum_{r_2\neq\{i,r_1,j\}}\cdots\sum_{r_{k-1}\neq\{i,r_1,\ldots,r_{k-2},j\}} a_{ir_1}a_{r_1r_2}\cdots a_{r_{k-1}j}$$

The definition of a path restricts the first index r_1 to $N - 2$ possible values, the second r_2 to $N - 3$, etc.. such that the total possible number of paths is

$$\prod_{l=1}^{k-1}(N-1-l) = \frac{(N-2)!}{(N-k-1)!}$$

whereas the total possible number of walks clearly is N^{k-1}.

A graph is connected if, for each pair of nodes, there exists a walk or, equivalently, if there exists some integer $k > 0$ for which $\left(A^k\right)_{ij} \neq 0$ for each i, j. The lowest integer k for which $\left(A^k\right)_{ij} \neq 0$ for each pair of nodes i, j is called the *diameter* of the graph G. Lemma B.1.1 demonstrates that the diameter equals the length of the longest shortest hop path in G.

B.2 The eigenvalues of the adjacency matrix

In this section, only general results of the eigenvalue spectrum of a graph G are treated. For special types of graphs, there exists a wealth of additional, but specific properties of the eigenvalues.

1. Since A is a real symmetric matrix, it has N real eigenvalues (Section A.2), which we order as $\lambda_1 \geq \lambda_2 \geq \cdots \geq \lambda_N$. Section A.1, art. **4** shows that, apart from a similarity transform, the set of eigenvalues with corresponding eigenvectors is unique. A similarity transform consists of a relabeling of the nodes in the graph that obviously does not alter the structure of the graph but merely expresses the eigenvectors in a different base. The classical Perron-Frobenius Theorem for non-negative square matrices (of which Theorem A.4.2 is a special case) states that λ_1 is a simple and positive root of the characteristic polynomial in (A.3) possessing the only eigenvector of A with non-zero components. Moreover, it follows from (A.34) that

$$\lambda_1 = \sup_{x \neq 0} \frac{x^T A x}{x^T x} = \max_{x \neq 0} \frac{\sum_{i=1}^{N} \sum_{j=1}^{N} a_{ij} x_i x_j}{\sum_{i=1}^{N} x_i^2}$$

The maximum is attained if and only if x is the eigenvector of A belonging to λ_1 and for any other vector $y \neq x$, $\lambda_1 \geq \frac{x^T A x}{x^T x}$ as shown in Section A.3. By choosing the vector $y = u = (1, 1, \ldots, 1)$, we have, with $\sum_{j=1}^{N} a_{ij} = d_i$ and (B.2),

$$\lambda_1 \geq \frac{1}{N} \sum_{i=1}^{N} \sum_{j=1}^{N} a_{ij} = \frac{1}{N} \sum_{i=1}^{N} d_i = \frac{2L}{N} \tag{B.5}$$

The stochastic matrix $P = \Delta^{-1} A$ where $\Delta = \text{diag}(d_1, d_2, \ldots, d_N)$ is the degree matrix has the characteristic polynomial $\det\left(\Delta^{-1} A - \lambda I\right) = \frac{\det(A - \lambda \Delta)}{\det \Delta}$ where $\det \Delta = \prod_{j=1}^{N} d_j$. Since the largest eigenvalue of a stochastic matrix equals $\lambda_1 = 1$ (Theorem A.4.2), for a regular graph where $d_j = r$, the largest eigenvalue equals $\lambda_1 = r$.

2. Since $a_{ii} = 0$, we have that $\text{trace}(A) = 0$. From (A.7),

$$(-1)^{N-1} c_{N-1} = \sum_{k=1}^{N} \lambda_k = 0 \tag{B.6}$$

3. *The Newton identities for polynomials.* Let $p_n(z)$ denote a polynomial of order n defined as

$$p_n(z) = \sum_{k=0}^{n} a_k(n)\, z^k = a_n(n) \prod_{k=1}^{n} (z - z_k(n)) \tag{B.7}$$

where $\{z_k(n)\}$ are the n zeros. It follows from (B.7) that $p_n(0) = a_0(n) = a_n(n) \prod_{k=1}^{n}(-z_k(n))$. The logarithmic derivative of (B.7) is

$$p_n'(z) = p_n(z) \sum_{k=1}^{n} \frac{1}{z - z_k(n)}$$

For $z > \max_k z_k(n)$ (which is always possible for polynomials, but not for functions), we have that

$$p_n'(z) = p_n(z) \sum_{k=1}^{n} \sum_{j=0}^{\infty} \frac{(z_k(n))^j}{z^{j+1}} = p_n(z) \sum_{j=0}^{\infty} \frac{Z_j(n)}{z^{j+1}}$$

where

$$Z_j(n) = \sum_{k=1}^{n} (z_k(n))^j$$

Thus

$$\sum_{k=1}^{n} k a_k(n)\, z^k = \sum_{k=0}^{n} a_k(n)\, z^k \sum_{j=0}^{\infty} Z_j(n) z^{-j} = \sum_{j=0}^{\infty} \sum_{k=0}^{n} a_k(n)\, Z_j(n) z^{k-j} \tag{B.8}$$

Let $l = k - j$, then $-\infty \le l \le n$. Also $j = k - l \ge 0$ such that $k \ge l$. Combined with $0 \le k \le n$, we have $\max(0, l) \le k \le n$. Thus,

$$\sum_{j=0}^{\infty} \sum_{k=0}^{n} a_k(n)\, Z_j(n) z^{k-j} = \sum_{l=-\infty}^{n} \sum_{k=\max(l,0)}^{n} a_k(n)\, Z_{k-l}(n) z^l$$

$$= \sum_{l=-\infty}^{-1} \sum_{k=0}^{n} a_k(n)\, Z_{k-l}(n) z^l + \sum_{l=0}^{n} \sum_{k=l}^{n} a_k(n)\, Z_{k-l}(n) z^l$$

Equating the corresponding powers of z in (B.8) yields

$$a_0(n)Z_{-l}(n) + \sum_{k=1}^{n} a_k(n) Z_{k-l}(n) = 0 \qquad l \leq 0$$

$$\sum_{k=l+1}^{n} a_k(n) Z_{k-l}(n) = (l-n)a_l(n)$$

The last set of equations for $0 \leq l < n$,

$$a_l(n) = -\frac{1}{n-l} \sum_{k=l+1}^{n} a_k(n) Z_{k-l}(n) \qquad (B.9)$$

are the *Newton identities* that relate the coefficients of a polynomial to the sum of the positive powers of the zeros. Applied to the characteristic polynomials (A.3) and (A.5) of the adjacency matrix with $z_k(n) = \lambda_k$, $a_k(n) = (-1)^N c_k$ and $c_{N-1} = 0$ (from (B.6)) yields for the first few values,

$$c_{N-2} = -\frac{1}{2} \sum_{k=1}^{N} \lambda_k^2$$

$$c_{N-3} = -\frac{1}{3} \sum_{k=1}^{N} \lambda_k^3$$

$$c_{N-4} = \frac{1}{8} \left(\left(\sum_{k=1}^{N} \lambda_k^2 \right)^2 - 2 \sum_{k=1}^{N} \lambda_k^4 \right)$$

4. From (A.4), the coefficient of the characteristic polynomial $c_{N-2} = \sum_{all} M_2$. Each principal minor M_2 has a principal submatrix of the form $\begin{bmatrix} 0 & x \\ y & 0 \end{bmatrix}$ with $x, y \in [0, 1]$. A minor M_2 is non-zero if and only if $x = y = 1$ in which case $M_2 = -1$. For each set of adjacent nodes, there exists such non-zero minor, which implies that

$$c_{N-2} = -L$$

From art. **3**, it follows that the number of links L equals

$$L = \frac{1}{2} \sum_{k=1}^{N} \lambda_k^2 \qquad (B.10)$$

5. Each principal submatrix $M_{3\times 3}$ is of the form

$$M_{3\times 3} = \begin{bmatrix} 0 & x & z \\ x & 0 & y \\ z & y & 0 \end{bmatrix}$$

with determinant $M_3 = \det M_{3\times 3} = 2xyz$, which is only non-zero for $x = y = z$. That form of $M_{3\times 3}$ corresponds with a subgraph of 3 nodes that are fully connected. Hence, $c_{N-3} = -2\times$ the number of triangles in G. From art. **3**, it follows that

$$\text{the number of triangles in } G = \frac{1}{6}\sum_{k=1}^{N}\lambda_k^3 \tag{B.11}$$

6. In general, from (A.4) and by identifying the structure of a minor M_k, any coefficient c_{N-k} can be expressed in terms of graph characteristics,

$$(-1)^N c_{N-k} = \sum_{\mathcal{G}\in G_k} (-1)^{cycles(\mathcal{G})} \tag{B.12}$$

where G_k is the set of all subgraphs of G with exactly k nodes and *cycles* $(\mathcal{G})$ is the number of cycles in a subgraph $\mathcal{G} \in G_k$. The minor M_k is a determinant of the $M_{k\times k}$ submatrix of A and defined as

$$\det M_k = \sum_p (-1)^{\sigma(p)} a_{1p_1} a_{2p_2} \cdots a_{kp_k}$$

where the sum is over all $k!$ permutations $p = (p_1, p_2, \ldots, p_k)$ of $(1, 2, \ldots, k)$ and $\sigma(p)$ is the parity of p, i.e. the number of interchanges of $(1, 2, \ldots, k)$ to obtain $(p_1, p_2, \ldots, p_k)$. Only if all the links $(1, p_1), (2, p_2), \ldots (k, p_k)$ are contained in G, $a_{1p_1} a_{2p_2} \ldots a_{kp_k}$ is non-zero. Since $a_{jj} = 0$, the sequence of contributing links $(1, p_1), (2, p_2), \ldots (k, p_k)$ is a set of disjoint cycles and $\sigma(p)$ depends on the number of those disjoint cycles. Now, $\det M_k$ is constructed from a specific set $\mathcal{G} \in G_k$ of k out of N nodes and in total there are $\binom{N}{k}$ such sets in G_k. Combining all contributions leads to the expression (B.12).

7. Since A is a symmetric 0-1 matrix, we observe that using (B.1),

$$\left(A^2\right)_{ii} = \sum_{k=1}^{N} a_{ik} a_{ki} = \sum_{k=1}^{N} a_{ik}^2 = \sum_{k=1}^{N} a_{ik} = d_i$$

Hence, with (A.16) or (B.10), (A.7) and basic law for the degree (B.2) is expressed as

$$\text{trace}(A^2) = \sum_{k=1}^{N} \lambda_k^2 = \sum_{k=1}^{N} d_k = 2L \tag{B.13}$$

Furthermore,

$$\sum_{i=1}^{N}\sum_{j=1;j\neq i}^{N}\left(A^2\right)_{ij} = \sum_{i=1}^{N}\sum_{j=1;j\neq i}^{N}\sum_{k=1}^{N}a_{ik}a_{kj} = \sum_{k=1}^{N}\sum_{i=1}^{N}a_{ki}\sum_{j=1;j\neq i}^{N}a_{kj}$$

$$= \sum_{k=1}^{N}\sum_{i=1}^{N}a_{ki}\left(d_k - a_{ki}\right) = \sum_{k=1}^{N}\left(d_k\sum_{i=1}^{N}a_{ki} - \sum_{i=1}^{N}a_{ki}\right)$$

or

$$\sum_{i=1}^{N}\sum_{j=1;j\neq i}^{N}\left(A^2\right)_{ij} = \sum_{k=1}^{N}d_k\left(d_k - 1\right) \qquad (B.14)$$

Lemma B.1.1 states that $\sum_{i=1}^{N}\sum_{j=1;j\neq i}^{N}\left(A^2\right)_{ij}$ equals twice the total number of two-hop walks with different source and destination nodes. In other words, the total number of connected triplets of nodes in G equals half (B.14).

8. The total number N_k of walks of length k in a graph follows from Lemma B.1.1 as

$$N_k = \sum_{i=1}^{N}\sum_{j=1}^{N}(A^k)_{ij}$$

Since any real symmetric matrix (Section A.2, art. **9**) can be written as $A = U\text{diag}(\lambda_j)U^T$ where U is an orthogonal matrix of the (normalized) eigenvectors of A, we have that $A^k = U\text{diag}(\lambda_j^k)U^T$ and $(A^k)_{ij} = \sum_{n=1}^{N}u_{in}u_{jn}\lambda_n^k$. Hence,

$$N_k = \sum_{i=1}^{N}\sum_{j=1}^{N}\sum_{n=1}^{N}u_{in}u_{jn}\lambda_n^k = \sum_{n=1}^{N}\left(\sum_{i=1}^{N}u_{in}\right)^2\lambda_n^k$$

9. Applying the Hadamard inequality for the determinant of any matrix $C_{n\times n}$,

$$|\det C| \le \prod_{j=1}^{n}\left(\sum_{i=1}^{n}|c_{ij}|^2\right)^{\frac{1}{2}}$$

yields, with $a_{ij} = a_{ji}$ and (B.1),

$$|\det A| \le \prod_{j=1}^{N}\left(\sum_{i=1}^{N}a_{ji}^2\right)^{\frac{1}{2}} = \prod_{j=1}^{N}\left(\sum_{i=1}^{N}a_{ji}\right)^{\frac{1}{2}} = \prod_{j=1}^{N}\sqrt{d_j}$$

Hence, with (A.6),

$$(\det A)^2 = \prod_{k=1}^{N} \lambda_k^2 \le \prod_{j=1}^{N} d_j \tag{B.15}$$

10. Applying the Cauchy–Schwarz inequality (5.17)

$$\left(\sum_{k=1}^{n} a_k b_k\right)^2 \le \sum_{k=1}^{n} a_k^2 \sum_{k=1}^{n} b_k^2$$

to the vector $(\lambda_2, \ldots, \lambda_N)$ and the 1 vector $(1, 1, \ldots, 1)$ gives

$$\left(\sum_{k=2}^{N} \lambda_k\right)^2 \le (N-1)\sum_{k=2}^{N} \lambda_k^2$$

Introducing (B.6) and (B.13)

$$\lambda_1^2 \le (N-1)\left(2L - \lambda_1^2\right)$$

leads to the bound for the largest (and positive) eigenvalue λ_1,

$$\lambda_1 \le \sqrt{\frac{2L(N-1)}{N}} \tag{B.16}$$

Alternatively, in terms of the average degree $d_a = \frac{1}{N}\sum_{j=1}^{N} d_j = \frac{2L}{N}$, the largest eigenvalue λ_1 is bounded by the geometric mean of the average degree and the maximum possible degree, $\lambda_1 \le \sqrt{d_a(N-1)}$. Combining the lower bound (B.5) and upper bound (B.16) yields

$$\frac{2L}{N} \le \lambda_1 \le \sqrt{\frac{2L(N-1)}{N}} \tag{B.17}$$

11. From the inequality (A.26) for Hölder q-norms, we find that, if

$$\sum_{k=1}^{N} |\lambda_k|^q < \Lambda^q$$

then

$$\sum_{k=1}^{N} |\lambda_k|^p < \Lambda^p$$

for $p > q > 0$. Since $\sum_{k=1}^{N} \lambda_k = 0$, not all λ_k can be positive and combined with $\left|\sum_{k=1}^{N} \lambda_k^p\right| \le \sum_{k=1}^{N} |\lambda_k|^p$, we also have that $\left|\sum_{k=1}^{N} \lambda_k^p\right| < \Lambda^p$. Applied to the case where $q = 2$ and $p = 3$ gives the following implication: if

$\sum_{k=2}^{N} \lambda_k^2 < \lambda_1^2$ then $\left| \sum_{k=1}^{N} \lambda_k^3 \right| < \lambda_1^3$. In that case, the number of triangles given in (B.11) is

$$\text{the number of triangles in } G = \frac{1}{6}\lambda_1^3 + \frac{1}{6}\sum_{k=2}^{N} \lambda_k^3 \geq \frac{1}{6}\lambda_1^3 - \frac{1}{6}\left| \sum_{k=2}^{N} \lambda_k^3 \right| > 0$$

Hence, if $\sum_{k=2}^{N} \lambda_k^2 < \lambda_1^2$, then the number of triangles in G is at least one. Equivalently, in view of (B.10), if $\lambda_1 > \sqrt{L}$ then the graph G contains at least one triangle.

12. A Theorem of Turan states that

Theorem B.2.1 *A graph G with N nodes and more than $\left[\frac{N^2}{4}\right]$ links contains at least one triangle.*

This theorem is a consequence of art. **7** and **11**. For, using $L > \frac{N^2}{4} \geq \left[\frac{N^2}{4}\right]$ which is equivalent to $N < 2\sqrt{L}$ in the bound on the largest eigenvalue (B.5),

$$\lambda_1 \geq \frac{2L}{N} > \frac{2L}{2\sqrt{L}} = \sqrt{L}$$

and $\lambda_1 > \sqrt{L}$ is precisely the condition in art. **11** to have at least one triangle.

13. The eigenvalues of the complete graph K_N are $\lambda_1 = N - 1$ and $\lambda_2 = \ldots = \lambda_N = -1$. This follows by computing the determinant in (A.2) in the same way as in Section B.1, art. **3**. Alternatively, the adjacency matrix of the complete graph is $J - I$ and, if $u^T = [1 \ 1 \ \cdots \ 1]$ is the all-one vector, then $J = u.u^T$. A direct computation yields

$$\det (J - I - \lambda I) = \det \left(u.u^T - (\lambda + 1) I \right) = (-(\lambda+1))^N \det \left(I - \frac{u.u^T}{\lambda + 1} \right)$$

Using (A.38) and $u^T u = N$,

$$\det (J - I - \lambda I) = (-(\lambda + 1))^N \left(1 - \frac{N}{\lambda + 1} \right)$$
$$= (-1)^{N-1} (\lambda + 1)^{N-1} (\lambda + 1 - N)$$

gives the eigenvalues of K_N. Since the number of links in K_N is $L = \frac{N(N-1)}{2}$, we observe that the equality sign in (B.16) can occur. Since $L \leq \frac{N(N-1)}{2}$ for any graph, the upper bound (B.16) shows that $\lambda_1 \leq N - 1$ for any graph.

14. The difference between the largest eigenvalue λ_1 and second largest λ_2 is never larger than N, i.e.

$$\lambda_1 - \lambda_2 \leq N \tag{B.18}$$

Since $\lambda_1 > 0$ as indicated by (B.17), it follows from (B.6) that

$$0 = \sum_{k=1}^{N} \lambda_k \leq \lambda_1 + \sum_{k=2}^{N} |\lambda_k| \leq \lambda_1 + (N-1)\,|\lambda_2|$$

such that

$$\lambda_2 \geq -\frac{\lambda_1}{N-1}$$

Hence,

$$\lambda_1 - \lambda_2 \leq \lambda_1 + \frac{\lambda_1}{N-1} = \frac{N\lambda_1}{N-1}$$

Art. **13** states that the largest possible eigenvalue is $\lambda_1 = N - 1$ of the complete graph which proves (B.18). Again, the equality sign in (B.18) occurs in case of the complete graph.

15. *Regular graphs.* Every node j in a regular graph has the same degree $d_j = r$ and relation (B.1) indicates that each row sum of A equals r.

Theorem B.2.2 *The maximum degree $d_{\max} = \max_{1 \leq j \leq N} d_j$ is an eigenvalue of the adjacency matrix A of a connected graph G if and only if the corresponding graph is regular (i.e. $d_j = d_{\max}$ for all j).*

Proof: If x is an eigenvector of A belonging to eigenvalue $\lambda = d_{\max}$ so is each vector kx for each complex k (Section A.1, art. **1**). Thus, we can scale the eigenvector x such that the maximum component, say $x_m = 1$, and $x_k \leq 1$ for all k. The eigenvalue equation $Ax = d_{\max}x$ for that maximum component x_m is

$$d_{\max}x_m = d_{\max} = \sum_{j=1}^{N} a_{mj}x_j$$

which implies that all $x_j = 1$ whenever $a_{mj} = 1$, i.e. when the node j is adjacent to node m. Hence, the degree of node m is $d_m = d_{\max}$. For any node j adjacent to m for which the component $x_j = 1$, a same eigenvalue relation holds and thus $d_j = d_{\max}$. Proceeding this process shows that every node $k \in G$ has same degree $d_k = d_{\max}$ because G is connected. Hence, $x = u$ where $u^T = [1 \ 1 \ \cdots \ 1]$. Conversely, if G is connected and regular, then $\sum_{j=1}^{N} a_{mj} = d_{\max}$ for each m such that u is the eigenvector belonging to eigenvalue $\lambda = d_{\max}$, and the only possible eigenvector (as follows from art. **1**). Hence, there is only one eigenvalue $d_{\max}$. $\qquad\square$

16. The characteristic polynomial of the complement G^c is

$$\det\left(A^c - \lambda I\right) = \det\left(J - A - (\lambda+1)\,I\right)$$
$$= (-1)^N \det\left(\left(A + (\lambda+1)\,I\right)\left(I - \left(A + (\lambda+1)\,I\right)^{-1}J\right)\right)$$
$$= (-1)^N \det\left(\left(A + (\lambda+1)\,I\right)\right)\det\left(I - \left(A + (\lambda+1)\,I\right)^{-1}u.u^T\right)$$

where we have used that $J = u.u^T$ and u is the all-one vector. Similar to the proof of Lemma A.4.4, we find

$$\det\left(A^c - \lambda I\right) = (-1)^N g\left(\lambda\right)\det\left(A + (\lambda+1)\,I\right) \qquad (\text{B.19})$$

where

$$g\left(\lambda\right) = 1 - u^T\left(A + (\lambda+1)\,I\right)^{-1}u$$

In general, $g\left(\lambda\right)$ is not a simple function of λ although a little more is known. For example, $g\left(\lambda\right) = 1 - \left\|\left(A + (\lambda+1)\,I\right)^{-\frac{1}{2}}u\right\|_2^2$ which shows that $g\left(\lambda\right) \in (-\infty, 1]$. Unlike in the proof of Lemma A.4.4, u is generally not an eigenvector of A and we can write (Section A.1, art. **8**)

$$\left(A + (\lambda+1)\,I\right)^{-1} = \frac{1}{\lambda+1}\sum_{k=0}^{\infty}\frac{(-1)^k A^k}{(\lambda+1)^k}$$

where the last sum $\sum_{k=0}^{\infty}A^k z^k$ can be interpreted as the matrix generating function of the number of walks of length k (see Section B.1, art. **5** and art. **8**). Since $A^k = U\operatorname{diag}\left(\lambda_j^k\right)U^T$ (Section A.2, art. **9**) where the orthogonal matrix U consists of eigenvectors u_j of A, the matrix product $u^T U = \left[\begin{array}{cccc} u.u_1 & u.u_2 & \cdots & u.u_N\end{array}\right] = \sqrt{N}\left[\begin{array}{cccc}\cos\alpha_1 & \cos\alpha_2 & \cdots & \cos\alpha_N\end{array}\right]$ where α_j is the angle between the eigenvector u_j and the all-one vector u. Hence, $u^T A^k u = u^T U\operatorname{diag}\left(\lambda_j^k\right)U^T u = N\sum_{j=1}^{N}\lambda_j^k\cos^2\alpha_j$ and, with $\sum_{k=0}^{\infty}\frac{(-\lambda_j)^k}{(\lambda+1)^k} = \frac{\lambda+1}{\lambda+1+\lambda_j}$, we can write

$$g\left(\lambda\right) = 1 - N\sum_{j=1}^{N}\frac{\cos^2\alpha_j}{\lambda+1+\lambda_j}$$

With (A.5), we have $c\left(-\lambda - 1\right) = \det\left(A + (\lambda+1)\,I\right) = \prod_{k=1}^{N}\left(\lambda_k + 1 + \lambda\right)$

and, hence,

$$\det(A^c - \lambda I) = \frac{(-1)^N}{N} \sum_{j=1}^{N} \left(\lambda + 1 + \lambda_j - N^2 \cos^2 \alpha_j\right) \prod_{k=1; k \neq j}^{N} (\lambda_k + 1 + \lambda)$$

(B.20)

which shows that the poles of $g(\lambda)$ are precisely compensated by the zeros of the polynomial $c(-\lambda - 1)$. Thus, the eigenvalues of A^c are generally different from $\{-\lambda_j - 1\}_{1 \leq j \leq N}$ where λ_j is an eigenvalue of A. Only if u is an eigenvector of A corresponding with λ_k, then $g(\lambda) = \frac{\lambda + 1 + \lambda_k - N}{\lambda + 1 + \lambda_k}$ and all eigenvalues of A^c belong to the set $\{-\lambda_j - 1\}_{1 \leq j \neq k \leq N} \cup \{N - 1 - \lambda_k\}$. According to art. **15**, u is only an eigenvector when the graph is regular.

B.3 The stochastic matrix $P = \Delta^{-1}A$

The stochastic matrix $P = \Delta^{-1}A$, introduced in Section B.2, art. **1**, characterizes a random walk on a graph. A random walk is described by a finite Markov chain that is time-reversible. Alternatively, a time-reversible Markov chain can be viewed as random walk on an undirected graph. Random walks on graphs have many applications in different fields (see e.g. the survey by Lovász (1993)); perhaps, the most important application is randomly searching or sampling.

The combination of Markov theory and algebra leads to interesting properties of $P = \Delta^{-1}A$. Section 9.3.1 and A.4.1 show that the left-eigenvector of P belonging to eigenvalue $\lambda = 1$ is the steady-state vector π (which is a $1 \times N$ row vector) and that the corresponding right-eigenvector is the all-one vector u, which essentially follows from (9.8) and which indicates that, at each discrete time step, precisely one transition occurs. These eigenvectors obey the eigenvalue equations $P^T \pi^T = \pi^T$ and $Pu = u$ and the orthogonality relation $\pi u = 1$ (Section A.1, art. **3**). If $d = (d_1, d_2, \ldots, d_N)$ is the degree vector, then the basic law for the degree (B.2) is written in vector form as $d^T u = 2L$, or, $\left(\frac{d}{2L}\right)^T u = 1$. Theorem 9.3.5 states that the steady-state eigenvector π is unique such that the equations $\pi u = 1$ and $\left(\frac{d}{2L}\right)^T u = 1$ imply that the steady-state vector is

$$\pi = \left(\frac{d}{2L}\right)^T$$

or

$$\pi_j = \frac{d_j}{2L}$$

(B.21)

In general, the matrix P is not symmetric, but, after a similarity transform $H = \Delta^{1/2}$, a symmetric matrix $R = \Delta^{1/2}P\Delta^{-1/2} = \Delta^{-1/2}A\Delta^{-1/2}$ is obtained whose eigenvalues are the same as those of P (Section A.1, art. **4**). The powerful property (Section A.2, art. **9**) of symmetric matrices shows that all eigenvalues are real and that $R = U^T\mathrm{diag}(\lambda_R)\,U$, where the columns of the orthogonal matrix U consist of the normalized eigenvectors v_k that obey $v_j^T v_k = \delta_{jk}$. Explicitly written in terms of these eigenvectors gives

$$R = \sum_{k=1}^{N} \lambda_k v_k v_k^T$$

where, with Frobenius Theorem A.4.2, the real eigenvalues are ordered as $1 = \lambda_1 \geq \lambda_2 \geq \cdots \geq \lambda_N \geq -1$. If we exclude bipartite graphs (where the set of nodes is $\mathcal{N} = \mathcal{N}_1 \cup \mathcal{N}_2$ with $\mathcal{N}_1 \cap \mathcal{N}_2 = \varnothing$ and where each link connects a node in $\mathcal{N}_1$ and in $\mathcal{N}_2$) or reducible Markov chains (Section A.4), then $|\lambda_k| < 1$, for $k > 1$. Section A.1, art. **4** shows that the similarity transform $H = \Delta^{1/2}$ maps the steady state vector π into $v_1 = H^{-1}\pi^T$ and, with (B.21),

$$v_1 = \frac{\Delta^{-1/2}\pi^T}{\left\|\Delta^{-1/2}\pi^T\right\|_2}$$

or

$$v_{1j} = \frac{\frac{\sqrt{d_j}}{2L}}{\sqrt{\sum_{j=1}^{N}\left(\frac{\sqrt{d_j}}{2L}\right)^2}} = \sqrt{\frac{d_j}{2L}} = \sqrt{\pi_j}$$

Finally, since $P = \Delta^{-1/2}R\Delta^{1/2}$, the spectral decomposition of the transition probability matrix of a random walk on a graph with adjacency matrix A is

$$P = \sum_{k=1}^{N} \lambda_k \Delta^{-1/2}v_k v_k^T \Delta^{1/2} = u\pi + \sum_{k=2}^{N} \lambda_k \Delta^{-1/2}v_k v_k^T \Delta^{1/2}$$

The n-step transition probability (9.10) is, with $\left(v_k v_k^T\right)_{ij} = v_{ki}v_{kj}$ and (B.21),

$$P_{ij}^n = \frac{d_j}{2L} + \sqrt{\frac{d_j}{d_i}} \sum_{k=2}^{N} \lambda_k^n v_{ki} v_{kj}$$

The convergence towards the steady state π_j can be estimated from

$$\left| P_{ij}^n - \pi_j \right| \leq \sqrt{\frac{d_j}{d_i}} \sum_{k=2}^{N} |\lambda_k^n| \, |v_{ki}| \, |v_{kj}| < \sqrt{\frac{d_j}{d_i}} \sum_{k=2}^{N} |\lambda_k^n|$$

Denoting by $\xi = \max(|\lambda_2|, |\lambda_N|)$ and by ξ_0 the largest element of the reduced set $\{|\lambda_k|\} \setminus \{\xi\}$ with $2 \le k \le N$, we obtain

$$\left|P_{ij}^n - \pi_j\right| < \sqrt{\frac{d_j}{d_i}} \xi^n + O\left(\xi_0^n\right)$$

B.4 Eigenvalues and connectivity

A graph G has k components (or clusters) if there exists a relabeling of the nodes such that the adjacency matrix has the structure

$$A = \begin{bmatrix} A_1 & O & \cdots & O \\ O & A_2 & & \vdots \\ \vdots & & \ddots & \\ O & & \cdots & A_k \end{bmatrix}$$

where the square submatrix A_m is the adjacency matrix of the connected component m. Disconnectivity is a special case of reducibility of a stochastic matrix defined in Section A.4 and expresses that no communication is possible between two states in a different component or cluster. Using (A.39) indicates that

$$\det(A - \lambda I) = \prod_{m=1}^{k} \det(A_m - \lambda_m I) \tag{B.22}$$

If A is a regular graph with degree r, so is each submatrix A_m. Since A_m is connected, Section B.2, art. **15** states that the largest eigenvalue of any A_m equals r. Hence, by (B.22), the multiplicity of the largest eigenvalue of A equals the number of components in the regular graph.

As shown in Section B.1, art. **2**, the Laplacian Q has non-negative eigenvalues of which at least one equals zero. In addition, the matrix

$$(N-1)I - Q = (N-1)I - \Delta + A$$

is non-negative with constant row sums all equal to $N - 1$. Although the matrix $(N-1)I - Q$ is not an adjacency matrix and does not represent a regular graph, the main argument in the proof of Theorem B.2.2 is the property of constant row sums and non-negative matrix elements. Hence, the multiplicity of the largest eigenvalue of $(N-1)I - Q$ is equal to the number of components of G. But the largest eigenvalue of $(N-1)I - Q$ is the smallest of $Q - (N-1)I$ and also of Q. Hence, we have proved

Theorem B.4.1 *The multiplicity of the smallest eigenvalue* $\lambda = 0$ *of the Laplacian* Q *is equal to the number of components in the graph* G

If Q has only 1 zero eigenvalue with corresponding eigenvector u (because $\sum_{k=1}^{N} q_{ik} = 0$ for each $1 \leq i \leq N$ is, in vector notation, $Qu = 0$), then the graph is connected; it has only 1 component. Theorem B.4.1 also implies that, if the second smallest eigenvalue $\mu_Q = \lambda_{N-1}^{(Q)}$ of Q is zero, the graph G is disconnected. Since all eigenvectors of a matrix are linearly independent, the eigenvector x_Q of μ_Q must satisfy $x_Q^T u = 0$ since u is the eigenvector belonging to $\lambda = 0$. By requiring this additional constraint and choosing the scaling of the eigenvector such that $x^T x = 1$, we obtain similar to (A.35) that

$$\mu_Q = \min_{\|x\|_2^2 = 1 \text{ and } x^T u = 0} x^T Q x$$

The second smallest eigenvalue μ_Q has many interesting properties that characterize how strongly a graph G is connected. It is interesting to mention the inequality (Cvetkovic *et al.*, 1995, p. 265)

$$\kappa(G) \geq \mu_Q \geq 2\lambda(G) \left(1 - \cos \frac{\pi}{N}\right) \tag{B.23}$$

where $\kappa(G)$ and $\lambda(G)$ are the vertex and edge connectivity respectively.

B.5 Random matrix theory

Random matrix theory investigates the eigenvalues of an $N \times N$ matrix A whose elements a_{ij} are random variables with a given joint distribution. Even in case all elements a_{ij} are independent, there does not exist a general expression for the distribution of the eigenvalues. However, in some particular cases (such as Gaussian elements a_{ij}), there exist nice results. Moreover, if the elements a_{ij} are properly scaled, in various cases the spectrum in the limit $N \to \infty$ seems to converge rapidly to a deterministic limit distribution. The fascinating results of random matrix theory and applications from nuclear physics to the distributions of the non-trivial zeros of the Riemann Zeta function are discussed by Mehta (1991).

Random matrix theory immediately applies to the adjacency matrix of the random graph $G_p(N)$ where each element a_{ij} is 1 with probability p and zero with probability $1 - p$.

B.5.1 The spectrum of the random graph $G_p(N)$

Let λ denote an arbitrary eigenvalue of the adjacency matrix of the random graph $G_p(N)$. Clearly, λ is a random variable with mean $E[\lambda] = \frac{1}{N}\sum_{k=1}^{N}\lambda_k = 0$ because of (B.6). In addition, the variance $\text{Var}[\lambda] = E[\lambda^2] = \frac{1}{N}\sum_{k=1}^{N}\lambda_k^2$ and from (B.10)

$$\text{Var}[\lambda] = \frac{2L}{N} = p(N-1)$$

This results implies that, for fixed p and large N, the eigenvalues of $G_p(N)$ grow as $O\left(\sqrt{N}\right)$, with the exception[1] of the largest eigenvalue λ_1.

The number of links L in $G_p(N)$ is binomially distributed with mean $E[L] = p\frac{N(N-1)}{2}$. Taking the expectation of the bounds (B.17) on the largest eigenvalue gives

$$\frac{2}{N}E[L] \le E[\lambda_1] \le \sqrt{\frac{2(N-1)}{N}}E\left[\sqrt{L}\right]$$

Using (2.12) yields

$$E\left[\sqrt{L}\right] = \sum_{k=0}^{\binom{N}{2}}\sqrt{k}\,\text{Pr}[L=k] = \sum_{k=0}^{\binom{N}{2}}\binom{\binom{N}{2}}{k}\sqrt{k}p^k(1-p)^{\binom{N}{2}-k}$$

Unfortunately, the sum cannot be expressed in closed form, but

$$\frac{1}{\sqrt{\binom{N}{2}}}\sum_{k=0}^{\binom{N}{2}}\binom{\binom{N}{2}}{k}\sqrt{k}p^k(1-p)^{\binom{N}{2}-k} \le \sqrt{p}$$

with equality for $N \to \infty$. In summary, for any N and p,

$$p(N-1) \le E[\lambda_1] \le \sqrt{p}(N-1) \tag{B.24}$$

The degree distribution (15.11) of the random graph is a binomial distribution with mean $E[D_{\text{rg}}] = p(N-1)$ and $\text{Var}[D_{\text{rg}}] = (N-1)p(1-p)$. The inequality (5.13) indicates that the degree D_{rg} converges exponentially fast to zero the mean $E[D_{rg}]$ for fixed p and large N, which means that the random graphs tends to a regular graph with high probability. Section B.2, art. **1** states that $\lambda_1 \to p(N-1)$ with high probability. Comparison with the bounds (B.24) indicates that the upper bound is less tight than the lower bound and that the upper bound is only sharp when $p \to 1$, i.e. for the complete graph. Section B.2, art. **13** shows that only for the complete graph the upper bound is indeed exactly attained.

---

[1] It is known that, for large N, the second largest eigenvalue of $G_p(N)$ grows as $O\left(N^{\frac{1}{2}+\epsilon}\right)$.

B.5.2 Wigner's Semicircle Law

Wigner's Semicircle Law is the fundamental result in the spectral theory of large random matrices.

Theorem B.5.1 (Wigner's Semicircle Law) *Let A be a random $N \times N$ real symmetric matrix with independent and identically distributed elements a_{ij} with $\sigma^2 = Var[a_{ij}]$ and denote by $\lambda^{(A_N)}$ an eigenvalue of the set of the N real eigenvalues of the scaled matrix $A_N = \frac{A}{\sqrt{N}}$. The probability density function $f_{\lambda^{(A_N)}}(x)$ of $\lambda^{(A_N)}$ tends for $N \to \infty$ to*

$$\lim_{N \to \infty} f_{\lambda^{(A_N)}}(x) = \frac{1}{2\pi\sigma^2}\sqrt{4\sigma^2 - x^2} \, 1_{|x| \leq 2\sigma} \qquad (B.25)$$

Since Wigner's first proof (Wigner, 1955) of this Theorem and his subsequent generalizations (Wigner, 1957, 1958) many proofs have been published. However, none of them is short and easy enough to include here. Wigner's Semicircle Law illustrates that, for sufficiently large N, the distribution of the eigenvalues of $\frac{A}{\sqrt{N}}$ does not depend anymore on the probability distribution of the elements a_{ij}. Hence, Wigner's Semicircle Law exhibits a universal property of a class of large, real symmetric matrices with independent random elements. Mehta (1991) suspects that, for a much broader class of large random matrices, a mysterious yet unknown law of large numbers must be hidden. The scaling of A by $\frac{1}{\sqrt{N}}$ can be understood from the previous Section B.5.1. The adjacency matrix of the random graph satisfies the conditions in Theorem B.5.1 with $\sigma^2 = p(1-p)$ and its eigenvalues (apart from the largest) grow as $O\left(\sqrt{N}\right)$. In order to obtain the finite limit distribution (B.25) scaling by $\frac{1}{\sqrt{N}}$ is necessary.

The spectrum of $G_p(50)$ together with the properly rescaled Wigner's Semicircle Law (B.25) is plotted in Fig. B.2. Already for this small value of N, we observe that Wigner's Semicircle Law is a reasonable approximation for the intermediate p-region. The largest eigenvalue λ_1 for finite N, which is distributed around $p(N-1)$ as demonstrated above and shown in Fig. B.2 but which is not incorporated in Wigner's Semicircle Law, influences the average $E[\lambda] = \frac{1}{N}\sum_{k=1}^{N}\lambda_k = 0$ and causes the major bulk of the pdf around $x = 0$ to shift leftward compared to Wigner's Semicircle Law, which is perfectly centered around $x = 0$.

The complement of $G_p(N)$ is $(G_p(N))^c = G_{1-p}(N)$, because a link in $G_p(N)$ is present with probability p and absent with probability $1-p$ and $(G_p(N))^c$ is also a random graph. For large N, there exists a large range of p values for which both $p \geq p_c$ and $1 - p \geq p_c$ such that both $G_p(N)$

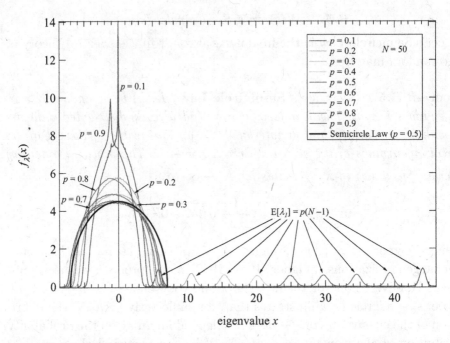

Fig. B.2. The probability density function of an eigenvalue in $G_p(50)$ for various p. Wigner's Semicircle Law, rescaled and for $p = 0.5$ ($\sigma^2 = \frac{1}{4}$), is shown in bold. We observe that the spectrum for p and $1 - p$ is similar, but slightly shifted. The high peak for $p = 0.1$ reflect disconnectivity, while the high peak at $p = 0.9$ shows the tendency to the spectrum of the complete graph where $N - 1$ eigenvalues are precisely -1.

and $(G_p(N))^c$ are connected almost surely. Figure B.2 shows that the normalized spectra of $G_p(N)$ and $G_{1-p}(N)$ are, apart from a small shift and ignoring the largest eigenvalue, almost identical. Equation (B.20) indicates that the spectrum of a graph and its complement tends to each other if $\cos \alpha_j \to 0$ (except for the largest eigenvalue which will tend to u). This seems to suggest that $G_p(N)$ and $G_{1-p}(N)$ are tending to a regular graph with degree $p(N-1)$ and $(1-p)(N-1)$ and that these regular graphs (even for small N) have nearly the same spectrum (apart from the largest eigenvalue $p(N-1)$ and $(1-p)(N-1)$ respectively): $\frac{\lambda_{1-p}}{\sqrt{N}} \simeq -\frac{\lambda_p}{\sqrt{N}} - \frac{1}{\sqrt{N}}$ where λ_p is an eigenvalue of $G_p(N)$.

Figure B.3 shows the probability density function $f_\lambda(x)$ of the eigenvalues of the adjacency matrix A of $G_p(N)$ with $N = 100$ together with the eigenvalues of the corresponding matrix A_U where all one elements in the adjacency matrix of $G_p(100)$ are replaced by i.i.d uniform random variables on $[0,1]$. Wigner's Semicircle Law provides an already better approximation

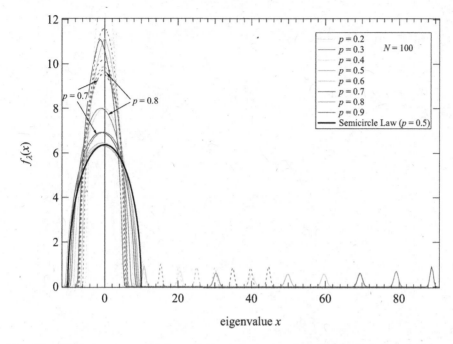

Fig. B.3. The spectrum of the adjacency matrix of $G_p(100)$ (full lines) and of the corresponding matrix with i.i.d. uniform elements (dotted lines). The small peaks at higher values of x are due to λ_1.

than for $N = 50$. Since the elements of A_U are always smaller (with probability 1) than those of A, the matrix norm $\|A_U\|_2 < \|A\|_2$, which implies by Section B.2, art. **1** that $\lambda_1(A_U) < \lambda_1(A)$. In addition, relation (B.13) shows that $\sum_{k=1}^{N} \lambda_k^2(A_U) < 2L$ such that $\text{Var}[\lambda(A_U)] < \text{Var}[\lambda(A)]$, which is manifested by a narrower and higher peaked pdf centered around $x = 0$.

Appendix C
Solutions of problems

C.1 Probability theory (Chapter 2)

(i) Using the general formula (2.12) for a non-zero random variable X, we have

$$E\left[\log X\right] = \sum_{k=1}^{\infty} \log k \Pr\left[X = k\right]$$

while (2.18), $\varphi_X(z) = \sum_{k=0}^{\infty} \Pr\left[X = k\right] z^k$, shows that we need to express $\log k$ in terms of z^k. A possible solution starts from the double integral with $0 < a \le b$,

$$\int_a^b dx \int_0^\infty dt e^{-tx} = \int_0^\infty dt \int_a^b dx e^{-tx}$$

where the reversal of integration is justified by absolute convergence (Titchmarsh, 1964, Section 1.8). Since $\int_0^\infty dt e^{-tx} = \frac{1}{x}$, the left-hand side integral equals

$$\int_a^b dx \int_0^\infty dt e^{-tx} = \log \frac{b}{a}$$

while the integral at the right hand side is

$$\int_0^\infty dt \int_a^b dx e^{-tx} = \int_0^\infty \frac{e^{-ta} - e^{-tb}}{t} dt$$

hence,

$$\log k = \int_0^\infty \frac{e^{-t} - e^{-tk}}{t} dt$$

Multiplying both sides by $\Pr\left[X = k\right]$ and summing over k, we obtain (reversal in operators is justified on absolute convergence),

$$\sum_{k=1}^{\infty} \log k \Pr\left[X = k\right] = \int_0^\infty \frac{dt}{t} \sum_{k=1}^{\infty} \left(e^{-t} - e^{-tk}\right) \Pr\left[X = k\right]$$

$$= \int_0^\infty \frac{dt}{t} \left(e^{-t} \sum_{k=1}^{\infty} \Pr\left[X = k\right] - \sum_{k=1}^{\infty} e^{-tk} \Pr\left[X = k\right]\right)$$

which finally gives with (2.18)

$$E\left[\log X\right] = \int_0^\infty \frac{e^{-t} - \varphi_X(e^{-t}) + \varphi_X(0)}{t} dt$$

493

(ii) (a) The pdf of the k-th smallest order statistic follows from (3.36) for an exponential distribution as

$$f_{X_{(k)}}(x) = m\alpha\binom{m-1}{k-1}\left(1 - e^{-\alpha x}\right)^{k-1} e^{-\alpha(m-k+1)x}$$

The probability generating function (2.37) is

$$\varphi_{X_{(k)}}(z) = E\left[e^{-zX_{(k)}}\right] = m\alpha\binom{m-1}{k-1}\int_0^\infty \left(1 - e^{-\alpha t}\right)^{k-1} e^{-(z+\alpha(m-k+1))t} dt$$

Let $u = e^{-\alpha t}$ and $\beta = z + \alpha(m-k+1)$, then the integral reduces to the well-known Beta function (Abramowitz and Stegun, 1968, Section 6.2.)

$$\int_0^\infty \left(1 - e^{-\alpha t}\right)^{k-1} e^{-\beta t} dt = \frac{1}{\alpha}\int_0^1 (1-u)^{k-1} u^{\beta/\alpha-1} dt = \frac{1}{\alpha}B\left(k, \beta/\alpha\right)$$

$$= \frac{1}{\alpha}\frac{\Gamma(k)\Gamma(\beta/\alpha)}{\Gamma(k+\beta/\alpha)}$$

Hence,

$$\varphi_{X_{(k)}}(z) = \frac{m!}{(m-k)!}\frac{\Gamma\left(\frac{z}{\alpha}+m+1-k\right)}{\Gamma\left(\frac{z}{\alpha}+m+1\right)} = \frac{m!}{(m-k)!}\prod_{j=0}^{k-1}\frac{1}{\frac{z}{\alpha}+m-j}$$

The mean follows from $E\left[X_{(k)}\right] = -L'_{w_{(k)}}(0)$ where L_X is the logarithm of the generating function (2.41) as

$$E\left[X_{(k)}\right] = \frac{1}{\alpha}\sum_{j=0}^{k-1}\frac{1}{m-j} \tag{C.1}$$

(b) For a polynomial probability density function $f_X(x) = \alpha x^{\alpha-1}1_{x\in[0,1]}$ with $\alpha > 0$, we have with (3.36) for $x \in [0,1]$ that

$$f_{X_{(k)}}(x) = \alpha m\binom{m-1}{k-1}x^{\alpha k-1}(1-x^\alpha)^{m-k}$$

with mean

$$E\left[X_{(k)}\right] = \alpha m\binom{m-1}{k-1}\int_0^1 x^{\alpha k}(1-x^\alpha)^{m-k}\,dx$$

$$= m\binom{m-1}{k-1}\int_0^1 t^{k+\frac{1}{\alpha}-1}(1-t)^{m-k}\,dt = \frac{m!}{(k-1)!}\frac{\Gamma\left(k+\frac{1}{\alpha}\right)}{\Gamma\left(m+1+\frac{1}{\alpha}\right)}$$

If $\alpha \to \infty$, then $E\left[X_{(k)}\right] = 1$, while for $\alpha \to 0$, $E\left[X_{(k)}\right] = 0$. For a uniform distribution where $\alpha = 1$, the result is $E\left[X_{(k)}\right] = \frac{k}{m+1}$. Indeed, the m independently chosen uniform random variables divide, after ordering, the line segment $[0,1]$ into $m+1$ subintervals. The length L of each subinterval has a same distribution, which more easily follows by symmetry if the line segment is replaced by a circle of unit perimeter. Since the length L of each subinterval is equal in distribution, one can consider the first subinterval $[0, X_{(1)}]$ whose length L exceeds a value $x \in (0,1)$ if and only if all m uniform random variables belong to $[x, 1]$. The latter event has probability equal to $(1-x)^m$ such that $\Pr[L > x] = (1-x)^m$ and, with (2.35), $E[L] = \frac{1}{m+1}$.

(iii) If X were a discrete random variable, then $\Pr[X = k] \approx \frac{n_k}{n}$, where n_k is the number of values in the set $\{x_1, x_2, \ldots, x_n\}$ that is equal to k. For a continuous random variable X, the values are generally real numbers ranging from $x_{\min} = \min_{1 \le j \le n} x_j$ until $x_{\max} = \max_{1 \le j \le n} x_j$. We first construct a histogram H from the set $\{x_1, x_2, \ldots, x_n\}$ by choosing a bin size $\Delta x = \frac{x_{\max} - x_{\min}}{m}$, where m is the number of bins (abscissa points). The choice of $1 < m < n$ is in general difficult to determine. However, most computer packages allow us to experiment with m and the human eye proves sensitive enough to make a good choice of m: if m is too small, we loose details, while a high m may lead to high irregularities due to the stochastic nature of X. Once m is chosen, the histogram consists of the set $\{h_0, h_1, \ldots, h_{m-1}\}$ where h_j equals the number of X values in the set $\{x_1, x_2, \ldots, x_n\}$ that lies in the interval $[x_{\min} + j\Delta x, x_{\min} + (j + 1)\Delta x]$ for $0 \le j \le m - 1$. By construction, $\sum_{j=0}^{m-1} h_j = n$.

The histogram H approximates the probability density function $f_X(x)$ after dividing each value h_j by $n\Delta x$ because

$$1 = \int_{x_{\min}}^{x_{\max}} f_X(x)dx = \lim_{\Delta x \to 0} \sum_{j=0}^{\frac{x_{\max}-x_{\min}}{\Delta x}-1} f_X(\xi_j)\, \Delta x \approx \sum_{j=0}^{m-1} \frac{h_j}{n\Delta x}\Delta x = 1$$

where in the Riemann sum ξ_j denotes a real number $\xi_j \in [x_{\min} + j\Delta x, x_{\min} + (j + 1)\Delta x]$. Alternatively from (2.31) we obtain

$$f_X(\xi_j) = \lim_{\Delta x \to 0} \frac{\Pr[\xi_j < X \le \xi_j + \Delta x]}{\Delta x} \approx \frac{\Pr[x_{\min} + j\Delta x < X \le x_{\min} + (j + 1)\Delta x]}{\Delta x}$$

such that

$$f_X(\xi_j) \approx \frac{h_j}{n\Delta x}$$

which reduces to the discrete case where $\Delta x = 1$.

(iv) The density of mobile nodes in the circle with radius r equals $\nu = \frac{N}{\pi r^2}$. Let R denote the (random) position of a mobile node. The probability that there is a mobile node between distance x and $x + dx$ (and $x \le r$) is

$$\Pr[x \le R \le x + dx] = \frac{2\pi x dx}{\pi r^2}$$

From (2.31), the pdf of R equals $f_R(x) = \frac{2x}{r^2}1_{x \le r}$ and the distribution function follows by integration as $F_R(x) = \frac{x^2}{r^2}1_{x \le r} + 1_{x > r}$. The (random) position $R_{(m)}$ of the m-th nearest mobile node to the center is given by (3.36)

$$f_{R_{(m)}}(x) = N f_R(x)\binom{N-1}{m-1}(F_R(x))^{m-1}(1 - F_R(x))^{N-m}$$

Written in terms of the density ν for $x \le r$,

$$f_{R_{(m)}}(x) = 2x\pi\nu\binom{N-1}{m-1}\left(\frac{\pi\nu x^2}{N}\right)^{m-1}\left(1 - \frac{\pi\nu x^2}{N}\right)^{N-1-(m-1)}$$

we recognize, apart from the prefactor $2x\pi\nu$, a binomial distribution (3.3) with $p = \frac{\pi\nu x^2}{N}$. Similar to the derivation of the law of rare events in Section 3.1.4, this binomial distribution tends, for large N but constant density ν, to a Poisson distribution with $\lambda = \pi\nu x^2$. Hence, asymptotically, the pdf of the position $R_{(m)}$ of the m-th nearest mobile node to the center is, for $x \le r$,

$$f_{R_{(m)}}(x) = 2x\pi\nu\frac{(\pi\nu x^2)^{m-1}}{(m-1)!}e^{-\pi\nu x^2}$$

(v) We use the law of total probability (2.46) first assuming that W is discrete,

$$\Pr[V - W \le x] = \sum_k \Pr[V - W \le x | W = k]\Pr[W = k]$$

and, by independence, $\Pr[V - W \le x | W = k] = \Pr[V \le k + x]$. Hence,

$$\Pr[V - W \le x] = \sum_k \Pr[V \le k + x]\Pr[W = k]$$

If W is continuous, the general formula is

$$\Pr[V - W \le x] = \int_{-\infty}^{\infty} \Pr[V \le x + y] \frac{d\Pr[W \le y]}{dy} dy \qquad (C.2)$$

from which the pdf follows by differentiation

$$f_{V-W}(x) = \int_{-\infty}^{\infty} f_V(x + y) f_W(y) dy$$

This resembles the convolution integral (2.62). If both V and W have the same distribution, direct integration of (C.2) yields

$$\Pr[V \le W] = \Pr[W \le V] = \frac{1}{2}$$

This equation confirms the intuitive result that two independent random variables with same density function have equal probability to be larger or smaller than the other.

C.2 Correlation (Chapter 4)

(i) In two dimensions, formula (4.2) becomes

$$E\left[e^{-z_1 X - z_2 Y}\right] = \exp\left[-z_2\mu_Y - \mu_X z_1 + \frac{\sigma_X^2}{2}z_1^2 + z_1 z_2 \rho \sigma_Y \sigma_X + \frac{\sigma_Y^2}{2}z_2^2\right]$$

$$= \exp\left[-z_2\mu_Y - \mu_X z_1 + \frac{1}{2}(\sigma_X z_1 + z_2\rho\sigma_Y)^2 + \frac{\sigma_Y^2(1-\rho^2)}{2}z_2^2\right]$$

Hence, with (4.21) the joint probability distribution is

$$f_{XY}(x, y; \rho) = \frac{1}{(2\pi i)^2} \int_{c_1-i\infty}^{c_1+i\infty} \int_{c_2-i\infty}^{c_2+i\infty} e^{z_1(x-\mu_X)+z_2(y-\mu_Y)}$$

$$\times e^{\frac{1}{2}(\sigma_X z_1 + z_2\rho\sigma_Y)^2 + \frac{\sigma_Y^2(1-\rho^2)}{2}z_2^2} dz_1 dz_2$$

$$= \frac{1}{2\pi i} \int_{c_2-i\infty}^{c_2+i\infty} e^{\frac{\sigma_Y^2(1-\rho^2)}{2}z_2^2} e^{z_2(y-\mu_Y)} dz_2$$

$$\times \frac{1}{2\pi i} \int_{c_1-i\infty}^{c_1+i\infty} e^{z_1(x-\mu_X)} e^{\frac{1}{2}(\sigma_X z_1 + z_2\rho\sigma_Y)^2} dz_1$$

Evaluating the last integral, denoted by L, yields

$$L = \frac{1}{2\pi i} \int_{c_1-i\infty}^{c_1+i\infty} e^{z_1(x-\mu_X)} \exp\left[\frac{1}{2}(\sigma_X z_1 + z_2\rho\sigma_Y)^2\right] dz_1$$

$$= \frac{1}{2\pi} \int_{-\infty}^{\infty} e^{(c_1+it)(x-\mu_X)} \exp\left[\frac{1}{2}(\sigma_X(c_1 + it) + z_2\rho\sigma_Y)^2\right] dt$$

$$= \frac{1}{2\pi} e^{c_1(x-\mu_X)} \int_{-\infty}^{\infty} e^{it(x-\mu_X)} \exp\left[\frac{-\sigma_X^2}{2}\left(t - i\left[c_1 + \frac{z_2\rho\sigma_Y}{\sigma_X}\right]\right)^2\right] dt$$

Since the integrand is an entire function, the contour can be shifted, which allows substitution as in real analysis. Thus, let $u = t - i\left[c_1 + \frac{z_2 \rho \sigma_Y}{\sigma_X}\right]$, then

$$
\begin{aligned}
L &= \frac{1}{2\pi} e^{c_1(x-\mu_X)} \int_{-\infty}^{\infty} e^{i\left(u + i\left[c_1 + \frac{z_2 \rho \sigma_Y}{\sigma_X}\right]\right)(x-\mu_X)} \exp\left[-\frac{\sigma_X^2}{2} u^2\right] du \\
&= \frac{1}{2\pi} e^{-\frac{z_2 \rho \sigma_Y}{\sigma_X}(x-\mu_X)} \int_{-\infty}^{\infty} \exp\left[-\frac{\sigma_X^2}{2}\left(u^2 + 2iu \frac{(x-\mu_X)}{\sigma_X^2}\right)\right] du \\
&= \frac{1}{2\pi} e^{-\frac{z_2 \rho \sigma_Y}{\sigma_X}(x-\mu_X)} \exp\left[-\frac{(x-\mu_X)^2}{2\sigma_X^2}\right] \int_{-\infty}^{\infty} \exp\left[-\frac{\sigma_X^2}{2}\left(u + i\frac{(x-\mu_X)}{\sigma_X^2}\right)^2\right] du
\end{aligned}
$$

By substituting $t = u + i\frac{(x-\mu_X)}{\sigma_X^2}$, the integral becomes

$$
\int_{-\infty}^{\infty} \exp\left[-\frac{\sigma_X^2}{2} t^2\right] dt = 2 \int_{0}^{\infty} \exp\left[-\frac{\sigma_X^2}{2} t^2\right] dt = \frac{\sqrt{2}}{\sigma_X} \int_{0}^{\infty} e^{-w} w^{-1/2} dw
$$

$$
= \frac{\sqrt{2}}{\sigma_X} \Gamma\left(\frac{1}{2}\right) = \frac{\sqrt{2\pi}}{\sigma_X}
$$

where we have used the Gamma function (Abramowitz and Stegun, 1968, Chapter 6). Hence,

$$
L = \frac{e^{-\frac{z_2 \rho \sigma_Y}{\sigma_X}(x-\mu_X)}}{\sigma_X \sqrt{2\pi}} \exp\left[-\frac{(x-\mu_X)^2}{2\sigma_X^2}\right]
$$

and

$$
f_{XY}(x,y;\rho) = \frac{\exp\left[-\frac{(x-\mu_X)^2}{2\sigma_X^2}\right]}{\sigma_X \sqrt{2\pi}} \frac{1}{2\pi i} \int_{c_2 - i\infty}^{c_2 - i\infty} e^{\frac{\sigma_Y^2(1-\rho^2)}{2} z_2^2 - \left(\mu_Y + \frac{\rho \sigma_Y}{\sigma_X}(x-\mu_X)\right) z_2} e^{z_2 y} dz_2
$$

The last integral is recognized with (3.22) as the inverse Laplace transform of a Gaussian with variance $\sigma^2 = \sigma_Y^2\left(1 - \rho^2\right)$ and mean $\mu = \mu_Y + \frac{\rho \sigma_Y}{\sigma_X}(x - \mu_X)$. Thus

$$
f_{XY}(x,y;\rho) = \frac{\exp\left[-\frac{(x-\mu_X)^2}{2\sigma_X^2}\right]}{\sigma_X \sqrt{2\pi}} \frac{\exp\left[-\frac{\left(y - \mu_Y - \frac{\rho \sigma_Y}{\sigma_X}(x-\mu_X)\right)^2}{2\sigma_Y^2(1-\rho^2)}\right]}{\sigma_Y \sqrt{1-\rho^2}\sqrt{2\pi}}
$$

which finally leads to the joint Gaussian density function (4.4). Hence, the linear combination method leads to exact results for Gaussian random variables.

C.3 Poisson process (Chapter 7)

(i) Let Y be a binomial random variable with parameters N and p, where N is a Poisson random variable with parameter λ. The probability density function of Y is obtained by applying the law of total probability (2.46),

$$
\Pr[Y = k] = \sum_{n=0}^{\infty} \Pr[Y = k | N = n] \Pr[N = n]
$$

With (3.3) and (3.9), we have

$$\Pr[Y = k] = \sum_{n=0}^{\infty} \binom{n}{k} p^k q^{(n-k)} \frac{\lambda^n e^{-\lambda}}{n!} = \frac{p^k}{k!} e^{-\lambda} \sum_{n=k}^{\infty} \frac{q^{(n-k)} \lambda^n}{(n-k)!}$$

$$= \frac{p^k \lambda^k}{k!} e^{-\lambda} \sum_{n=0}^{\infty} \frac{q^n \lambda^n}{n!} = \frac{(p\lambda)^k}{k!} e^{-\lambda + q\lambda}$$

Since $q = 1 - p$, we arrive at $\Pr[Y = k] = \frac{(p\lambda)^k}{k!} e^{-\lambda p}$, which means that Y is a Poisson random variable with mean $p\lambda$. If a sufficient sample of test strings defined above is sent and received, the average number of "one bits" at receiver divided by the average number of bits at the sender gives the probability p (if errors occur indeed independently).

(ii) Since the counting process of a sum of a Poisson process is again a Poisson counting process with rate equal to $\sum_{j=1}^{4} \lambda_j$, the average number of packets of the four classes in the router's buffers during interval T is $\lambda = T \sum_{j=1}^{4} \lambda_j$. Hence, the probability density function for the total number N of arrivals is $\Pr[N = n] = \frac{\lambda^n}{n!} e^{-\lambda}$.

(iii) Theorem 7.3.4 states the $N(t)$ is a Poisson counting process with rate $\lambda_1 + \lambda_2$. Then,

$$\Pr[X_1(t) = 1 | X(t) = 1] = \frac{\Pr[\{X_1(t) = 1\} \cap \{X(t) = 1\}]}{\Pr[X(t) = 1]}$$

$$= \frac{\Pr[\{X_1(t) = 1\} \cap \{X_2(t) = 0\}]}{\Pr[X(t) = 1]}$$

$$= \frac{\Pr[X_1(t) = 1] \Pr[X_2(t) = 0]}{\Pr[X(t) = 1]} = \frac{\lambda_1}{\lambda_1 + \lambda_2}$$

since the Poisson random variables X_1 and X_2 are independent. As an application we can consider a Poissonean arrival flow of packets at a router with rate λ. If the packets are marked randomly with probability $p = \frac{\lambda_1}{\lambda}$, the resulting flow consists of two types, those marked and those not. Each of these flows is again a Poisson flow, the marked flow with rate $\lambda_1 = p\lambda$ and the non-marked flow with $\lambda_2 = (1 - p)\lambda$. Actually, this procedure leads to a decomposition of the Poisson process into two independent Poisson processes and leads to the reverse of Theorem 7.3.4.

(iv) (a) Applying the solution of previous exercise immediately gives $\frac{\lambda_1}{\lambda_1 + \lambda_2 + \lambda_3}$

(b) Since the three Poisson processes are independent, the total number of cars on the three lanes, denoted by X, is also a Poisson process (Theorem 7.3.4) with rate $\lambda = \lambda_1 + \lambda_2 + \lambda_3$. Hence, $\Pr[X = n] = \frac{\lambda^n}{n!} e^{-\lambda}$.

(c) Let us denote the Poisson process in lane j by X_j. Then, using the independence between the X_j,

$$\Pr[X_1 = n, X_2 = 0, X_3 = 0] = \Pr[X_1 = n] \Pr[X_2 = 0] \Pr[X_3 = 0]$$

$$= \frac{\lambda_1^n}{n!} e^{-\lambda_1} e^{-\lambda_2} e^{-\lambda_3} = \frac{\lambda_1^n}{n!} e^{-\lambda}$$

(v) (a) The player relies on the fact that during the time there is exactly one arrival. Since the game rules mention that he should identify the last signal in $(0, T)$, signals arriving during $(0, s)$ do not influence his chance to win because of the memoryless property of the Poisson process. The number of arrivals in the interval (s, T) obeys a Poisson distribution with parameter $\lambda(T - s)$. The probability that precisely one signal arrives in the interval (s, T) is $\Pr[N(T) - N(s) = 1] = \lambda(T - s) e^{-\lambda(T-s)}$.

(b) Maximizing this winning probability with respect to s (by equating the first derivative to zero) yields

$$\frac{d}{ds} \Pr[N(T) - N(s) = 1] = -\lambda e^{-\lambda(T-s)} + \lambda^2 (T - s) e^{-\lambda(T-s)} = 0$$

with solution $\lambda(T - s) = 1$ or $s = T - 1/\lambda$. This maximum (which is readily verified by checking that $\frac{d^2}{ds^2} \Pr[N(T) - N(s) = 1] < 0$) lies inside the allowed interval $(0, T)$. The maximum probability of winning is $\Pr[N(T) - N(T - 1/\lambda) = 1] = 1/e$.

(vi) (a) We apply the general formula (7.1) for the pdf of a Poisson process with mean $E[X(t)] = \lambda t = 1$. Then, $\Pr[X(t+s) - X(s) = 0] = e^{-\lambda t} = \frac{1}{e}$.

(b) $\Pr[X(t+s) - X(s) > 10] = 1 - \Pr[X(t+s) - X(s) \leq 10] = 1 - \frac{1}{e}\sum_{k=0}^{10}\frac{1}{k!}$.

(c) Each minute is equally probable as follows from Theorem 7.3.3.

(vii) This exercise is an application of randomly marking in a Poisson flow as explained in solution (iii) above. The total flow of packets can be split up into an ACK stream, a Poisson process N_1 with rate $p\lambda = 3s^{-1}$ and a data flow, an independent Poisson process N_2 with rate $(1-p)\lambda = 7s^{-1}$. Then,

(a) $\Pr[N_1 > 1] = 1 - \Pr[N_1 = 0] = 1 - e^{-3}$

(b) The average number is $E[N_1 + N_2|N_1 = 5] = E[N_1|N_1 = 5] + E[N_2|N_1 = 5] = 5 + E[N_2] = 5 + 7 = 12$ packets.

(c) $\Pr[N_1 = 2|N_1 + N_2 = 8] = \frac{\Pr[N_1=2, N_1+N_2=8]}{\Pr[N_1+N_2=8]} = \frac{\frac{3^2}{2!}e^{-3}\frac{7^6}{6!}e^{-7}}{\frac{10^8}{8!}e^{-10}} \approx 29.65\%$

(viii) (a) Since the three Poisson arrival processes are independent, the total number of requests will also be a Poisson process with the parameter $\lambda = \lambda_1 + \lambda_2 + \lambda_3 = 20$ requests/hour (Theorem 7.3.4). The expected number of requests during an 8-hour working day is $E[N] = \lambda t = 20 \times 8 = 160$ requests.

(b) If we denote arrival processes of requests with different ADSL problems each with a random variable X_i for $i = 1, 2,$ and 3, then due to their mutual independence

$$\Pr[X_1 = 0, X_2 = k, X_3 = 0] = \Pr[X_1 = 0]\Pr[X_2 = k]\Pr[X_3 = 0]$$
$$= e^{-\lambda_1 t}\frac{(\lambda_2 t)^k e^{-\lambda_2 t}}{k!}e^{-\lambda_3 t}$$

from which $\Pr[X_1 = 0, X_2 = 3, X_3 = 0] = e^{-\frac{8}{3}}\frac{\left(\frac{6}{3}\right)^3 e^{-\frac{6}{3}}}{3!}e^{-\frac{6}{3}} = 1.7 \times 10^{-3}$.

(c) If we denote the total number of requests by X then $\Pr[X = 0] = e^{-\lambda t} = e^{-\frac{20}{4}} = 6.7 \times 10^{-3}$.

(d) The precise time is irrelevant for Poisson processes, only the duration of the interval matters. Here intervals are overlapping and we need to compute the probability

$$p = \Pr[\{X(0.2) = 1\} \cap \{X(0.5) - X(0.1) = 2\}]$$
$$= \sum_{k=0}^{1}\Pr[\{X(0.1) = k\} \cap \{X(0.2) - X(0.1) = 1 - k\} \cap \{X(0.5) - X(0.2)\} = 1 + k]$$
$$= \sum_{k=0}^{1}\Pr[X(0.1) = k]\Pr[X(0.2) - X(0.1) = 1 - k]\Pr[X(0.5) - X(0.2) = 1 + k]$$
$$= \sum_{k=0}^{1}e^{-2}\frac{(2)^k}{k!}e^{-2}\frac{(2)^{1-k}}{(1-k)!}e^{-6}\frac{(6)^{1+k}}{(1+k)!} = 48e^{-10} = 2.18 \times 10^{-3}$$

(e) Given that at the moment $t + s$ there are $k + m$ requests, the probability that there were k requests at the moment t is

$$\Pr[X(t) = k|X(t+s) = k+m] = \frac{\Pr[\{X(t) = k\} \cap \{X(t+s) = k+m\}]}{\Pr[X(t+s) = k+m]}$$
$$= \frac{\Pr[X(t) = k]\Pr[X(t+s) - X(t) = m]}{\Pr[X(t+s) = k+m]}$$
$$= \frac{\frac{(\lambda t)^k e^{-\lambda t}}{k!}\frac{(\lambda s)^m e^{-\lambda s}}{m!}}{\frac{(\lambda(t+s))^{k+m}e^{-\lambda(t+s)}}{(k+m)!}}$$
$$= \binom{k+m}{k}\left(\frac{t}{t+s}\right)^k\left(\frac{s}{t+s}\right)^m$$

(ix) (a) The number of attacks that are arriving to the PC is a Poisson random variable $X(t)$ with rate $\lambda = 6$. The probability that exactly one ($k = 1$) attack during one ($t = 1$) hour follows from (7.1) as $\Pr[X(1) = 1] = 6e^{-6}$.

(b) Applying (7.2), the expected amount of time that the PC has been on is $t = \frac{E[X(t)]}{\lambda} = \frac{60}{6} = 10$ hours.

(c) The arrival time of the fifth attack is denoted by T. Given that there are six attacks in one hour ($t = 1$), we compute the probability $\Pr[T < t|X(1) = 6]$ that either five attacks arrive in the interval $(0, t)$ and one arrives in $(t, 1)$ or all six attacks arrive in $(0, t)$ and none arrives in the interval $(t, 1)$. Hence, for $0 \le t < 1$,

$$
\begin{aligned}
F_T(t) &= \Pr[T < t|X(1) = 6] \\
&= \frac{\Pr[\{X(t) = 5\} \cap \{X(1) = 6\}] + \Pr[\{X(t) = 6\} \cap \{X(1) = 6\}]}{\Pr[X(1) = 6]} \\
&= \frac{\Pr[X(t) = 5]\Pr[X(1) - X(t) = 1] + \Pr[X(t) = 6]\Pr[X(1) - X(t) = 0]}{\Pr[X(1) = 6]} \\
&= \frac{((\lambda t)^5/5!)e^{-\lambda t}\lambda(1 - t)e^{-\lambda(1-t)} + ((\lambda t)^6/6!)e^{-\lambda t}e^{-\lambda(1-t)}}{(\lambda^6/6!)e^{-\lambda}} = 6t^5 - 5t^6
\end{aligned}
$$

The probability that the fifth attack will arrive between 1:30 p.m. and 2 p.m. is $F_T(1) - F_T\left(\frac{1}{2}\right) = 1 - \frac{7}{64} = \frac{57}{64}$.

(d) The expectation of T given $X(1) = 6$ follows from (2.33) as $E[T|X(1)] = \int_0^1 xf_T(x)dx$ where $f_T(t) = \frac{dF_T(t)}{dt}$ derived in (c). Alternatively, the expectation can be computed from (2.35), $E[T|X(1)] = \int_0^1 \left(1 - (6x^5 - 5x^6)\right) dx = \frac{5}{7}$. Hence the expected arrival time of the fifth attack between 1 p.m. and 2 p.m. is about 1:43 p.m.

(x) Let X and X_j denote the lifetime of system and subsystem j respectively. For a series of subsystems with independent lifetimes X_j is the event $\{X > t\} = \cap_{j=1}^n\{X_j > t\}$ and $\Pr[X > t] = \prod_{j=1}^n \Pr[X_j > t]$. Recall with (3.32) that $\Pr[X > t] = \Pr[\min_{1 \le j \le n} X_j > t]$. Using the definition of the reliability function (7.5) then yields

$$
R_{\text{series}}(t) = \prod_{j=1}^n R_j(t)
$$

(xi) The probability that the system S shown in Fig. 7.6 fails is determined by the subsystem with longest lifetime or $X = \max_{1 \le j \le n} X_j$. Invoking relation (3.33) combined with the definition of the reliability function (7.5) leads to

$$
R_{\text{parallel}}(t) = 1 - \prod_{j=1}^n (1 - R_j(t))
$$

C.4 Renewal theory (Chapter 8)

(i) The equivalence $\{N(t) > n\} \iff \{W_n \le t\}$ indicates

$$
\Pr\left[W_{N(t)} \le x\right] = \sum_{n=0}^{\infty} \Pr\left[\{W_n \le x\} \cap \{W_{n+1} > t\}\right]
$$

$$
= \Pr\left[W_0 \le x, W_1 > t\right] + \sum_{n=1}^{\infty} \Pr\left[W_n \le x, W_{n+1} > t\right]
$$

The convention $W_0 = 0$ reduces $\Pr[W_0 \leq x, W_1 > t] = \Pr[W_1 > t] = \Pr[\tau_1 > t] = 1 - F_\tau(t)$. Furthermore, by the law of total probability,

$$\Pr[W_n \leq x, W_{n+1} > t] = \int_0^\infty \Pr[W_n \leq x, W_{n+1} > t | W_n = u] \frac{d\Pr[W_n \leq u]}{du} du$$

$$= \int_0^x \Pr[W_{n+1} > t | W_n = u] d\Pr[W_n \leq u]$$

A renewal process restarts after each renewal from scratch (due to the stationarity and the independent increments of the renewal process). This implies that $\Pr[W_{n+1} > t | W_n = u] = \Pr[\tau_{n+1} > t - u] = 1 - F_\tau(t - u)$ because the interarrival times are i.i.d. random variables. Combined,

$$\Pr[W_{N(t)} \leq x] = \Pr[\tau > t] + \sum_{n=1}^\infty \int_0^x \Pr[\tau > t - u] d\Pr[W_n \leq u]$$

$$= \Pr[\tau > t] + \int_0^x \Pr[\tau > t - u] d\left(\sum_{n=1}^\infty \Pr[W_n \leq u]\right)$$

With the basic equivalence (8.6) and the definition (8.7) of the renewal function $m(t)$, we arrive at

$$\Pr[W_{N(t)} \leq x] = \Pr[\tau > t] + \int_0^x \Pr[\tau > t - u] \, dm(u)$$

This equation holds for all x. If $x = t$, we can use the renewal equation,

$$\int_0^t \Pr[\tau > t - u] \, dm(u) = m(t) - \int_0^t \Pr[\tau \leq t - u] \, dm(u)$$

$$= m(t) - m(t) + F_\tau(t)$$

which indeed confirms $\Pr[W_{N(t)} \leq t] = 1$.

(ii) The generating function of the number of renewals in the interval $[0, t]$ is with (8.10)

$$\varphi_{N(t)}(z) = E\left[z^{N(t)}\right] = \sum_{k=0}^\infty \Pr[N(t) = k] z^k$$

$$= \Pr[N(t) = 0] + \int_0^t \left(\sum_{k=1}^\infty \Pr[N(t - s) = k - 1] z^k\right) f_\tau(s) ds$$

$$= \Pr[N(t) = 0] + z \int_0^t \left(\sum_{k=0}^\infty \Pr[N(t - s) = k] z^k\right) f_\tau(s) ds$$

$$= \Pr[N(t) = 0] + z \int_0^t \varphi_{N(t-s)}(z) \, dF_\tau(s)$$

From (8.6), we have that $\Pr[N(t) = 0] = 1 - F_\tau(t)$ and

$$\varphi_{N(t)}(z) = 1 - F_\tau(t) + z \int_0^t \varphi_{N(t-s)}(z) \, dF_\tau(s)$$

By derivation with respect to z, we arrive at the differential-integral equation for the derivative of the generating function,

$$\varphi'_{N(t)}(z) = \int_0^t \varphi_{N(t-s)}(z) \, dF_\tau(s) + z \int_0^t \varphi'_{N(t-s)}(z) \, dF_\tau(s)$$

$$= \frac{\varphi_{N(t)}(z) - 1 + F_\tau(t)}{z} + z \int_0^t \varphi'_{N(t-s)}(z) \, dF_\tau(s)$$

which reduces to the renewal equation (8.9) for $z = 1$ since $\varphi'_{N(t)}(1) = m(t)$. The second derivative

$$\varphi''_{N(t)}(z) = 2\int_0^t \varphi'_{N(t-s)}(z)\, dF_\tau(s) + z\int_0^t \varphi''_{N(t-s)}(z)\, dF_\tau(s)$$

$$= \frac{2}{z}\varphi'_{N(t)}(z) - \frac{2}{z}\int_0^t \varphi_{N(t-s)}(z)\, dF_\tau(s) + z\int_0^t \varphi''_{N(t-s)}(z)\, dF_\tau(s)$$

evaluated at $z = 1$, is

$$\varphi''_{N(t)}(1) = 2m(t) - 2F_\tau(t) + \int_0^t \varphi''_{N(t-s)}(1)\, dF_\tau(s)$$

The variance $\mathrm{Var}[N(t)]$ follows from (2.27) as

$$\mathrm{Var}[N(t)] = \varphi''_{N(t)}(1) + \varphi'_{N(t)}(1) - \left(\varphi'_{N(t)}(1)\right)^2$$

$$= 3m(t) - m^2(t) - 2F_\tau(t) + \int_0^t \varphi''_{N(t-s)}(1)\, dF_\tau(s)$$

(iii) Every time an IP packet is launched by TCP, a renewal occurs and the reward is that 2000 km are travelled, in each renewal, thus $R_n = 2000$ km. The speed in a trip that suffers from congestion is, on average, 40 000 km/s, while the speed without congestion experience is 120 000 km/s. Since congestion only occurs in 1/5 cases, the average length (in s) of a renewal period is

$$E[\tau] = \frac{2000}{120000} \times \frac{4}{5} + \frac{2000}{40000} \times \frac{1}{5} = \frac{7}{300}$$

The average speed of an IP packet (in km/s) then follows from (8.20) as

$$\lim_{t \to \infty} \frac{R(t)}{t} = \frac{E[R]}{E[\tau]} = \frac{2000}{\frac{7}{300}} = 85714.3$$

(iv) Every transmission of an ATM cell is a renewal with average length of the renewal interval equal to $E[\tau] = N/r$, where $1/r$ is the mean interarrival time for a voice sample. If τ_n is the time between the n-th and $n+1$-th arrival of sample, then the average total cost per ATM cell transmission equals

$$E[R] = E\left[\sum_{n=1}^N nc \times \tau_n\right] + K = c\sum_{n=1}^N nE[\tau_n] + K$$

$$= \frac{c}{r}\frac{N(N-1)}{2} + K$$

Hence, the average cost per unit time incurred in UMTS is $\frac{E[R]}{E[\tau]} = c\frac{(N-1)}{2} + \frac{Kr}{N}$.

(v) (a) The replacement of a router is a renewal process where the time at which router R_j is replaced is $W_j = \min(X_j, T)$, and

$$R_j = \begin{cases} A, & \text{if } X_j \le T \\ B, & \text{if } X_j > T \end{cases}$$

The average cost per renewal period is $E[R] = A\Pr[X_j \le T] + B\Pr[X_j > T]$ and the average length of a renewal interval equals

$$E[X] = \int_0^\infty \Pr[W_j > t]\, dt = \int_0^\infty \Pr[\min(X_j, T) > t]\, dt = \int_0^T \Pr[X_j > t]\, dt$$

The time average cost rate of the policy CHANGEROUTER is $C = \frac{E[R]}{E[X]}$.

(b) For $A = 10000$, $B = 7000$, $\Pr[X_j \leq T] = 1 - e^{-\alpha T}$ with mean life time $\dfrac{1}{\alpha} = 10$ years and $T = 5$, we have

$$E[R] = A\Pr[X_j \leq T] + B\Pr[X_j > T] = 10000 \times \left(1 - e^{-1/2}\right) + 7000 \times e^{-1/2} \simeq 8200$$

and

$$E[X] = \int_0^T \Pr[X_j > t]\,dt = \int_0^5 e^{-0.1t}\,dt = 10 \times \left(1 - e^{-1/2}\right) \simeq 4$$

such that time average cost rate of the policy CHANGEROUTER is $C \simeq \dfrac{8200}{4} = 2050$.

C.5 Discrete-time Markov chains (Chapter 9)

(i) (a) The Markov chain is drawn in Fig. C.1

Fig. C.1. Three-state Markov chain.

(b) The steady-state vector π is computed via (9.24). The sequence P^{2^k} rapidly convergences to yield three correct digits after four multiplications

$$P^2 = \begin{bmatrix} 0.800 & 0.160 & 0.040 \\ 0.640 & 0.320 & 0.040 \\ 0.640 & 0.160 & 0.200 \end{bmatrix} \quad P^4 = \begin{bmatrix} 0.768 & 0.168 & 0.046 \\ 0.742 & 0.211 & 0.046 \\ 0.742 & 0.186 & 0.072 \end{bmatrix}$$

$$P^8 = \begin{bmatrix} 0.762 & 0.190 & 0.048 \\ 0.761 & 0.191 & 0.048 \\ 0.761 & 0.190 & 0.048 \end{bmatrix} \quad P^{16} = \begin{bmatrix} 0.762 & 0.190 & 0.048 \\ 0.762 & 0.190 & 0.048 \\ 0.762 & 0.190 & 0.048 \end{bmatrix}$$

from which we find that the row vector in P^{16} equals $\pi = \begin{bmatrix} 0.762 & 0.190 & 0.048 \end{bmatrix}$. The second method consists in solving the set (9.25) by Cramer's method. Hence,

$$\det M = \begin{vmatrix} -0.2 & 0.8 & 0.0 \\ 0.2 & -1.0 & 0.8 \\ 1 & 1 & 1 \end{vmatrix} = 0.84 \quad \pi_1 = \frac{\begin{vmatrix} 0 & 0.8 & 0.0 \\ 0 & -1 & 0.8 \\ 1 & 1 & 1 \end{vmatrix}}{\det M} = 0.762$$

$$\pi_2 = \frac{\begin{vmatrix} -0.2 & 0 & 0.0 \\ 0.2 & 0 & 0.8 \\ 1 & 1 & 1 \end{vmatrix}}{\det M} = 0.19 \quad \pi_1 = \frac{\begin{vmatrix} -0.2 & 0.8 & 0 \\ 0.2 & -1 & 0 \\ 1 & 1 & 1 \end{vmatrix}}{\det M} = 0.048$$

The third method relies on the specific structure of the Markov chain, a discrete birth and dead process or general random walk with constant $p_k = p$ and $q_k = q$. Applying formula

(11.9), taking into account that $N = 2$, $\rho = \frac{0.2}{0.8} = \frac{1}{4}$ yields,

$$\pi_1 = \frac{1 - \frac{1}{4}}{1 - \frac{1}{4^3}} = \frac{16}{21} = 0.762$$

$$\pi_2 = \pi_1 \rho = \frac{4}{21} = 0.190$$

$$\pi_3 = \pi_2 \rho = \frac{1}{21} = 0.048$$

(ii) The Markov chain is shown in Fig. C.2. The state 1 is an absorbing state. From (9.23),

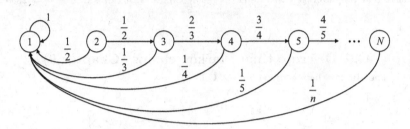

Fig. C.2. A recurrent Markov chain with positive drift.

the steady-state vector components are found as

$$\pi_1 = \sum_{k=1}^{N} \frac{\pi_k}{k}$$

$$\pi_2 = 0$$

$$\pi_j = \frac{j-2}{j-1} \pi_{j-1} \qquad j \geq 2$$

or $\pi_1 = 1$ and $\pi_j = 0$ for $j > 1$. Hence, the steady-state vector exists, and is different from $\pi = 0$, which demonstrates that the Markov chain is positive recurrent for any number of states N. However, the drift for $j > 1$ because $j = 1$ is absorbing is

$$E[X_{k+1} - X_k | X_k = j] = 1 - \frac{1}{j} - \frac{1}{j} = 1 - \frac{2}{j}$$

which is always positive for $j > 2$. Hence, given an initial state $j > 2$, the Markov chain will, on average, move to the right (higher states).

(iii) (a) The Markov chain is shown in Fig. C.3.

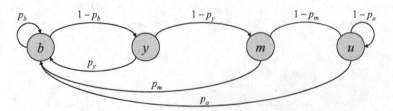

Fig. C.3. Markov chain of the growth process of trees in a forest during a period of 15 years.

(b) The evolution of the Markov process is defined by

$$
\begin{bmatrix} b[k+1] \\ y[k+1] \\ m[k+1] \\ u[k+1] \end{bmatrix} = \begin{bmatrix} p_b & p_y & p_m & p_o \\ 1-p_b & 0 & 0 & 0 \\ 0 & 1-p_y & 0 & 0 \\ 0 & 0 & 1-p_m & 1-p_o \end{bmatrix} \cdot \begin{bmatrix} b[k] \\ y[k] \\ m[k] \\ u[k] \end{bmatrix}
$$

(c) The number of trees in each category after 15 years (one period) is

$$
\begin{bmatrix} b[1] \\ y[1] \\ m[1] \\ u[1] \end{bmatrix} = \begin{bmatrix} 0.1 & 0.2 & 0.3 & 0.4 \\ 0.9 & 0 & 0 & 0 \\ 0 & 0.8 & 0 & 0 \\ 0 & 0 & 0.7 & 0.6 \end{bmatrix} \cdot \begin{bmatrix} 5000 \\ 0 \\ 0 \\ 0 \end{bmatrix} = \begin{bmatrix} 500 \\ 4500 \\ 0 \\ 0 \end{bmatrix}
$$

and after 30 years (two periods)

$$
\begin{bmatrix} b[2] \\ y[2] \\ m[2] \\ u[2] \end{bmatrix} = \begin{bmatrix} 0.1 & 0.2 & 0.3 & 0.4 \\ 0.9 & 0 & 0 & 0 \\ 0 & 0.8 & 0 & 0 \\ 0 & 0 & 0.7 & 0.6 \end{bmatrix} \cdot \begin{bmatrix} 500 \\ 4500 \\ 0 \\ 0 \end{bmatrix} = \begin{bmatrix} 950 \\ 450 \\ 3600 \\ 0 \end{bmatrix}
$$

(d) The steady-state vector π obeys equation (9.22) or, equivalently, (9.25). Applying a variant of (9.25), we have

$$
\begin{bmatrix} 1 \\ 0 \\ 0 \\ 0 \end{bmatrix} = \begin{bmatrix} 1 & 1 & 1 & 1 \\ 1-p_b & -1 & 0 & 0 \\ 0 & 1-p_y & -1 & 0 \\ 0 & 0 & 1-p_m & -p_o \end{bmatrix} \cdot \begin{bmatrix} \pi_b \\ \pi_y \\ \pi_m \\ \pi_o \end{bmatrix}
$$

The determinant is $\det P = -p_o - (1-p_b)(1+2p_o - p_y - p_m + p_y p_m - p_o p_y)$ and via Cramer's method we have

$$
\pi_b = \frac{1}{\det P} \det \begin{bmatrix} 1 & 1 & 1 & 1 \\ 0 & -1 & 0 & 0 \\ 0 & 1-p_y & -1 & 0 \\ 0 & 0 & 1-p_m & -p_o \end{bmatrix}
$$

$$
= \frac{\det \begin{bmatrix} -1 & 0 & 0 \\ 1-p_y & -1 & 0 \\ 0 & 1-p_m & -p_o \end{bmatrix}}{\det P} = \frac{-p_o}{\det P}
$$

With the numerical values given in (c), $\pi_b = 0.25773$. After a similar calculation for the other categories, the total number of trees in steady growth is

$$
\begin{bmatrix} 5000\pi_b \\ 5000\pi_y \\ 5000\pi_m \\ 5000\pi_o \end{bmatrix} \simeq \begin{bmatrix} 1289 \\ 1160 \\ 928 \\ 1624 \end{bmatrix}
$$

(iv) (a) The clustered error pattern is modeled as a two-state discrete Markov chain. When a bit is received incorrectly, the system is in state 0 else it is in state 1. The Markov chain is shown in Fig. 9.2, where $p = 1 - 0.95 = 0.05$ and $q = 1 - 0.999 = 0.001$. The transition probability matrix is $P = \begin{bmatrix} 0.95 & 0.05 \\ 0.001 & 0.999 \end{bmatrix}$.

(b) There is only one communicating class because both states 0 and 1 are reachable from each other. The Markov chain is therefore irreducible.

(c) The steady-state vector follows from (9.37) as

$$
\pi = \begin{bmatrix} \frac{50}{51} & \frac{1}{51} \end{bmatrix} = \begin{bmatrix} 0.0196 & 0.9804 \end{bmatrix}
$$

The fraction of correctly received bits in the long run is 98.04% and the fraction of incorrectly received bits is 1.96%.

(d) After repair, the system operates correctly in 99.9% of the cases, which implies that

$\pi_1 = 0.999$ and $\pi_0 = 0.001$. Formula (9.37) indicates that $\frac{p}{p+q} = 0.999$ and $\frac{q}{p+q} = 0.001$ or $p = 999q$. The test sequence shows that

$$\Pr[X_0 = 1, \ldots, X_{11} = 1] = (1-q)^{10} \Pr[X_0 = 1] = (1-q)^{10} = 0.9999$$

which leads to $q \simeq 10^{-5}$ and thus, $p = 0.0999$. A correctly (incorrectly) received bit is followed by a next correctly (incorrectly) received bit with probability $1 - q = 0.99999$, respectively $1 - p = 0.10001$.

C.6 Continuous-time Markov processes (Chapter 10)

(i) (a) The failure rate for each processor is $\lambda = 0.001$ per hour. The repair rate is $\mu = 0.01$ per hour. The Markov chain is shown in Fig. C.4.

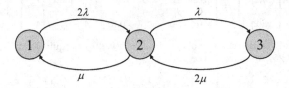

Fig. C.4. The Markov chain for the three states: (1) both processors work, (2) one processor is damaged and (3) both processors are damaged.

(b) The infinitesimal generator is

$$Q = \begin{pmatrix} -2\lambda & 2\lambda & 0 \\ \mu & -(\lambda + \mu) & \lambda \\ 0 & 2\mu & -2\mu \end{pmatrix}$$

If the state probability vector is denoted by $s(t)$, we can also write $s(t)Q = \frac{d}{dt}(s(t))$, or

$$[s_1(t) \quad s_2(t) \quad s_3(t)] \cdot \begin{pmatrix} -2\lambda & 2\lambda & 0 \\ \mu & -(\lambda + \mu) & \lambda \\ 0 & 2\mu & -2\mu \end{pmatrix} = [s_1'(t) \quad s_2'(t) \quad s_3'(t)]$$

(c) The steady-state $\pi = \lim_{t \to \infty} s(t)$ obeys the equation (10.19)

$$[\pi_1 \quad \pi_2 \quad \pi_3] \cdot \begin{pmatrix} -2\lambda & 2\lambda & 0 \\ \mu & -(\lambda + \mu) & \lambda \\ 0 & 2\mu & -2\mu \end{pmatrix} = [0 \quad 0 \quad 0]$$

Since $\pi_1 + \pi_2 + \pi_3 = 1$, we find that $\pi_1 = \left(\frac{\mu}{\lambda + \mu}\right)^2$, $\pi_2 = \frac{2\lambda\mu}{(\lambda + \mu)^2}$ and $\pi_3 = \left(\frac{\lambda}{\lambda + \mu}\right)^2$. From the balance equation, we know that the probability flux from state 1 to state 2 should precisely equal that in the opposite direction such that $2\lambda\pi_1 = \mu\pi_2$ and similar for the transitions $2 \to 3$, $\lambda\pi_1 = 2\mu\pi_2$. Using $\pi_1 + \pi_2 + \pi_3 = 1$ leads faster to the solution. With $\lambda = 0.001$ and $\mu = 0.01$, the values are $\pi_1 = 0.8264$, $\pi_2 = 0.1653$ and $\pi_3 = 0.0083$.

(d) The availability in case (i) is $\pi_1 = 0.8264$. The availability in case (ii) is $\pi_1 + \pi_2 = 0.9917$.

(ii) (a) In state 0, both servers are damaged, state 1 refers to one server down and one operating while in state 2, both servers are operating. The corresponding Markov chain is shown in Fig. C.5.

(b) The infinitesimal generator $Q = \begin{bmatrix} -\lambda_B & 0 & \lambda_B \\ \mu_F + \mu_E & -\mu_F - \mu_E - \lambda & \lambda \\ \mu_E & 2\mu_H & -\mu_E - 2\mu_H \end{bmatrix}$

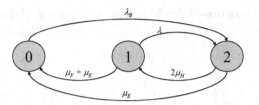

Fig. C.5. The Markov chain is specified by $\lambda = \dfrac{1}{15}\mathrm{h}^{-1} = 6.66 \times 10^{-2}\ \mathrm{h}^{-1}$, $\lambda_B = \dfrac{1}{20}\mathrm{h}^{-1} = 5 \times 10^{-2}\ \mathrm{h}^{-1}$, $\mu_H = 3 \times 10^{-4}\ \mathrm{h}^{-1}$, $\mu_F = 7 \times 10^{-4}\ \mathrm{h}^{-1}$ and $\mu_E = 6 \times 10^{-5}\ \mathrm{h}^{-1}$.

(c) The steady-state vector π obeys (10.19). The solution of $\pi Q = 0$ is

$$
\begin{bmatrix} \pi_0 & \pi_1 & \pi_2 \end{bmatrix}
\begin{bmatrix}
-\lambda_B & 0 & \lambda_B \\
\mu_F + \mu_E & -\mu_F - \mu_E - \lambda & \lambda \\
\mu_E & 2\mu_H & -\mu_E - 2\mu_H
\end{bmatrix}
= \begin{bmatrix} 0 & 0 & 0 \end{bmatrix}
$$

Since this linear set of equation is undetermined, we remove an arbitrary equation and add the normalization condition $\sum_{j=0}^{2} \pi_j = 1$,

$$
\begin{bmatrix} \pi_0 & \pi_1 & \pi_2 \end{bmatrix}
\begin{bmatrix}
-\lambda_B & 0 & 1 \\
\mu_F + \mu_E & -\mu_F - \mu_E - \lambda & 1 \\
\mu_E & 2\mu_H & 1
\end{bmatrix}
= \begin{bmatrix} 0 & 0 & 1 \end{bmatrix}
$$

The steady-state probabilities are

$$
\pi_2 = \frac{\lambda_B\,(\mu_F + \mu_E + \lambda)}{(\mu_E + \lambda_B)\,(\mu_F + \mu_E + \lambda) + 2\mu_H\,(\mu_F + \mu_E + \lambda_B)} = 0.9898
$$

$$
\pi_1 = \frac{2\mu_H}{(\mu_F + \mu_E + \lambda)}\pi_2 = 0.0088
$$

$$
\pi_0 = 1 - \pi_2 - \pi_1 = 0.0013
$$

(d) Theorem 10.2.3 states that the average lifetime of state j is $E[\tau_j] = \dfrac{1}{q_i}$. This yields

$$
E[\tau_0] = \frac{1}{q_0} = \frac{1}{\lambda_B} = 20\ \mathrm{h}
$$

$$
E[\tau_1] = \frac{1}{q_1} = \frac{1}{\mu_F + \mu_E + \lambda} = 14.9\ \mathrm{h}
$$

$$
E[\tau_2] = \frac{1}{q_2} = \frac{1}{\mu_E + 2\mu_H} = 1515\ \mathrm{h}
$$

(e) A repair takes place when the system transfers from state 1 to 2. When the system jumps from state 0 to state 2, two repairs take place. The fraction of time during which both servers are damaged is π_0 and the fraction of time in which one server is operating is π_1. The rate of repairs will be the rate of changing from state 1 to 2, plus two times the rate of changing from state 0 to state 2:

$$
f_r = \pi_1 q_{12} + 2\pi_0 q_{02} = \pi_1 \lambda + 2\pi_0 \lambda_B = 7.17\ 10^{-4}
$$

If we denote with X the random variable of the number of total failures over the period of 1 year, then the average value of X will be

$$
E[X] = f_r \times 24 \times 365 = 6.28
$$

C.7 Continuous-time Markov processes (Chapter 11)

(i) In both cases we apply the general formulae (11.15) and (11.16) for the steady-state of a general birth and death process.

(a) Using the notation $\rho = \frac{\lambda}{\mu}$, we first compute

$$\prod_{m=0}^{j-1} \frac{\lambda_m}{\mu_{m+1}} = \prod_{m=0}^{j-1} \frac{\lambda}{(m+1)\,\mu} = \frac{\lambda^j}{\mu^j} \prod_{m=1}^{j} \frac{1}{m} = \frac{\rho^j}{j!}$$

Then, with (11.16),

$$\pi_j = \frac{\frac{\rho^j}{j!}}{1 + \sum_{j=1}^{\infty} \frac{\rho^j}{j!}} = \frac{\rho^j}{j!} e^{-\rho} \qquad\qquad j \geq 0$$

which demonstrates that the steady-state probability that the birth and death process is in state j is Poisson distributed with mean ρ.

(b) Similarly, we first compute with $\lambda_m = \frac{\lambda}{(m+1)}$ and $\mu_m = \mu$,

$$\prod_{m=0}^{j-1} \frac{\lambda_m}{\mu_{m+1}} = \prod_{m=0}^{j-1} \frac{\lambda}{(m+1)\mu} = \frac{\rho^j}{j!}$$

which leads to precisely the same steady-state as in (a). Indeed, the steady-state is only a function of the ratios $\frac{\lambda_m}{\mu_{m+1}}$, which are the same in both (a) and (b).

(ii) All stations in slotted ALOHA operate independently and each has probability $p_t = 0.12$ to transmit in a timeslot. A station is successful in one slot with probability $p_s = p_t(1-p_t)^{N-1}$ where the number of stations $N = 8$. Thus, $p_s = 0.049$. The waiting time W to transmit one packet is a geometric random variable with parameter p_s from which (Section 3.1.3) the mean $E\,[W] = \frac{1}{p_s}$. Alternatively, $E\,[W]$ obeys the equation

$$E\,[W] = p_s + (1 - p_s)\,(1 + E\,[W])$$

because the average waiting time equals 1 timeslot with probability p_s plus 1 timeslot increased with the average waiting time with probability $1 - p_s$. Solving that equation again yields $E\,[W] = \frac{1}{p_s} = 20.39$ timeslots. The average transmission time for 7 packets is $7E\,[W] = 142.7$ timeslots.

C.8 Queueing (Chapters 13 and 14)

(i) Let us denote the number of packets in the server by N_x. Since a router either serves 0 or 1 packet, the problem states that $\Pr\,[N_x = 1] = 0.8$, and also that $E\,[N_x] = 0.8$. For any queue, it holds that $N_S = N_Q + N_x$ and $T = w + x$, the number in the system equals the number in the buffer and the number that is being served. From Little's Theorem (13.21), it follows with $E\,[x] = \frac{1}{\mu}$ that

$$E\,[N_x] = \frac{\lambda}{\mu}$$

or $\lambda = \mu E\,[N_x]$. Substituted into Little's law for the waiting time in the buffer, $E\,[N_Q] = \lambda E\,[w]$, and using $E\,[N_Q] = 3.2$ gives

$$E\,[w] = \frac{E\,[N_Q]}{E\,[N_x]} \frac{1}{\mu} = \frac{4}{\mu}$$

(ii) In a M/M/m/m queue, the number of busy servers equals the number (of packets) in the system N_S. From (14.16) and the definition (2.11), the average number of busy servers equals

$$E\left[N_S\right] = \sum_{j=0}^{m} j \Pr\left[N_S = j\right] = \frac{1}{\sum_{j=0}^{m} \frac{\lambda^j}{j!\mu^j}} \sum_{j=1}^{m} \frac{\lambda^j}{(j-1)!\mu^j}$$

The sum can be rewritten as

$$\sum_{j=1}^{m} \frac{\lambda^j}{(j-1)!\mu^j} = \frac{\lambda}{\mu} \sum_{j=0}^{m-1} \frac{\lambda^j}{j!\mu^j} = \frac{\lambda}{\mu} \left[\sum_{j=0}^{m} \frac{\lambda^j}{j!\mu^j} - \frac{\lambda^m}{m!\mu^m} \right]$$

such that

$$E\left[N_S\right] = \frac{\lambda}{\mu} \left[1 - \frac{\frac{\lambda^m}{m!\mu^m}}{\sum_{j=0}^{m} \frac{\lambda^j}{j!\mu^j}} \right] = \frac{\lambda}{\mu} \left(1 - \Pr\left[N_S = m\right]\right)$$

where the last probability is recognized as the Erlang B formula (14.17).

(iii) (a) Since the average service rate $\mu = 2$ s^{-1}, the average response time (average system time) follows from (14.2) as

$$E\left[T_{\mathrm{M/M/1}}\right] = \frac{1}{2 - \lambda}$$

(b) If $E\left[T_{\mathrm{M/M/1}}\right] = 2.5$ s, then it follows from (a) that $\lambda = 1.6$ s^{-1}. Hence, the number of jobs/s that can be processed for a given average response time of 2.5s equals 1.6 jobs/s

(c) A 10% increase in arrival rate corresponds to $\lambda = 1.76$ s^{-1} and from (a) we obtain that $E\left[T_{\mathrm{M/M/1}}\right] = \frac{1}{0.24} = 4.17$ s, which is with respect to 2.5s an increase in average response time of 67%.

(iv) We know that when the average call holding time is $1/\mu = 10$ min, the time blocking probability $P_{\mathrm{Bt}} = \frac{1}{10}$. Additionally, for Poisson call arrivals, the time blocking probability P_{Bt} equals the call blocking probability P_{B} on the PASTA property. The number of channels is $m = 2$. The arrival intensity λ can be calculated from the Erlang B formula (14.17)

$$P_{\mathrm{B}} = \frac{r^2/2}{1 + r + r^2/2}$$

where $r = \lambda/\mu$. Solving this equation for $r = 2\rho$ and taking into account that the traffic intensity $\rho \in [0, 1]$ yields $\rho = \frac{P_{\mathrm{B}} + \sqrt{2P_{\mathrm{B}} - P_{\mathrm{B}}^2}}{1 - P_{\mathrm{B}}}$. For $P_{\mathrm{B}} = \frac{1}{10}$, we have that $r = \frac{1+\sqrt{19}}{9}$ from which $\lambda = \frac{1+\sqrt{19}}{90}$. The blocking probability (14.17) corresponding to an average call holding time $1/\mu = 15$ min for which $r = \lambda/\mu = \frac{1 + \sqrt{19}}{90} \times 15$ is $P_{\mathrm{Bt}} \approx 0.174$.

(v) The queue 1 is a M/M/1 queue. By Burke's Theorem 14.1.1, the departure process from queue 1 is a Poisson with rate λ. By assumption, this departure process which is the arrival process to the second queue is also independent of the service process at queue 2. Therefore, queue 2, viewed in isolation, is also a M/M/1 queue. We know that the queueing processes in both queues are stable because the load $\rho_1 = \frac{\lambda}{\mu_1} < 1$ and $\rho_2 = \frac{\lambda}{\mu_2} < 1$. The steady-state distribution of the number of customers in queue 1 and queue 2 follow from (14.1) as $\Pr[n$ at queue 1$] = \rho_1^n(1-\rho_1)$ and $\Pr[m$ at queue 2$] = \rho_2^m(1-\rho_2)$. The number of customers presently in queue 1 is independent of the sequence of earlier arrivals at queue 2 and therefore also of the number of customers presently in queue 2. This implies that

$$\Pr[n \text{ at queue } 1, m \text{ at queue } 2] = \Pr[n \text{ at queue } 1]\Pr[m \text{ at queue } 2] = \rho_1^n(1-\rho_1)\rho_2^m(1-\rho_2)$$

(vi) The average system times $E[T]$ for the three different queueing systems are immediate. From (14.2), for system A, we have

$$E\left[T_A\right] = \frac{1}{k\mu\left(1 - \rho\right)}$$

For each of the k subqueues of system B, (14.2) gives

$$E\left[T_B\right] = \frac{1}{\mu\left(1-\rho\right)}$$

while for the M/M/k queue, (14.14) yields

$$E\left[T_C\right] = \frac{k\left(1-\rho\right) + \Pr\left[N_S \geq k\right]}{k\mu\left(1-\rho\right)}$$

Clearly, $E\left[T_B\right] = kE\left[T_A\right]$ shows that by replacing k small systems by one larger system with same processing capability, the average system time decreases by a factor k.
From the relation $E\left[T_C\right] = f(k,\rho)E\left[T_A\right]$ with $f(k,\rho) = k\left(1-\rho\right) + \Pr\left[N_S \geq k\right]$, it is more complicated to decide where $f(k,\rho)$ is larger or smaller than 1. The extreme values of $f(k,\rho)$ are known: $f(k,0) = k$ and $f(k,1) = 1$. Since $\frac{\partial f(k,\rho)}{\partial \rho} = -k + \frac{\partial \Pr[N_S \geq k]}{\partial \rho}$ and $\frac{\partial \Pr[N_S \geq k]}{\partial \rho} > 0$, it cannot be concluded that $\frac{\partial f(k,\rho)}{\partial \rho}$ is monotonously decreasing from k to 1 in which case we would have $f(k,\rho) > 1$. Assuming k real, we observe that $\frac{\partial f(k,\rho)}{\partial k} = (1-\rho) + \frac{\partial \Pr[N_S \geq k]}{\partial k} > 0$ for all $\rho < 1$, which implies that $f(2,\rho) < f(3,\rho) < \cdots$ and allows us to concentrate only on $f(2,\rho)$. Numerical results show that $f(2,\rho) < 1$ if $\rho \geq 0.85$, but $f(3,\rho) \geq 1$. This leads us to the conclusion that for $k > 2$, system A always outperforms system C; only if $k = 2$ and in the heavy traffic regime $\rho \geq 0.85$, system C leads to a slightly shorter average system time of maximum 1.7%. Hence, by replacing $k > 2$ processing units (servers) by one with same processing capability, always lowers the total time spent in the system. Of course, all conclusions only apply to systems that can be well modeled as M/M/m queueing systems. To first order a computing device (processor) may be regarded as a M/M/1 queue. Then, the analysis shows that replacing an old processor by a k times faster one is faster (on average) than installing k old processors in parallel.

(vii) The waiting process for aeroplanes is modeled as a M/D/1 queue because the arrival process is Poissonean with rate $\lambda = \frac{1}{10}$ arrivals/minute, it consists of a single queue as 1 aeroplane can land at a time and the service process (the landing process) takes precisely $x = 5$ minutes (constant service time). Thus, $E[x] = 5$ minutes, $\mathrm{Var}[x] = 0$ and $\rho = \frac{5}{10}$. Since the M/D/1 process is a special case of the M/G/1, we can apply the general formula (14.28) for the average waiting time in the queue of an M/G/1 system

$$E[w] = \frac{\lambda E[x^2]}{2(1-\rho)} = \frac{\frac{1}{10}\cdot 5^2}{2\cdot\frac{1}{2}} = 2.5 \text{ minutes}$$

(viii) (a) We know that the arrival intensity of new calls to the cell is $\lambda_f = 20$ calls/min. Let λ_h denote the arrival rate of the handover calls. The average time spent by a call in the cell is $E\left[T\right] = 1.64$ minutes and the average number of ongoing calls is $E\left[N\right] = 52$. Furthermore, the blocking rate is $P_B = 0.02$. The total arrival rate of calls that are carried by the base station is

$$\lambda_{\mathrm{carried}} = (1 - P_B)\,\lambda_{\mathrm{offered}} = (1 - P_B)\,(\lambda_f + \lambda_h)$$

Little's formula (13.21) states that $E\left[N\right] = \lambda_{\mathrm{carried}}\,E\left[T\right]$. Note that only the carried calls have an influence on the state of the system. We can solve the asked λ_h from these two equations as

$$\lambda_h = \frac{E\left[N\right]}{E\left[T\right](1 - P_B)} - \lambda_f = 12.35 \text{ calls/minute}$$

(b) The arrival intensity of lost calls is $\lambda_{\mathrm{lost}} = P_B\,(\lambda_f + \lambda_h) = 0.647$ calls/minute. If only the new calls are blocked, the asked blocking rate is

$$P_{Bf} = \frac{\lambda_{\mathrm{lost}}}{\lambda_f} = 3.24\%$$

(ix) (a) The derivation is given in the study of the M/G/1 queue in Section 14.3.1 where $A(z)$ given by (14.25) should be replace by $E\left[z^N\right]$.

b) Use in (14.25) the Laplace transform of an exponential random variable with mean $\frac{1}{\mu}$ given in (3.16)

$$\varphi_x(s) = \frac{\mu}{\mu + s}$$

One obtains

$$E\left[z^N\right] = \varphi_x\left(\lambda\left(1 - z\right)\right) = \frac{\mu}{(1 - z)\lambda + \mu} = \frac{\mu/(\lambda + \mu)}{1 - (\lambda/(\lambda + \mu))z}$$

$$= \left(1 - \frac{\lambda}{\lambda + \mu}\right)\sum_{k=0}^{\infty}\left(\frac{\lambda}{\lambda + \mu}\right)^k z^k$$

from which the probability density function

$$\Pr\left[N = k\right] = \left(1 - \frac{\lambda}{\lambda + \mu}\right)\left(\frac{\lambda}{\lambda + \mu}\right)^k$$

follows. Thus, $\Pr\left[N = k\right]$ is recognized as a geometric random variable with mean $E\left[N\right] = \frac{\lambda}{\mu}$.

(x) The queueing system is modeled by a birth and death process. The death rate is obvious and equal to μ. The arrival rate into state j equals the arrival rate of customers λ multiplied by the probability of really going to that state j which is $\frac{1}{j+1}$. Hence, $\lambda_j = \frac{\lambda}{j+1}$. The steady-state equation of this birth and death process is a Poisson process with rate $\rho = \frac{\lambda}{\mu}$ as derived above in Section C.7 solution (i).

(xi) *The M/M/m/m/s queue (The Engset formula)*. The arrival rate in Engset model is proportional to the size of the "still demanding subgroup" and the number of arrivals is exponential. The holding time of a line is exponentially distributed with mean $\frac{1}{\mu}$.

The Engset model is described as a birth–death process where each state k refers to the size of the "served subgroup". Since the total of customers is s, the "still demanding subgroup" consists of $s - k$ members. The birth rate is $\lambda_j = \alpha(s - k)$ and the death rate $\mu_k = k\mu$. The proportionality factor α can be interpreted as the arrival rate per "still demanding customer". The Markov graph is depicted in Fig. C.6. Application of the general birth–death

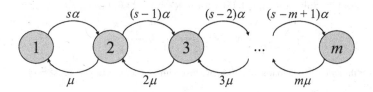

Fig. C.6. The Markov chain of the Engset loss model.

formulae for the steady-state vector (11.16) or $\Pr\left[N_S = j\right]$ yields, with $r = \frac{\alpha}{\mu}$,

$$\pi_j = \frac{\prod_{k=1}^{j}\frac{\alpha(s-k+1)}{k\mu}}{1 + \sum_{n=1}^{m}\prod_{k=1}^{n}\frac{\alpha(s-k+1)}{k\mu}} = \frac{\binom{s}{j}r^j}{\sum_{n=0}^{m}\binom{s}{n}r^n}$$

The computation of the blocking probability is more complex than for the Erlang B formula, because the arrival process is not a Poisson process. Indeed, due to the finite number of customers s, the largest number of possible arrivals is finite and the arrival rate depends on the state. Hence, the PASTA property cannot be applied. For a small time interval Δt, the blocking probability P_B equals the ratio of the $p_b(\Delta t)$, the probability of blocking in Δt, over $p_a(\Delta t)$, the probability of an arrival in Δt. Since the arrival rates depend on the state, the probability of an arrival in Δt is not equal to π_m as for the Erlang B model. Instead, we have

$$p_a(\Delta t) = \Delta t \sum_{n=0}^{m} (s-n)\alpha \Pr[N_S = n] = \alpha\Delta t \frac{\sum_{n=0}^{m} (s-n)\binom{s}{n} r^n}{\sum_{n=0}^{m} \binom{s}{n} r^n}$$

$$= \alpha s\Delta t \frac{\sum_{n=0}^{m} \binom{s-1}{n} r^n}{\sum_{n=0}^{m} \binom{s}{n} r^n}$$

Furthermore, blocking is only caused if $N_S = m$ and if at least one of the $s - m$ customers of the "still demanding group" generates an arrival. However, since the interval Δt can be made arbitrarily small[1], a generation of more than 1 arrival has probability $o(\Delta t)$ such that it suffices to consider only one call attempt. Hence,

$$p_b(\Delta t) = \Delta t(s-m) \Pr[N_S = m] = \alpha s\Delta t \frac{\binom{s-1}{m} r^m}{\sum_{n=0}^{m} \binom{s}{n} r^n}$$

The Engset call blocking probability $P_B = \frac{p_b(\Delta t)}{p_a(\Delta t)}$ becomes

$$P_B = \frac{\binom{s-1}{m} r^m}{\sum_{n=0}^{m} \binom{s-1}{n} r^n} \tag{C.3}$$

Observe that $P_B = \Pr[N_S = m]$ in a system with $s-1$ instead of s customers: an entering customer observes a system with $s - 1$ customers ignoring himself. At last, if we denote $\lambda = \alpha s$, the Engset call blocking formula (C.3) can be rewritten as

$$P_B = \frac{\frac{1}{m!}\left(\frac{\lambda}{\mu}\right)^m}{\sum_{n=0}^{m} \frac{(s-1-m)! s^{m-n}}{n!(s-1-n)!} \left(\frac{\lambda}{\mu}\right)^n}$$

The ratio $\frac{(s-1-m)!}{(s-1-n)!}$ is a polynomial in s of degree $n-m$ such that $\lim_{s\to\infty} \frac{(s-1-m)! s^{m-n}}{(s-1-n)!} = 1$. In conclusion, if $\lambda = \alpha s$ and $s \to \infty$, the Engset call blocking probability reduces to the Erlang B formula (14.17).

(xii) Although for a M/D/1 the exact expression of the overflow probability (14.44) exists, this series converges slowly for high traffic intensities $\rho = \lambda$ so that fast executable expressions are desirable. Substituting (14.45) with $\rho = \lambda$ into (14.67) gives

$$clr(\lambda) \approx \frac{1-\rho}{\rho} \frac{\frac{1-\rho}{\rho\zeta-1}\zeta^{-K}}{1 - \frac{1-\rho}{\rho\zeta-1}\zeta^{-K}}$$

For sufficiently high loads $\rho > 0.8$, we use the approximation $\zeta \approx \lambda^{-2}$ of Section 5.7 to obtain

$$clr_{M/D/1/K} \simeq \frac{(1-\rho)\rho^{2K}}{1-\rho^{2K+1}} \tag{C.4}$$

Comparing with (14.20) in the M/M/1/K queue,

$$clr_{M/M/1/K} \simeq \frac{(1-\rho)\rho^K}{1-\rho^{K+1}}$$

the M-server (in continuous-time) needs *approximately twice* as much buffer places to guarantee the same cell loss ratio as in the corresponding D-server (in discrete-time). Further combining (14.3) and (14.38) shows that

$$\mu_{M/M/1} E\left[w_{M/M/1}\right] = 2\mu_{M/D/1} E\left[w_{M/D/1}\right]$$

or, the average waiting time in the queue (normalized to the average service time) for the

---

[1] Similar arguments are used in Chapter 7 when studying the Poisson process.

M/M/1 queue is *exactly twice* as long as for the M/D/1 queue. The variability of the service in the M-server causes these rather large differences in performance. Furthermore, the simple formula (C.4) is particularly useful to engineer ATM buffers or to dimension simple queueing networks. If the number of individual flows that constitute the aggregate flow are large enough and none of the individual flows is dominant, the aggregate arrival process is quite well approximated by a Poisson process. Given as a QoS requirement a stringent cell loss ratio clr^*, the input flow ρ can be limited such that $clr_{M/D/1/K} < clr^*$. Alternatively, the buffer size K can be derived from (C.4) subject to $clr_{M/D/1/K} = clr^*$ for an aggregate Poisson input flow $\lambda = 0.9$. As long as the input flow is limited to $\lambda < 0.9$, the thus found buffer size K always guarantees a cell loss ratio below clr^* provided the input flow can be approximated as a Poisson arrival process.

C.9 General characteristics of graphs (Chapter 15)

(i) In one dimension ($d = 1$), the hopcount h_N of the shortest path between two uniformly chosen points x_A and x_B equals the distance between x_A and x_B. We allow the hopcount to be zero which is reflected by the small h while capital H refers to the case where the source A and the destination B are different. Thus, $\Pr[|x_A - x_B| = k] = \frac{1}{Z} 1_{k=0} + \frac{2(Z-k)}{Z^2} 1_{1 \le k \le Z-1}$ with corresponding generating function

$$\varphi_Z(x) = \sum_{k=0}^{Z-1} \Pr[|x_A - x_B| = k] x^k = \frac{Z - Zx^2 + 2x(x^Z - 1)}{Z^2(x-1)^2}$$

Since the nodes are uniformly chosen, all coordinate dimensions are independent and the generating function of the hopcount of the shortest path in a d-lattice is[2] $\varphi_Z^d(x)$. From (2.26) and (2.27), the average number of hops is immediate as $E[h_N] = \frac{d}{3Z}(Z^2 - 1)$ and the variance as $\mathrm{Var}[h_N] = \frac{d(Z^2-1)(Z^2+2)}{18Z^2}$. The total number of nodes in the d-lattice is $N = Z^d$ such that, for large N, we obtain

$$E[h_N] \simeq \frac{d}{3} N^{1/d}$$

and

$$\mathrm{Var}[h_N] \simeq \frac{d}{18} N^{2/d}$$

both increasing in $d > 1$ (for constant N) as in N (for constant d). For a two-dimensional lattice, the average hopcount scales as $O\left(\sqrt{N}\right)$.

(ii) Using the definition (15.6) of the clustering coefficient and applying the law of total probability (2.46) yields

$$\Pr\left[c_{G_p(N)} \le x\right] = \sum_{k=0}^{N-1} \Pr\left[\frac{2y}{d_v(d_v - 1)} \le x \Big| d_v = k\right] \Pr[d_v = k]$$

The degree distribution $\Pr[d_v = k]$ in the random graph is given by (15.11) and

$$\Pr\left[\frac{2y}{d_v(d_v-1)} \le x \Big| d_v = k\right] = \Pr\left[y \le \binom{k}{2}x \Big| d_v = k\right] = \sum_{j=0}^{\left[\binom{k}{2}x\right]} \binom{\binom{k}{2}}{j} p^j (1-p)^{\binom{k}{2}-j}$$

---

[2] If the sizes of the hypercube are not identical, the pgf is $\displaystyle\prod_{j=1}^{d} \varphi_{Z_j}(x)$.

because y is the number of links between the $d_v = k$ neighbors of v, which is binomially distributed with parameter p. Combined gives

$$\Pr\left[c_{G_p(N)} \le x\right] = \sum_{k=0}^{N-1} \binom{N-1}{k} p^k (1-p)^{N-1-k} \sum_{j=0}^{\left[\binom{k}{2}x\right]} \binom{\binom{k}{2}}{j} p^j (1-p)^{\binom{k}{2}-j}$$

The average $E\left[c_{G_p(N)}\right]$ is computed via (2.35) as

$$E\left[c_{G_p(N)}\right] = \int_0^1 \Pr\left[c_{G_p(N)} > x\right] dx$$

$$= \sum_{k=0}^{N-1} \binom{N-1}{k} p^k (1-p)^{N-1-k} \int_0^1 \sum_{j=\left[\binom{k}{2}x\right]+1}^{\binom{k}{2}} \binom{\binom{k}{2}}{j} p^j (1-p)^{\binom{k}{2}-j} dx$$

Let $t = \binom{k}{2}x$, then

$$\int_0^1 \sum_{j=\left[\binom{k}{2}x\right]}^{\binom{k}{2}} \binom{\binom{k}{2}}{j} p^j (1-p)^{\binom{k}{2}-j} dx = \frac{1}{\binom{k}{2}} \int_0^{\binom{k}{2}} \sum_{j=[t]+1}^{\binom{k}{2}} \binom{\binom{k}{2}}{j} p^j (1-p)^{\binom{k}{2}-j} dt$$

$$= \frac{1}{\binom{k}{2}} \sum_{t=0}^{\binom{k}{2}} \sum_{j=t+1}^{\binom{k}{2}} \binom{\binom{k}{2}}{j} p^j (1-p)^{\binom{k}{2}-j}$$

Reversing the t- and j- sum yields

$$\int_0^1 \sum_{j=\left[\binom{k}{2}x\right]}^{\binom{k}{2}} \binom{\binom{k}{2}}{j} p^j (1-p)^{\binom{k}{2}-j} dx = \frac{1}{\binom{k}{2}} \sum_{j=1}^{\binom{k}{2}} \sum_{t=0}^{j-1} \binom{\binom{k}{2}}{j} p^j (1-p)^{\binom{k}{2}-j}$$

$$= \frac{1}{\binom{k}{2}} \sum_{j=1}^{\binom{k}{2}} j \binom{\binom{k}{2}}{j} p^j (1-p)^{\binom{k}{2}-j} = p$$

Hence, we find that $E\left[c_{G_p(N)}\right] = p$. Along the same lines, we find that the generating function $\varphi_c(z)$ of the clustering coefficient $c_{G_p(N)}$ is

$$\varphi_c(z) = \sum_{k=0}^{N-1} \binom{N-1}{k} p^k (1-p)^{N-1-k} \left(1 - p + pe^{-\frac{z}{\binom{k}{2}}}\right)^{\binom{k}{2}}$$

The variance is computed from (2.43) as

$$\mathrm{Var}\left[c_{G_p(N)}\right] = (p-p^2) \sum_{k=2}^{N-1} \binom{N-1}{k} \frac{p^k (1-p)^{N-1-k}}{\binom{k}{2}}$$

(iii) The probability $\Pr[H_N = 2]$ is determined by the intersection of two independent events. First, there is no direct path between node A and B. This event has a chance proportional to $1-p$. Second, there is at least one path with two hops. All $N-2$ possible two-hops paths between A and B have the structure $(A \to j)(j \to B)$ and they have no links in common, i.e. they are mutually independent and independent from the direct link. The probability of the second event equals $1 - P_2$, where P_2 is the probability that there is no path with two hops. Hence, we have that $\Pr[H_N = 2] = (1 - p)(1 - P_2)$ and it remains to compute P_2. The event of no path with two hops is

$$\left(\cup_{j=1}^{N-2} 1_{(A \to j)(j \to B)}\right)^c = \cap_{j=1}^{N-2} 1_{((A \to j)(j \to B))^c}$$

such that

$$P_2 = \Pr\left[\left(\cup_{j=1}^{N-2} 1_{(A \to j)(j \to B)}\right)^c\right] = \Pr\left[\cap_{j=1}^{N-2} 1_{((A \to j)(j \to B))^c}\right]$$

$$= \prod_{j=1}^{N-2} \Pr\left[1_{((A \to j)(j \to B))^c}\right] = \prod_{j=1}^{N-2} \left(1 - \Pr\left[1_{((A \to j)(j \to B))}\right]\right) = \left(1 - p^2\right)^{N-2}$$

which demonstrates (15.26).

C.10 The uniform recursive tree (Chapter 16)

(i) The relative error r, defined as 1 minus the simulated value over the exact value at hop k given in (16.14), versus the number of hops k is shown in Fig. C.7. The insert in Fig. C.7 illustrates that, on a linear scale, the difference between simulation and theory (full line) is not distinguishable for $n \geq 10^5$ iterations. The average $E[r_n]$ and standard deviation

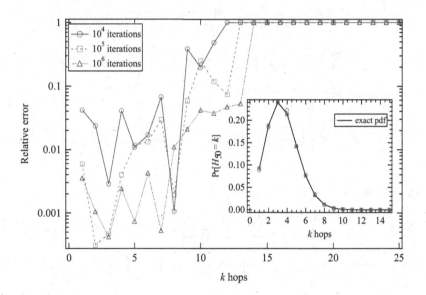

Fig. C.7. The relative error of the simulations of the hopcount in the complete graph with exponential link weight versus the hopcount for $10^4, 10^5$ and 10^6 iterations.

$\sigma[r_n]$ of the relative error for n iterations versus the hops k are

$$\begin{aligned} E[r_{10^4}] &= 0.12 & \sigma[r_{10^4}] &= 0.17 \\ E[r_{10^5}] &= 0.047 & \sigma[r_{10^5}] &= 0.073 \\ E[r_{10^6}] &= 0.017 & \sigma[r_{10^6}] &= 0.02 \end{aligned}$$

where the range of k values has been limited for $n = 10^4$ to 10 hops, for $n = 10^5$ to 11 hops, and for $n = 10^6$ to 12 hops. For larger hops, the simulations return zeros because the tail probability $\Pr[H_N > k]$ decreases as $O(1/k!)$ and simulating such a rare event requires on average at least as many simulations as $(\Pr[H_N = k])^{-1}$. The table roughly shows that the average error over the non-zero returned values decreases as $O\left(\frac{1}{\sqrt{n}}\right)$, which is in agreement with the Central Limit Theorem 6.3.1. Each iteration of the simulation can be regarded as an independent trial and the histogram sums in a particular way the number of these trials.

(ii) Using (2.43), we have

$$\mathrm{Var}\,[W_N] = \varphi''_{W_N}(0) - \left(\varphi'_{W_N}(0)\right)^2 = \varphi''_{W_N}(0) - (E\,[W_N])^2 \ ,$$

$$= \frac{1}{N-1} \sum_{k=1}^{N-1} \frac{d^2}{dz^2} \prod_{n=1}^{k} \frac{n(N-n)}{z+n(N-n)}\bigg|_{z=0} - \left(\frac{1}{N-1}\sum_{n=1}^{N-1}\frac{1}{n}\right)^2$$

where $E\,[W_N]$ is given in (16.18). The derivatives of the product

$$g\,(z) = \prod_{n=1}^{k} \frac{n(N-n)}{z+n(N-n)}$$

are elegantly computed via the logarithmic derivative $\frac{dg(z)}{dz} = g\,(z)\,\frac{d\log g(z)}{dz}$. The second derivative is $\frac{d^2 g(z)}{dz^2} = g\,(z)\left(\frac{d\log g(z)}{dz}\right)^2 + g\,(z)\,\frac{d^2 \log g(z)}{dz^2}$. With

$$\frac{d\log g\,(z)}{dz} = \frac{d}{dz}\sum_{n=1}^{k}\log\frac{n(N-n)}{z+n(N-n)} = -\sum_{n=1}^{k}\frac{1}{z+n(N-n)}$$

$$\frac{d^2 \log g\,(z)}{dz^2} = \sum_{n=1}^{k}\frac{1}{(z+n(N-n))^2}$$

we obtain since $g(0) = 1$,

$$\mathrm{Var}\,[W_N] = \frac{1}{N-1}\sum_{k=1}^{N-1}\left(\sum_{n=1}^{k}\frac{1}{n(N-n)}\right)^2 + \frac{1}{N-1}\sum_{k=1}^{N-1}\sum_{n=1}^{k}\frac{1}{n^2(N-n)^2} - \left(\frac{\sum_{n=1}^{N-1}\frac{1}{n}}{N-1}\right)^2$$

$$(\mathrm{C.5})$$

The first sum is

$$\sum_{k=1}^{N-1}\left(\sum_{n=1}^{k}\frac{1}{n(N-n)}\right)^2 = \sum_{k=1}^{N-1}\sum_{n=1}^{k}\frac{1}{n(N-n)}\sum_{j=1}^{k}\frac{1}{j(N-j)} = \sum_{n=1}^{N-1}\frac{\sum_{k=n}^{N-1}\sum_{j=1}^{k}\frac{1}{j(N-j)}}{n(N-n)}$$

and, with $\sum_{k=n}^{N-1}\sum_{j=1}^{k}\frac{1}{j(N-j)} = \sum_{k=1}^{N-1}\sum_{j=1}^{k}\frac{1}{j(N-j)} - \sum_{k=1}^{n-1}\sum_{j=1}^{k}\frac{1}{j(N-j)}$,

$$\sum_{k=1}^{N-1}\left(\sum_{n=1}^{k}\frac{1}{n(N-n)}\right)^2 = \sum_{n=1}^{N-1}\frac{1}{n(N-n)}\left(\sum_{k=1}^{N-1}\sum_{j=1}^{k}\frac{1}{j(N-j)} - \sum_{k=1}^{n-1}\sum_{j=1}^{k}\frac{1}{j(N-j)}\right)$$

$$= \sum_{n=1}^{N-1}\frac{1}{n(N-n)}\left(\sum_{j=1}^{N-1}\frac{1}{j(N-j)}\sum_{k=j}^{N-1}1 - \sum_{j=1}^{n-1}\frac{1}{j(N-j)}\sum_{k=j}^{n-1}1\right)$$

$$= \sum_{n=1}^{N-1}\frac{1}{n(N-n)}\sum_{j=1}^{N-1}\frac{1}{j} - \sum_{n=1}^{N-1}\frac{1}{(N-n)}\sum_{j=1}^{n-1}\frac{1}{j(N-j)}$$

$$+ \sum_{n=1}^{N-1}\frac{1}{n(N-n)}\sum_{j=1}^{n-1}\frac{1}{(N-j)}$$

Furthermore, since $\sum_{n=1}^{k} \frac{1}{n(N-n)} = \frac{1}{N} \sum_{n=1}^{k} \frac{1}{n} + \frac{1}{N} \sum_{n=N-k}^{N-1} \frac{1}{n}$, we have

$$\sum_{n=1}^{N-1} \frac{1}{(N-n)} \sum_{j=1}^{n-1} \frac{1}{j(N-j)} = \frac{1}{N} \sum_{n=1}^{N-1} \frac{1}{(N-n)} \sum_{k=1}^{n-1} \frac{1}{k} + \frac{1}{N} \sum_{n=1}^{N-1} \frac{1}{(N-n)} \sum_{k=N-n+1}^{N-1} \frac{1}{k}$$

$$= \frac{1}{N} \sum_{j=1}^{N-1} \frac{1}{j} \sum_{k=1}^{N-j-1} \frac{1}{k} + \frac{1}{N} \sum_{j=1}^{N-1} \frac{1}{j} \sum_{k=j+1}^{N-1} \frac{1}{k}$$

and

$$\sum_{n=1}^{N-1} \frac{1}{n(N-n)} \sum_{j=1}^{n-1} \frac{1}{(N-j)} = \frac{1}{N} \sum_{n=1}^{N-1} \left(\frac{1}{n} + \frac{1}{N-n} \right) \sum_{j=1}^{n-1} \frac{1}{(N-j)}$$

$$= \frac{1}{N} \sum_{j=1}^{N-1} \frac{1}{j} \sum_{k=N-j+1}^{N-1} \frac{1}{k} + \frac{1}{N} \sum_{j=1}^{N-1} \frac{1}{j} \sum_{k=j+1}^{N-1} \frac{1}{k}$$

Hence,

$$\sum_{k=1}^{N-1} \left(\sum_{n=1}^{k} \frac{1}{n(N-n)} \right)^2 = \frac{2}{N} \left(\sum_{n=1}^{N-1} \frac{1}{n} \right)^2 - \frac{1}{N} \sum_{j=1}^{N-1} \frac{1}{j} \sum_{k=1}^{N-j-1} \frac{1}{k} + \frac{1}{N} \sum_{j=1}^{N-1} \frac{1}{j} \sum_{k=N-j+1}^{N-1} \frac{1}{k}$$

$$= \frac{2}{N} \left(\sum_{n=1}^{N-1} \frac{1}{n} \right)^2 - \frac{1}{N} \sum_{j=1}^{N-1} \frac{1}{j} \left(\sum_{k=1}^{N-1} \frac{1}{k} - \sum_{k=N-j}^{N-1} \frac{1}{k} \right)$$

$$+ \frac{1}{N} \sum_{j=1}^{N-1} \frac{1}{j} \sum_{k=N-j+1}^{N-1} \frac{1}{k}$$

$$= \frac{1}{N} \left(\sum_{n=1}^{N-1} \frac{1}{n} \right)^2 + \frac{1}{N} \sum_{j=1}^{N-1} \frac{1}{j} \frac{1}{N-j} + \frac{2}{N} \sum_{j=1}^{N-1} \frac{1}{j} \sum_{k=N-j+1}^{N-1} \frac{1}{k}$$

$$= \frac{1}{N} \left(\sum_{n=1}^{N-1} \frac{1}{n} \right)^2 + \frac{2}{N^2} \sum_{j=1}^{N-1} \frac{1}{j} + \frac{2}{N} \sum_{j=1}^{N-1} \frac{1}{j} \sum_{k=N-j+1}^{N-1} \frac{1}{k}$$

Substituted into (C.5) yields

$$\text{Var}\left[W_N\right] = -\frac{\left(\sum_{n=1}^{N-1} \frac{1}{n} \right)^2}{(N-1)^2 N} + \frac{2 \sum_{j=1}^{N-1} \frac{1}{j}}{(N-1) N^2} + \frac{2 \sum_{j=1}^{N-1} \frac{1}{j} \sum_{k=N-j+1}^{N-1} \frac{1}{k}}{N(N-1)}$$

$$+ \frac{\sum_{k=1}^{N-1} \sum_{n=1}^{k} \frac{1}{n^2(N-n)^2}}{N-1}$$

Further,

$$\sum_{k=1}^{N-1} \sum_{n=1}^{k} \frac{1}{n^2(N-n)^2} = \sum_{n=1}^{N-1} \frac{1}{n^2(N-n)^2} \sum_{k=n}^{N-1} 1 = \sum_{n=1}^{N-1} \frac{1}{n^2(N-n)}$$

The partial fraction expansion of $\frac{1}{n^2(N-n)} = \frac{1}{N^2 n} + \frac{1}{N n^2} + \frac{1}{N^2(N-n)}$ such that

$$\sum_{k=1}^{N-1} \sum_{n=1}^{k} \frac{1}{n^2(N-n)^2} = \frac{2}{N^2} \sum_{n=1}^{N-1} \frac{1}{n} + \frac{1}{N} \sum_{n=1}^{N-1} \frac{1}{n^2}$$

Combined,

$$\text{Var}\left[W_N\right] = -\frac{\left(\sum_{n=1}^{N-1}\frac{1}{n}\right)^2}{(N-1)^2\,N} + \frac{4\sum_{n=1}^{N-1}\frac{1}{n}}{(N-1)\,N^2} + \frac{2}{N(N-1)}\sum_{j=1}^{N-1}\frac{1}{j}\sum_{k=N-j+1}^{N-1}\frac{1}{k} + \frac{\sum_{n=1}^{N-1}\frac{1}{n^2}}{N(N-1)}$$

Invoking the identity

$$\sum_{j=1}^{N-1}\frac{1}{N-j}\sum_{k=j}^{N-1}\frac{1}{k} = \sum_{n=1}^{N-1}\frac{1}{n^2}$$

(which can be verified by induction) yields

$$\sum_{j=1}^{N-1}\frac{1}{j}\sum_{k=N-j+1}^{N-1}\frac{1}{k} = \sum_{j=1}^{N-1}\frac{1}{N-j}\sum_{k=j+1}^{N-1}\frac{1}{k} = \sum_{j=1}^{N-1}\frac{1}{N-j}\left(\sum_{k=j}^{N-1}\frac{1}{k}-\frac{1}{j}\right)$$

$$= \sum_{j=1}^{N-1}\frac{1}{N-j}\sum_{k=j}^{N-1}\frac{1}{k} - \sum_{j=1}^{N-1}\frac{1}{j\,(N-j)} = \sum_{n=1}^{N-1}\frac{1}{n^2} - \frac{2}{N}\sum_{n=1}^{N-1}\frac{1}{n}$$

Finally, we arrive at (16.19).

(iii) The limit for $N \to \infty$ of the probability generating function (16.17) of the weight W_N of the shortest path

$$\varphi_{W_N}(z) = E\left[e^{-zW_N}\right] = \frac{1}{N-1}\sum_{k=1}^{N-1}\prod_{n=1}^{k}\frac{n(N-n)}{z+n(N-n)}$$

will be derived from which the distribution then follows by taking the inverse Laplace transform. Since

$$z + n(N-n) = \left(\sqrt{\left(\frac{N}{2}\right)^2+z}+\frac{N}{2}-n\right)\left(\sqrt{\left(\frac{N}{2}\right)^2+z}-\left(\frac{N}{2}-n\right)\right)$$

we have with $y = \sqrt{\left(\frac{N}{2}\right)^2+z}$,

$$\prod_{n=1}^{k}\frac{n(N-n)}{z+n(N-n)} = \frac{k!\,(N-1)!}{(N-k-1)!}\prod_{n=1}^{k}\frac{1}{\left(y+\frac{N}{2}-n\right)}\prod_{n=1}^{k}\frac{1}{\left(y-\frac{N}{2}+n\right)}$$

The products can be written in terms of the Gamma function,

$$\prod_{n=1}^{k}\frac{1}{\left(y+\frac{N}{2}-n\right)} = \frac{\Gamma\left(y+\frac{N}{2}-k\right)}{\Gamma\left(y+\frac{N}{2}\right)}$$

$$\prod_{n=1}^{k}\frac{1}{\left(y-\frac{N}{2}+n\right)} = \frac{\Gamma\left(y-\frac{N}{2}+1\right)}{\Gamma\left(y-\frac{N}{2}+k+1\right)}$$

Thus,

$$\varphi_{W_N}(z) = (N-2)!\,\frac{\Gamma\left(y-\frac{N}{2}+1\right)}{\Gamma\left(y+\frac{N}{2}\right)}\sum_{k=1}^{N-1}\frac{\Gamma\left(k+1\right)}{\Gamma(N-k)}\frac{\Gamma\left(y+\frac{N}{2}-k\right)}{\Gamma\left(y-\frac{N}{2}+k+1\right)}$$

Let the number of nodes be even $N = 2M$ such that $y = \sqrt{M^2 + z} \sim M + \frac{z}{2M}$ (provided $|z| < 2M$). The sum, denoted by S, can be split as

$$
\begin{aligned}
S &= \sum_{k=1}^{M} \frac{\Gamma(k+1)}{\Gamma(2M-k)} \frac{\Gamma(y+M-k)}{\Gamma(y-M+k+1)} + \sum_{k=M+1}^{2M-1} \frac{\Gamma(k+1)}{\Gamma(2M-k)} \frac{\Gamma(y+M-k)}{\Gamma(y-M+k+1)} \\
&= \sum_{j=0}^{M-1} \frac{\Gamma(M-j+1)}{\Gamma(M+j)} \frac{\Gamma(y+j)}{\Gamma(y-j+1)} + \sum_{k=1}^{M-1} \frac{\Gamma(M+k+1)}{\Gamma(M-k)} \frac{\Gamma(y-k)}{\Gamma(y+k+1)} \\
&= \sum_{j=-(M-1)}^{M-1} \frac{\Gamma(y+j)}{\Gamma(M+j)} \frac{\Gamma(M-j+1)}{\Gamma(y-j+1)}
\end{aligned}
$$

and

$$
\varphi_{W_{2M}}(z) = (2M-1)! \frac{\Gamma(y-M+1)}{\Gamma(y+M)} \frac{1}{2M-1} \sum_{j=-(M-1)}^{M-1} \frac{\Gamma(y+j)}{\Gamma(M+j)} \frac{\Gamma(M-j+1)}{\Gamma(y-j+1)}
$$

For large M,

$$
(2M-1)! \frac{\Gamma(y-M+1)}{\Gamma(y+M)} \sim \Gamma(2M) \frac{\Gamma\left(\frac{z}{2M}+1\right)}{\Gamma\left(2M+\frac{z}{2M}\right)} \sim (2M)^{-\frac{z}{2M}} \Gamma\left(\frac{z}{2M}+1\right)
$$

which suggests that we consider $z \to 2Mz$ since then, using (Abramowitz and Stegun, 1968, Section 6.1.47),

$$
\varphi_{W_{2M}}(2Mz) \sim (2M)^{-z} \Gamma(z+1) \frac{1}{2M-1} \sum_{j=-(M-1)}^{M-1} \left(1+O\left(\frac{1}{M}\right)\right)\left(1+O\left(\frac{1}{M}\right)\right)
$$

$$
\sim (2M)^{-z} \Gamma(z+1)
$$

Hence,

$$
\lim_{N\to\infty} N^z \varphi_{W_N}(Nz) = \Gamma(z+1)
$$

or equivalently,

$$
\lim_{N\to\infty} E\left[e^{-(NW_N - \log N)z}\right] = \Gamma(z+1) \tag{C.6}
$$

The inverse Laplace transform of $\Gamma(z+1)$ is a Gumbel distribution (3.37) and we arrive at the asymptotic distribution for the weight of the shortest path (16.20).

Since $\Pr[NW_N - \log N \leq y] = \Pr\left[W_N \leq \frac{y+\log N}{N}\right]$ from which after substitution of $x = \frac{y+\log N}{N}$ and ignoring the limit $N \to \infty$, it follows that $\Pr[W_N \leq x] = e^{-Ne^{-Nx}}$, the probability density function is found after derivation as

$$
\tilde{f}_{W_N}(x) = N^2 e^{-N(e^{-Nx}+x)} \tag{C.7}
$$

The goodness of this asymptotic distribution (C.7) for finite N is illustrated in Fig. C.8. Observe from Fig. C.8 that $f_{W_N}(0) = 1$ while $\tilde{f}_{W_N}(0) = N^2 e^{-N} \approx 0$. Since $\varphi_{W_N}(z) = \int_0^\infty e^{-zt} f_{W_N}(t)\, dt$ is a single-sided Laplace transform, integrating by parts yields $z\varphi_{W_N}(z) = f_{W_N}(0) + \int_0^\infty e^{-zt} f'_{W_N}(t)\, dt$ provided $f'_{W_N}(t)$ exists for all $t \geq 0$. Hence, we find a well-known limit criterion of single-sided Laplace transforms,

$$
f_{W_N}(0) = \lim_{z\to\infty} z\varphi_{W_N}(z) \tag{C.8}
$$

Applied to (16.17) leads to $f_{W_N}(0) = 1$ for all finite N and applied to the scaled link weight where the mean is $\frac{1}{a}$ such that $\varphi_{W_{N;a}}(z) = \varphi_{W_N}(az)$ gives $f_{W_N}(0) = a$. The interpretation of this property is related to the choice of the link weights. The shortest

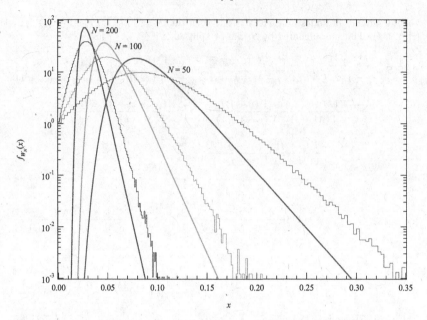

Fig. C.8. The pdf of the weight of the shortest path for various N. Each simulation consists of 10^6 iterations. The bold curves represent the finite N-equivalent (C.7) of the asymptotic result.

path includes almost surely the smallest link weights of $F_w(x) = f_w(0)x + O\left(x^2\right)$ since $F_w(0) = 0$. Both the exponential (with parameter 1) and uniform distribution are regular with $f_w(0) = 1$. Since the smallest values of the weight of the shortest path W_N in K_N occur for direct links a.s., the distribution of W_N around zero is dominated by the distribution of the link weight w around zero. The contribution cannot be due a n-hop shortest path with $n > 1$ since such a path exists of the sum of n exponentials, which has a probability density around $x = 0$ of the form $O\left(x^{n-1}\right)$. Indeed, see (3.24) or apply (C.8) to the pgf of a sum of n exponentials $\varphi_{S_n}(z) = \prod_{k=1}^{n} \frac{\alpha_k}{z+\alpha_k}$.

(iv) When the intermediate nodes of the shortest path between a source A and a destination B are removed from K_N, we obtain again a complete graph with $N - h_N + 1$ nodes. The resulting graph contains link weights that are not perfectly exponentially distributed anymore nor are they perfectly independent, because we have removed a special set of nodes and not a random set. But, since we have removed at each node of the shortest path, apart from the shortest link, $N - 3$ other links, we assume that the dependence between K_N and the reduced graph is ignorably small. Under these assumptions, the shortest node-disjoint path in K_N is a shortest path in K_{N-H_N+1} with exponential link weight with mean 1. The distribution of hopcount h_N^{nd} of that shortest node-disjoint path is

$$\Pr\left[h_N^{nd} = k\right] \simeq \sum_{j=0}^{N-1} \Pr\left[h_{N-j+1} = k | h_N = j\right] \Pr\left[h_N = j\right]$$

The hopcount H_N of the shortest path in the complete graph K_N with independent exponential link weights with mean 1 is given in (16.8). With the assumption that

$$\Pr\left[h_{N-j+1} = k | h_N = j\right] = \Pr\left[h_{N-j+1} = k\right]$$

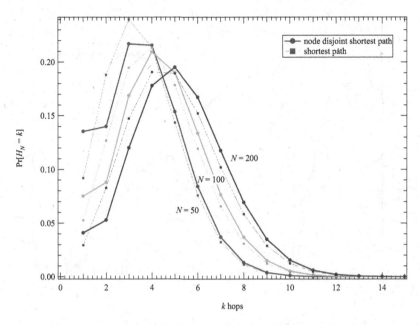

Fig. C.9. Both the pdf of the hopcount of the shortest path (thin line) and the shortest node-disjoint path (bold line).

we obtain

$$\Pr\left[h_N^{nd} = k\right] \simeq \frac{(-1)^{k+1}}{N!} \sum_{j=0}^{N-1} \frac{S_{N-j+1}^{(k+1)} S_N^{(j+1)}}{(N-j+1)!}$$

For large N, we can use the Poisson approximation (16.13)

$$\Pr\left[h_N^{nd} = k\right] \approx \frac{1}{Nk!} \sum_{j=0}^{N-1} \frac{(\log(N-j+1))^k}{N-j+1} \frac{(\log N)^j}{j!}$$

Since $(\log(N-j+1))^k = \log^k N - \frac{k(j-1)}{N} \log^{k-1} N + O\left(\frac{\log^{k-1}}{N^2}\right)$ and $\frac{1}{N-j+1} = \frac{1}{N} + O\left(\frac{1}{N^2}\right)$, we have to highest order in N,

$$\Pr\left[h_N^{nd} = k\right] \approx \frac{1}{Nk!} \sum_{j=0}^{N-1} \left(\frac{\log^k N}{N} + O\left(\frac{\log^{k-1} N}{N^2}\right)\right) \frac{(\log N)^j}{j!}$$

$$\approx \frac{1}{k!} \left(\frac{\log^k N}{N} + O\left(\frac{\log^{k-1} N}{N^2}\right)\right) \approx \Pr\left[h_N = k\right]$$

For large N, we expect approximately that the hopcount of the shortest and that of the shortest node-disjoint path have about the same distribution. The validity of the assumption is illustrated in Fig. C.9 for relatively small values of $N = 50, 100$, and 200. Each simulation consisted of $n = 10^6$ iterations. The corresponding weight of the shortest and node-disjoint shortest path are drawn in Fig. C.10. The weight of the node-disjoint shortest path is evidently always larger than that of the shortest path in the same graph. Nevertheless, for large N, the simulations suggest that both pdfs tend to each other.

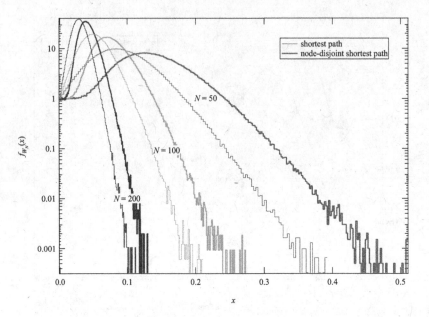

Fig. C.10. Pdf of the weight of the shortest path (thin line) and the node-disjoint shortest path (bold line).

C.11 The efficiency of multicast (Chapter 17)

(i) Using (17.25), we obtain

$$g_{N,k}(2) = N - 1 - \sum_{j=0}^{D-1} k^{D-j} \frac{\left(N - 1 - \frac{k^{j+1}-1}{k-1}\right)\left(N - 2 - \frac{k^{j+1}-1}{k-1}\right)}{(N-1)(N-2)}$$

$$= \frac{(2N-3)DN}{(N-1)(N-2)} + \frac{(2N-3)D}{(N-1)(N-2)(k-1)} - \frac{(2N-3)}{(N-2)(k-1)} +$$

$$- \frac{N(N-1-2D)}{(N-1)(N-2)(k-1)} - \frac{(N-1-2D)}{(N-1)(N-2)(k-1)^2} - \frac{1}{(N-2)(k-1)^2}$$

or, for large N,

$$g_{N,k}(2) \sim 2D - \frac{3}{k-1} + O\left(\frac{\log_k N}{N}\right)$$

the effective power exponent $\beta^*(N)$ as defined in (17.32), equals for the k-ary tree and large N,

$$\beta^*(N) \sim \frac{\log\left[\frac{2D - \frac{3}{k-1}}{D - \frac{1}{k-1}}\right]}{\log 2}$$

$$= 1 + \log_2\left[1 - \frac{1}{2(k-1)\left(\log_k N + \log_k(1 - 1/k) - \frac{1}{k-1}\right)}\right]$$

$$= 1 + \log_2\left[1 - \frac{1}{2(k-1)E[H_N]}\right] \sim 1 - \frac{1}{(\log 4)(k-1)E[H_N]}$$

which shows, for large N, that $\beta^*(N) < 1$, but that $\beta^*(N) \to 1$ if $k \to \infty$.

Bibliography

Abramowitz, M. and Stegun, I. A. (1968). *Handbook of Mathematical Functions.* (Dover Publications, Inc., New York).

Allen, A. O. (1978). *Probability, Statistics, and Queueing Theory.* Computer Science and Applied Mathematics, (Academic Press, Inc., Orlando).

Almkvist, G. and Berndt, B. C. (1988). Gauss, Landen, Ramanuyan, the Arithmic-Geometric Mean, Ellipses, π and the Ladies Diary. *American Mathematical Monthly* **95**, 585–608.

Anick, D., Mitra, D., and Sondhi, M. M. (1982). Stochastic theory of a data-handling system with multiple sources. *The Bell System Technical Journal 61*, 8 (October), 1871–1894.

Anupindi, R., Chopra, S., Deshmukh, S. D., Van Mieghem, J. A., and Zemel, E. (2006). *Managing Business Flows. Principles of Operations Management*, 2nd edn. (Prentice Hall, Upper Saddle River).

Barabasi, A.-L. (2002). *Linked, The New Science of Networks.* (Perseus, Cambridge, MA).

Baran, P. (2002). The beginnings of packet switching - some underlying concepts: The Franklin Institute and Drexel University seminar on the evolution of packet switching and the Internet. *IEEE Communications Magazine*, 2–8.

Berger, M. A. (1993). *An Introduction to Probabiliy and Stochastic Processes.* (Springer-Verlag, New York).

Bertsekas, D. and Gallager, R. (1992). *Data Networks*, 2nd edn. (Prentice-Hall International Editions, London).

Billingsley, P. (1995). *Probability and Measure*, 3rd edn. (John Wiley & Sons, New York).

Bisdikian, C., Lew, J. S., and Tantawi, A. N. (1992). On the tail approximation of the blocking probability of single server queues with finite buffer capacity. *Queueing Networks with Finite Capacity, Proc. 2nd Int. Conf.*, 267–280.

Bollobas, B. (2001). *Random Graphs*, 2nd edn. (Cambridge University Press, Cambridge, UK).

Borovkov, A. A. (1976). *Stochastic Processes in Queueing Theory.* (Springer-Verlag, New York).

Boyd, S. and Vandenberghe, L. (2004). *Convex Optimization.* (Cambridge University Press, Cambridge).

Brockmeyer, E., Halstrom, H. L., and Jensen, A. (1948). *The Life and Works of A. K. Erlang.* (Academy of Technical Sciences, Copenhagen).

523

Chalmers, R. C. and Almeroth, K. C. (2001). Modeling the branching characteristics and efficiency gains in global multicast trees. *IEEE INFOCOM2001, Alaska.*

Chen, L. Y. (1975). Poisson approximation for dependent trials. *The Annals of Probability 3*, 3, 534–545.

Chen, W.-K. (1971). *Applied Graph Theory.* (North-Holland Publishing Company, Amsterdam).

Chuang, J. and Sirbu, M. A. (1998). Pricing multicast communication: A cost-based approach. *Proceedings of the INET'98.*

Cohen, J. W. (1969). *The Single Server Queue.* (North-Holland Publishing Company, Amsterdam).

Cohen-Tannoudji, C., Diu, B., and Laloë, F. (1977). *Mécanique Quantique.* Vol. I and II. (Hermann, Paris).

Comtet, L. (1974). *Advanced Combinatorics*, revised and enlarged edn. (D. Riedel Publishing Company, Dordrecht, Holland).

Cormen, T. H., Leiserson, C. E., and Rivest, R. L. (1991). *An Introduction to Algorithms.* (MIT Press, Boston).

Cvetkovic, D. M., Doob, M., and Sachs, H. (1995). *Spectra of Graphs, Theory and Applications*, third edn. (Johann Ambrosius Barth Verlag, Heidelberg).

Dorogovtsev, S. N. and Mendes, J. F. F. (2003). *Evolution of Networks, From Biological Nets to the Internet and WWW.* (Oxford University Press, Oxford).

Embrechts, P., Klüppelberg, C., and Mikosch, T. (2001a). *Modelling Extremal Events for Insurance and Finance*, 3rd edn. (Springer-Verlag, Berlin).

Embrechts, P., McNeil, A., and Straumann, D. (2001b). *Correlation and Dependence in Risk Management: Properties and Pitfalls.* Risk Management: Value at Risk and Beyond, ed. M. Dempster and H. K. Moffatt, (Cambridge University Press, Cambridge, UK).

Erdös, P. and Rényi, A. (1959). On random graphs. *Publicationes Mathematicae Debrecen 6*, 290–297.

Erdös, P. and Rényi, A. (1960). On the evolution of random graphs. *Magyar Tud. Akad. Mat. Kutato Int. Kozl. 5*, 17–61.

Feller, W. (1970). *An Introduction to Probability Theory and Its Applications*, 3rd edn. Vol. 1. (John Wiley & Sons, New York).

Feller, W. (1971). *An Introduction to Probability Theory and Its Applications*, 2nd edn. Vol. 2. (John Wiley & Sons, New York).

Floyd, S. and Paxson, V. (2001). Difficulties in simulating the internet. *IEEE Transactions on Networking 9*, 4 (August), 392–403.

Fortz, B. and Thorup, M. (2000). Internet traffic engineering by optimizing OSPF weights. *IEEE INFOCOM2000.*

Frieze, A. M. (1985). On the value of a random minimum spanning tree problem. *Discrete Applied Mathematics 10*, 47–56.

Gallager, R. G. (1996). *Discrete Stochastic Processes.* (Kluwer Academic Publishers, Boston).

Gantmacher, F. R. (1959a). *The Theory of Matrices.* Vol. I. (Chelsea Publishing Company, New York).

Gantmacher, F. R. (1959b). *The Theory of Matrices.* Vol. II. (Chelsea Publishing Company, New York).

Gauss, C. F. (1821). Theoria combinationis observationum erroribus minimus obnoxiae. Pars prior. *Gauss Werke 4*, 3–26.

Gilbert, E. N. (1956). Enumeration of labelled graphs. *Canadian Journal of Mathematics* **8**, 405–411.

Gnedenko, B. V. and Kovalenko, I. N. (1989). *Introduction to Queuing Theory*, second edn. (Birkhauser, Boston).

Golub, G. H. and Loan, C. F. V. (1983). *Matrix Computations*. (North Oxford Academic, Oxford).

Goulden, I. P. and Jackson, D. M. (1983). *Combinatorial Enumeration*. (John Wiley & Sons, New York).

Grimmett, G. R. (1989). *Percolation*. (Springer-Verlag, New York).

Grimmett, G. R. and Stirzacker, D. (2001). *Probability and Random Processes*, 3rd edn. (Oxford University Press, Oxford).

Hardy, G. H. (1948). *Divergent Series*. (Oxford University Press, London).

Hardy, G. H., Littlewood, J. E., and Polya, G. (1999). *Inequalities*, 2nd edn. (Cambridge University Press, Cambridge, UK).

Hardy, G. H. and Wright, E. M. (1968). *An Introduction to the Theory of Numbers*, 4th edn. (Oxford University Press, London).

Harris, T. E. (1963). *The Theory of Branching Processes*. (Springer-Verlag, Berlin).

Harrison, J. M. (1990). *Brownian Motion and Stochastic Flow Systems*. (Krieger Publishing Company, Malabar, Florida).

van der Hofstad, R., Hooghiemstra, G., and Van Mieghem, P. (2001). First passage percolation on the random graph. *Probability in the Engineering and Informational Sciences (PEIS)* **15**, 225–237.

van der Hofstad, R., Hooghiemstra, G., and Van Mieghem, P. (2002a). The flooding time in random graphs. *Extremes 5*, 2 (June), 111–129.

van der Hofstad, R., Hooghiemstra, G., and Van Mieghem, P. (2002b). On the covariance of the level sizes in recursive trees. *Random Structures and Algorithms* **20**, 519–539.

van der Hofstad, R., Hooghiemstra, G., and Van Mieghem, P. (2005). Distances in random graphs with finite variance degree. *Random Structures and Algorithms 27*, 1 (August), 76–123.

van der Hofstad, R., Hooghiemstra, G., and Van Mieghem, P. (2006a). Size and weight of shortest path trees with exponential link weights. *Combinatorics, Probability and Computing*.

van der Hofstad, R., Hooghiemstra, G., and Van Mieghem, P. (2006b). The weight of the shortest path tree. *Random Structures and Algorithms*.

Hooghiemstra, G. and Koole, G. (2000). On the convergence of the power series algorithm. *Performance Evaluation* **42**, 21–39.

Hooghiemstra, G. and Van Mieghem, P. (2005). On the mean distance in scale free graphs. *Methodology and Computing in Applied Probability (MCAP)* **7**, 285–306.

Jamin, S., C. Jin, A. R. Kurc, D. R., and Shavitt, Y. (2001). Constrained mirror placement on the internet. *IEEE INFOCOM'01*.

Janic, M., Kuipers, F., Zhou, X., and Van Mieghem, P. (2002). Implications for QoS provisioning based on traceroute measurements. *Proceedings of 3nd International Workshop on Quality of Future Internet Services, QofIS2002 ed. B. Stiller et al., Zurich, Switzerland, Springer Verlag LNCS 2511*, 3–14.

Janson, S. (1995). The minimal spanning tree in a complete graph and a functional limit theorem for trees in a random graph. *Random Structures and Algorithms 7*, 4 (December), 337–356.

Janson, S. (2002). On concentration of probability. *Contemporary Combinatorics, ed. B. Bollobás, Bolyai Soc. Math. Stud. 10, János Bolyai Mathematical Society, Budapest*, 289–301.

Janson, S., Knuth, D. E., Luczak, T., and Pittel, B. (1993). The birth of the giant component. *Random Structures and Algorithms 4*, 3, 233–358.

Karlin, S. and Taylor, H. M. (1975). *A First Course in Stochastic Processes*, 2nd edn. (Academic Press, San Diego).

Karlin, S. and Taylor, H. M. (1981). *A Second Course in Stochastic Processes*. (Academic Press, San Diego).

Kelly, F. P. (1991). Special invited paper: Loss networks. *The Annals of Applied Probability 1*, 3, 319–378.

Kleinrock, L. (1975). *Queueing Systems*. Vol. 1 – Theory. (John Wiley and Sons, New York).

Kleinrock, L. (1976). *Queueing Systems*. Vol. 2 – Computer Applications. (John Wiley and Sons, New York).

Krishnan, P., Raz, D., and Shavitt, Y. (2000). The cache location problem. *IEEE/ACM Transactions on Networking 8*, 5 (October), 586–582.

Kuipers, F. A. and Van Mieghem, P. (2003). The Impact of Correlated Link Weights on QoS Routing. *IEEE INFOCOM03*.

Lanczos, C. (1988). *Applied Analysis*. (Dover Publications, Inc., New York).

Langville, A. N. and Meyer, C. D. (2005). Deeper inside PageRank. *Internet Mathematics 1*, 3 (Februari), 335–380.

Le Boudec, J.-Y. and Thiran, P. (2001). *Network Calculus, A Theory of Deterministic Queuing Systems for the Internet*. (Springer Verlag, Berlin).

Leadbetter, M. R., Lindgren, G., and Rootzen, H. (1983). *Extremes and Related Properties of Random Sequences and Processes*. (Springer-Verlag, New York).

Leon-Garcia, A. (1994). *Probability and Random Processes for Electrical Engineering*, 2nd edn. (Addison-Wesley, Reading, Massachusetts).

van Lint, J. H. and Wilson, R. M. (1996). *A course in Combinatorics*. (Cambridge University Press, Cambridge, UK).

Lovász, L. (1993). Random Walks on Graphs: A Survey. *Combinatorics 2*, 1–46.

Markushevich, A. I. (1985). *Theory of functions of a complex variable*. Vol. I – III. (Chelsea Publishing Company, New York).

Mehta, M. L. (1991). *Random Matrices*, 2nd edn. (Academic Press, Boston).

Meyer, C. D. (2000). *Matrix Analysis and Applied Linear Algebra*. (Society for Industrial and Applied Mathematics (SIAM), Philadelphia).

Mitra, D. (1988). Stochastic theory of a fluid model of producers and consumers coupled by a buffer. *Advances in Applied Probability* **20**, 646–676.

Morse, P. M. and Feshbach, H. (1978). *Methods of Theoretical Physics*. (McGraw-Hill Book Company, New York).

Neuts, M. F. (1989). *Structured Stochastic Matrices of the M/G/1 Type and Their Applications*. (Marcel Dekker Inc., New York).

Norros, I. (1994). A storage model with self-similar input. *Queueing Systems 16*, 3-4, 387–396.

Pascal, B. (1954). *Oeuvres completes*. Bibliothèque de la Pléade, (Gallimard, Paris).

Paxson, V. (1997). End-to-end Routing Behavior in the Internet. *IEEE/ACM Transactions on Networking 5*, 5 (October), 601–615.

Phillips, G., Schenker, S., and Tangmunarunkit, H. (1999). Scaling of multicast trees: Comments on the chuang-sirbu scaling law. *ACM Sigcomm99*.

Pietronero, L. and Schneider, W. (1990). Invasion percolation as a fractal growth problem. *Physica A* **170**, 81–104.

Press, W. H., Teukolsky, S. A., Vetterling, W. T., and Flannery, B. P. (1992). *Numerical Recipes in C*, 2nd edn. (Cambridge University Press, New York).

Rainville, E. D. (1960). *Special Functions.* (Chelsea Publishing Company, New York).

Riordan, J. (1968). *Combinatorial Identities.* (John Wiley & Sons, New York).

Roberts, J. W. (1991). *Performance Evaluation and Design of Multiservice Networks.* Information Technologies and Sciences, vol. COST 224. (Commission of the European Communities, Luxembourg).

Robinson, S. (2004). The prize of anarchy. *SIAM News 37,* 5 (June), 1–4.

Ross, S. M. (1996). *Stochastic Processes*, 2nd edn. (John Wiley & Sons, New York).

Royden, H. L. (1988). *Real Analysis*, 3rd edn. (Macmillan Publishing Company, New York).

Sansone, G. and Gerretsen, J. (1960). *Lectures on the Theory of Functions of a Complex Variable.* Vol. 1 and 2. (P. Noordhoff, Groningen).

Schoutens, W. (2000). *Stochastic Processes and Orthogonal Polynomials.* (Springer-Verlag, New York).

Siganos, G., Faloutsos, M., Faloutsos, P., and Faloutsos, C. (2003). Power laws and the AS-level internet topology. *IEEE/ACM Transactions on Networking 11,* 4 (August), 514–524.

Smythe, R. T. and Mahmoud, H. M. (1995). A survey of recursive trees. *Theory of Probability and Mathematical Statistics* 51, 1–27.

Steyaert, B. and Bruneel, H. (1994). Analytic derivation of the cell loss probability in finite multiserver buffers, from infinite buffer results. *Proceedings of the second workshop on performance modelling and evaluation of ATM networks, Bradford UK*, 18.1–11.

Strogatz, S. H. (2001). Exploring complex networks. *Nature 410,* 8 (March), 268–276.

Syski, R. (1986). *Introduction to Congestion Theory in Telephone Systems*, 2nd edn. Studies in Telecommunication, vol. 4. (North-Holland, Amsterdam).

Titchmarsh, E. C. (1948). *Introduction to the Theory of Fourier Integrals*, 2nd edn. (Oxford University Press, Ely House, London W. I).

Titchmarsh, E. C. (1964). *The Theory of Functions.* (Oxford University Press, Amen House, London).

Titchmarsh, E. C. and Heath-Brown, D. R. (1986). *The Theory of the Zeta-function*, 2nd edn. (Oxford Science Publications, Oxford).

Van Mieghem, P. (1996). The asymptotic behaviour of queueing systems: Large deviations theory and dominant pole approximation. *Queueing Systems* **23**, 27–55.

Van Mieghem, P. (2001). Paths in the simple random graph and the Waxman graph. *Probability in the Engineering and Informational Sciences (PEIS)* **15**, 535–555.

Van Mieghem, P. (2004a). *Data Communications Networking.* (Delft University of Technology, Delft).

Van Mieghem, P. (2004b). The Probability Distribution of the Hopcount to an Anycast Group. Delft University of Technology, Report 2003605 (www.nas.ewi.tudelft.nl/people/Piet/teleconference) .

Van Mieghem, P. (2005). The limit random variable W of a branching process. Delft University of Technology, Report 20050206 (www.nas.ewi.tudelft.nl/people/Piet/teleconference).

Van Mieghem, P., Hooghiemstra, G., and van der Hofstad, R. (2000). A Scaling Law for the Hopcount in the Internet. Delft University of Technology, Report2000125 (www.nas.ewi.tudelft.nl/people/Piet/telconference).

Van Mieghem, P., Hooghiemstra, G., and van der Hofstad, R. (2001a). On the efficiency of multicast. *IEEE/ACM Transactions on Networking 9*, 6 (December), 719–732.

Van Mieghem, P., Hooghiemstra, G., and van der Hofstad, R. W. (2001b). Stochastic model for the number of traversed routers in internet. *Proceedings of Passive and Active Measurement: PAM-2001, April 23-24, Amsterdam.*

Van Mieghem, P. and Janic, M. (2002). Stability of a multicast tree. *Proceedings IEEE INFOCOM2002* **2**, 1099–1108.

Veres, A. and Boda, M. (2000). The chaotic nature of TCP congestion control. *IEEE INFOCOM'2000, Tel-Aviv, Israel.*

Walrand, J. (1998). *Communication Networks, A First Course*, 2nd edn. (McGraw-Hill, Boston).

Wästlund, J. (2005). Evaluation of Janson's constant for the variance in the random minimum spanning tree problem. *Linköping studies in Mathematics. Series editor: Bengt Ove Turesson* 7 (www.ep.liu.se/ea/lsm/2005/007).

Waxman, B. M. (1998). Routing of multipoint connections. *IEEE Journal on Selected Areas in Communications 6*, 9 (December), 1617–1622.

Whittaker, E. T. and Watson, G. N. (1996). *A Course of Modern Analysis*, Cambridge Mathematical Library edn. (Cambridge University Press, Cambridge, UK).

Wigner, E. P. (1955). Characteristic vectors of bordered matrices with infinite dimensions. *Annals of Mathematics 62*, 3 (November), 548–564.

Wigner, E. P. (1957). Characteristic vectors of bordered matrices with infinite dimensions ii. *Annals of Mathematics 65*, 2 (March), 203–207.

Wigner, E. P. (1958). On the distribution of the roots of certain symmetric matrices. *Annals of Mathematics 67*, 2 (March), 325–327.

Wilkinson, J. H. (1965). *The Algebraic Eigenvalue Problem.* (Oxford University Press, New York).

Wolff, R. W. (1982). Poisson arrivals see time averages. *Operations Research 30*, 2 (April), 223–231.

Wolff, R. W. (1989). *Stochastic Modeling and the Theory of Queues.* (Prentice-Hall International Editions, New York).

Index

k-ary tree, 387, 401, 414, 417, 423, 432, 522

adjacency matrix, 320, 471, 488
  eigenvalues, 475

Bayes' rule, 28, 197
Beneš' equation, 261, 301

cell loss ratio (clr), 309, 512
Central Limit Theorem, 104, 148, 366, 377,
    516
Chernoff bound, 88
Chuang–Sirbu scaling law, 404–407
complete graph, 319, 321, 327, 347, 349, 359,
    371, 373, 380, 392, 473, 481, 482, 488, 520
conditional distribution function, 28
conditional expectation, 34, 233, 341
conditional probability, 26
conditional probability density function, 28
correlation coefficient, 30, 61, 67, 69, 74, 119
covariance, 29, 71, 78
  matrix, 63, 66

degree graph, 323
degree of a node, 225, 322, 472
disjoint paths, 520
  Merger's Theorem, 327
distribution
  k-th order statistics, 53, 494
  Bernoulli, 37
  binomial, 38, 332, 488
  Cauchy, 54
  chi-square, 50
  Erlang, 48, 125, 274, 278
  exponential, 44, 75
  extremal, 106
  Fréchet, 107
  Gamma, 48, 51, 125
  Gaussian, 46, 103, 400
  geometric, 39, 272
  Gumbel, 54, 107, 400
  joint Gaussian, 64
  lognormal, 57, 77

Pareto, 56
Poisson, 40, 116, 129, 335
polynomial, 44, 348, 494
regular, 348, 362
uniform, 43, 74
Weibull, 55, 107, 132

Engset formula, 314, 511
Erlang B formula, 2, 280
Erlang C formula, 277
event, 10, 53
  mutually exclusive, 10

failure rate, 131
flooding time, 362

giant component, 335, 337, 339
Google, 224
graph connectivity, 325, 486
  edge connectivity, 326, 487
  vertex connectivity, 326, 487
graph metrics
  betweenness, 329
  clustering coefficient, 328, 346, 513
  diameter, 475
  distortion, 329
  expansion, 328
  hopcount, 329
  resilience, 329

histogram, 118, 494
hopcount, 340, 347, 354, 357, 387, 392, 403,
    409, 418, 420, 423, 431, 513, 520

incidence matrix, 471
inclusion-exclusion formula, 12, 335, 391
  sieve of Eratosthenes, 15
indicator function, 12, 17, 43, 321
inequality
  Boole, 15
  Cauchy-Schwarz, 90, 91, 480
  Chebyshev, 88
  Gauss, 92

Hölder, 90, 480
Jensen, 85, 342
Markov, 88
Minkowsky, 91
infinitesimal generator, 181

Laplacian (admittance matrix), 472, 486
law of rare events, 41, 128, 495
law of total probability, 27, 123, 142, 159, 204,
    205, 238, 255, 274, 278, 284, 295, 324,
    341, 367, 410, 422, 445, 495
level set of a tree, 352
Lindley's equation, 255
link weight, 320, 340, 341, 347, 349, 359, 362,
    373, 392, 406, 408
Little's law, 267, 273, 275, 281, 287, 297, 508,
    510

Markov chain
    absorbing states, 164
    communicating states, 162
    conservative, 181
    continuous-time, 179
    discrete-time, 158
    embedded, 186, 188
    hitting time, 163
    irreducible Markov chain, 161, 226
    periodic and aperiodic, 162
    transient and recurrent states, 165
mean time to failure, 131
memoryless property, 27, 40, 45, 125, 132, 185,
    351
Metcalfe's law, 320
minimum spanning tree (MST), 373, 399
modes of convergence, 99

Newton identities for polynomials, 477

order statistics, 52, 127

PageRank (Google), 224
phase transition, 335, 376
Poisson arrivals see time averages (PASTA),
    267, 274, 275, 283, 288, 312, 509, 511
Pollaczek-Khinchin equation, 286
power law, 325
probability density function (pdf), 16, 20, 22
    joint, 28, 32
probability generating function (pgf), 18
    logarithm of, 19, 25
    moment generating function, 25, 235
process
    arrival, 248, 270
    birth and death, 208, 304, 351
    branching, 229, 342
        geometric, 244
        Poisson, 246, 345
    counting, 263
    Markov, 180, 253, 349
        balance equation, 187
        Chapman-Kolmogorov equation, 180

forward and backward equation, 182, 195
    time reversibility, 196
nonhomogeneous Poisson, 129
Poisson, 120, 210
queueing, 250
renewal, 137
service, 249, 270
stochastic, 115
    modeling, 117
Yule, 212, 230

quality of service (QoS), 2, 249, 283, 309, 340,
    419

random graph, 330, 332, 337, 339, 346, 354,
    362, 373, 374, 377, 387, 392, 403, 404,
    406, 408, 410, 488, 513
random variable
    continuous, 20, 59
    discrete, 16, 58
    expectation, 17, 22
    indepedent, 97, 104
    independent, 28, 29, 32, 34, 47, 49, 51, 78
    normalized, 31, 93, 400
random vector, 62
random walk, 202, 484
redundancy level, 325
regular graphs, 322, 328, 475, 482, 486, 488
reliability function, 131
renewal
    alternating renewal process, 153
    Blackwell's Renewal Theorem, 146
    Elementary Renewal Theorem, 145, 170
    inspection paradox, 152
    Key Renewal Theorem, 146, 151
    renewal equation, 141
    renewal function, 140
    renewal process, 137
    renewal theory, 165

server placement problem, 419, 424, 429
shortest path, 340, 347
    tree (SPT), 387, 392, 399, 407, 419, 428
slotted Aloha, 219
stochastic matrix, 159, 450, 451, 455, 473

total variation distance, 42
transition probability matrix, 159, 190, 201
    spectral decomposition, 184

unfinished work, 256, 261, 263, 301
uniform recursive tree (URT), 354, 380, 392,
    404, 407, 411, 417, 419, 422, 424
uniformization, 189

Wald's identity, 34, 145, 154
web graph, 224, 323
Wigner's Semicircle Law, 489